Classification Scheme

1. Structure of Crystalline Solids
   - 1.1 Perfectly Periodic Structure
   - 1.2 Solid-State Phase Transformations
   - 1.3 Alloys. Metallurgy
   - 1.4 Microstructure (Magnetic Domains See 18; Ferroelectric Domains See 14.4.1)
   - 1.5 Films
   - 1.6 Surfaces
2. Non-Crystalline State
3. Crystal Growth
4. Bonding Properties
5. Mössbauer Spectroscopy
6. Lattice Dynamics. Phonons
7. Acoustic Properties
8. Thermal Properties
9. Diffusion
10. Defect Properties (Irradiation Defects See 11)
    - 10.1 Metals
    - 10.2 Non-Metals
11. Irradiation Effects (X-Ray Diffraction Investigations See 1 and 10)
12. Mechanical Properties (Plastic Deformations See 10)
    - 12.1 Metals
    - 12.2 Non-Metals
13. Electron States
    - 13.1 Band Structure
    - 13.2 Fermi Surfaces
    - 13.3 Surface and Interface States
    - 13.4 Impurity and Defect States
    - 13.5 Elementary Excitations (Phonons See 6)
      - 13.5.1 Excitons
      - 13.5.2 Plasmons
      - 13.5.3 Polarons
      - 13.5.4 Magnons
14. Electrical Properties. Transport Phenomena
    - 14.1 Metals. Semi-Metals
    - 14.2 Superconductivity. Superconducting Materials and Devices
    - 14.3 Semiconductors
      - 14.3.1 Films
      - 14.3.2 Surfaces and Interfaces
      - 14.3.3 Devices. Junctions (Contact Problems See 14.3.4)
      - 14.3.4 High-Field Phenomena, Space-Charge Effects, Inhomogeneities, Injected Carriers (Electroluminescence See 20.3; Junctions See 14.3.3)
    - 14.4 Dielectrics
      - 14.4.1 Ferroelectrics
15. Thermoelectric and Thermomagnetic Properties
16. Photoconductivity. Photovoltaic Effects
17. Emission of Electrons and Ions
    - 17.1 Field Emission Microscope Investigations
18. Magnetic Properties
    - 18.1 Paramagnetic Properties
    - 18.2 Ferromagnetic Properties
      - 18.2.1 Ferromagnetic Films
    - 18.3 Ferrimagnetic Properties
    - 18.4 Antiferromagnetic Properties

(Continued on cover three)

physica status solidi (a)

applied research

Volume 14 · Number 2 · Pages 355 to 704, K95 to K162, and A9 to A16

December 16, 1972

PSSA 14(2) 355—704, K95—K162, A9—A16 (1972)

AKADEMIE-VERLAG · BERLIN

Subscriptions and orders for single copies should be addressed to

AKADEMIE-VERLAG GmbH, 108 Berlin, Leipziger Straße 3 – 4 or to Buchhandlung KUNST UND WISSEN, Erich Bieber, 7 Stuttgart 1, Wilhelmstr. 4 – 6 or to Deutsche Buch-Export und -Import GmbH, 701 Leipzig, Postschließfach 160

Editorial Note:

"physica status solidi (a)" undertakes that an original paper *accepted for publication* before the 23rd of any month will be published within 50 days of this date unless the author requests a postponement. In special cases there may be some delay between *receipt* and *acceptance* of a paper due to the review and, if necessary, revision of the paper.

Schriftleiter und verantwortlich für den Inhalt: Professor Dr. Dr. h. c. P. Görlich, 102 Berlin, Neue Schönhauser Str. 20 bzw. 69 Jena, Humboldtstr. 26. Chefredakteur: Dr. H.-J. Hänsch. Redaktionskollegium: Prof. Dr. E. Gutsche, Dr. H.-J. Hänsch, Dr. H. Lange, Dr. S. Oberländer. Anschrift der Schriftleitung: 102 Berlin, Neue Schönhauser Str. 20, Fernruf: 4223380. Verlag: Akademie-Verlag GmbH, 108 Berlin, Leipziger Str. 3—4, Fernruf: 220441, Telex-Nr. 112020, Postscheckkonto: Berlin 35021. — Die Zeitschrift „physica status solidi (a)" erscheint jeweils am 16. des Monats. Bezugspreis eines Bandes 120,— M. Bestellnummer dieses Bandes: 1085/14. Jeder Band enthält zwei Hefte. Gesamtherstellung: VEB Druckerei „Thomas Müntzer" Bad Langensalza. — Veröffentlicht unter der Lizenznummer 1310a des Presseamtes beim Vorsitzenden des Ministerrates der Deutschen Demokratischen Republik.

Contents

Short Notes

Pre-printed Titles

Systematic List

Subject classification:	Corresponding papers begin on the following pages (pages given in italics refer to the principal subject classification):
1	483
1.1	*639, 663*
1.2	363
1.3	*431*
1.5	*473*, 605, K101
1.6	*K97*
3	*473, 605*
5	*689*
6	363, 423, *453*
8	*363, 639*
9	*521, 673, K95*
10.1	431, *463, 545*, 561, *689, K143*
10.2	363, *443, 489, 599, 613, 623, K101*
11	541, *605, K97*
12.2	*489, K125*
13	483
13.1	363, 423, 531
13.3	*573, K135*
13.4	*405, 439*, 453, *527*, 531, 565, 579, *655*
14	*517*
14.1	*561*
14.3	417, *439*, 453, *531, 541*
14.3.1	*K135*
14.3.3	*405, 457, 655, 695, K95*
14.3.4	*423, 681*, K113
14.4	*K105*
14.4.1	*501*
15	517
16	*527, 573, 681, K113, K121*
17	*K129, K149, K157*
17.1	*417*
18	363, *591*
18.2	423, 463, *483, K157*
18.3	*511*, K105, *K109, K153*
19	*511, K147*
20	*501*
20.1	*541, 565, 591, 643, K117*

The Author Index of Volume 14 Begins on Page 705

(It will be delivered together with Volume 15, Number 1.)

Review Article

phys. stat. sol. (a) **14**, 363 (1972)

Subject classification: 8; 1.2; 6; 10.2; 13.1; 18; 22

A. F. Ioffe Physico-Technical Institute, Academy of Sciences of the USSR, Leningrad

Thermal Properties of Rare-Earth Chalcogenides

By

I. A. Smirnov

Contents

1. Introduction

Rare-earth chalcogenides are new perspective materials. During the last ten years they started to attract more and more attention.

Compounds of this class are characterized by a wide range of properties. There are metals, semiconductors, and insulators among them. All of them have relatively high melting points, good thermal stability, interesting magnetic, optical, electric, and thermal properties. There is an outlook to use them in technics (high-temperature thermoelectric materials, operated magnetic elements, high-temperature termistors, laser materials, etc. [1, 2]).

However, the knowledge of rare-earth chalcogenides is unfortunately very insufficient. This review deals with thermal properties (thermal conductivity $\varkappa$, specific heat C, thermal expansion coefficient δ) of this group of compounds. The investigation of $\varkappa$, C, and δ not only gives information about their absolute values necessary for various technical calculations, but also the opportunity to obtain many additional data on the fundamental physical parameters of the substances.

The total solid state specific heat can be expressed as

$$C_{\text{tot}} = C^{\text{L}} + C_{\text{e}} + C_{\text{m}} + C_{\text{f}} + C_{\text{o}} + C_{\text{vac}} + C_{\text{nuc}} + C_{\text{loc}} + \cdots, \qquad (1)$$

where C^{L} is the crystal lattice specific heat, C_{e} the electron component, C_{m} the magnon specific heat due to the excitation of the spin system in magnetically ordered substances, C_{f} is the component associated with the transitions to

Table 1

Information on solids which can be obtained from the specific heat investigations

$$C_v^{\text{tot}} = C^{\text{L}} + C_{\text{e}} + C_{\text{m}} + C_{\text{f}} + C_{\text{vac}} + C_{\text{nuc}} + C_{\text{loc}} + \cdots$$

Term	Information
C_v^{tot}	entropy
C^{L}	Debye temperature
	crystal lattice dynamics
	influence of impurities on crystal lattice
C_{e}	effective mass of current carriers
	density of states at the Fermi level
C_{m}	phase transition temperature
	entropy changes at phase transformation
	spin magnetic moment
	exchange integral value
C_{f}	energy spectrum of a crystal
C_{vac}	vacancy formation energy
C_{nuc}	magnetic or quadrupole moment of the nucleus
	value of the magnetic field at a nucleus
	gradient of the electric field at a nucleus
C_{loc}	local vibrations of impurity atoms

higher energy levels (ionization of impurities, band-to-band transition, transitions between levels obtained as the result of splitting of atomic levels by inner or outer fields (Schottky effect)), C_0 is the specific heat connected with ordering processes (ferro- and antiferromagnetics, ferroelectric materials, structural ordering, etc.), C_{vac} is a component due to the formation of vacancies (noticeable at high temperatures, especially near the melting point), C_{nuc} a component connected with superfine splitting of nuclear levels (Schottky nuclear effect), and C_{loc} the specific heat connected with the local vibrations of impurity atoms.

The coefficient of thermal expansion for insulators, metals, and semiconductors is written as

$$\delta_{tot} = \delta_L + \delta_m + \delta_e , \tag{2}$$

where the δ_L, δ_m, and δ_e are lattice, magnon, and electron components of thermal expansion, respectively.

The total thermal conductivity also consists of a number of components:

$$\varkappa_{tot} = \varkappa_L + \varkappa_e + \varkappa_{bip} + \varkappa_{ph} + \varkappa_{ex} + \varkappa_m + \cdots , \tag{3}$$

where $\varkappa_L$ is the crystal lattice thermal conductivity, $\varkappa_e$ the electron component of the thermal conductivity, $\varkappa_{bip}$ the bipolar thermal conductivity appearing in semiconductors in the region of intrinsic conductivity due to the diffusion of electron–hole pairs, $\varkappa_{ph}$ is the photon component due to the heat transfer resulting from electromagnetic irradiation, $\varkappa_{ex}$ is the exciton component, and $\varkappa_m$ the magnon component of thermal conductivity (the latter occurs only in magnetic crystals).

Each of the components of C, δ, and $\varkappa$ bears certain (and often unique) information on the physical parameters of a material under investigation. This is vividly illustrated by Tables 1 to 3.

Presently there is a sufficiently large number of publications on thermal properties of rare-earth chalcogenides. It is high time to analyse critically and to systematize the accumulated results as well as to see what new information about the physical parameters of rare-earth chalcogenides can be obtained from measurements of thermal properties of these compounds.

2. Specific Heat of Rare-Earth Chalcogenides

Table 4 gives the list of rare-earth chalcogenides, the specific heat of which has been measured up to now. It was found sufficient to involve only four specific heat components from the right-hand side of equation (1) to analyse the available experimental data. Let us consider the main formulas for them.[1])

Crystal lattice specific heat

The crystal lattice specific heat can be expressed as below, taking into account only acoustic modes:

$$C_v^L = 9\,R \left(\frac{T}{\theta}\right)^3 \int\limits_0^{\theta/T} \frac{e^x\,x^4}{(e^x-1)^2}\,dx = 3\,R\,D\left(\frac{\theta}{T}\right), \tag{4}$$

---

1) A detailed development of the formulas can be found, e.g., in [31] to [33].

Table 2

Information on solids which can be obtained from the thermal expansion coefficient investigations

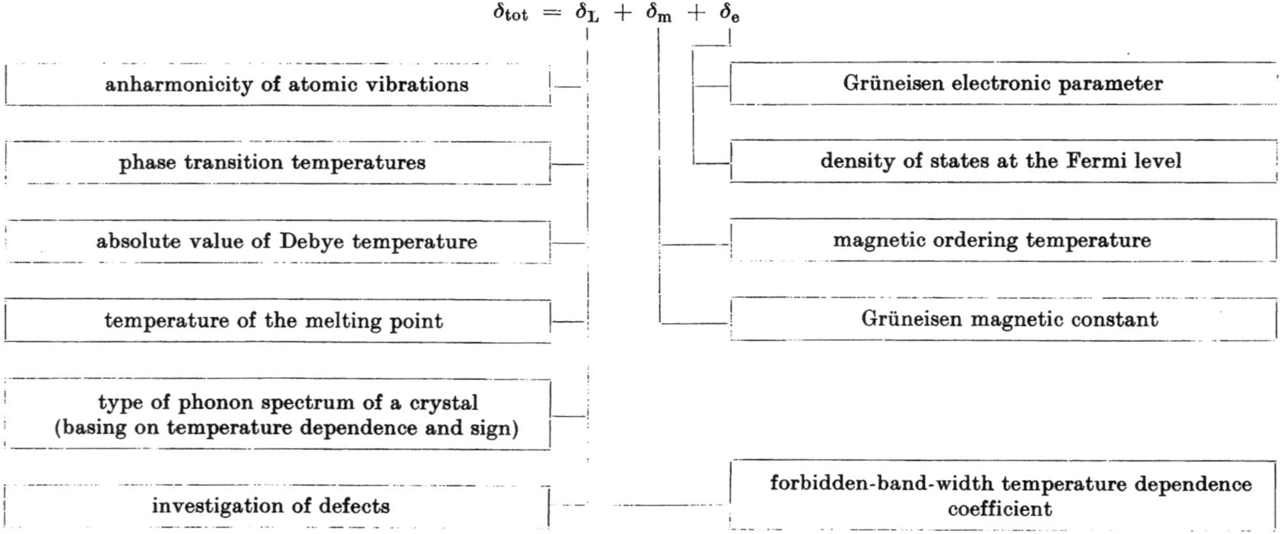

Table 3

Information on solids which can be obtained from thermal conductivity investigations

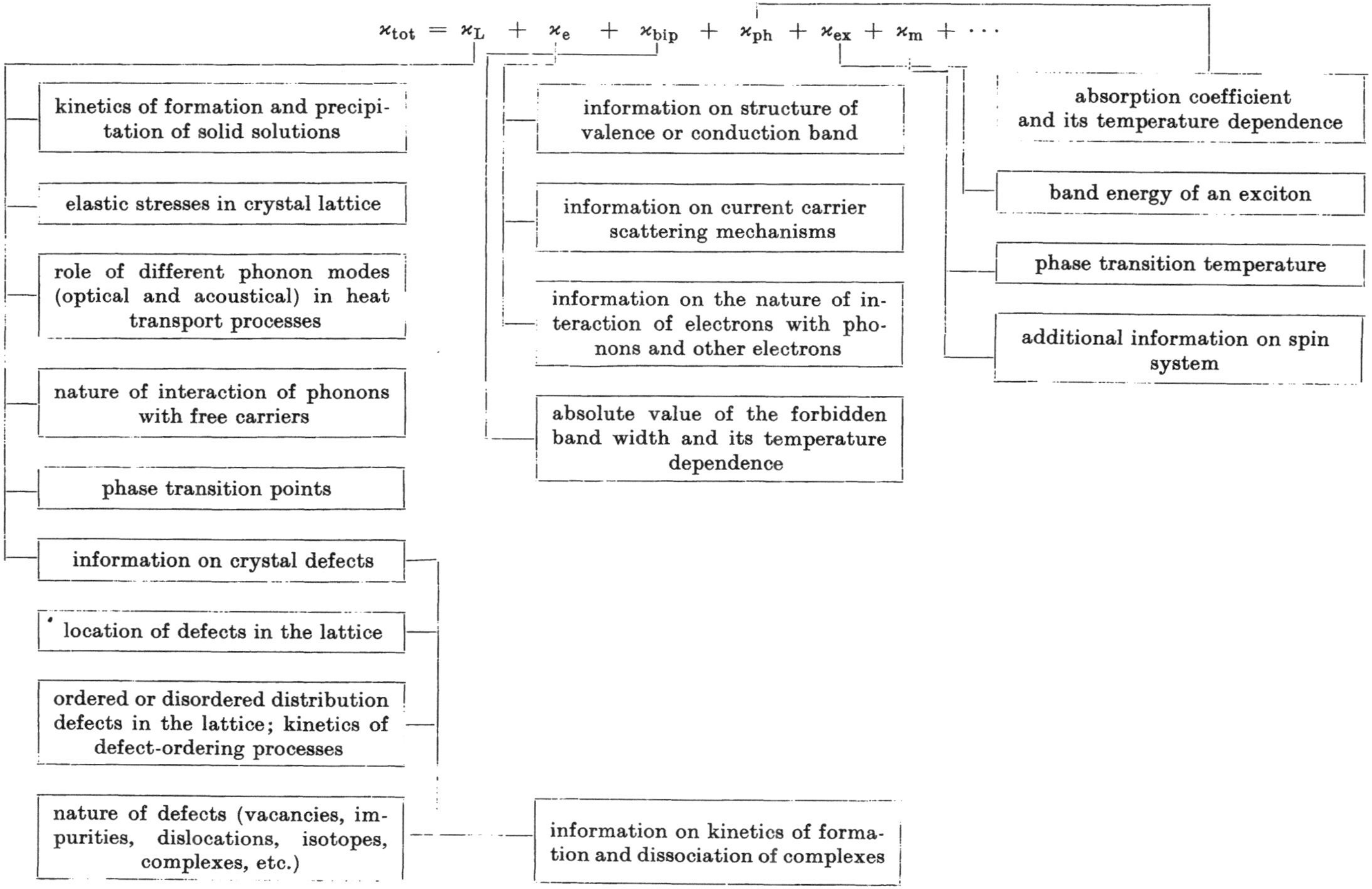

Table 4
Measured specific heat of rare-earth chalcogenides

compound	temperature range (°K)	references
MeX		
LaTe	90 to 400	[3]
	1.5 to 13	[37]
LaSe	90 to 400	[3]
	1.5 to 13	[37]
LaS	1.5 to 13	[37]
CeS	25 to 1200	[4, 5]
NdS	90 to 400	[3]
	in the vicinity of 8 °K	[6]
NdSe	in the vicinity of 10 °K	[6]
NdTe	in the vicinity of 10 °K	[6]
PrS	1.5 to 13	[37]
PrSe	1.5 to 13	[37]
PrTe	1.5 to 13	[37]
EuTe	0.5 to 20	[7 to 9]
EuSe	0.5 to 20	[7, 8, 10]
EuS	1 to 80	[7, 11 to 19]
EuS + Gd	1 to 80	[11, 17]
SmS	90 to 400	[20]
SmSe	90 to 400	[20]
TmTe	1.5 to 13	[37]
TmSe	1.5 to 13	[37]
TmS	1.5 to 13	[37]
Me_2X_3		
La_2Te_3	13 to 298	[21, 22]
La_2Se_3	13 to 298	[21, 23]
La_2S_3	13 to 298	[24]
Ce_2S_3	25 to 300	[4]
Nd_2Te_3	12 to 298	[21, 25]
Nd_2Se_3	12 to 298	[21, 26]
Nd_2S_3	13 to 298	[27]
Me_3X_4		
La_3Te_4	90 to 400	[28]
$La_{3-x}Se_4$	1 to 10	[29]
Ce_3Te_4	90 to 400	[28]
Nd_3Te_4	90 to 400	[28]
Pr_3Te_4	90 to 400	[28]
Nd_3S_4	in the vicinity of 46 °K	[30]

where R is the gas constant, $x = \hbar\omega/(k_0 T)$ the frequency, $D(\theta/T)$ the tabulated Debye function, and θ the Debye temperature.[2]) According to (4) C_v^L equals

[2]) Usually C_p, the specific heat at constant pressure, is measured experimentally. It is necessary to know C_v, specific heat at constant volume, for thermodynamic calculations. One of the following formulas is used to calculate C_v from C_p: $C_p - C_v = T\,V\,\beta^2/K_T$ and $C_p = C_v\,(1 - \gamma\,\beta\,T)$, where V is the molecular volume, K_T the isothermal compressibility, β the thermal volume expansion coefficient, and γ the Grüneisen constant.

zero at 0 °K, rises at low temperatures as T^3, and becomes constant at high temperatures (Dulong-Petit law) being equal to $3\,R\,n = 5.96\,n$ cal mol^{-1} deg^{-1} = = 24.94 n J mol^{-1} deg^{-1} (n number of atoms per molecule). If the optical modes are excited, the so-called Einstein thermal capacity is added to (4).

Electronic component of specific heat

The formula for the electronic specific heat in a free-electron model looks like

$$C_e = \gamma_0\, T = C^0\, m^*\, T\,, \tag{5}$$

where $C^0 = 4\,\pi^2\,k_0^2\,N^{1/3}/(3\,h^2)\{V\,[3\,N_a/(\pi\,V)]^{1/3}\}$, N_a is the number of electrons per atom, m^* the effective mas sof current carriers, N the Avagadro number, and k_0 the Boltzmann constant.

Magnon specific heat

The magnon specific heat can be represented as follows:

$$C_m^F = C^F\, R\left(\frac{k_0 T}{2\,J\,s}\right)^{3/2} \quad \text{for ferromagnetics}\,, \tag{6}$$

$$C_m^A = C^A\, R\left(\frac{k_0 T}{2\,J'\,s}\right)^{3} \quad \text{for antiferromagnetics}\,, \tag{7}$$

where J and J' are exchange integrals, C^F and C^A constants depending on the type of crystal structure, and s is the spin. The magnon specific heat changes at applying a magnetic field.[3])

Specific heat connected with transitions to higher energy levels

The specific heat due to band–band transitions for intrinsic conductivity in semiconductors and for band overlapping in semimetals is small and will not be taken into account in the analysis of experimental data. The Schottky effect (electronic transitions between various energy levels) is observed to give a considerable contribution to specific heat. The Schottky specific heat is expressed as follows for a two-level system:

$$C_{\text{Schott}} = R\left(\frac{\Delta_1}{k_0 T}\right)^2 \frac{g_1}{g_0}\, \frac{\exp(-\Delta_1/k_0 T)}{[1 + (g_1/g_0)\exp(-\Delta_1/k_0 T)]^2}\,, \tag{8}$$

where Δ_1 is the distance between ground states and first excited levels, and g_i ($i = 1, 2, \ldots$) is the degeneracy.

$$\text{At}\quad T \ll \frac{\Delta_1}{k_0 T} \qquad C_{\text{Schott}} = R\left(\frac{g_1}{g_0}\right)\left(\frac{\Delta_1}{k_0 T}\right)\exp\left(-\frac{\Delta_1}{k_0 T}\right), \tag{9}$$

$$\text{at}\quad T \gg \frac{\Delta_1}{k_0 T} \qquad C_{\text{Schott}} = R\, g_0\, g_1\,(g_0 + g_1)^{-2}\left(\frac{\Delta_1}{k_0 T}\right)^2. \tag{10}$$

The experimental data on specific heat will be considered separately for compounds of MeX, Me_3X_4, and Me_2X_3 type.

---

[3]) Formulas for C_m in a magnetic field are rather complicated. They can be found in [34] and [35].

2.1 MeX-type compounds

2.1.1 Rare-earth monochalcogenides with metallic conductivity

The references give data for twelve monochalcogenides of this group: LaTe, LaSe, LaS, PrTe, PrSe, PrS, CeS, NdTe, NdSe, NdS, TmSe, and TmS (see Table 4 and Fig. 1 and 2). These compounds crystallize in a NaCl-type lattice and are metals with current-carrier concentration equal to $\approx 10^{22}$ cm^{-3}. Therefore they are expected to have a considerable contribution of the electronic component to the total specific heat, i.e.,

$$C_v^{\text{tot}} = C_v^{\text{L}} + C_{\text{e}} + C_{\text{f}} = C_v^{\text{L}} + \gamma_0 T + C_{\text{f}} . \tag{11}$$

The specific heat of LaTe, LaSe, and NdS at medium temperatures has been investigated in [3]. The C_{f}-contribution to C_{tot} is absent for all three compounds. Some extra Schottky-type specific heat could be expected in NdS due to the availability of non-filled f-shell in Nd. However, such extra specific heat has not been revealed experimentally [3]. Possibly this is due to the small value of C_{f} within the investigated temperature range. C_{f} does not appear in lanthanum monochalcogenides since the f-levels in La are empty.

Since the specific heat measurements in [3] were taken at sufficiently high temperatures, there were some great difficulties in the separation of C_v^{L} and C_{e}. The successive approximation method[4]) was used to overcome them. For

Table 5

Values of the γ_0-parameter for some rare-earth metals and monochalcogenides

compound	$\gamma_0 \times 10^4$ (cal mol^{-1} deg^{-2})	references
La	21.6	[36]
Nd	20.5	[36]
Pr	17.4	[36]
LaTe	12.5	[3]
LaSe	10.5	[3]
NdS	8.5	[3]

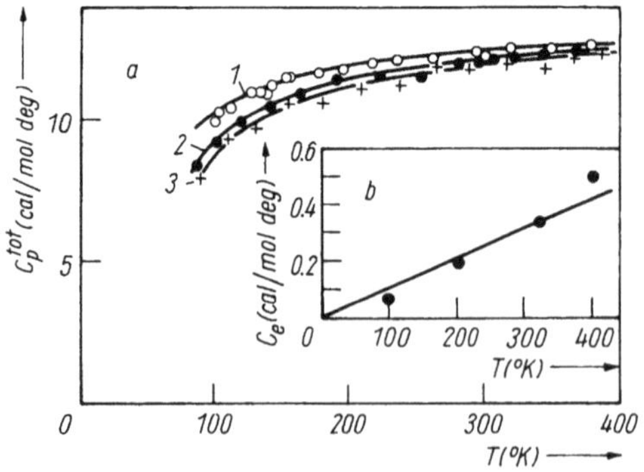

Fig. 1. Dependence of specific heat on temperature [3] a) $C_p^{\text{tot}} = f(T)$: (1) LaTe, (2) LaSe, (3) NdS; b) $C_{\text{e}} = f(T)$ for LaSe

[4]) The separation of C_v^{L} and C_{e} by the successive approximation method is described in [3].

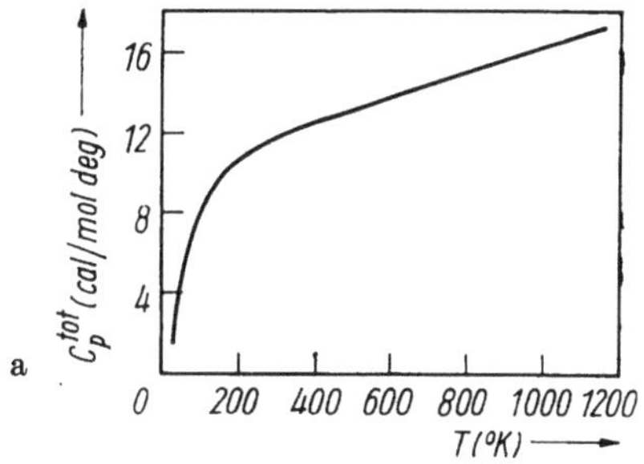

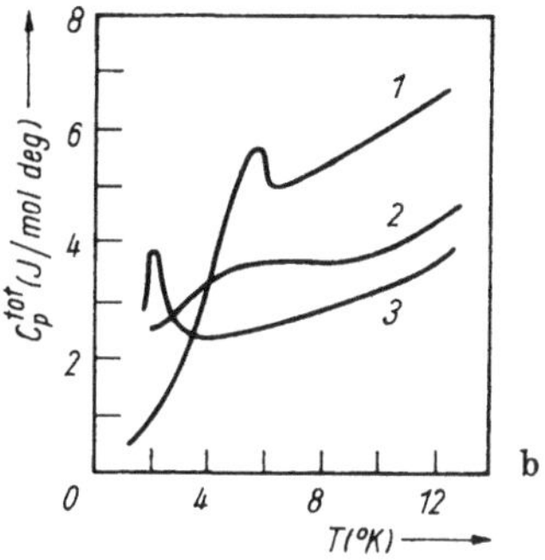

Fig. 2. Temperature dependence of specific heat. a) CeS [4, 5]; b) TmS (1), TmSe (2), TmTe (3) [37] (averaged values)

example, Fig. 1 b shows the electronic component of specific heat thus obtained for LaSe. The obtained values of the parameter γ_0 for rare-earth metals and monochalcogenides were of the same order of magnitude (see Table 5).

Formula (5) can be used for the calculation of the absolute value of the effective mass of current carriers from data for C_e. This was done in [3] for the three compounds NdS, LaSe, and LaTe; the following results were obtained:

compound	m^*
NdS	$2.9\ m_0$
LaSe	$3.1\ m_0$
LaTe	$3.3\ m_0$

(m_0 effective mass of the free electron).

It is possible to determine the Debye temperature θ from C_v^L with the help of formula (4). Fig. 3 gives values of θ thus calculated at 100 °K for NdS, LaSe, and LaTe depending on average atomic weight $\bar{A}$. Fig. 3 also contains experimental values of Debye temperatures for PbS, PbSe, and PbTe [38] which have the same NaCl-type lattice as the investigated monochalcogenides. All data lie on the same straight line which also contains the θ-values for LaS, PrSe, NdSe, and GdTe calculated by the Lindemann formula [39]

$$\theta = B\, T_m^{1/2}\, d^{1/3}\, \bar{A}^{-5/6}\,, \tag{12}$$

where B is a constant coefficient equal to 120 for compounds with NaCl-type lattice and d is the density. This fact gives, firstly, the possibility to approximately determine θ from the straight line of Fig. 3 for all other rare-earth monochalcogenides (NaCl-type lattice metals) and, secondly, to approximately evaluate from formula (12) their melting points basing on the obtained θ values. Since the error in the determination of θ is $\pm$ 5 deg, the melting point (T_m)

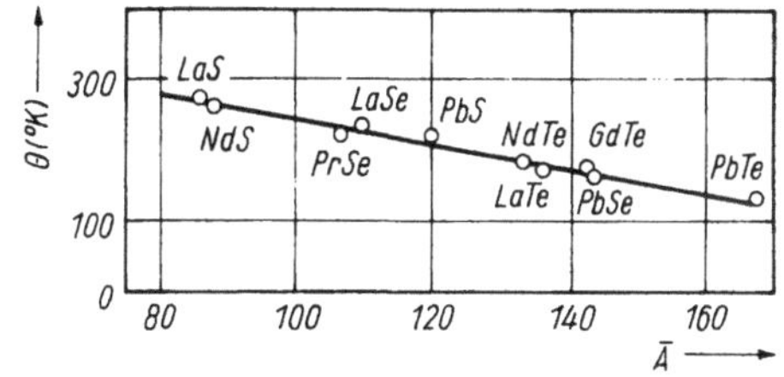

Fig. 3. Dependence of θ on $\bar{A}$ [3]

Table 6
Values of θ and T_m for rare-earth chalcogenides

compound	θ (°K)	T_m (°C) by formula (12)	T_m (°C) from references
LaS	266 (276)	2290	2325 [40], 2200 [41, 42], 1970 [41, 43]
LaSe	226 (231)	2250	1980 [45]
LaTe	184 (175)	2020	1800 [46], 1725 [47], 1720 [48, 49]
CeS	265	2220	2450 [50], 2200 [42], 2100 [44, 51]
CeSe	225	2180	—
CeTe	183	1970	—
PrS	264	2190	2230 [41]
PrSe	224	2150	2100 [53]
PrTe	182	1930	1950 [50, 54]
NdS	261	2130	2140 [41, 42, 44], 2200 [43]
NdSe	222	2100	—
NdTe	180	1870	2040 [48, 57], 2025 [58], 1755 [55]
GdS	251	2020	—
GdSe	210	1880	—
GdTe	170	1655	1825 [56], 1870 [52]
TbS	249	1970	—
TbSe	208	1840	—
TbTe	167	1585	—
DyS	247	1940	—
DySe	206	1790	—
DyTe	165	1530	—
HoS	244	1890	—
HoSe	204	1760	—
HoTe	163	1520	—
ErS	243	1870	—
ErSe	202	1710	1630 [57], 1750 [48]
ErTe	160	1445	1610 [48], 1525 [57]
TmS	241	1840	—
TmSe	200	1670	—
TmTe	159	1410	—

Figures in brackets for LaS, LaSe, and LaTe are θ-data obtained from specific heat measurements at helium temperatures [37].

from formula (12) can be obtained with the accuracy of $\pm$ 100 deg. There are no T_m-values for a number of rare-earth monochalcogenides, and T_m is determined with great errors for the others. So even a rough estimation of T_m on the basis of θ-data could be interesting. Values of θ and T_m are given in Table 6.

Magnetic ordering of antiferromagnetic type takes place in the neodymium monochalcogenides NdS, NdSe, and NdTe at (8.5 $\pm$ 0.5), 10.6, and 10.2 °K, respectively [59, 6]. The temperature dependence of specific heat at the given temperatures reveals clearly visible peaks [6].

The same picture takes place in case of tullium compounds (see Fig. 2b). An anomaly of C_p is observed at the Néel point at temperatures of 5.18, 1.85, and 0.21 °K for TmS, TmSe, and TmTe, respectively [37] (the latter compound is a semiconductor).

The component C_f gives a considerable contribution to specific heat in PrS, PrSe, and PrTe at low temperatures [37]. In [37] it is supposed that the contribution of C_e and C^L to C_p^{tot} will be similar in the pairs of compounds LaS and PrS, LaSe and PrSe, LaTe and PrTe and the difference of their total specific heat is equal to C_f (e.g., $C_p^{tot}(\mathrm{PrS}) - C_p^{tot}(\mathrm{LaS}) = C_f(\mathrm{PrS})$). The deviation of the straight lines $\ln(C_p^{tot}(\mathrm{PrX}) - C_p^{tot}(\mathrm{LaX}))$ from $1/T$ (where X = S, Se, Te) gave the way to determine the value of energy gaps (Δ_1) between the first excited and the ground state levels for all three compounds:

compound	Δ_1/k_0 (°K)
PrS	130 ± 20
PrSe	118 ± 10
PrTe	75 ± 5
	(79 ± 12)

(The figures in brackets for PrTe represent the data obtained from non-elastic neutron scattering measurements [37].)

2.1.2 Rare-earth monochalcogenides with non-metallic conductivity

2.1.2.1 SmS, SmSe

The current-carrier concentration in these compounds is $\approx 10^{19}\,\mathrm{cm}^{-3}$ so that their electronic component C_e is negligibly small. However, additional Schottky-type specific heat appears in SmS and SmSe due to thermal excitation of 4f electrons from the main state of multiplet $^8F_{1/2}$ to higher energy levels [20]. Therefore, in this case C_v^{tot} can be written in the following way:

$$C_v^{tot} = C_v^L + C_f\,. \tag{13}$$

Since there is no information on the influence of the lattice crystal field on the multiplet ^8F, C_f was calculated according to formula (13) on the assumption that levels in the compounds do not differ from the levels of the free ion Sm^{2+} [60]. The paper [20] took into account the contribution of three excited levels $F_{3/2}$, $F_{5/2}$, and $F_{7/2}$ located at distances of 326.64, 511.58, and 1489.16 cm^{-1}, respectively, from the main state $F_{1/2}$ [60]. It appeared that at 400 °K the filling of these levels was 30.4% and 1%, respectively. Calculated and experimental results for specific heats of SmS and SmSe are given in Fig. 4. The Debye temperatures calculated at 100 °K from the data of C_v^L for SmS and SmSe are equal to 269 and 206 °K, respectively.

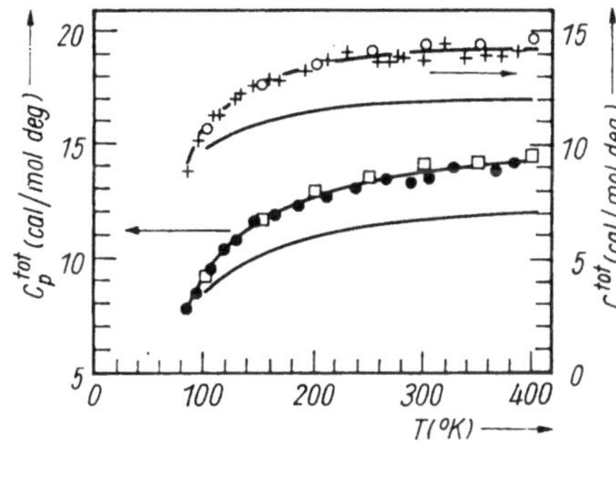

Fig. 4. Temperature dependence of specific heat of SmS and SmSe [20]. +, ● Experimental values; ○, □ calculated values $C_p^L + C_f$; solid lines calculated values C_p^L. The right-hand ordinate refers to SmSe and the left-hand one to SmS

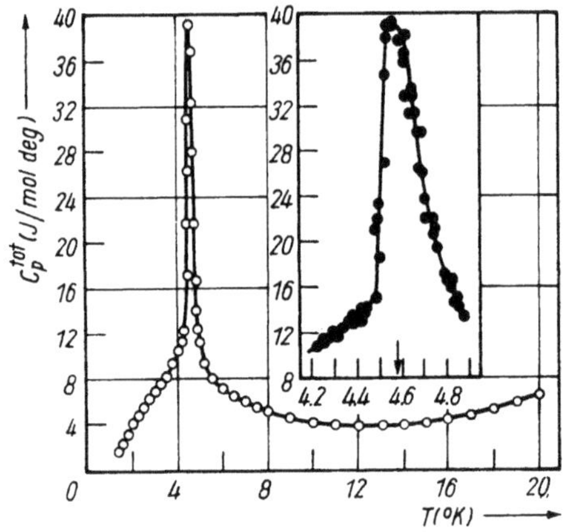

Fig. 5. Temperature dependence of specific heat of EuSe [8]. The dependence of $C_p(T)$ in the vicinity of T_N is shown in detail in the inset

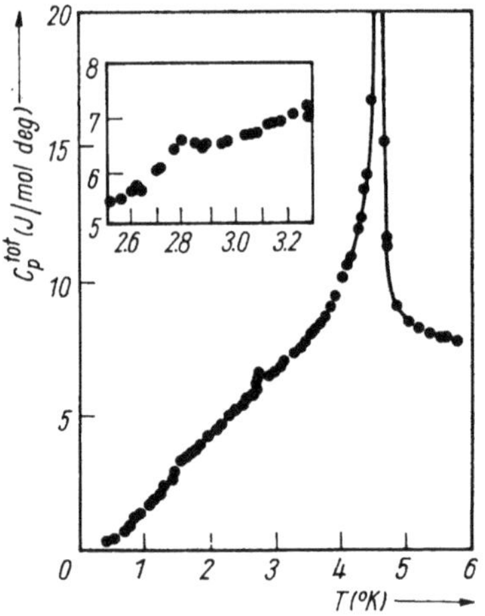

Fig. 6. Temperature dependence of specific heat of EuSe [10]. The vicinity of the peak at 2.8 °K is shown in an enlarged scale in the inset. (A small peak in the region of 2 °K is due to the presence of impurities in the specimen. This peak is absent in purer specimens)

2.1.2.2 EuTe, EuSe, EuS

The largest number of publications is dealing with these compounds. At low temperatures EuTe is antiferromagnetic, EuS is ferromagnetic, EuSe between 4 and 2.8 °K is antiferromagnetic and below 2.8 °K it is ferromagnetic (or ferrimagnetic) [61]. So the total specific heat of these compounds can be expressed as the sum of lattice and magnon components:

$$C_v^{\text{tot}} = C_v^{\text{L}} + C_{\text{m}} \,. \tag{14}$$

(Since europium monochalcogenides are insulators, the electronic component vanishes. C_{f} (see Section 3) also gives no contribution in these compounds.)

Analysing the data on specific heat of europium monochalcogenides one gets the following parameters:

a) magnetic ordering temperature (Curie temperature, T_{C}, and Néel temperature, T_{N}); b) Debye temperature; c) absolute values of exchange integrals and magnetic entropy.

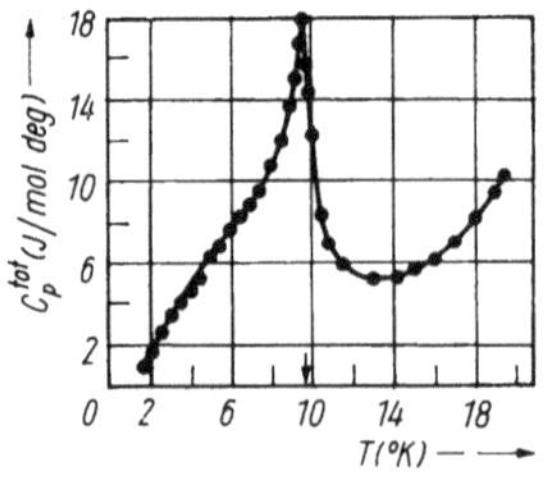

Fig. 7. Temperature dependence of specific heat of EuTe [8]

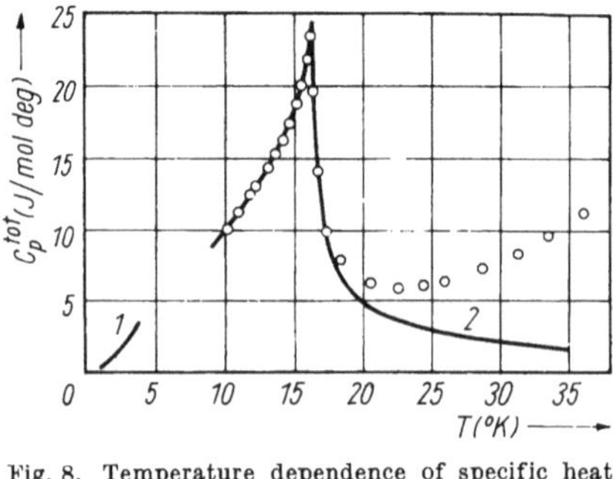

Fig. 8. Temperature dependence of specific heat of EuS. (1) Experimental results [11]; circles indicate the total specific heat [13]. (2) Magnetic specific heat (solid curve) [13]

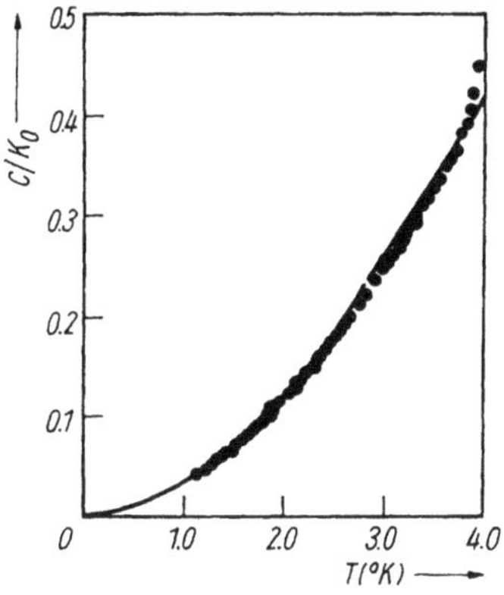

Fig. 9. Temperature dependence of specific heat of EuS. Full circles: experimental data [11]; solid curve: calculation [12]

Fig. 5 to 9 give the temperature dependence of specific heat for all europium monochalcogenides. Peaks in specific heat appear at magnetic ordering temperatures, the absolute values of which are given in Table 7 (column 2).

Also a specific heat anomaly for EuSe [10] was revealed at 2.8 °K. It is connected with the transition from antiferromagnetic to ferromagnetic ordering at lower temperatures.

Table 7

Values of T_C, T_N, and θ for europium monochalcogenides obtained from specific heat data

compounds	magnetic ordering temperature (°K)	T_C, T_N	θ (°K)
EuS	16.3 ± 0.1 [7,8] 16.2 ± 0.1 [13] 17.1 [11] 16.426, 16.427 [15]	T_C	≈ 200 [11] 204 [62] 208 [13]
EuSe	4.48 ± 0.03 [7, 8] 4.6 [10]	T_N	180 [8]
EuTe	9.64 ± 0.06 [7, 8] 9.81 ± 0.05 [9]	T_N	150 ± 10 [9] 140 [8]

The separation of the total specific heat into C_v^L and C_m is often done graphically. Besides, C_m can be calculated by (6) and (7) by selecting the values for exchange integrals. A better agreement of experiment and theory is obtained when taking into account the exchange interaction with the nearest (J_1) and second-nearest (J_2) neighbours [9, 12, 62, 63]. The absolute values of exchange integrals obtained in this way are given in Table 8. The experimental values of exchange integrals agree well with those calculated on the basis of the spin-wave theory for the Heisenberg ferromagnet [11, 62].

The calculations show that the contribution of the lattice specific heat to the total specific heat is negligibly small at low temperatures ($T < 10\cdots15$ °K), making $\approx 0.5\%$ to 1% of the total specific heat below 4 °K [11], so that in practice $C_m = C_v^{tot}$ (see Fig. 8). The lattice component becomes noticeable only from (15···20) °K and above. The Debye temperatures for all europium monochalcogenides were calculated from C_v^L at these temperatures (see Table 7, column 4).

The magnetic entropy can be determined from the magnetic specific heat and then compared to theoretical entropy. This was done in papers [8] and [9] (see Table 9).

Very interesting results for the theory of phase transitions can be obtained from the analysis of specific heat behaviour in the vicinity of magnetic ordering

Table 8

Values of exchange integrals for EuS and EuTe calculated from specific heat data

compound	exchange integrals J_1/k_0	exchange integrals J_2/k_0	references
EuS	0.20 ± 0.01	-0.08 ± 0.02	[63]
	0.20	−0.07	[63]
	0.172 ± 0.019	-0.012 ± 0.032	[12]
EuTe	0.02 ± 0.02	-0.12 ± 0.02	[9]

temperatures. A jump in the specific heat curve indicates that the observed phase transition is possibly a second-order transition [64].

Papers [15], [16], [65], and [66] consider the behaviour of magnetic specific heat of EuS in the vicinity of T_C. The temperature dependence of C_m in the vicinity of T_C can be written as follows [15]:

$$C_m = \left(\frac{A}{\alpha}\right)(\varepsilon^{-\alpha} - 1) + B \quad \text{for} \quad T \to T_C^+, \tag{15}$$

$$C_m = \left(\frac{A'}{\alpha'}\right)(\varepsilon^{-\alpha'} - 1) + B' \quad \text{for} \quad T \to T_C^-, \tag{16}$$

where $\varepsilon = (T - T_C)/T$. When $\alpha = 0$ formula (15) becomes $C_m = -a \ln \varepsilon + b$ (formula (16) can be written analogously at $\alpha' = 0$).

According to the analysis of C_m-data in paper [15] for the regions above and below T_C but in the vicinity of this temperature (T_C^+ and T_C^-), the following parameters were obtained for equations (15) and (16):

$T \to T_C^+$:	$T \to T_C^-$:
$\alpha = 0.00 \pm 0.03$	$\alpha' = -0.25 \pm 0.03$
$A = 4.21$	$A' = 7.59$
$B = -2.5$	$B' = 4.33$
$a = 9.47$	—
$b = -2.5$	—
$T_C = 16.427$ °K	$T_C = 16.426$ °K

(A, B, A', B', a, b are expressed in J mol^{-1} deg^{-1}).

Table 9

Experimental values of magnetic entropy for europium monochalcogenides obtained from specific heat data and comparison with theory

compound	experimental values of magnetic entropy (J mol^{-1} deg^{-1}) $S_m = \int_0^\infty \left(\frac{C_m}{T}\right) dT$	theoretical values of magnetic entropy (J mol^{-1} deg^{-1}) $S_m = R \ln (2s + 1)$
EuSe	16 [8]	17.3
EuTe	16.9 [9]	17.3

Fig. 10. Temperature dependence of specific heat of EuS [14]. (1) $H_{\text{intern}} = 0$; (2) 5820 Oe; (3) 10800 Oe. Circles: experimental data; solid curves: calculated data

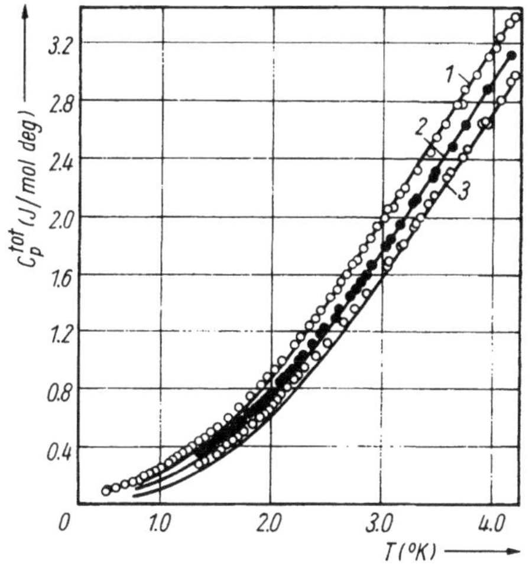

The fact that $\alpha = 0$ in EuS is also supported by the measurements of C_m in a magnetic field [15].

The specific heat of EuS was measured at low temperatures in a magnetic field [14, 15, 18, 19]. Fig. 10 shows the experimental results obtained in [14]. C_v^L is negligibly small in the temperature range under investigation. C_m decreases with the increase of the applied magnetic field [14].

The addition of a small amount of trivalent metal impurities to europium chalcogenides produces anomalies in magnetic properties and transport phenomena. A slight increase of T_C is also observed with increasing impurity concentration. It is interesting to observe the behaviour of gadolinium-doped EuS (see Fig. 11) [17]. When the gadolinium concentration increases, the transition temperature rises and the cooperative character of the transition gets considerably destroyed.

2.2 Me_3X_4-type compounds

Fig. 12 (curves 1 to 4) gives data on the temperature dependence of the specific heat of La_3Te_4, Ce_3Te_4, Pr_3Te_4, and Nd_3Te_4 [28]. These compounds crystallize in a Th_3P_4-type cubic lattice. The figure shows that the specific heat of La_3Te_4 differs from that of the other compounds both by its absolute value and its temperature

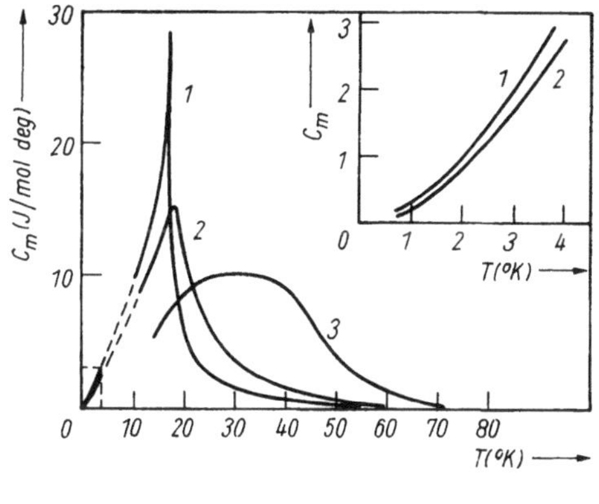

Fig. 11. Temperature dependence of specific heat of $Eu_{1-x}Gd_xS$ [17]. (1) $x = 0$; (2) 0.01; (3) 0.1. Low-temperature data from [11] are given in an enlarged scale in the inset

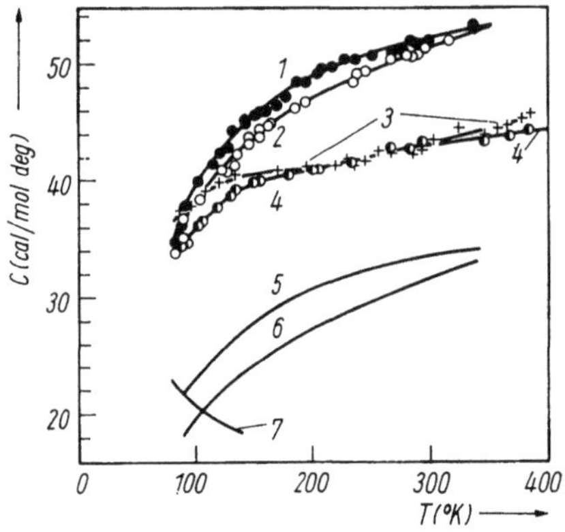

Fig. 12. Temperature dependence of total specific heat for a number of Me_3X_4-type compounds [28]. (1) La_3Te_4; (2) Ce_3Te_4; (3) Pr_3Te_4; (4) Nd_3Te_4. — Temperature dependence C_f: (5) Pr_3Te_4; (6) Ce_3Te_4; (7) Nd_3Te_4

dependence. For example, the specific heat of Ce_3Te_4 and Pr_3Te_4 at 340 °K is 25 to 28% higher than the specific heat of La_3Te_4. What is the reason of the observed difference? All the compounds studied have the same crystal structure and similar values of T_m and density. So it could be considered that C_v^L of all compounds is very similar. The electronic component of specific heat is very small due to the insignificant value of current carrier concentration ($\approx 10^{21}$ cm^{-3}). These compounds are paramagnetic in the considered temperature range, so their C_m equals zero. Therefore, the observed difference occurs most probably due to the Schottky component C_f. There are no 4f electrons in lanthanum, so there is no contribution of C_f to the specific heat of La_3Te_4 and the experimental value of C_v^{tot} can be described only by the lattice component.[5]) Ce, Pr, and Nd have non-filled 4f shells. Terms of trivalent ions of these elements are split (see Table 10) in the crystal field of the lattice into a number of levels. So the transitions of the 4f electrons between these levels lead to an increase in the specific heat in the chalcogenides of Nd, Pr, and Ce compared to that in La chalcogenide. Non-filled f-layers in rare-earth metals are located deep in the electron shell. Filled 5s and 5p levels screen the 4f layer from external influences. Due to the "screening" effect and the high localization of wave functions of 4f electrons the internal crystal field causes rather weak ($\approx$ 100 °K) splitting of electron levels in ions, so that the transitions between split levels will occur just at the considered temperatures. Thus C_v^{tot} for Me_3X_4-type Nd, Pr, and Ce chalcogenides can be represented by equation (13). Since it is very

Table 10

Splitting of the rare-earth ions in a field of cubic symmetry

ion	Ce^{3+}	Pr^{3+}	Nd^{3+}
term of ground state	$^2F_{5/2}$	$^3H_4$	$^4I_{9/2}$
number of levels in cubic symmetry field	2	4	3
statistical weight of the levels	2, 4	1, 2, 3, 3	2, 4, 4

[5]) At $T > 300$ °K the specific heat of La_3Te_4 insignificantly exceeds the values calculated by the Dulong-Petit law. Most probably this occurs due to the anharmonic contribution to the specific heat which appears at high temperatures.

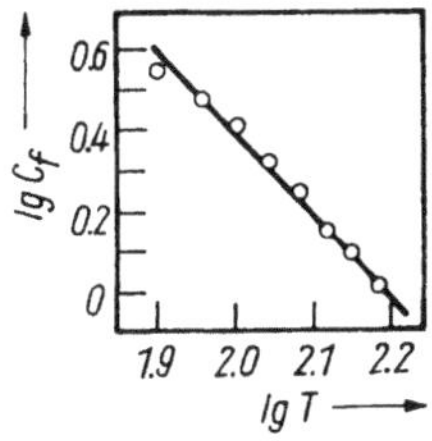

Fig. 13. Dependence of lg C_f on lg T for Nd_3Te_4 [28]

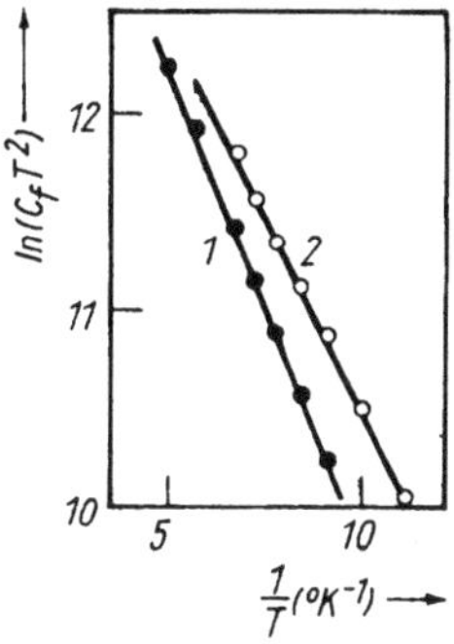

Fig. 14. Dependence of ln (C_f T^2) on $1/T$ for Pr_3Te_4 (1) and Ce_3Te_4 (2) [28]

complicated to separate the lattice component of specific heat by calculation in Nd_3Te_4, Ce_3Te_4, and Pr_3Te_4, it is better to suppose in a first approximation that C_v^L is the same as that of La_3Te_4 and that the differences of experimental values of specific heat of Nd, Pr, Ce tellurides and La_3Te_4 determine C_f (see Fig. 12, curves 5 to 7).

The shape of the curves 5 to 7 in Fig. 12 gives the idea that at experimental temperatures for Nd_3Te_4 we are at the right-hand (high-temperature) slope and for Pr_3Te_4 and Ce_3Te_4 at the left-hand (low-temperature) slope of the Schottky curve for C_f, since in the first case the dependence (10), $C_f \sim 1/T^2$ (see Fig. 14), and in the second case the dependence (9), $C_f \sim \exp [\Delta_1/(k_0 T)]$ (see Fig. 13), are true. The energy gaps between the ground and the nearest excited states in Ce_3Te_4 and Pr_3Te_4 calculated from Fig. 13 seem to equal (400 ± 30) °K and (500 ± 40) °K, respectively. The more thorough analysis of temperature dependence of C_f in Ce_3Te_4 and Pr_3Te_4 shows that at high temperatures the transitions of 4f electrons to higher energy levels start to influence.

The specific heat of $La_{2.91}[\ ]_{0.09}Se_4$ within the range of 1 to 10 °K was investigated in paper [29] ([ ] indicates a vacancy).

Formula (5) was used to determine the value of the effective mass

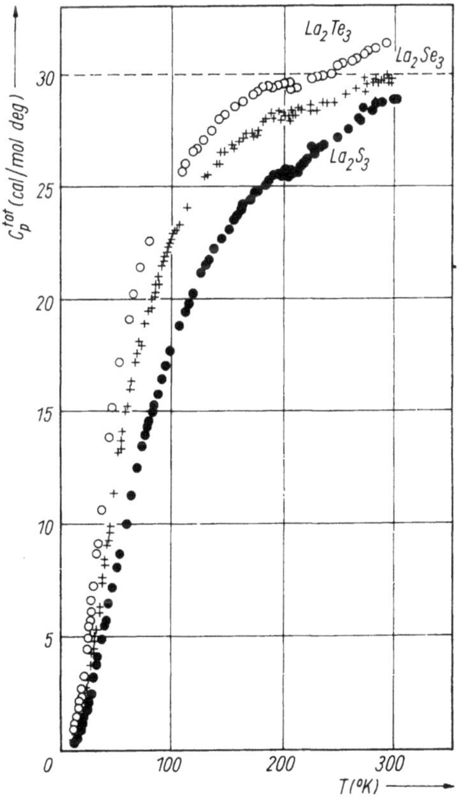

Fig. 15. Temperature dependence of heat capacity of La_2Te_3 [22], La_2Se_3 [23], and La_2S_3 [24]

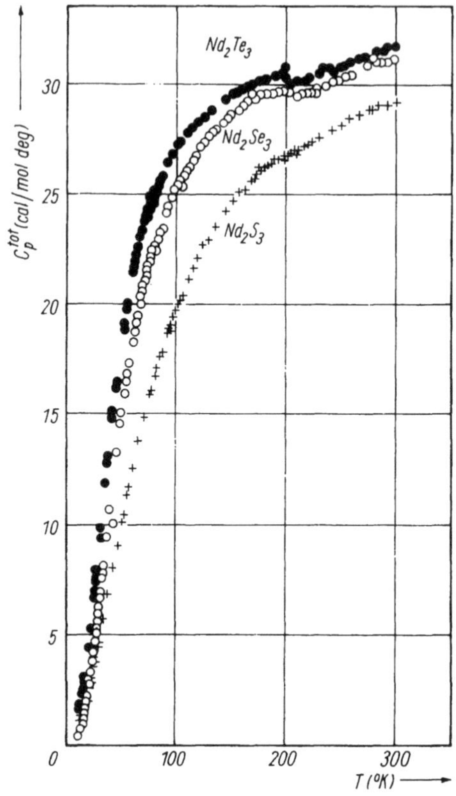

Fig. 16. Temperature dependence of specific heat of Nd_2Te_3 [25], Nd_2Se_3 [26], and Nd_2S_3 [27]

of electrons from the electronic component of specific heat:

$$m^* = (2.3 \pm 0.2)\, m_0 .$$

2.3 Me_2X_3-type compounds

Fig. 15, 16, and 17 give the experimental data on specific heat of Me_2X_3-type chalcogenides presently available. It seems that the absolute value of specific heat of La_2Te_3, La_2Se_3, and La_2S_3 is considerably smaller than that of Nd_2Te_3, Nd_2Se_3, and Nd_2S_3. The analysis has shown that as in the case of Me_3X_4-type chalcogenides the observed difference is connected with the presence of an additional Schottky-type specific heat component C_f in Nd chalcogenides [67]. Thus the total specific heat for Ln_2X_3 can be written as follows:

$$C_{tot} = C_v^L ,$$

which becomes $C_v^{tot} = C_v^L + C_f$ for Nd_2X_3. C_v^L for Nd_2X_3 was calculated in paper [67] in the following way: The Debye temperature of lanthanum chalcogenides was determined for a number of temperatures; then curves were plotted showing the dependence of θ on average atomic weight. The temperature dependence θ of Nd_2X_3 was determined with the help of these curves by the known average atomic weight of Nd chalcogenides, and on the basis of this temperature dependence the lattice specific heat was calculated. The difference between the experimental values of specific heat C_v^{tot} of Nd_2X_3 taken from Fig. 16 and those calculated for C_v^L determines C_f. It is given in Fig. 18a, b, c for the temperature range of 13 to 120 °K.

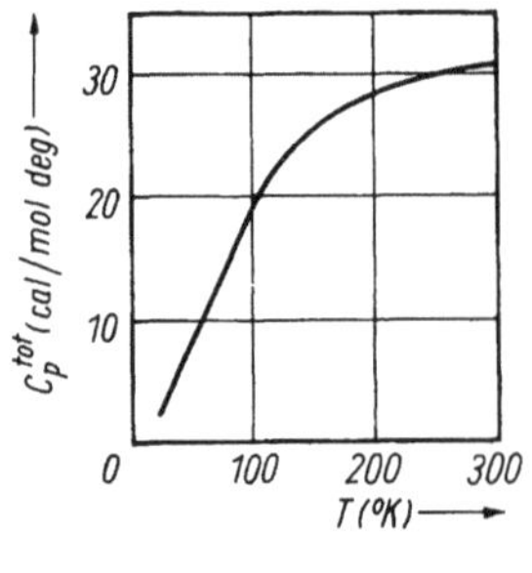

Fig. 17. Temperature dependence of specific heat of Ce_2S_3 [4]

A good agreement of experimental and theoretical data is obtained when the transitions to the first two excited levels are taken into account. In this case the equation for C_f differs a little from (8):

$$C_f = R \frac{\frac{g_1}{g_0}\left(\frac{\Delta_1}{k_0 T}\right)^2 \exp\left(-\frac{\Delta_1}{k_0 T}\right) + \frac{g_2}{g_0}\left(\frac{\Delta_2}{k_0 T}\right)^2 \exp\left(-\frac{\Delta_2}{k_0 T}\right)}{\left[1 + \frac{g_1}{g_0}\exp\left(-\frac{\Delta_1}{k_0 T}\right) + \frac{g_2}{g_0}\exp\left(-\frac{\Delta_2}{k_0 T}\right)\right]}, \tag{17}$$

where Δ_2 is the distance from the main level to the second excited levels.

The parameters at which C_f calculated by formula (17) agrees with the experiment best of all are given in Table 11.

Curves $C_p^{tot} = f(T)$ in the investigated rare-earth Me_2X_3-type chalcogenides show breaks and peaks (see Fig. 15, 16 and Table 12). The causes of these anomalies are not yet clear.

The experimentally measured specific heat of La_2X_3 equals the lattice component. This allows to calculate the Debye temperature (θ was determined at 70 °K):

compound	θ (°K)
La_2S_3	320
La_2Se_3	240
La_2Te_3	190

3. Thermal Conductivity of Rare-Earth Chalcogenides

Tables 13 to 15 give the list of rare-earth chalcogenides for which thermal conductivity is measured by now. It appears that in order to analyse the experimental data it is necessary to involve all thermal conductivity components (excluding $\varkappa_{bip}$) of equation (3). To facilitate the discussion of the experimental results it would be reasonable to separate the basic formulas for the thermal conductivity components most frequently used. The formula required for the discussion of details will be further given in sections directly dealing with experimental data.

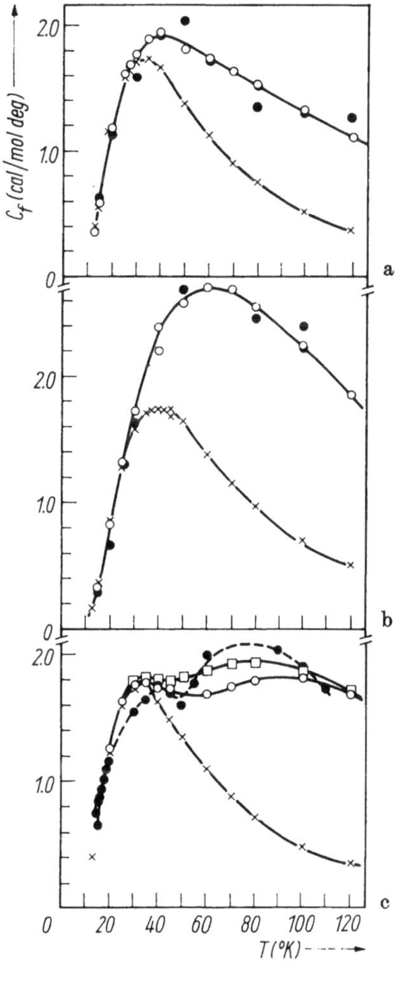

Fig. 18. Temperature dependence of C_f for a) Nd_2S_3, b) Nd_2Se_3, and c) Nd_2Te_3 [67]. ● $C_{f\,exp}$, × $C_{f\,calc}$, taking into account the first excited level; ○ $C_{f\,calc}$ with the account of the contribution of two excited levels $\Delta_1/k_0 = 76$ °K and $\Delta_2/k_0 = 325$ °K; □ the same at $\Delta_1/k_0 = 76$ °K and $\Delta_2/k_0 = 300$ °K

Table 11

Main parameters of excited 4f levels in Nd_2X_3 crystals

compound	Nd_2S_3	Nd_2Se_3	Nd_2Te_3
Δ_1/k_0 (°K)	78	93	76
g_1/g_0	1	1	1
Δ_2/k_0 (°K)	250	250	300 to 325
g_2/g_0	1	2	2

Table 12

Anomalies observed on $C_p^{\text{tot}} = f(T)$ curves for Me_2X_3 rare-earth chalcogenides

compound	crystal structure	T (°K) at which the anomaly is observed
La_2Te_3	Th_3P_4	≈ 204 [22]
La_2S_3	β-phase with impurity (< 2%) of α-phase	190 to 210 [24]
La_2Se_3	Th_3P_4	175 to 270 [23]
Nd_2Te_3	rhombic, Sb_2S_3-type	199 [25] 26 to 28 [25] 245 [25]
Nd_2S_3	β-phase with impurity of α-phase	189 to 199 [27]
Nd_2Se_3	Th_3P_4	160 to 240 [26]

α-phase: orthorhombic lattice; β-phase: tetragonal lattice.

Lattice thermal conductivity

Formula (18) is frequently used to calculate quantitatively the crystal lattice thermal conductivity in Debye approximation:

$$\varkappa_{\text{L}} = G\, T^3 \int\limits_0^{\theta/T} \frac{x^4\, \mathrm{d}x}{\tau^{-1} \sinh^2 (x/2)}, \tag{18}$$

where $G = [\hbar/(8\,\pi^2\, v)]\, (k_0/\hbar)^4$, $x = \hbar\omega/k_0 T$, ω is the phonon frequency, v the average sound velocity, and τ^{-1} the inverse relaxation time of phonons. If there are several scattering processes in a solid (e.g., phonon–phonon scattering, scattering by crystal boundaries, impurities, isotopes, dislocations, etc.)[6]), the total relaxation time can be represented in the form

$$\tau^{-1} = \sum \tau_j^{-1}\,. \tag{19}$$

[6]) Formula (19) is true only in case of addition of inverse relaxation times for U-processes (which conserve the total momentum in a crystal). The addition for U- and N-processes (N-processes conserve the total momentum) is effected with the help of a more complicated dependence [97].

Table 13

MeX-type rare-earth chalcogenides
for which thermal conductivity was measured

compound	temperature range T (°K)	references
LaTe	80 to 460	[68]
LaSe	300	[69, 97]
	80 to 460	[68]
LaS	300	[69, 70]
	80 to 460	[68]
CeTe	300	[69]
CeSe	300	[69]
CeS	300	[69, 70]
PrTe	80 to 460	[71]
	300	[69]
PrSe	80 to 460	[71]
	300	[69, 86]
PrS	80 to 460	[71]
NdTe	2 to 100	[6]
	300	[87]
NdSe	2 to 100	[6]
	300	[69]
NdS	2 to 100	[6]
	300	[69]
SmS	80 to 300	[71, 72]
	300	[73]
	300 to 1000	[74]
SmTe	300	[107]
EuS	1.5 to 4.2	[75 to 77]
GdTe	300	[56]
GdS	80 to 725	[78]
YbTe	300	[79, 106]
DyTe	300	[106]

It is necessary to note that at high temperature ($T > \theta$) the lattice thermal resistivity of an ideal crystal lattice for three-phonon processes varies in accordance with the law

$$W_L^0 = \frac{1}{\varkappa_L} = A\,T\,, \tag{20}$$

and any defects give an additional thermal resistivity which is independent of temperature, i.e.,

$$W_L^{\mathrm{impur}} = A\,T + B = W_L^0 + \Delta W\,. \tag{21}$$

Table 14

Me_3X_4-type rare-earth chalcogenides for which thermal conductivity was measured

compound	temperature range (°K)	references
La_3Te_4	80 to 460	[80, 92]
	300 to 1250	[93]
	300 to 600	[94]
	300	[81]
La_3Se_4	300	[83, 96]
La_3S_4	300	[83, 96]
	300 to 1250	[93]
Ce_3Te_4	300	[81, 96]
Ce_3Se_4	300	[93]
	80 to 600	[92]
Ce_3S_4	300	[81]
	300 and 1300	[95]
Pr_3Te_4	300	[54, 96]
Pr_3Se_4	300	[86, 93, 96]
Nd_3Te_4	300	[55, 58, 81, 87, 93, 94, 96]
Nd_3S_4	1.5 to 120	[30]
	80 to 700	[92]
Sm_3S_4	300	[107]
	80 to 700	[52]
Gd_3Te_4	300	[88]
Dy_3Te_4	300	[81, 106]

Electronic component of thermal conductivity

The electronic component of thermal conductivity is calculated by the Wiedemann-Franz law:

$$\varkappa_e = L_e \sigma T , \tag{22}$$

where L_e is the Lorenz number. L_e for semiconductors with parabolic band and elastic scattering of current carriers can be written as

$$L_e = \left[\frac{\left(r + \frac{7}{2}\right) F_{r+5/2}(\xi^*)}{\left(r + \frac{3}{2}\right) F_{r+1/2}(\xi^*)} - \frac{\left(r + \frac{5}{2}\right)^2 F^2_{r+3/2}(\xi^*)}{\left(r + \frac{3}{2}\right)^2 F_{r+1/2}(\xi^*)} \right] \left(\frac{k_0}{e}\right)^2 , \tag{23}$$

where F are Fermi integrals, $\xi^* = \xi/k_0T$ is the reduced Fermi level, and r is the scattering parameter (r is the exponent in dependence of the current-carrier relaxation time on energy). The Lorenz number in non-degenerate samples is equal to

$$L_e = \left(r + \frac{5}{2}\right) \left(\frac{k_0}{e}\right)^2 \tag{24}$$

and depends on the scattering parameters.

Table 15

Me_2X_3-type rare-earth chalcogenides for which thermal conductivity was measured

compound	temperature range (°K)	references
La_2Te_3	80 to 300	[80, 92]
	300	[81, 96]
La_2Se_3	300	[82, 83]
La_2S_3	300	[70, 83, 84, 96]
Ce_2Te_3	300	[81, 96]
Ce_2Se_3	300	[82]
Ce_2S_3	300	[70, 84]
	300 and 1200	[85]
Pr_2Te_3	300	[54, 81, 96]
Pr_2Se_3	300	[86, 96]
Pr_2S_3	300	[70, 84]
Nd_2Te_3	300	[58, 81, 87, 88, 96]
Nd_2Se_3	300	[82, 96]
Nd_2S_3	300	[70, 84]
	80 to 380	[89]
Sm_2Te_3	300	[105]
Gd_2Te_3	300	[56, 88, 90, 91]

Note: Paper [81] investigated not exactly Me_2X_3 compounds, but those slightly displaced to Me_3X_4.

In the case of strong degeneration ($\xi^* \gg 1$) in semiconductors, the Lorenz number, as for pure metals and metal alloys at $T > \theta$, is a universal constant:

$$L_0 = \frac{\pi^2}{3}\left(\frac{k_0}{e}\right)^2. \tag{25}$$

Usually ξ^* is determined from the experimental values of thermoelectric power.

In case of non-parabolic band, non-elastic current-carrier scattering mechanism, or the appearance of additional subbands with "heavy" or "light" current carriers the formulas for L_e become much more complicated [98]. A decrease of L_e is observed in the first two cases, while the latter case (depending on the location of the Fermi level with respect to the top of the additional subband) gives either an increase or decrease of the Lorenz number compared to L_e calculated by formula (23).

Photon component of thermal conductivity

In optically isotropic substances at small temperature gradients, when the absorption coefficient, K, does not depend on the wavelength and $1/K$ is small compared to the linear dimensions of a sample, it is possible to express $\varkappa_{ph}$ for the heat transfer by electromagnetic irradiation (photons) as follows [99]:

$$\varkappa_{ph} = \frac{16}{3} \frac{n^2 \sigma_0 T^3}{K}, \tag{26}$$

where n is the refractive index and $\sigma_0 = 1.38 \times 10^{-12}$ cal cm^{-2} deg^{-4} is the Stefan-Boltzmann constant. (Formulas for $\varkappa_{ph}$ for the cases when K depends on the wavelength and those for anisotropic crystals are given in paper [99].) Rare-earth compounds are expected to reveal the *resonance phonon scattering on paramagnetic ions* [104].

Orbital and spin moments of atoms and ions with completely filled shells are equal to zero, and the above-mentioned effect is not observed. The situation is quite different with ions with non-filled shells. When free, an ion with total moment $\boldsymbol{J}$ is in a state $(2\,J + 1)$-fold degenerated in the direction of $\boldsymbol{J}$. In the crystal the ion is located in the crystal electric field which influences the shell electrons. This field has the symmetry of the crystal and variously oriented ions may have different energy. So the removal of degeneration along orientations of $\boldsymbol{J}$ takes place: a system of energy levels appears instead of one energy level. Lattice vibration can move an ion from one level to another changing the orientation of $\boldsymbol{J}$. That causes the absorption of a phonon. Thus, such ions are phonon-scattering defects.

If the uncompensated shell is an outer one, then the crystal field influence is very strong and split levels are spaced at magnitudes of thousand degrees. As a rule this value exceeds the phonon energy and does not influence thermal conductivity. Rare-earth metals have uncompensated inner shells screened by outer shells from the crystal field. So the splitting is smaller, in the order of one hundred degrees only, and under usual conditions phonons can cause transitions between these levels.

The decrease of thermal conductivity due to this effect in case of two levels with energy E and $E + \Delta E$ can be written according to [104] as follows:

$$-\Delta\varkappa = \frac{G\,T^3\,\Delta T_{Res}}{4\left(T + \dfrac{D}{T_{Res}}\right)\sinh^2\dfrac{T_{Res}}{2\,T}\sinh\dfrac{T_{Res}}{T}}\left[1 - \frac{1}{1 + \dfrac{M(e^{T_{Res}/T} - 1)}{T_{Res}^2\,(T + D/T_{Res})}}\right], \tag{27}$$

where $T_{Res} = \Delta E/k_0$ is the resonance temperature, ΔT_{Res} the temperature resonance width, G, M, and D are constant. At high temperatures $(T \gg T_{Res})$ equation (27) looks as follows:

$$-\Delta\varkappa = \frac{\text{const}}{T_{Res}^2\,T^2}. \tag{28}$$

Thus, at high temperatures $-\Delta\varkappa$ is inversely proportional to T^2, but when the temperature decreases, $-\Delta\varkappa$ passes a maximum and then tends to zero due to freezing out the resonance phonons.

The decrease of thermal conductivity due to the above discussed effect (see Table 16) can be expected in all rare-earth chalcogenides (excluding La, Lu, and Eu compounds) which crystallize in a cubic lattice. The corresponding temperature range depends on the value of the level splitting energy. If ΔE and T_{Res} are small ($\approx$ 10 to 40 °K), then the condition $T \gg T_{Res}$ is true within the range of medium and high temperatures, equation (28) holds, and the resonance scattering effect is insignificant. In this case it is difficult to distinguish it from the

Table 16

Properties of trivalent rare-earth ions [108]

ion	term of ground state	splitting of ground level in a field of cubic symmetry			
		singlet	duplet	triplet	quadru-plet
1	2	3	4	5	6
La	$^1S_0$	1	—	—	—
Ce	$^2F_{5/2}$	—	1	—	1
Pr	$^3H_4$	1	1	2	—
Nd	$^4I_{9/2}$	—	1	—	2
Pm	$^5I_4$	1	1	2	—
Sm	$^6H_{5/2}$	—	1	—	1
Eu	$^7F_0$	1	—	—	—
Gd	$^8S_{7/2}$	—	2	—	1
Tb	$^7F_6$	2	1	3	—
Dy	$^6H_{15/2}$	—	2	—	3
Ho	$^5I_8$	1	2	4	—
Er	$^4I_{15/2}$	—	2	—	3
Tm	$^3H_6$	2	1	3	—
Yb	$^2F_{7/2}$	—	2	—	1
Lu	$^1S_0$	1	—	—	—

effect caused by impurities and defects, since in both cases the additional thermal resistivity is independent of temperature. At $T_{Res} \approx 100$ °K it is possible to reveal the resonance scattering effect within the medium temperature range.

Experimental data for thermal conductivity will be discussed separately for MeX-, Me_2X_3-, and Me_3X_4-type compounds.

3.1 MeX-type compounds

3.1.1 Rare-earth monochalcogenides with metallic conductivity

This class of compounds is very interesting for the investigation of thermal conductivity. Due to the high current-carrier concentration ($\approx 10^{22}$ cm^{-3}) in the thermal resistivity of the crystal lattice of these compounds an important role is played by the phonon scattering on conduction electrons, besides the main three-phonon scattering mechanism.

Only those phonons the wave vector $\boldsymbol{q}$ of which is less than $2\,\boldsymbol{k}_e$ (k_e is the absolute value of the electron wave vector) can interact with electrons. $\boldsymbol{k}_e$ in metals at the Fermi level is larger than the maximum wave vector of phonons, so all phonons can interact with electrons and the effect of phonon–electron scattering must occur at each temperature. There was no chance to observe this effect in pure metals at high temperatures due to the difficulty of separating the lattice component of thermal conductivity. Usually in such metals $\varkappa_e \gg \varkappa_L$ and $\varkappa_{tot} \approx \varkappa_e$ is measured. The situation is different in rare-earth metals and monochalcogenides with metallic character of conductivity. It is possible to separate the electronic and the lattice components of thermal conductivity due to the fact that these compounds have a small mobility of current carriers

a

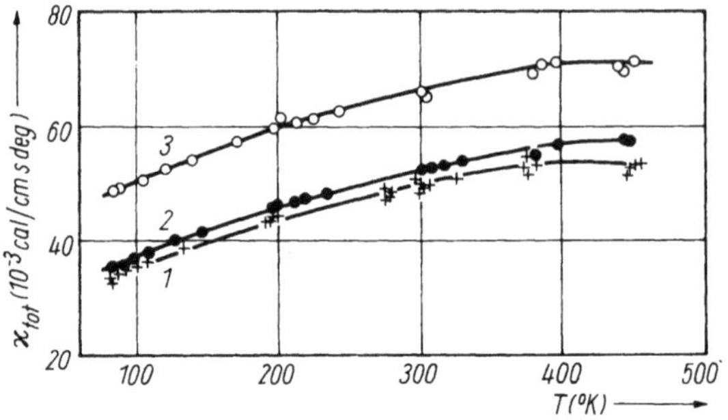

Fig. 19. a) Temperature dependence of a) $\varkappa_{tot}$ and b) ϱ of LaTe (1), LaSe (2), and LaS (3) [68]

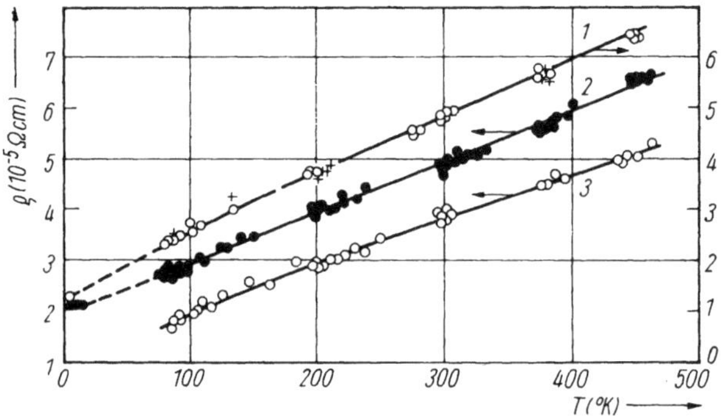

b

which leads to a decrease of σ and, consequently, of $\varkappa_e$ as well. Thus $\varkappa_L$ becomes a value comparable by its absolute value to $\varkappa_e$.

Fig. 19a, b, Fig. 20a, b, and Fig. 21a, b give the most reliable experimental results for thermal conductivity (as well as for σ or ϱ) of some rare-earth monochalcogenides. Let us discuss in detail the data for thermal conductivity of lanthanum and gadolinium sulphides [78].

Fig. 22 gives the values of thermal resistivity of the crystal lattice of these compounds. The electronic thermal conductivity was calculated by formula (22). For temperatures $T > \theta$, L_e was determined by formula (25). Difficulties appear for $T < \theta$. Here the Lorenz number is not yet a constant but, depending on the purity of the specimen, changes from zero to L_0 (see Fig. 23). L_e/L_0 for metals

a

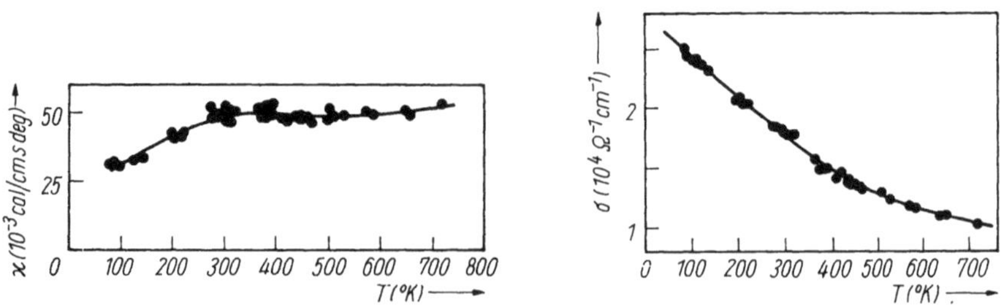

b

Fig. 20. Temperature dependence of a) $\varkappa_{tot}$ and b) σ of GdS [78]

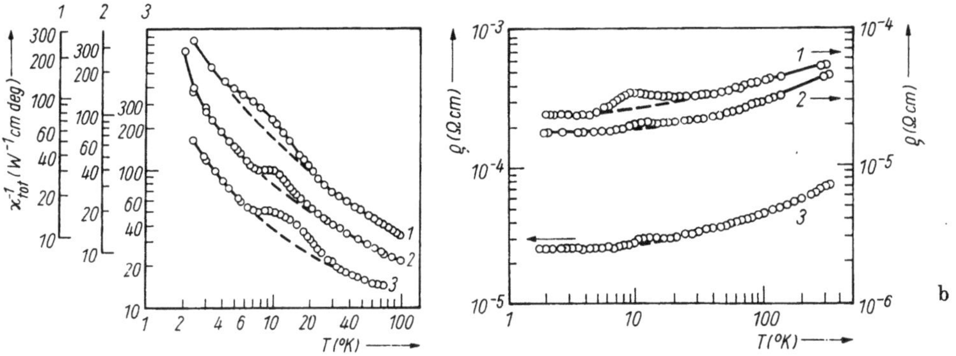

Fig. 21. Temperature dependences of total thermal resistivity a) $W_{tot} = 1/\varkappa_{tot}$ and b) ϱ of NdS (1), NdSe (2), and NdTe (3) [6]

can be written for temperatures above and below θ as follows [100]:

$$\frac{L_e}{L_0} = \frac{\dfrac{\varrho_0}{4 A_0} + \left(\dfrac{T}{\theta}\right)^5 J_5\left(\dfrac{\theta}{T}\right)}{\dfrac{\varrho_0}{4 A_0} + \chi\left(\dfrac{T}{\theta}, \dfrac{D}{\xi}\right)}, \tag{29}$$

where ϱ_0 is the residual resistance,

$$\chi\left(\frac{T}{\theta}, \frac{D}{\xi}\right) = \left(\frac{T}{\theta}\right)^5 \left\{\left[1 + \frac{3}{2\pi^2}\frac{\xi}{D}\left(\frac{\theta}{T}\right)^2\right] J_5\left(\frac{\theta}{T}\right) - \frac{1}{2\pi^2} J_7\left(\frac{\theta}{T}\right)\right\},$$

ξ is the Fermi energy, $J_5(\theta/T)$ and $J_7(\theta/T)$ are integrals of the type[7])

$$J_n(x) = \int_0^x \frac{z^n \, dz}{(e^z - 1)(1 - e^{-z})},$$

A_0 is a constant including the constant of electron–phonon interaction [101], and D is also a constant. It follows from the theory that $D/\xi = 2^{-1/3} N_a^{-2/3}$ (N_a is the number of free electrons per atom) and that at $T/\theta > 0.6$, $\varrho_i = \varrho_{tot} - \varrho_0 = A_0 T/\theta$ (ϱ_i is the part of resistivity caused by the electron scat-

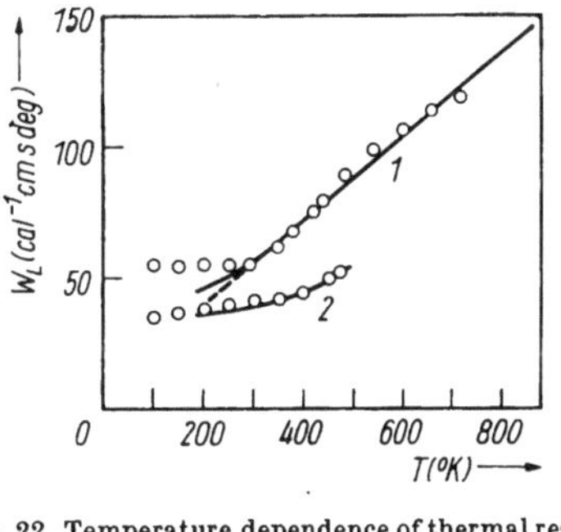

Fig. 22. Temperature dependence of thermal resistivity of crystal lattice of GdS (1) and LaS (2). Circles: experimental data; solid curves: theoretical data

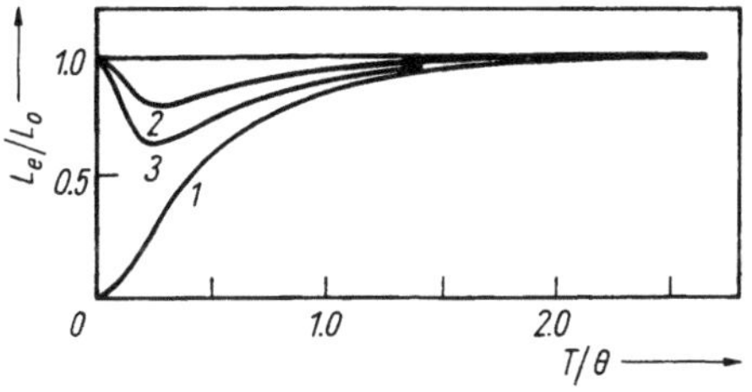

Fig. 23. Dependence of L_e/L_0 on T/θ (calculation by formula (29)). (1) Pure metals, (2) GdS, and (3) LaS

7) The integrals are tabulated in [102] and [103].

tering on lattice vibrations, ϱ_{tot} is the experimentally measured resistivity). The parameter $\varrho_0/(4\,A_0)$ is a measure of the purity of material. We get $\varrho_0/(4\,A_0) = 0$ for very pure metals.

The following results were obtained for the thermal resistivity of LaS and GdS. At temperatures exceeding 300 to 350 °K, for both compounds $W_L \sim T$, at lower temperatures the dependence of W_L on T becomes weaker. Such behaviour of the lattice thermal resistivity is not usual, since for the majority of compounds the law of $W_L \sim T$ is well observed up to temperatures of $(0.3\cdots0.5)\,\theta$. The above noticed peculiarity can be explained with the help of the additional mechanism of phonon scattering on conductivity electrons. Paper [104] shows that for the high-temperature region, simultaneously taking into account the three-phonon and phonon–electron scattering, formula (18) assumes the form

$$\varkappa_L = \frac{P\,\theta}{T}\left[1 - \frac{D}{\theta\,T}\ln\left(\frac{\theta\,T}{D} + 1\right)\right], \tag{30}$$

where P and D are constants determinable from the comparison of theoretical and experimental values of $\varkappa_L$ at two temperatures. The comparison for LaS and GdS was arranged at 300 and 450 °K. It resulted in the following values for the above-mentioned constants (P in cal cm^{-1} s^{-1} deg^{-1}, D in deg^2):

	LaS	GdS
P	0.089	0.0362
D	5.15×10^4	1.87×10^4

The theoretical dependence calculated by formula (30) is plotted in Fig. 22 by solid curves. The coincidence of theoretical and experimental results was very good for LaS. A noticeable difference between the calculated and experimental values of $\varkappa_L$ at $T < 300$ °K was observed for GdS. Such a difference may be caused by the additional phonon scattering on Gd paramagnetic levels split by the crystal field. This effect does not arise in principle in LaS.

Nd monochalcogenides show antiferromagnetic-type magnetic ordering in the region 8 to 10 °K. Fig. 21a, b give the temperature dependences of ϱ and $\varkappa_{tot}^{-1}$ for three Nd monochalcogenides. Both ϱ and $\varkappa_{tot}^{-1}$ pass through a maximum in the vicinity of the magnetic ordering temperature.

3.1.2 Rare-earth monochalcogenides with non-metallic conductivity

3.1.2.1 EuS

The thermal conductivity of europium sulphide was investigated only within the low-temperature range $T < T_c$, where $\varkappa_m \gg \varkappa_L$. $\varkappa_m$ (not taking into account the magnon–magnon scattering) can be written [77] as follows:

$$\varkappa_m = \frac{k_0}{3\,(2\,\pi)^3\,\hbar^2}\int\left(\frac{E}{k_0 T}\right)^2 \frac{e^{E/(k_0 T)}}{(e^{E/(k_0 T)} - 1)^2}\,(\nabla_k\,E)^2\,\tau(\boldsymbol{k})\,d^3\boldsymbol{k}\,, \tag{31}$$

where $E = E(\boldsymbol{k})$ is the energy of the spin modes with wave vector $\boldsymbol{k}$ and $\tau(\boldsymbol{k})$ is the relaxation time for such spin modes. The magnon thermal conductivity depends on the magnetic field through the function $E(\boldsymbol{k})$.

The experimental data for the thermal conductivity of EuS are given in Fig. 24. Solid curves represent the calculated values obtained under the supposition that heat is transferred only by magnons and that phonon scattering occurs only on crystal boundaries [77]. If there is no magnetic field, the experimental

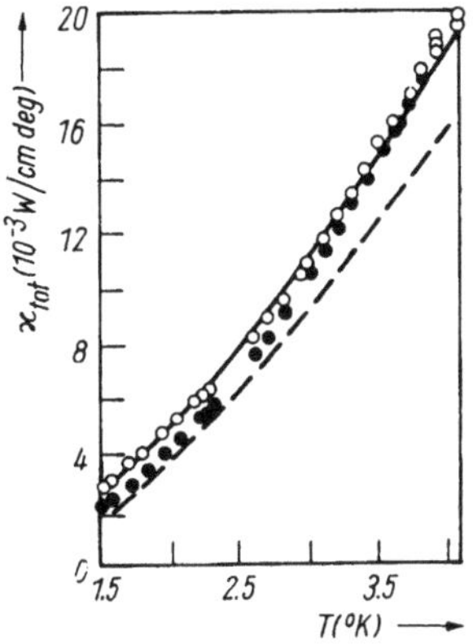

Fig. 24. Temperature dependence of thermal conductivity of EuS [77].
○ $H = 0$; ● $H = 9000$ Oe. Solid and dashed curves are theoretical data for $H = 0$ and 9000 Oe, respectively

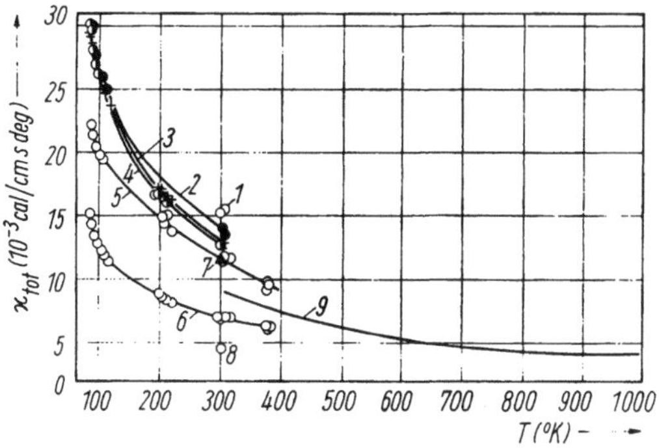

Fig. 25. Temperature dependence of thermal conductivity of SmS specimens [72]. Figures at the curves are the numbers of specimens; No. 8 is taken from [73], No. 9 from [74]

and theoretical data coincide very well, but they differ considerably with a magnetic field present. Experimentally the external magnetic field of 9000 Oe leads to a decrease of thermal conductivity approximately by 25% within the range of 1.5 to 2 °K and by 3% at 4 °K. The theory predicts that at these temperatures $\varkappa_m$ decreases by 31 and 11% respectively. This disagreement of experiments and theory is probably due to the fact that the role of the lattice component of thermal conductivity is not taken into account which increases with the temperature rise.

3.1.2.2 SmS

Fig. 25 gives all available data about thermal conductivity of SmS. The lattice thermal conductivity was calculated with the help of formulas (22) and (23); the experimental values of σ are given in Fig. 26. It was supposed in the calculation that $r = 0.5$ [109].

Fig. 27 gives the temperature dependence W_L for a number of investigated specimens of SmS. The lattice thermal resistivity looks analogous to (21). The additional thermal resistivity arises due to the presence of defects in the crystals. Fig. 28 gives $\varkappa_L$- and ΔW-values of investigated specimens de-

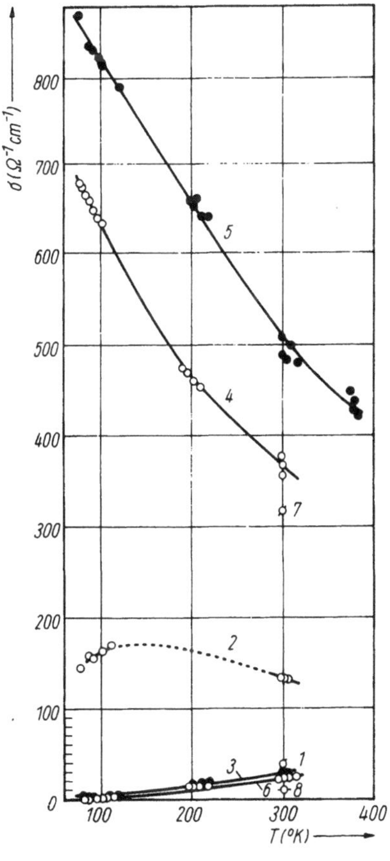

Fig. 26. Temperature dependence of electrical conductivity of SmS specimens [72]. Figures at the curves correspond to Fig. 25, data for specimen No. 9 are not available

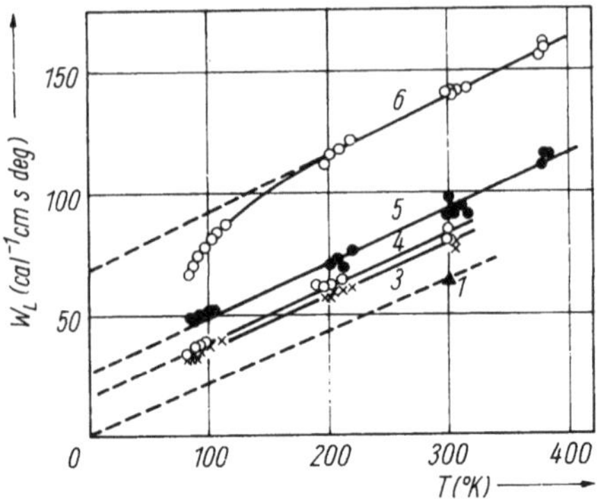

Fig. 27. Temperature dependence of the crystal lattice thermal resistivity of SmS specimens [72]. Figures at the curves have the same meaning as in Fig. 25

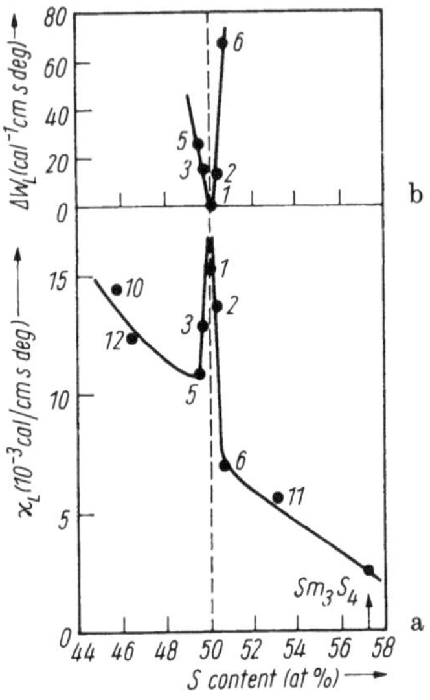

Fig. 28. Dependence of a) the crystal lattice thermal conductivity and b) the additional thermal resistivity of SmS on composition. Figures at the curves have the same meaning as in Fig. 25. 10 to 12 are the numbers of specimens

pending on the composition. $\varkappa_L$ passes through a maximum and ΔW through a minimum in the vicinity of the stoichiometric composition corresponding to SmS. The analysis of experimental results on thermal conductivity shows (the data on other physical parameters being involved, [72]) that in Sm-rich specimens (within the homogeneity range [(50···54) $\pm$ 0.5] at% Sm) excess Sm atoms take interstitial positions. In samples with Sm deficiency a visible amount of the Sm_3S_4 phase always exists, and this circumstance makes an interpretation of the experimental data difficult. The sharp decrease of $\varkappa_L$ for samples with Sm deficiency compared with the stoichiometric composition of SmS can be connected with the formation of defect associates, which precede the Sm_3S_4 phase precipitation, possibly of the form Sm^{3+} $[\ ]^{2-}$ Sm^{3+}.

Judging from the shape of the temperature dependence W_L, it is possible to conclude that within the investigated temperature range there is, probably, no phonon scattering on samarium paramagnetic ions in SmS or it is very small. Probably the splitting of the ground state level in the crystal field of SmS is insignificant and the inter-multiplet transitions will be ineffective, since there are no phonons which could cause the transition to a higher level in single- and double-phonon processes.

3.1.2.3 SmTe, YbTe

The literature gives rather scarce information on the thermal conductivity of SmTe and YbTe. At 300 °K $\varkappa_L$ of these compounds equals 2.6×10^{-3} [107] and $\approx 4.8 \times 10^{-3}$ cal cm^{-1} s^{-1} deg^{-1} [79, 106], respectively.

3.2 Me_2X_3- and Me_3X_4-type compounds

The compounds of this type are very interesting for the investigation of the lattice thermal conductivity. Me_2X_3- and Me_3X_4-type chalcogenides are indeed the limiting compounds of the same phase within which the same Th_3P_4 type of crystal lattice is retained and the lattice parameter is actually not changed. However, different from Me_3X_4, Me_2X_3 chalcogenides have a larger amount of structural vacancies since each ninth atom of their metal sublattice is vacant. During the transition to Me_3X_4 metal atoms gradually fill the vacancies in the lattice. At complete filling of the vacancies, the Me_3X_4 composition is obtained. The transition from a defect lattice to a defectless one greatly influences the absolute value of the lattice thermal conductivity.

Let us discuss first the data for the *thermal conductivity of Me_2X_3 chalcogenides*. The literature gives rather contradictory data of $\varkappa_L$[8]) of this group of compounds. Some papers give very high, the other, to the contrary, very small absolute values of $\varkappa_L$. Even if one considers that in all cases the measurements were taken similarly reliably, the difference in the absolute values of thermal conductivity may arise due to the following facts:

1. $\varkappa_L$ may be larger or smaller depending on the vacancy distribution in the investigated specimens (ordered or not ordered).

2. Impurity atoms of the metal can occupy the vacancies. In this case the concentration of active phonon-scattering lattice defects (vacancies) will decrease and the thermal conductivity will increase.

3. Me_2S_3-type sulphides are polymorphic. Each phase may have a different thermal conductivity. However, in many papers the structure of the investigated compounds was not determined.

4. Most of the data for $\varkappa_L$ refer to the temperature 300 °K. However, as shown in [89], at this temperature the phonon component gives a large contribution to the total thermal conductivity for a number of Me_2X_3 chalcogenides. So the absolute value of $\varkappa_L$ at 300 °K greatly depends on the degree of specimen purity.

The temperature dependence of $\varkappa_L$ was measured only for two compounds: Nd_2S_3 [89] and La_2Te_3 [80] (see Fig. 29 and 30). α- and β-phases were investigated in the first compound (with orthorhombic and tetragonal crystal lattice) and the γ-phase was investigated in La_2Te_3 (Th_3P_4-type lattice).

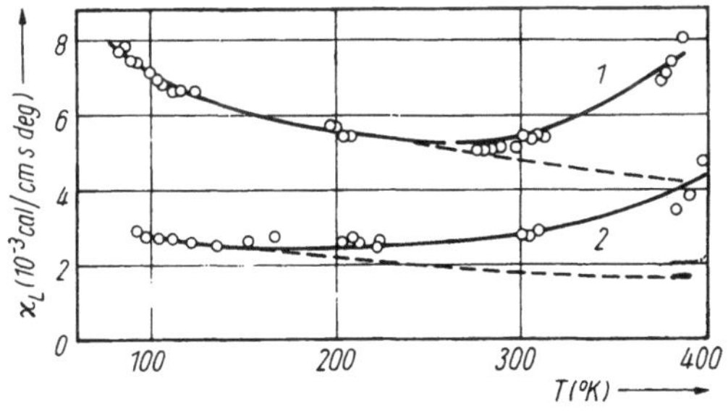

Fig. 29. Temperature dependence of thermal conductivity of Nd_2S_3. (1) α-phase; (2) β-phase [89]. Dashed lines are extrapolated curves of $\varkappa_L = A/T$

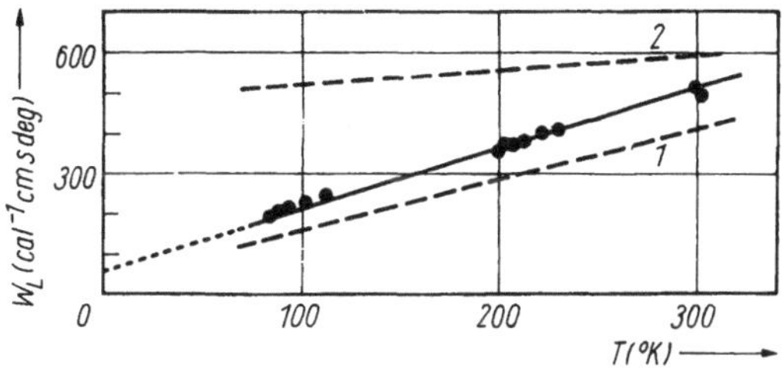

Fig. 30. Temperature dependence of the lattice thermal resistivity of La_2Te_3 [80]. Data for W_L of In_2Te_3 from [110] are given for comparison. (1) α-In_2Te_3 (ordered location of vacancies); (2) β-In_2Te_3 (statistical location of vacancies)

[8]) Since σ of Me_2X_3 rare-earth chalcogenides is small, the experimentally measured $\varkappa_{tot}$ equals $\varkappa_L$.

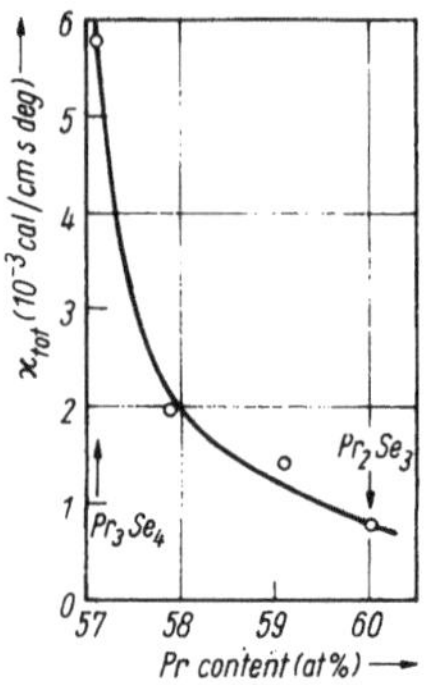

Fig. 31. Dependence of $\varkappa_{tot}$ on composition in a Pr_3Se_4–Pr_2Se_3 system at 300 °K [86]

at% Pr	σ (Ω^{-1} cm^{-1})
57.1	673
57.9	96.6
59.2	1.50×10^{-6}
60.0	9.37×10^{-7}

The absolute values of the lattice thermal conductivity of Nd_2S_3 and La_2Te_3 appeared to be very small. α- and β-phases of Nd_2S_3 at high temperatures have shown a deviation from the $\varkappa_L = A/T$ law.[9] Basing on the value and on the temperature dependence of the additional thermal conductivity it is possible to conclude that it is caused by the photon component ($\varkappa_{ph}$). The character of the temperature dependence and the absolute value W_L of La_2Te_3 are the same as those of the highly defective In_2Te_3 compound [110]; moreover, they are close to α-In_2Te_3 in which the vacancies are ordered.

During the transition from Me_2X_3 to Me_3X_4 chalcogenides the lattice thermal conductivity increases due to the gradual filling of the vacancies by metal atoms (see Fig. 31 and 32).

Interesting results were obtained in paper [92] for $\varkappa_L$ of the La_2Te_3–La_3Te_4 system. Fig. 32 gives the dependence of $\varkappa_L$ on σ at $T = 300$ °K. The crystal lattice thermal conductivity was determined by formulas (22) and (23) for $r = -0.5$.[10] What is unusual in the dependence of $\varkappa_L(\sigma_{300\,°K})$? It is the presence of a maximum in the region of $\sigma_{300\,°K} \approx (1000\cdots800)\ \Omega^{-1}$ cm^{-1}. It is natural to suppose that the composition corresponding to the maximum of $\varkappa_L$ is the stoichiometric compound of La_3Te_4, and the further decrease of $\varkappa_L$ accompanying

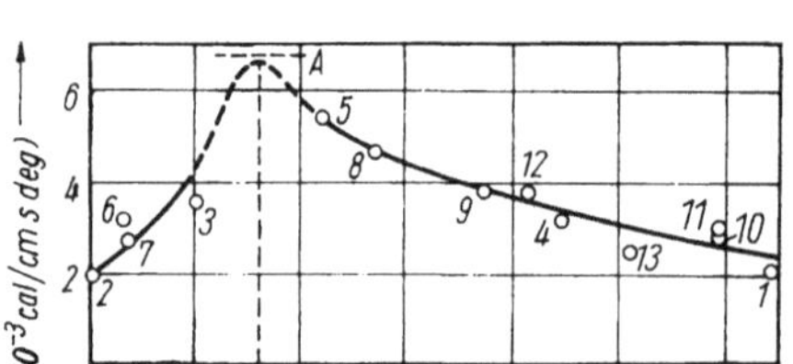

Fig. 32. Dependence of $\varkappa_L$ on σ for the La_2Te_3–La_3Te_4 system. 1 to 3, 12, and 13 are from [92], 4 to 7 from [81], 8 to 10 from [93], and 11 from [94]

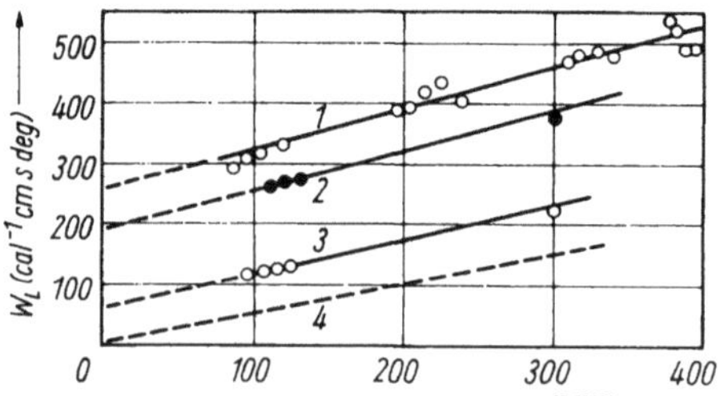

Fig. 33. Dependence of the lattice thermal resistivity on temperature for La_3Te_4 [92]. (1) For a specimen with $\sigma_{300\,°K} = 3200\ \Omega^{-1}$ cm^{-1}; (2) 2600 Ω^{-1} cm^{-1}; (3) 1400 Ω^{-1} cm^{-1}; (4) for a specimen of stoichiometric composition

---

9) The dependence $\varkappa_L = A/T$ for the high-temperature region is marked by a dashed line in Fig. 29.

10) The choice of this or that mechanism of scattering is not much influencing the final result since for compounds close to La_2Te_3 the contribution of $\varkappa_e$ to $\varkappa_{tot}$ is insignificant, and $L_e = L_0 = \pi^2/3\ (k_0/e)^2$ for the whole temperature range for La_3Te_4.

Fig. 34. Dependence of ΔW on $\sigma_{300\,°K}$ for La_3Te_4 [92]. Figures at the curves correspond to the straight lines of Fig. 33

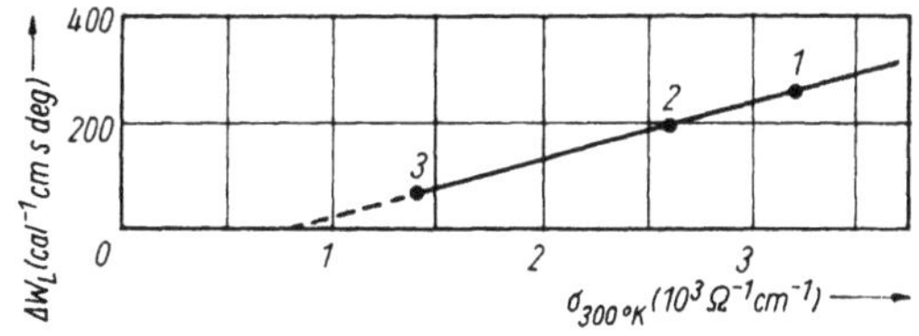

the increase of σ is explained by the formation of new defects. Excessive lanthanum at interstitial positions can play the role of such defects.[11]) The conclusion that the stoichiometric La_3Te_4 has small values of $\sigma_{300\,°K}$ is supported by the data obtained from the temperature dependence of the lattice thermal conductivity [92] of La_3Te_4 specimens with various values of $\sigma_{300\,°K}$. As it is seen from Fig. 33, W_L for a specimen with $\sigma_{300\,°K} = 3200\ \Omega^{-1}\ cm^{-1}$ (specimen 1) obeys equation (21) which agrees well with the theory of thermal conductivity for specimens with defects at $T > \theta$. The following regularities were revealed for La_3Te_4 with smaller values of $\sigma_{300\,°K}$:

a) $W_L(T)$ is parallel to $W_L(T)$ of specimen No. 1.

b) ΔW drops with decreasing absolute value of $\sigma_{300\,°K}$ and equals zero at $\sigma_{300\,°K} \approx 800\ \Omega^{-1}\ cm^{-1}$ (see Fig. 34). The stoichiometric composition La_3Te_4 probably corresponds to this value of $\sigma_{300\,°K}$. $W_L(T)$ for the stoichiometric specimen can be represented by curve 4 of Fig. 33. $\varkappa_L$ at 300 °K is equal to 6.7×10^{-3} cal cm^{-1} s^{-1} deg^{-1} (dashed line A in Fig. 32).

Now let us discuss *the thermal conductivity of Me_3X_4 chalcogenides*. Fig. 35 gives all data available in literature for the temperature dependence of $\varkappa_{tot}$ for the compounds of this type. The behaviour of $\varkappa_{tot}$ at medium (up to 700 °K) and high (> 700 °K) temperatures is different. Since a sufficient amount of reliable experimental data is presently unavailable, it is difficult to evaluate the correctness or the errors of these results.

Interesting conclusions on the character of band structure of Nd_3S_4, La_3Te_4, and Ce_3Se_4 compounds were obtained in paper [112] on the basis of $\varkappa_L$-data

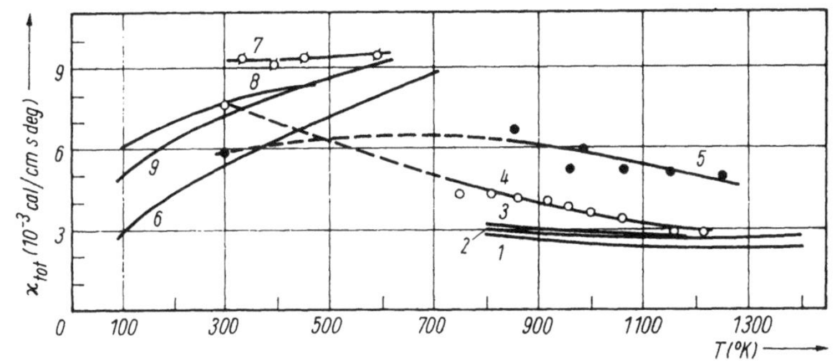

Fig. 35. Temperature dependence of $\varkappa_{tot}$ for a number of Me_3X_4-type rare-earth chalcogenides. (1) $CeS_{1.38}$ [111], (2) $CeS_{1.37}$ [111], (3) $(CeS)_{1.39}(SrS)_{0.07}$ [111], (4) La_3Te_4 [93], (5) La_3S_4 [93], (6) Nd_3S_4 [112], (7) La_3Te_4 [94], (8) La_3Te_4 [80, 112], (9) Ce_3Se_4 [112]

[11]) A similar dependence of $\varkappa_L$ on $\sigma_{300\,°K}$ was obtained for a number of other systems as well: Nd_2Te_3–Nd_3Te_4, Gd_2Te_3–Gd_3Te_4, Nd_2S_3–Nd_3S_4, Ce_2S_3–Ce_3S_4, Pr_2Se_3–Pr_3Se_4, Ce_2Se_3–Ce_3Se_4, Ce_2Te_3–Ce_3Te_4. In all cases the maximum of $\varkappa_L$ also coincided with small values of $\sigma_{300\,°K}$.

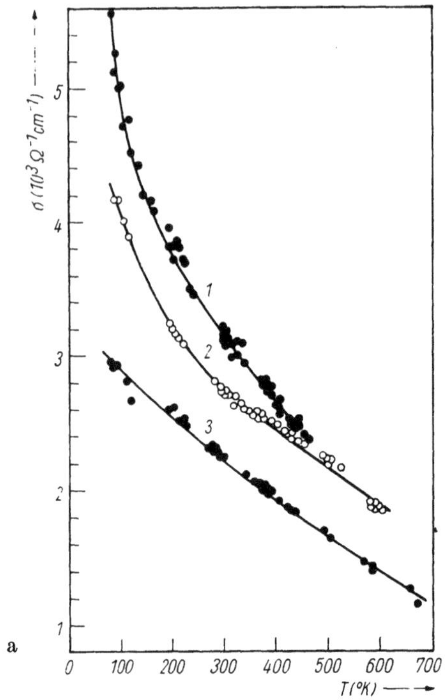

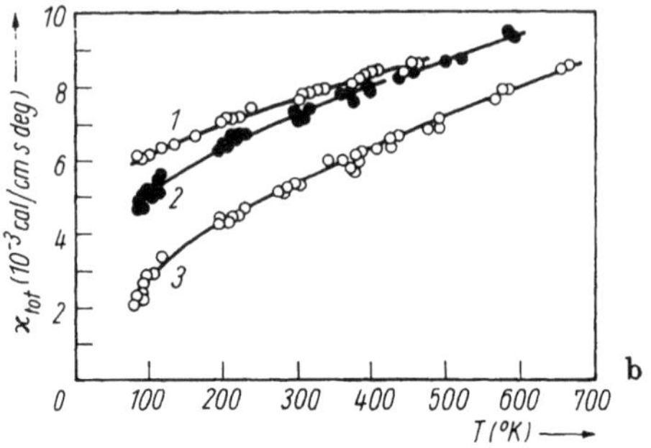

Fig. 36. Temperature dependence of a) σ and b) $\varkappa_{tot}$ of La_3Te_4 (1), Ce_3Se_4 (2), and Nd_3S_4 (3)

analysis. The temperature dependence of $\varkappa_{tot}$ and σ of the above listed compounds is given in Fig. 36. All specimens have n-type conductivity. Since there was no information on the type of conductivity band in these compounds, $\varkappa_L$ was primarily calculated by the usual standard formulas (22) and (23) with $r = -0.5$. The corresponding W_L-values are given in Fig. 37a. The behaviour of the lattice thermal resistivity is unusual for all three investigated compounds, i.e.:

1. It does not increase as equation (21) at high temperatures with increasing temperature[12]) as required by theory, but, contrarily, decreases.

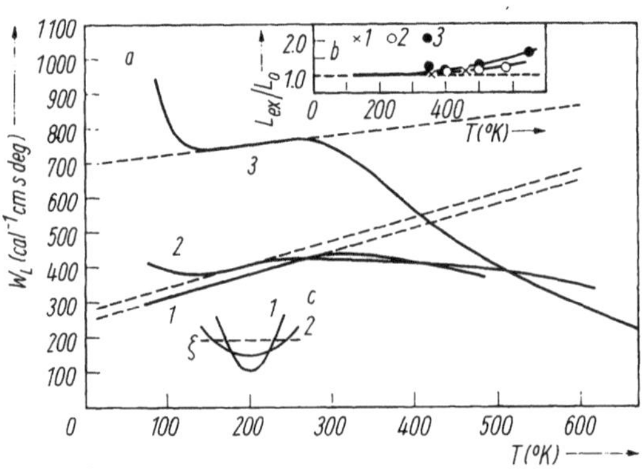

Fig. 37. Temperature dependence of a) the lattice thermal resistivity and b) the relation L_{ex}/L_0 for La_3Te_4 (1), Ce_3Se_4 (2), and Nd_3S_4 (3). c) Possible conduction band of the investigated compounds

[12]) An additional thermal resistivity ΔW, which is independent of temperature, appears due to the presence of impurities of non-controllable defects in crystals.

Table 17

Rare-earth chalcogenides for which the thermal expansion coefficient is measured

MeX	Me_2X_3	Me_3X_4
LaS [70, 120 to 122, 124, 126, 127]	La_2S_3 [70, 124 to 126, 129]	$La_{3-x}S_4$ [122]
LaTe [122]	La_2Se_3 [120, 124, 130]	$La_{3-x}Se_4$ [122]
CeS [70, 120, 121, 123, 124, 126, 127]	La_2Te_3 [131]	
PrS [70, 120, 121, 124, 126, 127]	Ce_2S_3 [70, 124 to 126, 129]	
NdS [70, 120, 121, 124, 126, 127]	Ce_2Se_3 [120, 124, 130]	
EuS [120, 124]	Pr_2S_3 [70, 124 to 126]	
EuSe [120, 124]	Pr_2Se_3 [120, 124, 130]	
	Nd_2Se_3 [70, 124 to 126]	
	Nd_2Te_3 [120, 124, 130]	
	Sm_2S_3 [124, 125, 130]	
	Sm_2Se_3 [120, 124, 130]	

2. Nd_3S_4 and Ce_3Se_4 specimens show some deviation from equation (21) at low temperatures, but this is connected with a W_L-increase. La_3Te_4 does not reveal such an effect.

Let us discuss separately high- and low-temperature regions.

High temperatures

Most probably the unusual behaviour of W_L at high temperatures is connected with an incorrect consideration of the Lorenz number in (22). In order to get W_L in this temperature range obeying dependence (21), it is necessary that L_e should exceed the values calculated by formula (23). This can be expected

1. in semiconductors within the region of intrinsic conductivity;
2. in semimetals;
3. in doped semiconductors with a complicated conduction band consisting of subbands with "heavy" and "light" current carriers;
4. in metals with a band considered in 3.

Cases 1 and 2 are unreal for the compounds under consideration. The fact that there is no intrinsic conductivity is illustrated by the temperature dependence of thermoelectric power and electrical conductivity. Case 2 does not suit our purpose by the following reasons: As it was mentioned above, Me_3X_4 and Me_2X_3 compounds are the limiting compositions of the same phase within which the type of crystal lattice is retained and the lattice parameter actually remains unchanged. It can be concluded that during the transition from Me_2X_3 compounds (which are semiconductors with a sufficiently wide forbidden gap E_g) to Me_3X_4 compounds E_g will either remain unchanged or, if changed, will change insignificantly.

Thus only cases 3 and 4 remain, which cannot be differed basing on the data for the Lorenz number. Presently the literature does not give any uniform point of view on the nature of Me_3X_4 compounds. Some authors rather consider them bad metals, the others consider them heavily doped semiconductors. It seems the second suggestion is more justified.

The Lorenz number in cases 3 and 4 may increase compared to (23) due to the appearance of inter-band scattering of "light" and "heavy" current carriers. In this case the Fermi level must be located in the second band closer to its bottom [113].[13])

Fig. 37b gives the values of L_{ex}/L_0. The value of L_{ex} was derived experimentally from the expression

$$\frac{\varkappa_{tot} - \varkappa_L^0}{\sigma T} = L_{ex}, \tag{32}$$

where $\varkappa_L^0$ is the value of the lattice thermal conductivity taken from the straight line of $W_L(T)$ obtained by extrapolation of the low-temperature linear sections to the high-temperature side (see dashed lines in Fig. 37a). L_{ex}/L_0 increases with temperature within reasonable limits.

So the data on the electronic component of thermal conductivity for La_3Te_4, Ce_3Se_4, and Nd_3S_4 indicate that there is a complicated conduction band in these compounds, which consists of two subbands with heavy and light electrons. The approximate band scheme of these compounds is given in Fig. 37c.

Table 18

Experimental values of thermal expansion coefficient for rare-earth monochalcogenides

compound	$\delta_{tot} \times 10^6$ (deg^{-1})	average meanings for temperature range (°K)
LaS	9.8 *) [121] 11.62 *) [122] 11.62 [80, 123 to 126]	300 to 673 293 to 573 300 to 1350
CeS	11.6 *) [121] 12.37 [123 to 127] 9.8 [128]	300 to 673 300 to 1350 293 to 1297
PrS	13.2 *) [121] 14.3 [123 to 127]	300 to 673 300 to 1350
NdS	14.3 *) [121] 15.35 ± 0.34 [123 to 127]	300 to 1350 300 to 1350
EuS	14.6 [120] 14.2 [124]	300 to 800 300 to 800
EuSe	18.6 [120] 18.2 [124]	300 to 800 300 to 800
LaTe	18.0 ± 3 *) [122]	293 to 573

*) Taken from X-ray measurements; the other data were obtained from dilatometric measurements.

13) For the first time this problem was theoretically considered in [113] and experimentally discovered in the heavily doped semiconductors p-PbTe [114], LaSb [115], PbTe–PbSe [116], p- and n-Bi_2Te_3 [117], and SnTe [118, 119].

Fig. 38. Temperature dependence of $\varkappa_{tot}$ (1), σ (2), and $d\varkappa_{tot}/dT$ (3) for Nd_3S_4 [30]

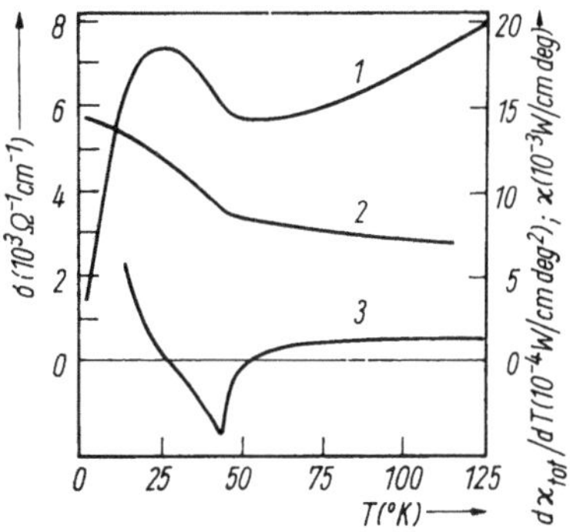

Low temperatures

The effect that W_L increases compared with the dependence (21) can be explained only by resonance scattering of phonons on Ce and Nd paramagnetic

Table 19

Experimental values of thermal expansion coefficients for Me_2X_3-type rare-earth chalcogenides

compound	$\delta_{tot} \times 10^6$ (deg^{-1})	average meanings for temperature range (°K)
La_2S_3	9.9 ± 0.67 [124 to 126, 129]	300 to 1350
Ce_2S_3	10.45 ± 0.67 [125, 126, 129] 13.2 [128]	300 to 1350 300 to 1350
Pr_2S_3	12.09 ± 0.4 [126] 11.3 [124, 125]	300 to 1350 300 to 1350
Nd_2S_3	12.9 ± 0.65 [124 to 126]	300 to 1350
Sm_2S_3	11.8 [120] 11.4 [124]	300 to 800 300 to 800
La_2Se_3	11.9 [120] 11.5 [124] 10.2 [130]	300 to 800 300 to 800 300 to 800
Ce_2Se_3	12.6 [120] 12.2 [124]	300 to 800 300 to 800
Pr_2Se_3	13.0 [120] 12.6 [124]	300 to 800 300 to 800
Nd_2Se_3	13.5 [120] 13.1 [124]	300 to 800 300 to 800
Sm_2Se_3	14.8 [120] 14.4 [124]	300 to 800 300 to 800
La_2Te_3	19.6*) [131]	293 to 573

*) Taken from X-ray measurements; the other data were obtained from dilatometric measurements.

Table 20

Experimental values of thermal expansion coefficients of Me_3X_4-type rare-earth chalcogenides [122]

compound	$\delta_{tot} \times 10^6$ (deg^{-1})	average meanings for temperature range (°K)
$La_{3-x}S_4$	9.9	293 to 573
$La_{3-x}Te_4$	18 + 2	293 to 573

Note: Data were obtained from X-ray measurements.

Table 21

Experimental values of thermal expansion coefficients of the Ce_2S_3–Ce_3S_4 system (average values for the temperature range 293 to 1273 °K)

composition	$\delta_{tot} \times 10^6$ (deg^{-1})
$CeS_{1.33}$	12.3 [132]
$CeS_{1.36}$	12.6 [132]
$CeS_{1.4}$	13.0 [132]
$CeS_{1.5}$	13.2 [128]

Note: Data were obtained from dilatometric measurements.

Table 22

Influence of impurities on δ_{tot} of $CeS_{1.4}$ (average values for the temperature range 293 to 1273 °K)

impurity (1.5 at%)	$\delta_{tot} \times 10^6$ (deg^{-1}) [132]
Zr	12.5
Nb	12.4
Mo	12.5
Hf	12.5
Ta	12.6
W	12.5

ion levels split by the lattice crystal field. As repeatedly mentioned above, this effect should not in principle be observed in lanthanides.

Fig. 38 gives the results of $\varkappa_{tot}$ of Nd_3S_4 for the low-temperature region [30]. The derived value of $d\varkappa_{tot}/dT$ has a sharp minimum in the region of 43 °K. Antiferromagnetic ordering in Nd_3S_4 is observed in the vicinity of this tempera ture.

4. Thermal Expansion Coefficient of Rare-Earth Chalcogenides

Unfortunately very few experimental data are presently available in literature on the investigation of thermal expansion coefficients of rare-earth chalcogenides. The available results mostly refer to the high-temperature range. δ_m and δ_e (see Table 2) were not investigated in these compounds. Table 17 gives an list and Tables 18 to 22 give the absolute values of δ_{tot} of rare-earth chalcogenides measured up to the present time.

5. Conclusions

The investigation of the thermal properties of rare-earth chalcogenides gives sufficiently vast information on the physical properties of these compounds. Using thermal methods it became possible

a) to determine the absolute value of the effective mass of current carriers for rare-earth monochalcogenides with metallic character of conductivity;

b) to determine the temperatures of magnetic ordering and the values of exchange integrals and magnetic entropy in europium monochalcogenides;

c) to determine the Debye temperature for a great number of compounds and to estimate the melting points;

d) to estimate the value of the ground state splitting in the cubic symmetry field for a number of rare-earth MeX, Me_3X_4, and Me_2X_3 chalcogenides;

e) to reveal the presence of a number of phase transitions in rare-earth chalcogenides;

f) to confirm the existence of strong phonon–electron interaction in rare-earth monochalcogenides with metallic character of conductivity;

g) to determine the type of the band structure for Me_3X_4-type compounds;

h) to determine the exact stoichiometric compositions for a number of Me_3X_4 and MeX chalcogenides;

i) to establish a number of defects typical for some rare-earth chalcogenides; etc.

However, this is far from the complete usage of all opportunities granted by the thermal methods while investigating rare-earth chalcogenides. It is only the start for the serious investigation of the thermal properties of compounds belonging to this group.

References

[1] G. V. Lashkarev, Metallovedenie, Izd. Nauka, Moskva 1971 (p. 388).
[2] S. Mathfessel, IEEE Trans. Magnetics **1**, 144 (1965).
[3] V. V. Tikhonov, A. V. Golubkov, and I. A. Smirnov, Fiz. tverd. Tela **8**, 3578 (1966).
[4] E. C. King and W. W. Weller, U.S. Bur. Mines Rep. Invest. 5483, 1–5 (1959).
[5] R. Mezaki, E. W. Tilleux, T. F. Jambois, and I. L. Margrave, III. ASME Symp. Thermophysical Properties, Purdue Univ. 1965 (p. 138).
[6] V. P. Zhuze, V. I. Novikov, V. M. Sergeeva, and S. S. Shalyt, Fiz. tverd. Tela **11**, 2192 (1969).
[7] G. Busch, P. Junod, R. G. Morris, and J. Muheim, Helv. phys. Acta **37**, 637 (1964).
[8] G. Busch, P. Junod, R. G. Morris, J. Muheim, and W. Stutius, Phys. Letters (Netherlands) **11**, 9 (1964).
[9] B. C. Passenheim and D. C. McCollum, Bull. Amer. Phys. Soc. **13**, 1645 (1968).
[10] H. W. White, D. C. McCollum, and J. Callaway, Phys. Letters (Netherlands) **25A**, 388 (1967).
[11] D. C. McCollum and J. Callaway, Phys. Rev. Letters **9**, 376 (1962).
[12] J. Callaway and D. C. McCollum, Phys. Rev. **130**, 1741 (1963).
[13] V. L. Moruzzi and D. T. Teaney, Solid State Commun. **1**, 127 (1963).
[14] B. C. Passenheim, D. C. McCollum, and J. Callaway, Phys. Letters (Netherlands) **23**, 634 (1966).
[15] B. J. C. van der Hoeven, Jr., D. T. Teaney, and V. L. Moruzzi, Phys. Rev. Letters **20**, 719 (1968).
[16] D. T. Teaney and V. L. Moruzzi, Coll. Internat. C.N.R.S. "Les éléments des terres rares", Vol. 2, Paris 1970 (p. 131).
[17] V. L. Moruzzi, D. T. Teaney, and B. J. C. van der Hoeven, Jr., Solid State Commun. **6**, 461 (1968).
[18] B. J. C. van der Hoeven, Jr., Coll. Internat. C.N.R.S. "Les éléments des terres rares", Vol. 2, Paris 1970 (p. 119).
[19] D. T. Teaney, B. J. C. van der Hoeven, Jr., and V. L. Moruzzi, Phys. Rev. Letters **20**, 722 (1968).
[20] A. V. Golubkov and V. V. Tikhonov, Fiz. tverd. Tela **9**, 954 (1967).
[21] I. E. Paukov, Raboty po fiz. tverd. tela, Vyp. 1, Izd. Nauka (Sibir. Otd.), Novosibirsk 1967 (p. 91).
[22] V. V. Nogteva, I. E. Paukov, and E. I. Yarembash, Zh. fiz. Khim. **43**, 2118 (1969).
[23] I. E. Paukov, V. V. Nogteva, and E. I. Yarembash, Zh. fiz. Khim. **40**, 3094 (1966).
[24] I. E. Paukov, V. V. Nogteva, and E. I. Yarembash, Zh. fiz. Khim. **43**, 2351 (1969).
[25] I. E. Paukov, V. V. Nogteva, and E. I. Yarembash, Zh. fiz. Khim. **42**, 998 (1968).
[26] V. V. Nogteva, I. E. Paukov, and E. I. Yarembash, Zh. fiz. Khim. **42**, 232 (1968).

[27] V. V. NOGTEVA, I. E. PAUKOV, and E. I. YAREMBASH, Zh. fiz. Khim. **43**, 2344 (1969).
[28] V. V. TIKHONOV, M. V. ROMANOVA, I. A. SMIRNOV, and V. M. SERGEEVA, Fiz. tverd. Tela **13**, 2023 (1971).
[29] F. HOLTZBERG, P. E. SEIDEN, and S. VON MOLNAR, Phys. Rev. **168**, 408 (1968).
[30] V. I. NOVIKOV and S. S. SHALYT, Fiz. tverd. Tela **12**, 3252 (1970).
[31] E. S. K. GOPAL, Specific Heats at Low Temperatures, Heywood Books, London 1966.
[32] A. I. ANSELM, Vvedenie v teoriyu poluprovodnikov, Fizmatgiz, Moskva/Leningrad 1962.
[33] H. M. ROSENBERG, Low Temperature Solid State, Clarendon Press, Oxford 1963.
[34] A. I. AKHIEZER, V. G. BARYAKHTAR, and S. V. PELETMINSKI, Spinovye volny, Izd. Nauka, Moskva 1967.
[35] J. E. KUNZLER, L. R. WALKER, and J. K. GALT, Phys. Rev. **119**, 1609 (1960).
[36] K. A. GSCHNEIDER, Proc. IV. Conf. Rare Earth Research, Phoenix, Arizona 1964, Gordon & Breach, New York/London/Paris 1964 (p. 153).
[37] E. BUCHER, A. C. GOSSARD, K. ANDRES, J. P. MAITA, and A. S. COOPER, Proc. VIII. Conf. Rare Earth Research, Reno (Nev.) 1970, Reno Metallurgy Res. Center, U. S. Bureau of Mines, 1970 (p. 74).
[38] D. H. PARKINSON and J. E. QUARRINGTON, Proc. Phys. Soc. **67**, 569 (1954).
[39] J. M. ZIMAN, Electrons and Phonons, Clarendon Press, Oxford 1960.
[40] E. O. CARTER, T. E. LEE, E. W. JOHNSON, E. G. RAUCH, and H. A. TICK, J. phys. Chem. **69**, 2684 (1965).
[41] M. C. PICON, J. FLAHAUT, M. GRUTTARD, and M. PATRIE, XVI. Internat. Congr. Pure Appl. Chem., Paris 1958, Butterworths, London/Washington 1958 (p. 685).
[42] M. PICON and M. PATRIE, C. R. Acad. Sci. (France) **242**, 1321 (1956).
[43] G. V. SAMSONOV, Sbornik "Khalkogenidy (svoistva, metody polucheniya i primeneniya)", Naukova Dumka, Kiev 1967 (p. 12).
[44] F. GAUME and F. MAHN, Bull. Soc. Chim. France, No. 11/12, 1862 (1956).
[45] E. I. YAREMBASH, E. S. VIGILEVA, R. R. KAGRAMANOVA, and L. KH. KRAVCHENKO, Izv. Akad. Nauk SSSR, Ser. neorg. Mater. **5**, 260 (1969).
[46] A. A. ELISEEV, E. I. YAREMBASH, V. G. KUZNETSOV, L. I. ANTONOVA, and Z. P. STOYANTSEVA, Izv. Akad. Nauk SSSR, Ser. neorg. Mater. **1**, 1027 (1965).
[47] T. H. RAMSEY, H. STEINFINK, and E. J. WEISS, Proc. III. Conf. Rare Earth Research, Cleawater (Fla.) 1963, Ed. K. S. VORRES, Vol. II, Gordon and Breach Science Publishers Inc., 1963 (p. 301).
[48] D. J. HAASE, H. STEINFINK, and E. WEISS, Proc. IV. Conf. Rare Earth Research, Phoenix, Arizona 1964, Gordon & Breach, New York/London/Paris 1964 (p. 535).
[49] A. A. ELISEEV, E. I. YAREMBASH, and V. S. VIGILEVA, Dokl. Akad. Nauk SSSR **153**, 1333 (1963).
[50] E. D. EASTMAN, L. BREWER, L. A. BROMLEY, P. V. GILLES, and N. L. LOFGREN, J. Amer. Chem. Soc. **72**, 2248 (1950).
[51] J. FLAHAUT and M. GUITTARD, C. R. Acad. Sci. (France) **242**, 1318 (1956).
[52] I. A. SMIRNOV. L. S. PARFENEVA, V. KHUSNUTDINOVA, and V. M. SERGEEVA, Fiz. tverd. Tela **14**, 2783 (1972).
[53] G. V. LASHKAREV, Avtoreferat kand. diss., Inst. Problem Metallovedeniya, Kiev 1966.
[54] E. I. YAREMBASH and V. S. VIGILEVA, Izv. Akad. Nauk SSSR, Ser. neorg. Mater. **6**, 1572 (1970).
[55] N. KH. ABRIKOSOV and V. SH. ZARGARYAN, Izv. Akad. Nauk SSSR, Ser. neorg. Mater. **1**, 1462 (1965).
[56] V. SH. ZARGARYAN and N. KH. ABRIKOSOV, Izv. Akad. Nauk SSSR, Ser. neorg. Mater. **3**, 769 (1967).
[57] W. LIN, H. STEINFINK, and E. F. WEISS, J. inorg. nucl. Chem. **4**, 877 (1965).
[58] K. A. ZINCHENKO, N. P. LUZHNAYA, E. I. YAREMBASH, and A. L. ELISEEV, Izv. Akad. Nauk SSSR, Ser. neorg. Mater. **2**, 1747 (1966).
[59] V. E. ADAMYAN and G. M. LOGINOV, Zh. eksper. teor. Fiz. **51**, 1044 (1966).
[60] M. A. ELYASHEVICH, Spektra redkich zemel, GITTL, Moskva 1963.
[61] G. BUSH, J. appl. Phys. **38**, 1386 (1967).

[62] P. J. Wojtowicz, J. appl. Phys. **35**, 991 (1964).
[63] S. H. Charap and E. L. Boyd, Phys. Rev. **133**, 811 (1964).
[64] A. Z. Patashinskii and V. A. Pokrovskii, Zh. eksper. teor. Fiz. **46**, 994 (1964).
[65] A. Quattropani, Solid State Commun. **7**, 689 (1969).
[66] C. Domb and R. G. Bowers, J. Phys. C (Solid State Phys.) **2**, 755 (1969).
[67] V. V. Tikhonov and I. A. Smirnov, Fiz. tverd. Tela **13**, 2749 (1971).
[68] A. V. Golubkov, E. D. Devyatkova, V. P. Zhuze, V. M. Sergeeva, and I. A. Smirnov, Fiz. tverd. Tela **8**, 1761 (1966).
[69] V. P. Zhuze, A. V. Golubkov, E. V. Goncharova, and V. M. Sergeeva, Fiz. tverd. Tela **6**, 257 (1964).
[70] A. V. Rodzikovskaya and V. I. Marchenko, Sulfidy redkozemelnykh metallov i aktinoidov, Naukova Dumka, Kiev 1966.
[71] E. D. Devyatkova, V. P. Zhuze, A. V. Golubkov, V. M. Sergeeva, and I. A. Smirnov, Fiz. tverd. Tela **6**, 430 (1964).
[72] V. P. Zhuze, E. V. Goncharova, N. F. Kartenko, T. I. Komarova, L. S. Parfeneva, V. M. Sergeeva, and I. A. Smirnov, to be published.
[73] M. D. Hauston, Ceramic Age **77**, 50 (1964).
[74] Y. S. Toulonkian (Ed.), Thermophysical Properties of High Temperature Solid Materials, Vol. 5, Thermophys. Prop. Res. Center, Purdue Univ., 1967.
[75] R. L. Wild and D. C. McCollum, Bull. Amer. Phys. Soc., Ser. II, **8**, 208 (1963).
[76] R. L. Wild and D. C. McCollum, Bull. Amer. Phys. Soc., Ser. II, **9**, 599 (1964).
[77] D. C. McCollum, R. L. Wild, and J. Callaway, Phys. Rev. **136**, A426 (1964).
[78] V. Khusnutdinova, V. S. Oskotskii, I. A. Smirnov, and V. M. Sergeeva, phys. stat. sol. (b) **48**, 353 (1971).
[79] K. A. Zinchenko, A. A. Eliseev, and N. Kh. Abrikosov, Izv. Akad. Nauk SSSR, Ser. neorg. Mater. **6**, 1172 (1970).
[80] I. A. Smirnov, L. S. Parfeneva, and V. M. Sergeeva, to be published in Izv. Akad. Nauk SSSR, Ser. neorg. Mater. (1972).
[81] W. L. Mularz and S. J. Wolwik, Proc. III. Conf. Rare Earth Research, Cleawater (Fla.) 1963, Ed. K. S. Vorres, Gordon & Breach, New York/London 1964.
[82] G. V. Lashkarev and Yu. V. Paderno, Izv. Akad. Nauk SSSR, Ser. neorg. Mater. **1**, 1791 (1965).
[83] R. R. Kagramanova, I. S. Lisker, and E. I. Yarembash, Azerb. khim. Zh., No. 4, 46 (1967).
[84] V. I. Marchenko and N. G. Barantseva, Inzh. fiz. zhurnal **6**, 120 (1964).
[85] Yu. Goryachev and T. G. Kutsenok, Fiz. Tekh. Poluprov. **3**, 277 (1969).
[86] I. S. Lisker, B. Sh. Malkovich, E. I. Yarembash, and V. I. Kalitin, Izv. Akad. Nauk SSSR, Ser. neorg. Mater. **3**, 5 (1967).
[87] N. Kh. Abrikosov and V. Sh. Zargaryan, Izv. Akad. Nauk SSSR, Ser. neorg. Mater. **3**, 280 (1967).
[88] V. Sh. Zargaryan. Avtoreferat kand. diss., IMET, Moskva 1967.
[89] I. A. Smirnov, L. S. Parfeneva, V. M. Sergeeva, and T. B. Zhukova, Fiz. tverd. Tela **14**, 2471 (1972).
[90] D. F. Miller and R. Ts. Haims, Fizicheskie svoistva redkozemelnykh metallov i ikh soedinenii, Izd. in. Lit., Moskva 1962 (p. 24).
[91] I. Miller, F. Reid, and R. Himes, J. Electrochem. Soc. **106**, 1043 (1962).
[92] I. A. Smirnov, L. S. Parfeneva, and V. M. Sergeeva, Fiz. tverd. Tela **14**, 1050 (1972).
[93] V. P. Zhuze, O. A. Golikova, V. M. Sergeeva, and I. M. Rudnik, Fiz. tverd. Tela **13**, 811 (1971).
[94] V. P. Zhuze, V. M. Sergeeva, and O. A. Golikova, Fiz. tverd. Tela **11**, 2568 (1969).
[95] F. M. Ryan, I. N. Greenberg, F. L. Carter, and R. C. Miller, J. appl. Phys. **33**, 864 (1962).
[96] E. I. Yarembach, Coll. Internat. C.N.R.S. "Les éléments des terres rares", Vol. 1, Paris 1970 (p. 471).
[97] J. Callaway, Phys. Rev. **113**, 1046 (1959).

[98] YU. I. RAVICH, B. A. EFIMOVA, and I. A. SMIRNOV, Metody issledovaniya poluprovodnikov v primenenii k khalkogenidam svintsa PbTe, PbSe, PbS, Izd. Nauka, Moskva 1968. YU. I. RAVICH, B. A. EFIMOVA, and I. A. SMIRNOV, Semiconducting Lead Chalcogenides, Plenum Press, New York/London 1970.
[99] E. D. DEVYATKOVA, B. YA. MOYZHES, and I. A. SMIRNOV, Fiz. tverd. Tela **1**, 613 (1959).
[100] A. H. WILSON, The Theory of Metals, Cambridge University Press, 1954 (p. 290).
[101] J. L. OLSEN and H. M. ROSENBERG, Adv. Phys. **2**, 28 (1953).
[102] L. COLQUITT, J. appl. Phys. **36**, 2454 (1965).
[103] E. H. SONDHEIMER, Proc. Roy. Soc. **A203**, 75 (1950).
[104] V. S. OSKOTSKII and I. A. SMIRNOV, Fiz. tverd. Tela **13**, 1635 (1971).
[105] A. KARABEKOV, Avtoreferat kand. diss., Inst. Neorg. i Fiz. Khimii, Akad. Nauk KirgSSR, Frunze 1970.
[106] K. A. ZINCHENKO, Avtoreferat kand. diss., Inst. Tonkoi Khimich. Tekh., Moskva 1971.
[107] E. G. TYURIN, Avtoreferat kand. diss., Inst. Stali i Splavov, Moskva 1971.
[108] C. H. DIEKE, Spectra and Energy Levels of Rare Earth Ions in Crystals, Interscience Publishers, New York/London/Sydney/Toronto 1969.
[109] A. V. GOLUBKOV, E. V. GONCHAROVA, V. P. ZHUZE, and I. G. MANOILOVA, Fiz. tverd. Tela **7**, 2430 (1965).
[110] A. I. ZASLAVSKII, V. M. SERGEEVA, and I. A. SMIRNOV, Fiz. tverd. Tela **2**, 2885 (1960).
[111] F. L. CARTER, R. C. MILLER, and F. M. RYAN, Adv. Energy Conv. **1**, 165 (1961).
[112] I. A. SMIRNOV, L. S. PARFENEVA, V. YA. KHUSNUTDINOVA, and V. M. SERGEEVA, Fiz. tverd. Tela **14**, 844 (1972).
[113] N. V. KOLOMOETS, Fiz. tverd. Tela **8**, 999 (1966).
[114] I. A. SMIRNOV, M. N. VINOGRADOVA, N. V. KOLOMOETS, and L. M. SYSOEVA, Fiz. tverd. Tela **9**, 2638 (1967); **9**, 3004 (1967).
[115] E. V. GONCHAROVA, V. P. ZHUZE, V. V. ZHDANOVA, T. B. ZHUKOVA, I. A. SMIRNOV, and E. V. SHADRICHEV, Fiz. tverd. Tela **10**, 1321 (1968).
[116] V. A. SAAKYAN and V. V. TIKHONOV, Fiz. tverd. Tela **10**, 1600 (1968).
[117] I. A. SMIRNOV, E. V. SHADRICHEV, and V. A. KUTASOV, Fiz. tverd. Tela **11**, 3311 (1969).
[118] D. H. DAMON, J. appl. Phys. **37**, 3181 (1966).
[119] N. A. ERASOVA and V. I. KAIDANOV, Fiz. Tekh. Poluprov. **2**, 33 (1968).
[120] G. V. LASHKAREV, E. M. DUDNIK, and YU. B. PADERNO, Khimicheskaya svyaz v poluprovodnikakh i termodinamika, Izd. "Nauka i Tekhnika", Minsk 1966 (p. 291).
[121] N. N. ZHURAVLEV, A. A. STEPANOVA, and M. P. SHEBATINOV, Kristallografiya **9**, 116 (1964).
[122] A. A. ELISEEV, V. G. KUZNETSOV, and E. I. YAREMBASH, Izv. Akad. Nauk SSSR, Ser. neorg. Mater. **3**, 1267 (1967).
[123] E. EASTMAN, L. BREWER, L. BROWLEY, P. GILLES, and N. LOFGREN, J. Amer. Chem. Soc. **72**, 2248 (1950).
[124] E. M. DUDNIK, G. V. LASHKAREV, YU. B. PADERNO, and V. A. OBOLONCHIK, Izv. Akad. Nauk SSSR, Ser. neorg. Mater. **2**, 980 (1966).
[125] V. I. MARCHENKO, Sbornik "Voprosy teorii i primeneniya redkozemelnykh metallov", Izd. Nauka, Moskva 1964.
[126] V. I. MARCHENKO and G. V. SAMSONOV, Fiz. Metallov i Metallovedenie **15**, 631 (1963).
[127] V. I. MARCHENKO and G. V. SAMSONOV, Zh. neorg. Khim. **8**, 2035 (1963).
[128] K. A. GSCHNEIDER, Rare Earth Alloys, D. van Nostrand Company, Inc., Princeton, New Jersey, Toronto/New York/London 1961.
[129] G. V. SAMSONOV and V. I. MARCHENKO, Dokl. Akad. Nauk SSSR **152**, 671 (1963).
[130] V. A. OBOLONCHIK and G. V. LASHKAREV, Selenidy i telluridy redkozemelnykh metallov i aktinoidov, Naukova Dumka, Kiev 1966.
[131] A. A. ELISEEV, E. I. YAREMBASH, V. G. KUZNETSOV, E. S. VIGILEVA, A. A. RESHCHIKOVA, and L. I. ANTONOVA, Zh. neorg. Khim. **9**, 876 (1964).
[132] YU. M. GORYACHEV and T. G. KUTSENOK, Khalkogenidy, Vyp. 2, Naukova Dumka, Kiev 1970.

(Received May 9, 1972)

Original Papers

phys. stat. sol. (a) **14**, 405 (1972)

Subject classification: 13.4 and 14.3.3; 22.1.2

Department of Electrical Engineering and Materials Research Laboratory, University of Illinois, Urbana

Thermal Ionization Rates and Energies of Holes at the Double Acceptor Zinc Centers in Silicon[1])

By

J. M. Herman III and C. T. Sah

The thermal emission rates and activation energies of holes trapped at the neutral and singly ionized zinc centers are measured using the dark capacitance and current transient techniques in n^+–p junctions. The thermal activation energies of holes from these centers are 316 and 617 meV, respectively. Detailed electric field dependences of the thermal emission rates are obtained using a differential bias voltage technique. The thermal emission rate of holes from neutral zinc centers exhibits a very large field dependence, while that from singly ionized zinc centers is small. A second zinc-related center, which is attributed to the previously reported acceptorlike, zinc–boron complex, is studied. The thermal activation energy of trapped holes from this center is 167 meV.

Die thermischen Emissionsraten und die Ionisationsenergien von Löchern, die von einfach geladenen und neutralen Zinkzentren eingefangen sind, werden gemessen. Dabei wurden die Kapazitätsmethode und die Analyse der zeitlichen Änderung eines Strompulses in n^+–p-Dioden verwendet. Die thermischen Ionisationsenergien von Löchern dieser Zentren betragen 316 und 617 meV. Die genaue Feldabhängigkeit der thermischen Emissionsraten wurden mit Hilfe einer veränderlichen Vorspannung ermittelt. Die thermische Emissionsrate für Löcher von neutralen Zinkzentren zeigt eine starke Feldabhängigkeit, während die von einfach ionisierten Zinkzentren schwach ist. Ein zweites Zentrum, das vom Zink herrührt und das dem kürzlich berichteten Bor–Zink-Komplex zugeschrieben wird, wird untersucht. Die thermische Ionisationsenergie von eingefangenen Löchern dieses Zentrums ist 167 meV.

Table of Symbols

A	junction area (10^{-4} cm^2)
A_{ps}	defined by equation (10)
B	heating rate
$C(t)$, C	time dependent and steady state high frequency capacitance (pF)
e_n^t, e_p^t	thermal emission rates of electrons and holes
e_{n1}^t, e_{n2}^t	thermal emission rates of electrons from singly and doubly ionized zinc centers
e_{p0}^t, e_{p1}^t	thermal emission rates of holes from neutral and singly ionized zinc centers
$\mathcal{E}$, $\mathcal{E}_{\vdash\!\dashv}$	electric field and electric field interval
E_V, E_C, E_F	energy of valence band edge, conduction band edge, and Fermi level
$E_{1/2}$, $E_{3/2}$	effective energy levels for zinc in silicon

[1]) Work supported in part by the Advanced Research Projects Agency and the Air Force Office of Scientific Research.

$\Delta E_{1/2}$, $\Delta E_{3/2}$	thermal activation energy of holes for zinc, defined by equation (10)
F	number of negative charges per zinc center (0 to 2)
$i(t)$	time dependent junction current
k_B	Boltzmann constant
$K_S \varepsilon_0$	dielectric constant of semiconductor
m	defined by equation (10)
n, p	electron and hole concentration
n_T, p_T	trapped electron and hole concentration
N_I, N_D, N_A, N_{TT}	total ionized impurity, donor, acceptor, and trap concentrations
N_0, N_1, N_2	concentration of neutral, singly ionized, and doubly ionized zinc centers
N_{Zn}, N_B, N_{ZnB}	zinc, boron, and zinc–boron complex concentration
q	electronic charge
τ	time constant
t	time
T, T_1, T_2	junction temperatures
T_M	peak TSCAP derivative junction temperature
μ_n, μ_p	electron and hole conductivity mobilities
V	junction voltage
V_D	junction diffusion voltage
V_R, V_1	reverse biased junction voltages
W	depletion region width
x	space coordinate in the junction

1. Introduction

The electrical properties of zinc in silicon were first investigated by Fuller and Morin in 1956 [1]. They measured the diffusion coefficient, solid solubility, and a Hall effect thermal activation energy of 0.31 eV for holes at neutral zinc centers. A year later Carlson [2] observed a Hall effect thermal activation energy of 0.55 eV for electrons at doubly ionized zinc centers and established the double acceptor nature for substitutional zinc in silicon from mobility and solubility data.

It is the purpose of this paper to report the application of the recently developed dark junction capacitance and current transient techniques [3] to measure the thermal emission rates and activation energies of trapped holes at neutral and singly ionized zinc centers in silicon. The electric field dependences of the thermal emission rates are measured using a new differential bias voltage technique. A second zinc-related center (zinc–boron complex) is also investigated.

2. Device Fabrication

Both p^+–n and n^+–p abrupt junctions with nearly constant impurity concentration profiles are prepared using the conventional silicon planar technology. The cross sectional view of the n^+–p gated diode is given in Fig. 1. The shallow majority impurities are phosphorous donors and boron acceptors. Zinc diffusions were carried out in sealed quartz ampoules evacuated to 10^{-6} Torr. The zinc vapor pressure in the ampoule was kept at 1 atm or less at the diffusion temperature by placing an appropriate amount of zinc in the ampoules. The

Fig. 1. Diode structure

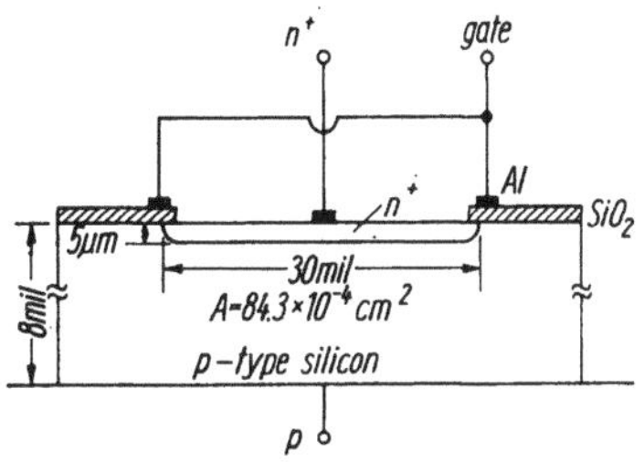

silicon slices were placed on a small quartz boat in each ampoule. The zinc diffusion times were 12 h. The substrate majority impurity concentration and the zinc diffusion temperature were selected to give a zinc concentration equal to approximately 10% of the substrate concentration. This assumes that solid solubility [1, 4] is reached for the diffusion times used [5].

An n-type silicon control slice was also placed in each zinc diffusion ampoule. The zinc concentration was determined by measuring the initial and final room temperature resistivity and conductivity type of the control slice. From the resistivity measurements, the free electron or hole concentrations can be determined assuming $\mu_n = 1350$ and $\mu_p = 480\ cm^2/Vs$. The zinc concentration is then determined using Fig. 2 which was generated from the neutrality equation

$$N_D + p(E_F) = n(E_F) + N_{Zn}\, F(E_F)\,. \tag{1}$$

Here $F(E_F)$, the number of negative charges per zinc center, is determined from the equilibrium statistics for zinc in silicon and is given by

$$F(E_F) = \frac{N_1 + 2\,N_2}{N_0 + N_1 + N_2}\,, \tag{2a}$$

$$F(E_F) = \frac{1 + 2\exp\left[(E_F - E_{3/2})/k_B T\right]}{1 + \exp\left[(E_F - E_{3/2})/k_B T\right] + \exp\left[(E_{1/2} - E_F)/k_B T\right]}\,. \tag{2b}$$

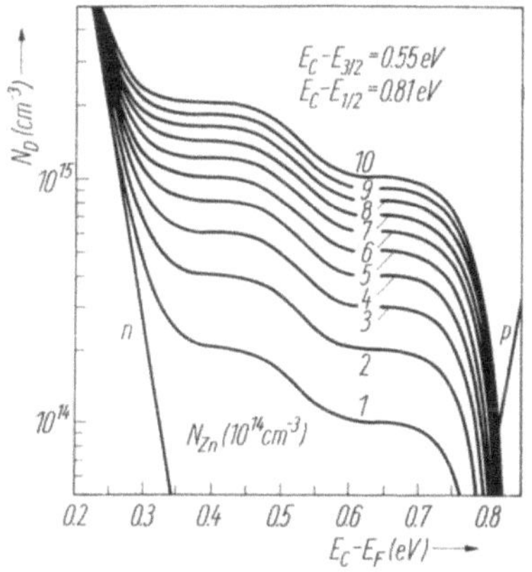

Fig. 2. E_F, p, and n in n-type zinc-doped silicon at 300 °K ($N_{Zn} = 10^{14}\ cm^{-3}$)

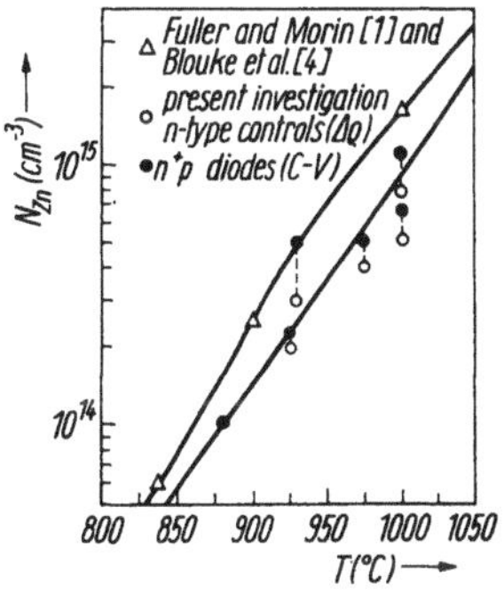

Fig. 3. Zinc concentrations determined from n-type controls ($\Delta\varrho$) and n^+–p diodes (C–V) for the zinc diffusion temperatures shown. Dashed lines denote the same zinc diffusion

The zinc concentrations determined from the resistivity change $\Delta\varrho$ are shown in Fig. 3. For comparison, zinc concentrations determined from reverse bias C–V measurements of the zinc-diffused n^+–p junctions are shown. These two agree quite well. Also shown in Fig. 3 is the solid solubility curve reported for zinc in silicon [1, 4]. The zinc concentrations obtained in this investigation are approximately a factor of 2 lower for the temperature range shown.

Three main fabrication problems were overcome in order to provide reproducible and uniform devices. First, the unwanted deep level centers were removed by a high temperature phosphorous glass gettering step introduced prior to the zinc diffusion. This was also employed in the sulfur work [6]. Secondly, the "soft" reverse current–voltage characteristics due to zinc precipitates, frequently observed after a rapid cooling by quenching at the end of the zinc diffusion, were eliminated using a relatively slow pull rate of one to three seconds from the zinc diffusion temperature to room temperature. The pull rate could not be too slow, for then the zinc would out diffuse, leaving little or no zinc near the metallurgical junction and space charge layer to be studied. The third problem was the clustering of the zinc on the silicon surface during the zinc diffusion, especially the surface where the contacts to the device were to be made. This was overcome by removing the oxide on the back surface of the device prior to the zinc diffusion, and opening the top contact holes in the oxide only after the zinc diffusion.

3. Measurement Techniques and Results

3.1 Thermal scanning of traps

A survey of the electrically active impurity and defect centers present in each sample can be obtained from a thermal scan of the junction current, thermally stimulated current (TSC), or a thermal scan of the high frequency junction capacitance (TSCAP) [7 to 11]. In this study the TSCAP method was employed where the time derivative of the capacitance, $\mathrm{d}C/\mathrm{d}t$, is recorded on a strip-chart recorder as a function of the sample temperature with a constant sample heating rate, $T = B\,t$, from 77 °K. Two TSCAP scans are made for each device so that the capacitance change or $\mathrm{d}C/\mathrm{d}t$ due to the thermal activation of trapped carriers is unambiguously defined. For the first scan the junction is cooled to 77 °K under reverse bias V_R, zero-biased, returned to V_R, and then heated to room temperature at a constant heating rate. The second scan is the same except the junction is not zero-biased before the heat is applied. Each peak of the $\mathrm{d}C/\mathrm{d}t$ curve that occurs only for the first scan corresponds to the thermal activation of a trap level whose energy can be estimated from the peak temperature T_M [10]. The derivative technique has been applied extensively to survey defects generated by keV to MeV ions [12] and by 1 MeV electrons [13] in silicon.

Experimental data for typical n^+–p zinc-diffused and control diodes are shown in Fig. 4. The zinc-diffused diode exhibits two peaks which can be associated with the two acceptor levels of zinc. The small peak at 127 °K for the control device is attributed to the donor level of the residue gold in silicon [10, 11]. This residue gold is gettered out by zinc in the zinc-diffused diodes as indicated by the absence of the 127 °K peak in Fig. 4. Gettering by a surface layer of metal on a silicon surface in liquid phase has been reported previously [14].

Fig. 4. Experimentally differentiated TSCAP curves for zinc-diffused and control n⁺–p diodes ($V_{np}(t = 0) = 0$; solid lines). The dashed lines are the curves that result when the levels are not filled with holes ($V_{np}(t = 0) = V_R = 30$ V). (a) n⁺–p Zn-doped device (B21-Z6-2); (b) n⁺–p control device (B21-C-8), heating rate ≈0.4 deg/s

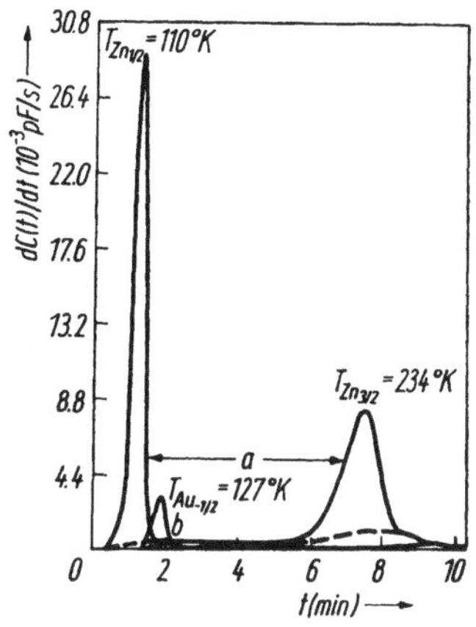

A thermal scan of the p⁺–n junctions for both the zinc-diffused and control devices revealed only one deep level center, which was previously reported [15], with a peak temperature of 185 °K ($B \approx 0.4$ deg/s). The concentration of this center was always less than $10^{12}/\text{cm}^3$.

3.2 *Charge concentrations*

After the device has been thermally scanned and the deep level traps located, the charge concentration for each trap is determined from the steady state dark junction capacitance versus reverse bias voltage measurements at a temperature between each of the peak derivative TSCAP temperatures. For an abrupt p–n junction with one side lowly doped compared with the other side, the total ionized center concentration (shallow majority impurity plus ionized traps) on the lowly doped side is given by

$$N_I(W, T) = 6.03 \times 10^6 A^2 \left[\frac{C^3}{\left|\frac{dC}{dV}\right|}\right], \tag{3}$$

where

$$W = 1.04 \frac{A}{C}. \tag{4}$$

The trap concentration with a peak dC/dt at T_M is then given by

$$N_{TT}(W, T_M) = |N_I(W, T_2) - N_I(W, T_1)|, \tag{5}$$

where $T_1 < T_M < T_2$. The charge state of the trap, whether donor-like or acceptor-like, can also be determined using this method by making a comparison of the control and impurity-diffused devices, assuming there are no traps introduced in the upper half of the band gap which have a comparable concentration to those in the lower half of the gap. This analysis technique has been employed previously for sulfur centers in silicon [6] and in great detail to sort out the electron-irradiated defects in silicon [13].

Fig. 5 and 6 are the experimental results for the n⁺–p and p⁺–n diodes, respec-

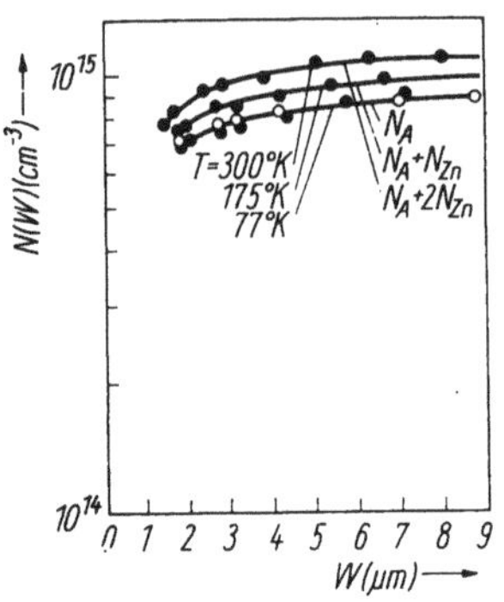

Fig. 5. Concentration profiles of zinc-diffused and control n$^+$–p diodes at the temperatures indicated. $N_{Zn} = 10^{14}$ cm^{-3}, ○ n$^+$–p control device (B21-C-8), ◉ n$^+$–p Zn-doped device (B21-Z6-2)

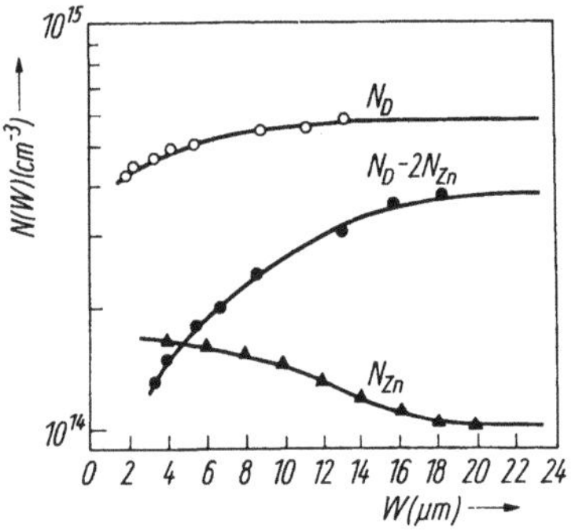

Fig. 6. Concentration profiles of zinc-diffused (◉) and control (○) p$^+$–n diodes at room temperature (300 °K). The resulting zinc concentration profile for the zinc-diffused diode is shown (▲)

tively. Three different temperatures were used for the zinc-diffused n$^+$–p diode in Fig. 5. At each temperature the capacitance and $\Delta C/\Delta V$ were measured experimentally for a number of reverse bias voltages, always starting at zero bias. Then using equation (3) the effective ionized impurity concentrations were determined. At 77 °K $N_I \approx N_A$, since the holes were frozen-in at the zinc acceptor centers. At the temperature of 175 °K the zinc centers are singly ionized, since $e^t_{p0} \gg e^t_{n1}$ and e^t_{p0} is large, so that $N_I \approx N_A + N_1 \approx N_A + N_{Zn}$. At room temperature e^t_{p1} and e^t_{n2} are large and $e^t_{p1} \gg e^t_{n2}$, so that most of the zinc centers are in the doubly ionized state, giving $N_I \approx N_A + N_1 + 2\,N_2 \approx$ $\approx N_A + 2\,N_2 \approx N_A + 2\,N_{Zn}$. The trap charge concentration using equation (5) around each peak $\mathrm{d}C/\mathrm{d}t$ is just N_{Zn}, the zinc concentration.

The same type measurements were made at room temperature on zinc-diffused and control p$^+$–n diodes. The total ionized impurity concentrations were determined using equation (3) and are shown in Fig. 6 along with half of the difference between these total concentrations, which is just the zinc concentration for the zinc-diffused p$^+$–n diode. It is clear from Fig. 6 that zinc is rejected by the boron rich p$^+$ region.

One other zinc-related center was detected in n$^+$–p diodes with boron substrate concentrations ranging from 10^{14} to 10^{16}/cm^3. This center, which gave rise to a shallow level with respect to the valence band, is identified with the acceptorlike, zinc–boron complex previously reported [1, 2]. The charge concentration for this center was measured at 77 °K using

$$N_{ZnB} \approx \frac{2\,\Delta C\,N_B}{C}\,. \tag{6}$$

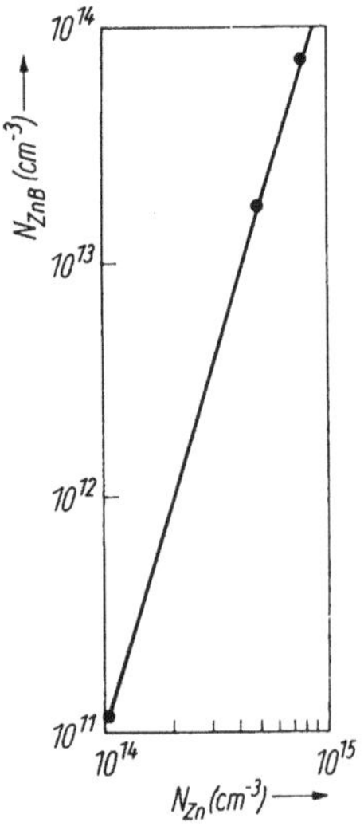

Fig. 7. Shallow level, zinc–boron center concentration versus zinc concentration for $N_{Zn}/N_B < 0.4$ (samples B21-Z6-2, B22-Z7-7, B23-Z8-7)

For $N_{Zn}/N_B < 0.4$, the concentration of this center depended on the zinc concentration in the device as $N_{ZnB} \sim N_{Zn}^3$ shown in Fig. 7, indicating that the complex may be of the form Zn_3B from the simple mass action law.

3.3 Thermal emission rates

The thermal emission rates of trapped holes, e_p^t, are obtained at constant sample temperatures over a wide temperature range using the dark junction capacitance (1 MHz) and current transients [3]. For the n^+–p diodes employed, the trapped hole concentration at a reverse bias V_R, after initially filling the traps with holes in the p-side by zero biasing the junction, is

$$p_T(t) = N_{TT} \left\{ \left[\frac{e_n^t}{e_n^t + e_p^t} \right] + \left[\frac{e_p^t}{e_n^t + e_p^t} \right] \exp\left(-\frac{t}{\tau} \right) \right\}, \tag{7}$$

where $\tau = (e_n^t + e_p^t)^{-1}$. The total ionized charge concentration (deep levels plus shallow majority impurity) is $N_I(t) = N_A + n_T(t) = N_{TT} = p_T(t)$ for an acceptor-like trap. Thus, if $N_{TT} \ll N_A$, then the capacitance transient is given by

$$\Delta C(t) = C(\infty) - C(t) \approx \frac{C(\infty)}{2} \left[\frac{N_{TT}}{N_{TT} + N_A} \right] \exp\left(- e_p^t\, t \right), \tag{8}$$

where $C^2(\infty) = [q\, K_s\, \varepsilon_0\, A^2/2\,(V + V_D)]\,(N_A + N_{TT})$ and $e_p^t \gg e_n^t$ is used. Thus, both the trap concentration N_{TT} and the thermal emission rate, e_p^t are obtained. A similar measurement for larger emission rates at higher temperatures is made for the junction current transient using

$$\Delta i(t) \approx \left[\frac{q\, A\, W\, N_{TT}\, e_p^t}{2} \right] \exp\left(- e_p^t\, t \right). \tag{9}$$

Results of the transient measurements on the two zinc acceptor levels are given in Fig. 8 and 9 using an equipment previously described [6, 15]. The

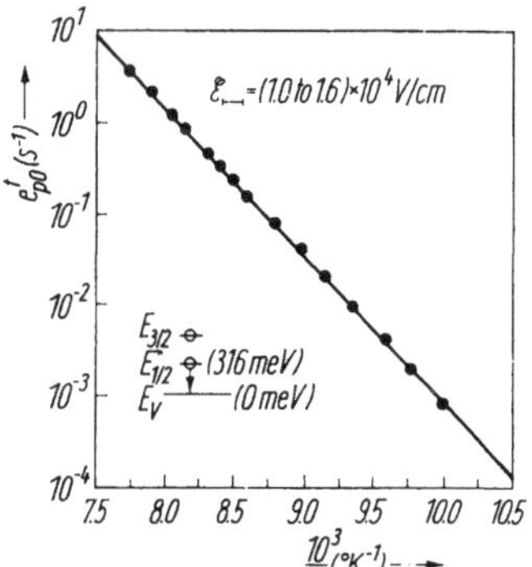

Fig. 8. Thermal emission rate of holes from neutral zinc centers versus reciprocal temperature. B21-Z6-2

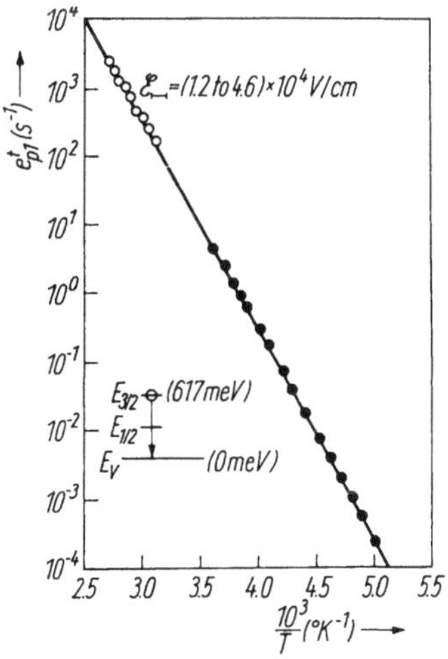

Fig. 9. Thermal emission rate of holes from singly ionized zinc centers versus reciprocal temperature B21-Z6-2, ○ current, ● capacitance

Table 1

Least squares fit parameters for different powers of T in the pre-exponential factor. Tolerances indicated are 70% fiducial limits

$e^{t}_{ps}(\mathscr{E}_{\vdash\dashv}) = A_{ps}\,(T/300)^m \exp\,(-\Delta E_{s+1/2}/k_B T)$

$\mathscr{E}_{\vdash\dashv} = (1.0 \text{ to } 1.6)\times 10^4$ V/cm

m	A_{p0} (10^{12} s^{-1})	$\Delta E_{1/2}$ (meV)
−2	7.64 ± 10%	336 ± 1
−1	7.41 ± 10%	326 ± 1
0	7.19 ± 10%	316 ± 1
1	6.98 ± 10%	306 ± 1
2	6.78 ± 10%	297 ± 1

$\mathscr{E}_{\vdash\dashv} = (1.2 \text{ to } 4.6)\times 10^4$ V/cm

m	A_{p1} (10^{12} s^{-1})	$\Delta E_{3/2}$ (meV)
−2	5.36 ± 4%	664 ± 1
−1	2.16 ± 1%	641 ± 1
0	0.868 ± 4%	617 ± 1
1	0.350 ± 5%	594 ± 1
2	0.141 ± 1%	571 ± 1

maximum sample temperature error is ±0.2 °K. High electric field dependence was observed for the hole emission rate from neutral zinc centers (e^{t}_{p0}, $E_V +$ + 316 meV), while little field dependence was observed for the hole emission rate from singly ionized zinc centers (e^{t}_{p1}, E_V + 617 meV). Thus, a technique is used, as described in the next section, to obtain the emission rates in a narrow range of electric field as indicated in Fig. 8 and 9.

The data of Fig. 8 and 9 are least squares fit to [6, 16]

$$e^{t}_{ps} = A_{ps}\left(\frac{T}{300}\right)^m \exp\left(-\frac{\Delta E_{s+1/2}}{k_B T}\right) \tag{10}$$

for $m = -2, -1, 0, 1$, and 2. The results are given in Table 1. The tolerances indicated are the 70% fiducial limit where the numerical constant in the "Student's t distribution" is between 1.05 and 1.10.

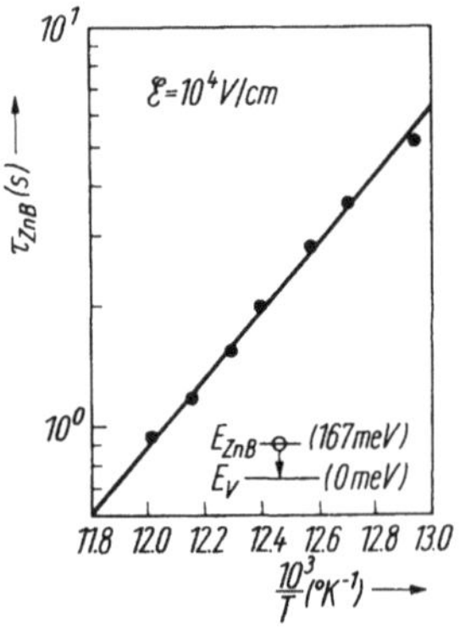

Fig. 10. Hole emission time constant for the shallow level, zinc–boron center versus reciprocal temperature. B22-Z7-7

The time constant or reciprocal hole emission rate, $\tau_{\mathrm{ZnB}} = 1/e^{\mathrm{t}}_{\mathrm{pZnB}}$, for the shallow level zinc–boron center was measured only over one order of magnitude from dark capacitance transients and is shown in Fig. 10 for an electric field of 10^4 V/cm. The data least squares fit to equation (10) gave $A(m = 0) = (1.5 \pm 2.0) \times 10^{10}$ and $\Delta E = (167 \pm 10)$ meV.

3.4 Electric field dependence

The electric field dependence of the thermal emission rates is obtained by a differential bias voltage technique in the following manner. First, the initial condition of the traps in the depletion region $0 \leqq x \leqq W$ is changed by only filling the traps in the region $x_1 \leqq x \leqq W$ with holes. This was achieved by switching the junction bias from V_{R} momentarily to V_1 and observing the $\Delta C(t)$ transient after returning to V_{R}. This is then repeated for $V_1 + \Delta V$, filling the traps with holes in $x_1 + \Delta x \leqq x \leqq W$. The difference between the two ΔC curves, $\Delta C(V_1, t) - \Delta C(V_1 + \Delta V, t)$, then gives the thermal emission rate of holes trapped in the region $x_1 \leqq x \leqq x_1 + \Delta x$ which corresponds to a well defined field interval, $\mathcal{E}(x_1 + \Delta x) \leqq \mathcal{E}_{\vdash\dashv}(x) \leqq \mathcal{E}(x_1)$. This technique is particularly simple to use for the capacitance transient if $N_{\mathrm{TT}} \ll N_{\mathrm{A}}$ and both N_{TT} and N_{A} are independent of position [17]. These conditions are approximately satisfied in the samples studied in this paper. It is evident that this differential bias voltage technique can also be applied to the dark current transient, but experimentally the current measurement is not as sensitive as the capacitance, and therefore the emission rates would be subject to a large amount of error.

These differential bias voltage methods have also been developed independently [11, 18] for probing the spatial variation of $N_{\mathrm{TT}}(x)$, where the field dependences of the thermal emission rates have been ignored.

The experimental data at fixed temperatures for the electric field dependence of the hole emission rate from neutral and singly ionized zinc centers are shown in Fig. 11. Fig. 12 presents the reciprocal temperature dependence of the hole emission rate from neutral zinc centers for fixed field intervals. There was no reciprocal temperature dependence observed for the hole emission rate from singly ionized zinc centers.

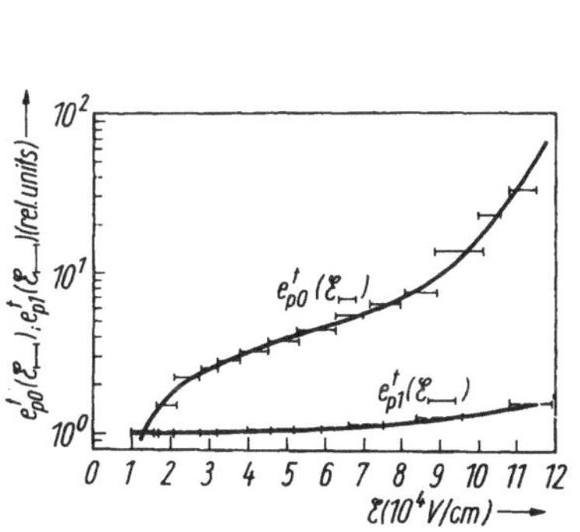

Fig. 11. Electric field dependence of the thermal emission rates. B21-Z6-2, ├──┤ $10^3/T = 9.600$, ├ – –┤ $10^3/T = 4.314$ °K^{-1}

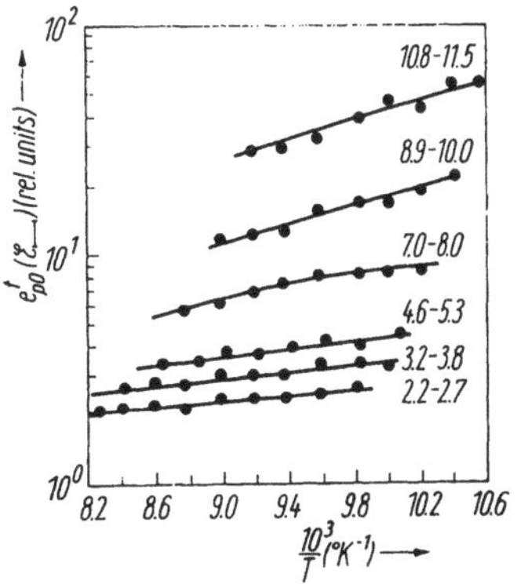

Fig. 12. Electric field dependence of the thermal emission rate of holes from neutral zinc centers versus reciprocal temperatures. B21-Z6-2; figures at the curves give the electric field interval in 10^4 V/cm

4. Discussion and Summary

The data presented in Fig. 4 indicate two zinc-related deep levels are present in the zinc-diffused n^+–p junctions, while that in Fig. 5 clearly identifies both of them with different charge states of the same center, since they have the same concentration profile, and as being acceptor-like. Thus, the double acceptor nature of substitutional zinc in silicon with three charge states (0, −1, −2) has been verified and unequivocally demonstrated. In both n^+–p and p^+–n zinc-diffused diodes no evidence of thermal emission for electrons from the doubly ionized zinc centers was detected. Therefore, it is concluded that $e^t_{p1} \gg e^t_{n2}$, and only the thermal emission rates and activation energies of holes from neutral and singly ionized zinc centers may be measured experimentally. The activation energies are shown in Fig. 8 and 9.

The electric field dependence of the thermal emission rate of holes from neutral zinc centers is very large. Fig. 11 and 12 indicate that in the range of fields shown more than one mechanism might be reponsible for the enhancement of the emission process, since two or more slopes exist in Fig. 11. This would suggest excited states play a role in the emission process. This was found to be the case for the helium-like double donor sulfur center in silicon [19]. The field dependence also increases with decreasing temperature, which would be consistent with the barrier lowering Poole-Frenkel model [20, 21]. The electric field dependence of the thermal emission rate of holes from singly ionized zinc centers was small, with just a slight field enhancement of the emission process.

A shallow level center with respect to the valence band was detected in n^+–p zinc-diffused diodes. The center is identified with the previously reported [1, 2] zinc–boron acceptor-like complex. The concentration of this center dependend on the cube of the zinc concentration in boron-doped substrates for $N_{Zn}/N_B < 0.4$, suggesting that it is a Zn_3B complex. The thermal emission rate of holes from this center was measured, which gave a thermal activation energy of 167 meV. Kornilov [22] explained the previously reported [1, 2] slopes of the Hall effect curves, attributed to the shallow levels of zinc-related complexes, as different degrees of compensation of the boron concentration by donors introduced during the high temperature zinc diffusion. In this investigation no measurable compensation of the boron substrate is observed. This is shown in Fig. 5 where the substrate concentration is the same for both the zinc-diffused and control devices.

Acknowledgements

The authors wish to thank T. H. Ning, L. L. Rosier, J. W. Walker, and L. D. Yau for many helpful discussions throughout the course of this work.

References

[1] C. S. Fuller and F. J. Morin, Phys. Rev. **105**, 379 (1957).
[2] R. O. Carlson, Phys. Rev. **108**, 1390 (1957).
[3] C. T. Sah, L. Forbes, L. L. Rosier, and A. F. Tasch, Jr., Solid State Electronics **13**, 759 (1970).
[4] M. M. Blouke, N. Holonyak, Jr., B. G. Streetman, and H. R. Zwicher, J. Phys. Chem. Solids **31**, 173 (1970).

[5] S. M. ZALAR, J. appl. Phys. **41**, 3458 (1970).
[6] L. L. ROSIER and C. T. SAH, Solid State Electronics **14**, 41 (1971).
[7] L. R. WEISBERG and H. SCHADE, J. appl. Phys. **39**, 5149 (1968).
[8] H. SCHADE and D. HERRICK, Solid State Electronics **12**, 857 (1969).
[9] L. FORBES and C. T. SAH, Solid State Electronics **14**, 182 (1971).
[10] C. T. SAH, W. W. CHAN, H. S. FU, and J. W. WALKER, Appl. Phys. Letters **20**, 193 (1972).
[11] M. G. BUEHLER, Solid State Electronics **15**, 69 (1972).
[12] H. S. FU, W. W. CHAN, and C. T. SAH, Bull. Amer. Phys. Soc. **17**, 307 (1972); to be published.
[13] J. W. WALKER and C. T. SAH, phys. stat. sol. (a) **11**, 513 (1972).
[14] S. J. SILVERMAN and J. B. SINGLETON, J. Electrochem. Soc. **105**, 591 (1958).
[15] L. D. YAU and C. T. SAH, phys. stat. sol. (a) **6**, 561 (1971).
[16] C. T. SAH, L. FORBES, L. L. ROSIER, A. F. TASCH, JR., and A. B. TOLE, Appl. Phys. Letters **15**, 145 (1969).
[17] C. T. SAH and H. S. FU, phys. stat. sol. (a) **11**, 297 (1972); **14**, 59 (1972).
[18] H. G. GLOVER, IEEE Trans. Electron Devices **19**, 138 (1972).
[19] T. H. NING and C. T. SAH, Phys. Rev. **B4**, 3482 (1971).
[20] J. FRENKEL, Phys. Rev. **54**, 647 (1938).
[21] J. L. HARKE, J. appl. Phys. **39**, 4871 (1968).
[22] B. V. KORNILOV, Soviet Phys. — Semicond. **2**, 1480 (1969).

(Received June 26, 1972)

phys. stat. sol. (a) 14, 417 (1972)
Subject classification: 17.1; 14.3; 22.1.1

Institute of Physics, Leningrad State University

Gold Adsorption on Clean Germanium Field Emitters

By
I. L. SOKOLSKAYA†, N. V. MILESHKINA, and R. Z. BAKHTIZIN[1])

Gold adsorption on clean surfaces of p-type and n-type Ge crystals has been studied by FEM technique. A structural selectivity of adsorption and desorption is observed. Contrary to what might be expected gold adsorption accesible at all for FEM coverages does not essentially change the field emission current. The straightening of the *i–U* characteristics for p-type Ge is observed for several monolayers of Au.

Goldadsorption auf sauberen Oberflächen von p-leitenden und n-leitenden Ge-Kristallen wurden mit der FEM-Technik untersucht. Ein strukturelles Trennvermögen von Adsorption und Desorption wird beobachtet. Im Gegensatz zum erwarteten Verhalten ändert Goldadsorption, die für FEM-Bedeckungen überhaupt möglich ist, den Feldemissionsstrom nicht wesentlich. Linearität der *i–U*-Charakteristiken für p-leitendes Ge wird für einige Monoschichten von Au beobachtet.

1. Introduction

The study of the metal–semiconductor contact at the first stages of metal adsorption [1] is of special interest as it helps to understand the processes which take place at the semiconductor–metal boundary. Some useful information may be obtained when studying the adsorption of thin metal layers on a semiconductor substrate. There is a very small number of works researching the metal adsorption on atomically clean surfaces of even such widely used semiconductors as Ge and Si. At the present time adsorption of some alkali metals on Ge and Si [2, 3] has been only studied as well as that of barium, titanium, and oxygen by field emission microscopy (FEM) technique [4 to 6]. The method of FEM is quite convenient for such studies as it gives the possibility of continuous visual surface control during the experiment, and allows to observe adsorption on one or several faces of the same single crystal simultaneously. Furthermore, the non-linear *i–U* characteristics of p-type semiconductor field emitters seem to contain some supplementary information on the emitting surface [6].

This study concerns the gold adsorption on the atomically clean surface of germanium single crystals as well as on germanium layers on tungsten. In comparison with the investigated electropositive adsorbates [2 to 5] gold has a greater work function than germanium ($\varphi_{Au} = 5.3$ eV [8 to 10]), and a decrease of the field emission current might be expected at adsorption.

2. Experimental

Emitter blanks oriented in the ⟨011⟩ and ⟨111⟩ were prepared from Sb-doped n-type Ge single crystals with resistivity 30 Ωcm and Ga-doped p-type Ge with resistivities 3.5 and 30 Ωcm. The tips were etched in CP-4a. The emitters were cleaned by dc field desorption.

[1]) Present address: Ufa State University, Ufa, USSR.

The experiments were carried out in a Müller-type microscope (FEM). The probe hole technique was used for the study of emission from different faces. Necessary tip positioning was achieved by attaching the sample holder to the stem of the tube via stainless steel bellows. The gold source was a small golden ball on a previously outgassed tungsten arc. The temperature of the source at the time of evaporating was kept constant, which has been controlled by the optical pyrometre, and because of this the quantity of deposited gold has been taken to be proportional to the time of the evaporation. The coverage of the surface by gold (θ in monolayers) has been calculated by the Vekshinskii method [12] from the measurement of the total evaporation time.

While outgassing the Au source in the process of pumping out the formation of amalgam is excepted. As it was shown by means of a tungsten emitter the evaporated gold films had the same work function as that of solid samples [8 to 11].

3. Results

The field emission of the germanium emitter surface cleaned by field desorption and covered by gold at room temperature is shown in Fig. 1. It may be clearly seen that irrespective of the emitter orientation an increase of the emission from the faces {001} and {011} takes place after adsorption, but at the same time the contribution to the total emission current from the {111} faces is decreased. With increased adsorbate thickness the crystallization of gold on germanium surface is observed except on {111} faces (Fig. 2a). The structural selectivity is observed also at the field desorption of gold from germanium. It can be seen from the emission patterns of desorption (Fig. 2b to d) that, if the coverage considerably exceeds one monolayer, the Au adatoms seem to migrate at high field in the direction of {111} faces and then are desorbed (Fig. 2e). At the higher electric field gold is desorbed from {100} and {110} faces (Fig. 2f). The desorption field for thin gold layers does not exceed that for germanium evaporation, which fact may point to the comparatively weak binding of Au adatoms to the substrate. Heating the tip after field desorp-

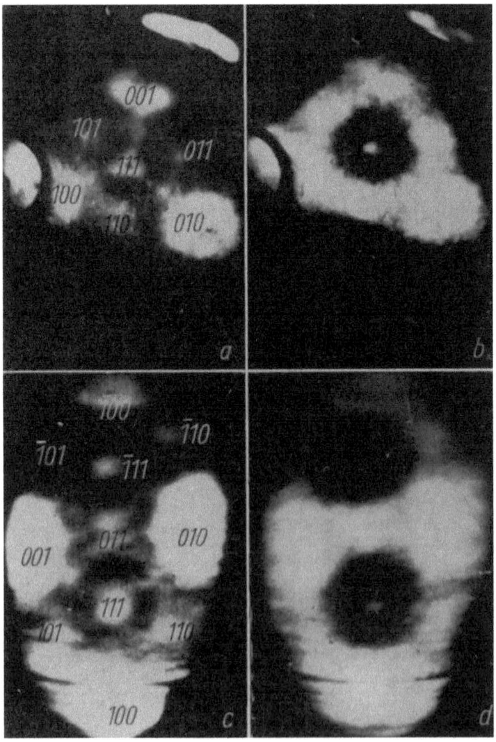

Fig. 1. FEM patterns of Au deposited on Ge at 300 °K. a) clean ⟨111⟩ oriented Ge field emitter; b) the same tip after depositing Au (θ = 1.0); c) clean ⟨011⟩ oriented Ge field emitter; d) the same tip after depositing Au (θ = 1.0)

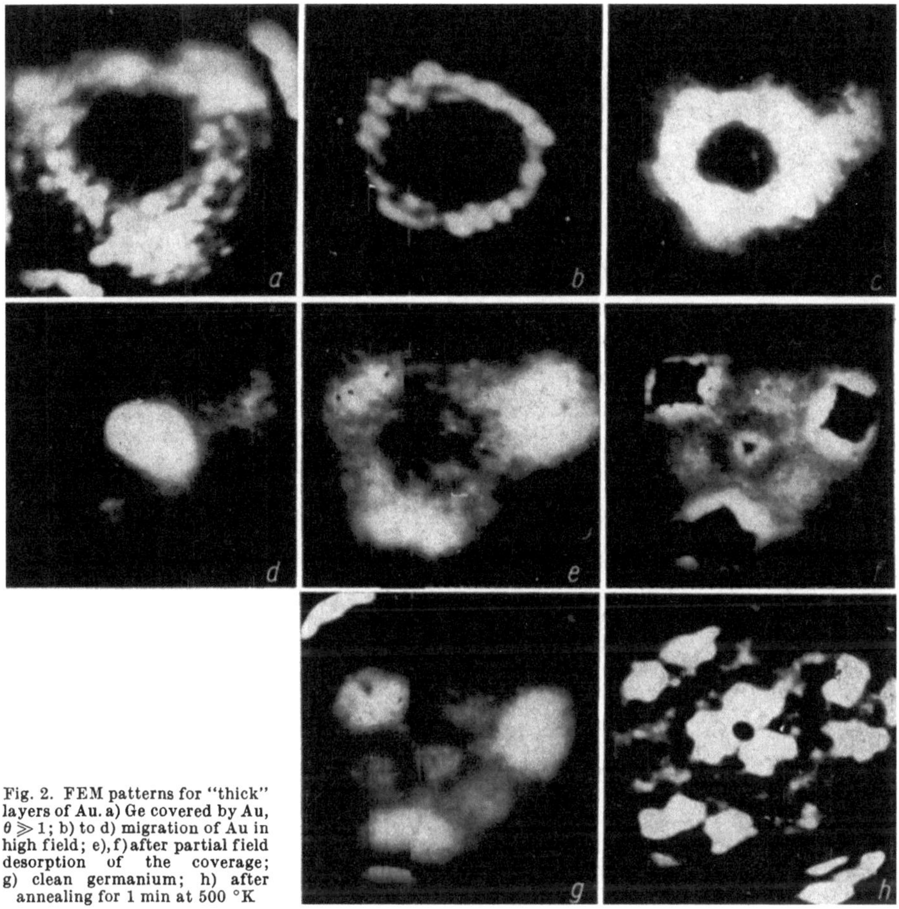

Fig. 2. FEM patterns for "thick" layers of Au. a) Ge covered by Au, $\theta \gg 1$; b) to d) migration of Au in high field; e), f) after partial field desorption of the coverage; g) clean germanium; h) after annealing for 1 min at 500 °K

tion (500 °K; 1 min) resulted in the appearance of contrast emission pattern (Fig. 2f) which may serve as an additional evidence for the atomically cleaning of the initial Ge surface [13].

The i–U characteristics of p-type and n-type germanium field emitters obtained at gold adsorption are shown on Fig. 3 and 4, respectively. Contrary to what might be expected gold adsorption at all investigated degrees of coverage does not essentially decrease the field emission current. Both for p-type and n-type Ge only a slight decrease of the emission is observed for low coverage (curves 2, Fig. 3 and Fig. 2). At higher coverage the emission current

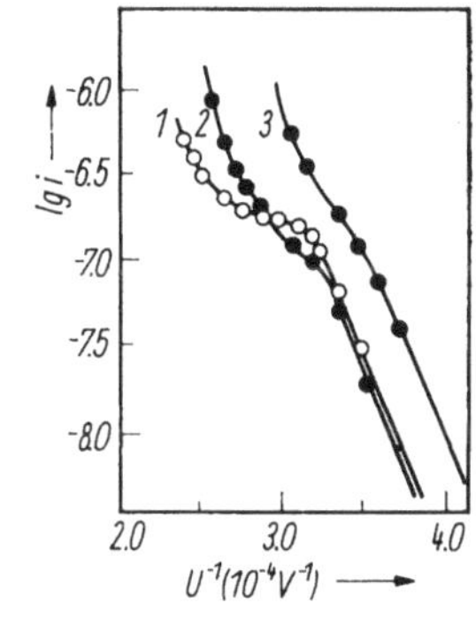

Fig. 3. i–U characteristics of p-type Ge. (1) Initial i–U characteristic of 30 Ωcm p-type Ge; (2) after depositing Au, $\theta = 0.5$; (3) $\theta = 1.5$

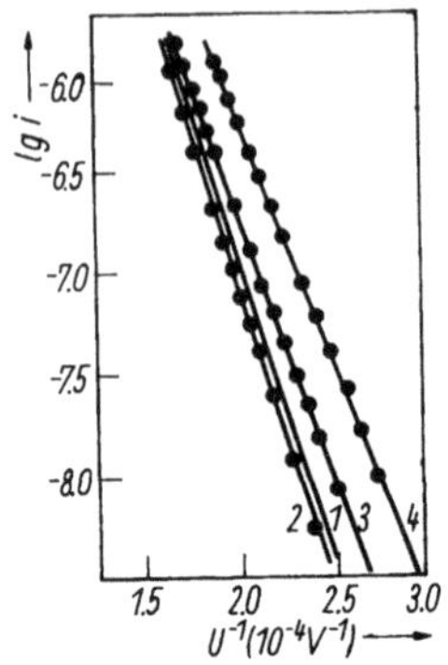

Fig. 4. i–U characteristics for the case of n-type Ge. (1) Clean germanium; (2) after Au adsorption, $\theta = 0.7$; (3) $\theta = 1.5$; (4) $\theta = 3.0$

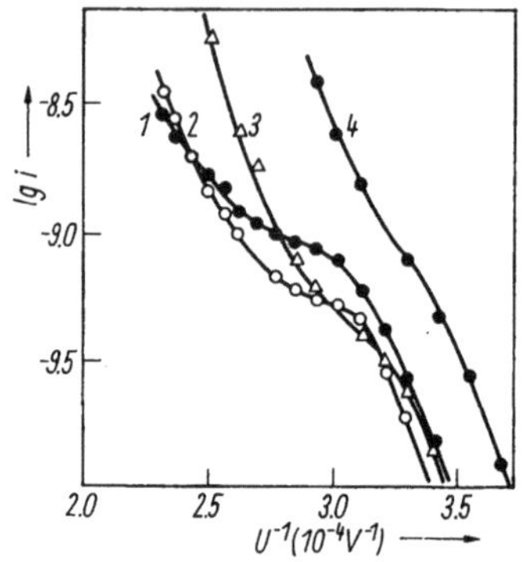

Fig. 5. i–U characteristics for Au on the (100) face of Ge field emitter. (1) Initial curve for 3.5 Ωcm p-type Ge; (2) after depositing Au, $\theta = 0.3$; (3) $\theta = 0.5$; (4) $\theta = 1.5$

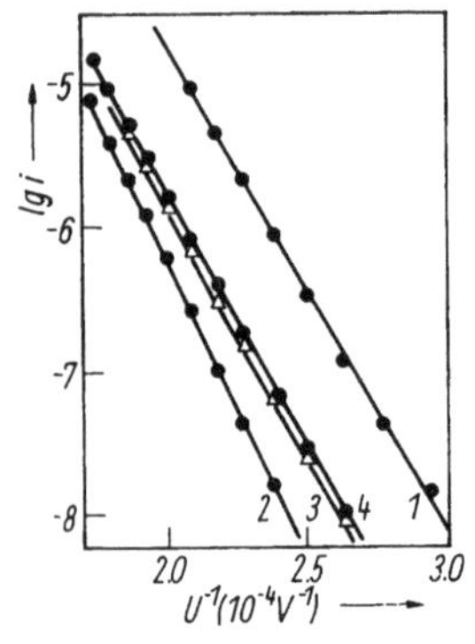

Fig. 6. Fowler-Nordheim characteristics. (1) Clean tungsten ($\varphi = 4.5$ eV); (2) Au monolayer on W field emitter ($\varphi = 5.3$ eV); (3) W covered by Ge, $\theta = 1$; (4) after Au adsorption on W covered by Ge, $\theta = 1$

increases and the slope of F-N plots decreases (Fig. 3, curve 3; and Fig. 4 curves 3 and 4). These results refer to the current of the whole tip and cannot correspond to the behaviour of gold at different parts of the emitting surface. As it may be seen in Fig. 5 the effect of gold adsorption on the (100) face is the same as that on the total current (Fig. 3).

For checking up this effect the gold adsorption is observed on tungsten covered with germanium monolayers. As it is noted in [14], the emission properties of these emitters are non-metallic essentially, and the adsorption properties of such layers are determined by the origin of adatoms and the interaction between adatoms and substrate atoms. The layer of gold on the pure tungsten emitter decreases the emission current and increases the work function up to 5.3 eV (Fig. 6, curve 2). The same layer on a germanium film deposited on tungsten does not change the emission characteristics significantly (Fig. 6, curve 4)

Opposite to the cases of barium [4] and oxygen [6] for gold adsorption non-linear i–U characteristics of the p-type Ge emitter are observed even at the coverage $\theta > 1$ (Fig. 3 and 5). Only for several monolayers of gold a straightening of the i–U characteristics is obtained. The removal of gold by field desorption could at least partially restore the initial non-linearity, but at small ($\theta = 0.5$) coverages only. It seems, the irreversible changes occur due to the noticeable diffusion of gold into the germanium substrate when the adsorbed film becomes thicker.

4. Discussion

As it was pointed out above, gold adsorption both at p-type and n-type germanium does not essentially decrease the emission as well as increase the work function, though the work function of gold films ($\varphi_{\text{Au}} = 5.3$ eV) exceeds that of germanium ($\varphi_{\text{Ge}} = 4.8$ eV [5]). One of the possible explanations of this anomaly might be the mercury adsorption on gold layers that according to Huber [8] decreases the work function to $\varphi = 4.7$ eV. However, in our experiments such a possibility was excluded, as it was mentioned above.

Another reason of such a behaviour of gold on the germanium surface might be the solution of gold in germanium already at room temperature (as gold and germanium are known to form the eutectic at $T = 629$ °K). But special experiments performed with cooled emitters produced the same results.

It may be suggested as well that gold adatoms are weakly bound to the germanium substrate (the fact is in accordance with the results of field desorption of thin gold films) and have a tendency to crystallize in thin films already. Then the observed change of field emission characteristics might be referred to clusters of gold atoms which give rise to locally increased fields. At higher coverage this effect grows. However it is difficult to suppose the constancy of these local fields at different cases resulting in the stability of the work function of $\theta \leqq 1$. Furthermore, the control tungsten emitter shows that the degree of "granularity" of gold films on germanium at $\theta \leqq 1$ does not result in the emission characteristics.

Such an anomalous behaviour of gold on germanium as well as on germanium films on tungsten substrates may be explained by the specific interaction of gold with the germanium substrate: it seems, that for the monolayer there are no metallic bonds among the gold atoms as in the solid. So the work function of germanium and the emission current after adsorption of thin gold layers almost does not change (curves 2, Fig. 3 and 4). It is possibly useful to take into account the "island" structure of gold layers in the investigations of coverages accessible for FEM.

The changes of the i–U characteristics of p-type Ge which is observed at gold adsorption (Fig. 3 and 5) is not due to the increase in the conductivity of the lateral cone surface, shunting the current in the volume of the semiconductor emitter [4], as in this case one could have observed the increase in the "saturation" current (as well as in the case of increasing temperature or light intensity [16]) because of the additional carriers supply. Besides, as has been shown in [17], the shunting effect of gold layers on germanium occurs only in the case of coverages $\theta \geqq 2$, while the changes of the emission characteristics may be observed already at $\theta < 1$.

We suppose that these changes may be referred to the change in the density of the surface states of the clean germanium surface at gold adsorption, which results in a decrease in the screening of the external field. Then according to [7] the straightening of the non-linear i–U characteristics of germanium at gold adsorption (Fig. 3 and 5) may be explained by the increase in the field generation rate of carriers in the depletion region of the p-type field emitter. This conception is developed in [6].

Acknowledgements

We are indebted to T. P. Andreeva for her assistance and to Dr. V. M. Ushakov and Dr. P. G. Shlyakhtenko for useful discussions.

References

[1] W. MÖNCH, phys. stat. sol. **40**, 257 (1970); Surface Sci. **21**, 443 (1970).
[2] F. G. ALLEN and G. W. GOBELI, Phys. Rev. **144**, 558 (1966).
[3] P. W. PALMBERG and W. T. PERIA, Surface Sci. **6**, 57 (1967).
[4] R. E. WEBER and W. T. PERIA, Surface Sci. **14**, 13 (1969).
I. L. SOKOLSKAYA, V. G. IVANOV, and G. N. FURSEY, phys. stat. sol. **21**, 789 (1967).
[5] R. Z. BAKHTIZIN and I. L. SOKOLSKAYA, Fiz. tverd. Tela **12**, 2815 (1970).
[6] N. V. MILESHKINA and R. Z. BAKHTIZIN, Surface Sci. **29**, 644 (1972).
[7] A. F. YATSENKO, phys. stat. sol. (a) **1**, 333 (1970).
[8] E. E. HUBER, Appl. Phys. Letters **8**, 169 (1966).
[9] J. C. RIVIERE, Appl. Phys. Letters **8**, 172 (1966).
[10] A. A. HOLSCHER, Surface Sci. **4**, 89 (1966).
[11] N. V. MILESHKINA and I. L. SOKOLSKAYA, Proc. LGU, Voprosy elektr. tverd. Tela **3**, 11 (1970).
[12] S. A. VEKSHINSKII, Novyi metod metallograficheskogo issledovaniya splavov, Gostekhizdat, Moscow/Leningrad 1948 (p. 28).
[13] F. G. ALLEN, J. Phys. Chem. Solids **19**, 87 (1961).
[14] N. V. MILESHKINA and I. L. SOKOLSKAYA, Fiz. tverd. Tela **5**, 2501 (1963); **7**, 1043 (1965).
[15] J. R. ARTHUR, JR., J. Phys. Chem. Solids **25**, 683 (1964).
[16] J. R. ARTHUR, JR., J. appl. Phys. **36**, 3221 (1965).
[17] A. I. KLIMOVSKAYA and O. V. SNITKO, Fiz. tverd. Tela **8**, 611 (1966).

(Received September 18, 1972)

phys. stat. sol. (a) **14**, 423 (1972)

Subject classification: 14.3.4; 6; 13.1; 18.2; 22.8

Technische Universität Dresden, Sektion Physik, Arbeitsgruppe Theoretische Physik I (a), and Akademie der Wissenschaften der DDR, Zentralinstitut für Festkörperphysik und Werkstofforschung, Institutsteil für magnetische Werkstoffe, Jena (b)

Transport Properties of a Non-Degenerate Ferromagnetic Semiconductor in High Electric Fields

III. Low Temperature Region

By

S. ROTH (a) and W. HAUBENREISSER (b)

On the basis of a theory, given in a preceding paper [1], the high field transport properties of $CdCr_2Se_4$ are calculated for a temperature below the Curie temperature. The scattering at the magnetic moments is calculated in the spin wave approximation. Scattering at optical and acoustical phonons is taken into account, and the influence of the effective mass of the conduction electrons is studied. A negative differential conductivity is found.

Auf der Grundlage einer Theorie, die in einer früheren Arbeit [1] dargelegt wurde, werden die Transporteigenschaften von $CdCr_2Se_4$ bei hohen elektrischen Feldern für eine Temperatur unterhalb der Curie-Temperatur berechnet. Die Streuung an den magnetischen Momenten wird in der Spinwellen-Näherung behandelt. Außerdem wird die Streuung an optischen und akustischen Phononen berücksichtigt und der Einfluß der effektiven Masse der Leitungselektronen untersucht. Eine negative differentielle Leitfähigkeit wird gefunden.

1. Introduction

In two preceding papers [1, 2] (referred to as I and II) we gave a theory for the calculation of high-field transport properties of a non-degenerate ferromagnetic semiconductor and applied this theory to calculate the high-field transport properties of $CdCr_2Se_4$ above the Curie temperature and for different magnetic fields. A negative differential conductivity (ndc) was found at large electric fields [2]. To investigate whether at low temperatures ndc occurs also, we performed calculations for a temperature below the Curie temperature. These calculations are presented in this paper.

2. The Approximate Solution of Boltzmann's Equation

As outlined in paper I we use a displaced Maxwellian (equation (I.13)) for the distribution function of the conduction electrons in the two subbands. The parameters in this distribution function are determined by the conservation laws for electron number, wave vector, and energy in each subband.

In the low temperature case we have to substitute the collision integrals for optical phonons (equation (I.10)), for acoustical phonons (equation (I.11)), and for magnons (equation (I.9)) into the equations obtained from the conservation laws (equations (I.15) to (I.17)). For simplicity we made the further approximation that the energy of magnons is independent of the wave vector of magnons

$$\hbar\,\omega_{\boldsymbol{k}} = \hbar\,\omega_0\,. \tag{1}$$

This approximation is supported by some calculations of Woolsey and White [3] who calculated the magnon lifetime due to electron–magnon interaction for a ferromagnetic semiconductor. They found that only magnons within a very small range of wave vectors are damped due to magnon–electron interaction. This indicates that only magnons with a certain wave vector and therefore with a certain energy interact with the conduction electrons. This justifies our approximation.

If we now substitute the relevant collision integrals (equations (I.9) to (I.11)) into the conservation laws (equations (I.15) to (I.17)) we get a set of equations similar to that for the high temperature region (equations (II.1) to (II.3)). This set of equations is given in the Appendix and was solved in the same manner as outlined in paper II, Section 2.

3. The Material Parameters

As in paper II we apply the theory to $CdCr_2Se_4$. We need in this calculation several material parameters, some of which are not known. The parameters which are known are the static and the optical dielectric constants ($\varepsilon_0 = 10.28$ and $\varepsilon_\infty = 9.0$) and the optical phonon frequency ($\hbar\,\Omega_0 \approx 0.03$ eV) [4]. The interaction constant J is taken to be the same as in paper II [5]. The gap between the two subbands Δ is taken from measurements of the red shift of the absorption edge [6]. The gap should be a little bit smaller than the difference between the measured absorption edge and that extrapolated from the paramagnetic state [7]. We obtain the value $\Delta \approx 0.3$ eV. The lattice constant of the cubic cell is 10.75 Å. The elementary lattice cell is four times smaller than the cubic one and contains 14 atoms, i.e. it has the composition $(CdCr_2Se_4)_2$. Two parameters which occur in the coupling constant of the acoustical phonons are unknown. We took the values $C = 5$ eV for the deformation potential and $u = 3000$ m/s for the velocity of sound. The effective mass m^* of the conduction electrons in $CdCr_2Se_4$ is not known, too. As in paper II we performed calculations for $m^*/m_0 = 1$ and $m^*/m_0 = 0.1$. To investigate the influence of the energy of magnons $\hbar\,\omega_0$, which is a constant in our approximation, we calculated the transport properties for three magnon energies ($\hbar\,\omega_0 = 1$, 3, and 5 meV). The lattice temperature T is taken to be $T = 80$ K and the Curie temperature of $CdCr_2Se_4$ is $T_c = 129$ K [8].

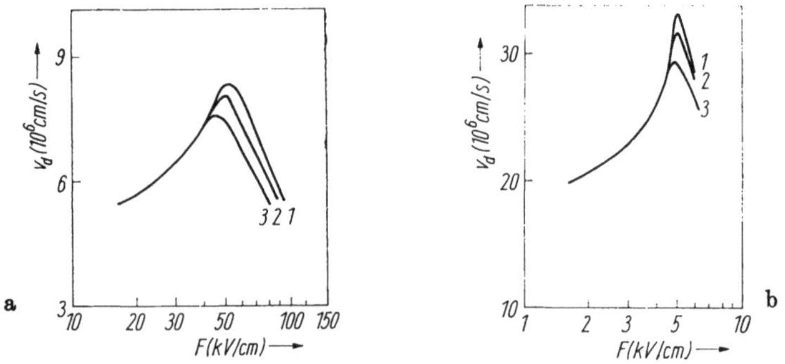

Fig. 1. Drift velocity as a function of an applied dc electric field for different magnon energies: (1) $\hbar\omega_0 = 5$meV; (2) 3 meV; (3) 1 meV. a) $m^*/m_0 = 1$, b) $m^*/m_0 = 0.1$

4. Discussion of the Results

The results of our calculation are given in Fig. 1a, 2a, 3a, and 4a for $m^*/m_0 = 1$ and in Fig. 1b, 2b, 3b, and 4b for $m^*/m_0 = 0.1$. In Fig. 1 (a und b) the field dependence of the drift velocity is plotted. Fig. 2 (a and b) shows the drift mobility as a function of the electric field, and Fig. 3 and 4 (a and b) show the field dependence of the difference between the effective temperature of electrons in the (+)-subband and the lattice temperature $T_e^+ - T$ and the ratio of the number of electrons in the (+)-subband and in the (—)-subband n^+/n^-, respectively.

From Fig. 1 (a and b) we see that there is a negative differential conductivity at high electric fields. In comparison to the high-temperature case the critical fields are considerably smaller. The ndc is (as in the high-temperature case [2]) due to the "switching on" of the magnetic interband scattering with increasing effective temperature of electrons. Those electrons in the (+)-subband which are scattered by magnetic scattering into the (—)-subband have to overcome the energy gap Δ. Therefore high electron energies and thus a high effective temperature of electrons are necessary to make the magnetic interband scattering efficient. This is indicated by the fact that all the curves for different magnon energies coincide at fields for which the mean energy of electrons is not high enough to make the magnetic interband scattering contribute to the resistivity. Only the μ^--curves are at low fields different for different magnon energies. But in this case the magnetic interband scattering is always present because an electron which is scattered from the (—)-subband to the (+)-subband has not to overcome the energy gap Δ. A similar mechanism to explain the ndc in CdTe has been assumed by Canali et al. [9] who believe that the ndc in CdTe

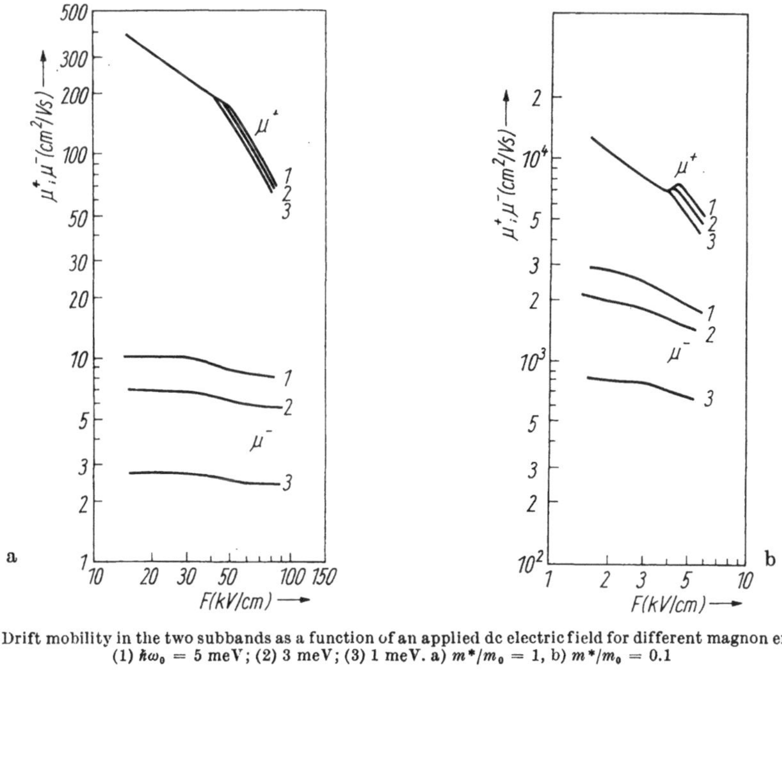

Fig. 2. Drift mobility in the two subbands as a function of an applied dc electric field for different magnon energies: (1) $\hbar\omega_0 = 5$ meV; (2) 3 meV; (3) 1 meV. a) $m^*/m_0 = 1$, b) $m^*/m_0 = 0.1$

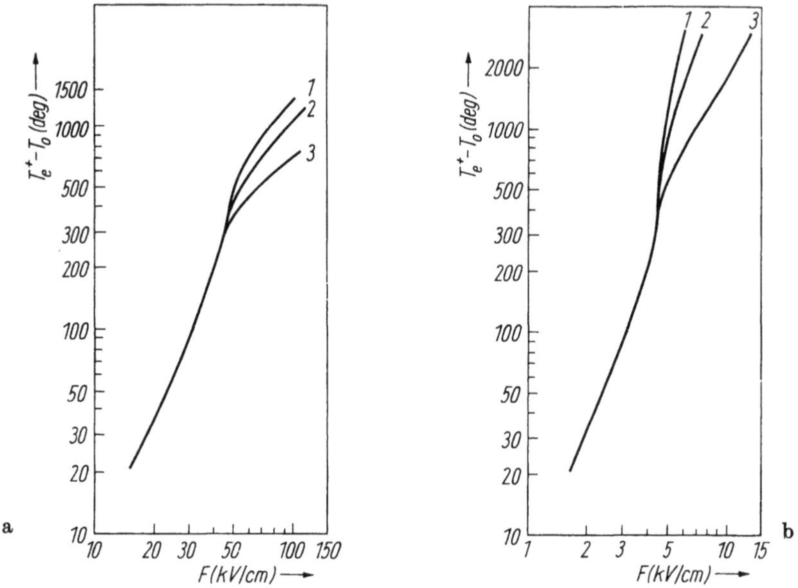

Fig. 3. Difference between the effective temperature of electrons in the (+)-subband and the lattice temperature as a function of an applied dc electric field for different magnon energies: (1) $\hbar\omega_0 = 5$ meV; (2) 3 meV; (1) 1 meV. a) $m^*/m_0 = 1$, b) $m^*/m_0 = 0.1$

is due to "switching on" of the intervalley scattering rather than to the larger effective mass of the electrons in the secondary valleys.

A low magnon energy leads to a lower critical field, because the magnon scattering is proportional to $[\exp(\hbar\,\omega_0/k_B T) - 1]^{-1}$ (equations (A2) and (A3)) and therefore is more efficient for lower magnon energies. At higher magnon energies the ndc is more striking because the effective temperature of electrons is increasing more rapidly in this case, which induces a more rapid switching on of the magnetic interband scattering (see Fig. 3 (a and b)).

From Fig. 3 a and b we see that the assumption of Haas [7] namely that the difference between the effective temperature of electrons and the lattice temperature ($T_e^+ - T$) is proportional to the square of the electric field is valid only below the critical field where magnon scattering is negligible and where the effec-

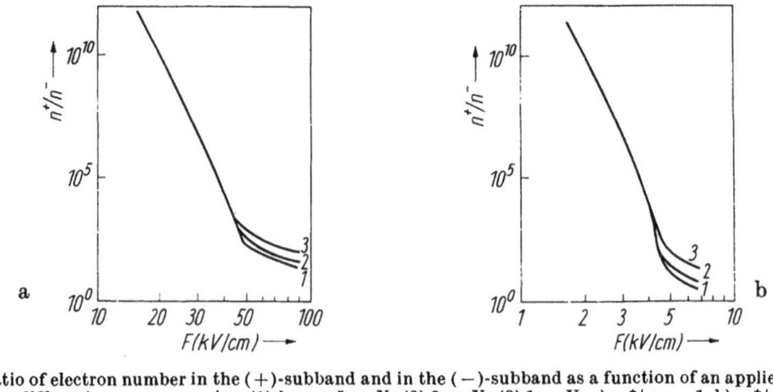

Fig. 4. Ratio of electron number in the (+)-subband and in the (−)-subband as a function of an applied dc electric field for different magnon energies: (1) $\hbar\omega_0 = 5$ meV; (2) 3 meV; (3) 1 meV. a) $m^*/m_0 = 1$, b) $m^*/m_0 = 0.1$

tive temperature of electrons is smaller than $\hbar\, \Omega_0/k_B$. If the mean energy of electrons $T_e^+\, k_B$ is of the order of the optical phonon energy $\hbar\, \Omega_0$ the optical phonon scattering becomes small [10] and the effective temperature of electrons T_e^+ increases more rapidly with increasing electric field. But with increasing T_e^+ and with decreasing magnon energy $\hbar\, \omega_0$ the magnetic scattering becomes more efficient as already has been discussed. Therefore, after switching on the magnetic scattering the effective temperature of electrons increases less rapidly with increasing electric field and with decreasing magnon energies. On the other hand for higher magnon energies there are more electrons in the higher (−)-subband due to the higher effective temperature of electrons.

The μ^+-curves for $\hbar\, \omega_0 = 3$ and 5 meV in Fig. 2b have a maximum. This is due to the fact that the optical phonon scattering is decreasing whilst the magnetic scattering is increasing with increasing electric field. For high magnon energies and low effective mass of the conduction electrons the optical phonon scattering is stronger in comparison with the magnetic scattering (see equations (A.2) and (A.3)). Therefore, this effect of decreasing optical phonon scattering can be seen best in these cases. On the other hand this indicates that we are near the dielectric breakdown voltage.

5. Conclusions

In $CdCr_2Se_4$ an ndc behaviour at high electric fields can be expected also below the Curie temperature. The calculated critical fields are much lower than in the high-temperature case. Therefore, it could be possible to observe an ndc in $CdCr_2Se_4$ at low temperatures easier than at high temperatures. As at high temperatures a smaller effective mass of electrons lowers the critical field in the low temperature region, too.

Acknowledgements

The authors are indebted to Prof. Dr. G. Heber, Prof. Dr. F. Voigt, and Dr. J. Monecke for helpful discussions and stimulating interest.

Appendix

If we substitute the relevant collision integrals (equations (I.9) to (I.11)) into the conservation laws for electron number, wave vector, and energy in each subband we get the following set of equations. Electron number conservation gives (equation (I.15))

$$\frac{n^-}{n^+} = \frac{e^{-\frac{\Delta - \hbar\omega_0}{k_B T_e^+}}}{e^{\frac{\hbar\omega_0}{k_B T}}} \frac{e^{-\left(\frac{d^+}{t^+}\right)^2}}{e^{-\left(\frac{d^-}{t^-}\right)^2}} \frac{\int\limits_0^\infty dx\, e^{-x} \sqrt{x} \sqrt{x + \frac{\Delta - \hbar\, \omega_0}{k_B T_e^+}}\; V\left(2\sqrt{x + \frac{\Delta - \hbar_0 \omega_0}{k_B T_c^+}}\, \frac{d^+}{t^+}\right)}{\int\limits_0^\infty dx\, e^{-x} \sqrt{x} \sqrt{x + \frac{\Delta - \hbar\, \omega_0}{k_B T_e^-}}\; V\left(2\sqrt{x}\, \frac{d^-}{t^-}\right)}. \tag{A.1}$$

Energy conservation (equation (I.17)) gives

$$\frac{\hbar e}{m^*} \boldsymbol{F d}^+ = \left(\frac{2 m^*}{\hbar^2 \pi}\right)^{3/2} \frac{J^2 v}{\hbar\left(e^{\frac{\hbar\omega_0}{k_B T}} - 1\right)} \times$$

$$\times \left\{\frac{n^+}{n^-} e^{\frac{\hbar\omega_0}{k_B T}} (k_B T_e^-)^{3/2} e^{-\left(\frac{d^-}{t^-}\right)^2} \int_0^\infty dx \sqrt{x} \left(x + \frac{\Delta - \hbar \omega_0}{k_B T_e^-}\right)^{3/2} V\left(2 \frac{d^-}{t^-} \sqrt{x}\right) e^{-x} - \right.$$

$$- (k_B T_e^+)^{3/2} e^{-\frac{\Delta - \hbar \omega_0}{k_B T_e^+}} e^{-\left(\frac{d^+}{t^+}\right)^2} \int_0^\infty dx \sqrt{x} \left(x + \frac{\Delta - \hbar \omega_0}{k_B T_e^+}\right)^{3/2} \times$$

$$\left. \times V\left(2 \frac{d^+}{t^+} \sqrt{x + \frac{\Delta - \hbar \omega_0}{k_B T_e^+}}\right) e^{-x}\right\} -$$

$$- N_0 \Omega_0^2 e^2 \left(\frac{1}{\varepsilon_\infty} - \frac{1}{\varepsilon_0}\right) \sqrt{\frac{2 m^*}{\pi k_B T_e^+}} e^{-\left(\frac{d^+}{t^+}\right)^2} \int_0^\infty dx \ln \left[\frac{\sqrt{x + \frac{\hbar \Omega_0}{k_B T_e^+}} + \sqrt{x}}{\sqrt{x + \frac{\hbar \Omega_0}{k_B T_e^+}} - \sqrt{x}}\right] \times$$

$$\times \left\{e^{\frac{\hbar \Omega_0}{k_B}\left(\frac{1}{T} - \frac{1}{T_e^+}\right)} V\left(2 \frac{d^+}{t^+} \sqrt{x + \frac{\hbar \Omega_0}{k_B T_e^+}}\right) - V\left(2 \frac{d^+}{t^+} \sqrt{x}\right)\right\} e^{-x} \tag{A.2a}$$

and

$$\frac{\hbar e}{m^*} \boldsymbol{F d}^- = \left(\frac{2 m^*}{\hbar^2 \pi}\right)^{3/2} \frac{J^2 v}{\hbar\left(e^{\frac{\hbar\omega_0}{k_B T}} - 1\right)} \times$$

$$\times \left\{\frac{n^+}{n^-} (k_B T_e^+)^{3/2} e^{-\frac{\Delta - \hbar \omega_0}{k_B T_e^+}} e^{-\left(\frac{d^+}{t^+}\right)^2} \int_0^\infty dx\, x^{3/2} \sqrt{x + \frac{\Delta - \hbar \omega_0}{k_B T_e^+}} \times \right.$$

$$\times V\left(2 \frac{d^+}{t^+} \sqrt{x + \frac{\Delta - \hbar \omega_0}{k_B T_e^+}}\right) e^{-x} -$$

$$\left. - e^{\frac{\hbar\omega_0}{k_B T}} (k_B T_e^-)^{3/2} e^{-\left(\frac{d^-}{t^-}\right)^2} \int_0^\infty dx\, x^2 \sqrt{x + \frac{\Delta - \hbar \omega_0}{k_B T_e^-}} V\left(2 \frac{d^-}{t^-} \sqrt{x}\right) e^{-x}\right\} -$$

$$- N_0 \Omega_0^2 e^2 \left(\frac{1}{\varepsilon_\infty} - \frac{1}{\varepsilon_0}\right) \sqrt{\frac{2 m^*}{\pi k_B T_e^-}} e^{-\left(\frac{d^-}{t^-}\right)^2} \int_0^\infty dx \ln \left[\frac{\sqrt{x + \frac{\hbar \Omega_0}{k_B T_e^-}} + \sqrt{x}}{\sqrt{x + \frac{\hbar \Omega_0}{k_B T_e^-}} - \sqrt{x}}\right] \times$$

$$\times \left\{e^{\frac{\hbar \Omega_0}{k_B}\left(\frac{1}{T} - \frac{1}{T_e^-}\right)} V\left(2 \frac{d^-}{t^-} \sqrt{x + \frac{\hbar \Omega_0}{k_B T_e^-}}\right) - V\left(2 \frac{d^-}{t^-} \sqrt{x}\right)\right\} e^{-x} . \tag{A.2b}$$

Wave vector conservation (equation (I.16)) gives

$$\frac{e}{\hbar}\boldsymbol{F} = -\frac{2}{3}\boldsymbol{d}^{+}\left(\frac{2\,m^{*}}{\hbar^{2}\,\pi}\right)^{3/2}\frac{J^{2}\,v}{\hbar\left(e^{\frac{\hbar\omega_0}{k_B T}}-1\right)}\sqrt{k_B T_e^{+}}\,e^{-\left(\frac{d^{+}}{t^{+}}\right)^{2}}\times$$

$$\times\, e^{-\frac{\Delta-\hbar\omega_0}{k_B T_e^{+}}}\int_0^{\infty}dx\,e^{-x}\sqrt{x}\left(x+\frac{\Delta-\hbar\omega_0}{k_B T_e^{+}}\right)^{3/2}U\left(2\frac{d^{+}}{t^{+}}\sqrt{x+\frac{\Delta-\hbar\omega_0}{k_B T_e^{+}}}\right)-$$

$$-\,\boldsymbol{d}^{+}\frac{8}{27}\left(\frac{2\,m^{*}}{\pi}\right)^{3/2}\frac{C^{2}\,k_B T}{\varrho\,\hbar^{4}\,u^{2}}\sqrt{k_B T_e^{+}}\,e^{-\left(\frac{d^{+}}{t^{+}}\right)^{2}}\int_0^{\infty}dx\,e^{-x}\,x^{2}\,U\left(2\frac{d^{+}}{t^{+}}\sqrt{x}\right)-$$

$$-\,\frac{2}{3}\boldsymbol{d}^{+}\,e^{2}\,N_0\,\Omega_0\left(\frac{1}{\varepsilon_\infty}-\frac{1}{\varepsilon_0}\right)\left(\frac{\hbar^{2}\,k_B T_e^{+}}{2\,m^{*}}\right)^{-1/2}e^{-\left(\frac{d^{+}}{t^{+}}\right)^{2}}\times$$

$$\times\int_0^{\infty}dx\,e^{-x}\left\{\sqrt{x+\frac{\hbar\,\Omega_0}{k_B T_e^{+}}}\sqrt{x}\left[U\left(2\frac{d^{+}}{t^{+}}\sqrt{x}\right)+e^{\frac{\hbar\,\Omega_0}{k_B}\left(\frac{1}{T}-\frac{1}{T_e^{+}}\right)}U\left(2\frac{d^{+}}{t^{+}}\sqrt{x+\frac{\hbar\,\Omega_0}{k_B T_e^{+}}}\right)\right]+\right.$$

$$+\,\frac{\hbar\,\Omega_0}{2\,k_B T_e^{+}}\ln\left[\frac{\sqrt{x+\frac{\hbar\,\Omega_0}{k_B T_e^{+}}}+\sqrt{x}}{\sqrt{x+\frac{\hbar\,\Omega_0}{k_B T_e^{+}}}-\sqrt{x}}\right]\left[e^{\frac{\hbar\,\Omega_0}{k_B}\left(\frac{1}{T}-\frac{1}{T_e^{+}}\right)}U\left(2\frac{d^{+}}{t^{+}}\sqrt{x+\frac{\hbar\,\Omega_0}{k_B T_e^{+}}}\right)-\right.$$

$$\left.\left.-\,U\left(2\frac{d^{+}}{t^{+}}\sqrt{x}\right)\right]\right\} \tag{A.3a}$$

and

$$\frac{e}{\hbar}\boldsymbol{F} = -\frac{2}{3}\boldsymbol{d}^{-}\left(\frac{2\,m^{*}}{\hbar^{2}\,\pi}\right)^{3/2}\frac{J^{2}\,v}{\hbar\left(e^{\frac{\hbar\omega_0}{k_B T}}-1\right)}\sqrt{k_B T_e^{-}}\,e^{-\left(\frac{d^{-}}{t^{-}}\right)^{2}}\times$$

$$\times\, e^{\frac{\hbar\omega_0}{k_B T}}\int_0^{\infty}dx\,e^{-x}\,x^{3/2}\sqrt{x+\frac{\Delta-\hbar\omega_0}{k_B T_e^{-}}}\;U\left(2\frac{d^{-}}{t^{-}}\sqrt{x}\right)-$$

$$-\,\boldsymbol{d}^{-}\frac{8}{27}\left(\frac{2\,m^{*}}{\pi}\right)^{3/2}\frac{C^{2}\,k_B T}{\varrho\,\hbar^{4}\,u^{2}}\sqrt{k_B T_e^{-}}\,e^{-\left(\frac{d^{-}}{t^{-}}\right)^{2}}\int_0^{\infty}dx\,e^{-x}\,x^{2}\,U\left(2\frac{d^{-}}{t^{-}}\sqrt{x}\right)-$$

$$-\,\frac{2}{3}\boldsymbol{d}^{-}\,e^{2}\,N_0\,\Omega_0\left(\frac{1}{\varepsilon_\infty}-\frac{1}{\varepsilon_0}\right)\left(\frac{\hbar^{2}\,k_B T_e^{-}}{2\,m^{*}}\right)^{-1/2}e^{-\left(\frac{d^{-}}{t^{-}}\right)^{2}}\times$$

$$\times\int_0^{\infty}dx\,e^{-x}\left\{\sqrt{x+\frac{\hbar\,\Omega_0}{k_B T_e^{-}}}\sqrt{x}\left[U\left(2\frac{d^{-}}{t^{-}}\sqrt{x}\right)+e^{\frac{\hbar\,\Omega_0}{k_B}\left(\frac{1}{T}-\frac{1}{T_e^{-}}\right)}U\left(2\frac{d^{-}}{t^{-}}\sqrt{x+\frac{\hbar\,\Omega_0}{k_B T_e^{-}}}\right)\right]+\right.$$

$$+\,\frac{\hbar\,\Omega_0}{2\,k_B T_e^{-}}\ln\left[\frac{\sqrt{x+\frac{\hbar\,\Omega_0}{k_B T_e^{-}}}+\sqrt{x}}{\sqrt{x+\frac{\hbar\,\Omega_0}{k_B T_e^{-}}}-\sqrt{x}}\right]\left[e^{\frac{\hbar\,\Omega_0}{k_B}\left(\frac{1}{T}-\frac{1}{T_e^{-}}\right)}U\left(2\frac{d^{-}}{t^{-}}\sqrt{x+\frac{\hbar\,\Omega_0}{k_B T_e^{-}}}\right)-\right.$$

$$\left.\left.-\,U\left(2\frac{d^{-}}{t^{-}}\sqrt{x}\right)\right]\right\}. \tag{A.3b}$$

Here v is the volume of the elementary lattice cell and T is the lattice temperature. The functions $U(x)$ and $V(x)$ are defined by

$$U(x) = \frac{3}{x^3} [x \cosh x - \sinh x]$$

and

$$V(x) = \frac{1}{x} \sinh x \,.$$

Finally we used the abbreviation $t^{\pm} = \sqrt{\frac{2\, m^* \, k_B T_e^{\pm}}{\hbar^2}}$ · All other notations are the same as in paper I.

References

[1] S. ROTH and W. HAUBENREISSER, phys. stat. sol. (a) **13**, 437 (1972).
[2] S. ROTH and W. HAUBENREISSER, phys. stat. sol. (a) **14**, 91 (1972).
[3] R. B. WOOLSEY and R. M. WHITE, Phys. Rev. **B1**, 4474 (1970).
[4] T. H. LEE, J. appl. Phys. **42**, 1441 (1971).
[5] C. HAAS, Phys. Rev. **168**, 531 (1968).
[6] G. HARBEKE and H. PINCH, Phys. Rev. Letters **17**, 1090 (1966).
[7] C. HAAS, IEEE Trans. Magnetics **5**, 487 (1969).
[8] S. METHFESSEL and D. C. MATTIS, Hdb. Phys., Vol. XVIII/1, Ed. S. FLÜGGE, Springer-Verlag, Freiburg 1968.
[9] C. CANALI, M. MARTINI, G. OTTOVIANI, and K. R. ZANIO, Phys. Rev. **B4**, 422 (1971).
[10] E. M. CONWELL, High Field Transport in Semiconductors, Academic Press, New York 1967.

(Received September 18, 1972)

phys. stat. sol. (a) **14**, 431 (1972)

Subject classification: 1.3; 10.1; 21.1; 21.1.1

Laboratoire de Métallurgie Physique, Faculté des Sciences, Poitiers

Energy of the Antiphase Area Created by Dislocation Slip in $L2_0$ Type Ordered Alloys

By

J. P. EYMERY, H. GAREM, and P. MOINE

The energy of the antiphase area created by the slip of n dislocations in an ordered alloy of $L2_0$-type is computed whose order state is characterized by the parameters S of Bragg and Williams [2] and σ of Bethe [3]. The results are then discussed and consequences are drawn about the deformation of such an alloy.

On calcule l'énergie de frontiére antiphase créée par le glissement de n dislocations dans un alliage ordonné du type $L2_0$ dont l'état d'ordre est caractérisé par les paramétres S de Bragg et Williams [2] et o de Bethe [3]. On discute ensuite les résultats obtenus et leurs conséquences sur la déformation d'un tel alliage.

1. Introduction

The creation of antiphase boundary trails by slipping dislocations has heavy consequences for the deformation of ordered alloys. Moine and Grosbras [1] have calculated the energy of the antiphase area created by the slip of one and two unit dislocation(s) of $\frac{1}{2}\langle 111\rangle$ Burgers vector in the case of a $L2_0$ type stoichiometric alloy A–B whose order state was characterized by the long-range order parameter S of Bragg and Williams [2] and by the short-range order parameter σ of Bethe [3]. These computations were more rigorous than those of Sumino [4] and Rudman [5] though they have been performed using the first-nearest neighbour approximation. In this paper we intend to calculate the energy of the antiphase area created by the slip of n dislocations using the same order parameters and the same method as Moine and Grosbras [1]. The results will be illustrated by some electron microscope photographs of Fe–Co–V alloy (49% Fe, 49% Co, 2% V) whose order–disorder transition temperature is about 725 °C.

2. Energy of the Antiphase Area Created by the Slip of n Dislocations

A slip along a (hkl) plane alters the number of A–A, B–B, A–B bonds connecting A and B atoms of the fixed plane $P_{(hkl)}$ relative to those of the parallel contiguous planes $P'_{(hkl)}$, $P''_{(hkl)}$ . . . which have slid. The method we use consists in computing the number of nearest neighbour A–A, B–B, and A–B atom pairs across the unit area of the considered slip plane before and after n successive slips. It has been shown (cf. [1]) that it is sufficient to know only the increment of the number of atom pairs A–A (let $\Delta N^{(n)}_{\mathrm{AA}}$ be this increment) to calculate the antiphase boundary energy created by the slip of n unit dislocations of $\frac{1}{2}\langle 111\rangle$ Burgers vector; the required energy $\Gamma^{(n)}_{(hkl)}$ may then be written as $\Gamma^{(n)}_{(hkl)} = 2\,v\,\Delta N^{(n)}_{\mathrm{AA}}$, where v is equal to $(V_{\mathrm{AA}} + V_{\mathrm{BB}})/2 - V_{\mathrm{AB}}$ (V_{AA}, V_{BB}, and V_{AB} are the energies of the A–A, B–B, and A–B atom pairs, respectively); according to Bethe's theory [3], v is given by $kT_c/4$ (k Boltzmann's constant, T_c critical ordering temperature. Let α and β be the two kinds of sites of the b.c.c. lattice. The initial number of A–A atom pairs (let $N_{\overline{\mathrm{AA}}}$ be this number) across the slip plane is given by

$$N_{\overline{\mathrm{AA}}} = N_{\overline{\mathrm{A}\alpha\mathrm{A}\beta}} + N_{\overline{\mathrm{A}\beta\mathrm{A}\alpha}},$$

where $N_{\overline{\mathrm{A}\alpha\mathrm{A}\beta}}$ is the number of bonds connecting an A atom on an α-site of $P_{(hkl)}$ to a nearest neighbour A atom on a β-site of $P'_{(hkl)}$, $P''_{(hkl)}$ . . .; the definition of $N_{\overline{\mathrm{A}\beta\mathrm{A}\alpha}}$ would be analogous. In what follows, the numbers $N_{\overline{\mathrm{A}\alpha\mathrm{A}\beta}}$ and $N_{\overline{\mathrm{A}\beta\mathrm{A}\alpha}}$ will be taken as x and $\hat{x}$, respectively; we shall have also to consider the numbers of atom pairs $N_{\overline{\mathrm{A}\alpha\mathrm{B}\beta}} = y$ and $N_{\overline{\mathrm{A}\beta\mathrm{B}\alpha}} = \hat{y}$. In the case of $L2_0$ type stoichiometric compounds these numbers are expressed as

$$x = \hat{x} = \frac{N Z}{8}(1 - \sigma),$$

$$y = \frac{N Z}{8}(1 + \sigma + 2 S),$$

$$\hat{y} = \frac{N Z}{8}(1 + \sigma - 2 S),$$

where N is the number of atoms per unit area of the plane $P_{(hkl)}$ and Z the number of bonds (per atom) connecting lattice points of $P_{(hkl)}$ with lattice points of $P'_{(hkl)}$, $P''_{(hkl)}$ . . .

The slip of an unit dislocation of $1/2\,\langle 111\rangle$ Burgers vector replaces, on one side of the slip plane, α-sites by β-sites and inversely. The number $N^{(1)}_{\overline{\mathrm{AA}}}$ of A–A atom pairs across the slip plane is then given by

$$N^{(1)}_{\overline{\mathrm{AA}}} = N^{(1)}_{\overline{\mathrm{A}\alpha\mathrm{A}\alpha}} + N^{(1)}_{\overline{\mathrm{A}\beta\mathrm{A}\beta}}.$$

The notations are analogous to the former ones; the exponent 1 means we take into account one slip. In what follows, $N^{(1)}_{\overline{\mathrm{A}\alpha\mathrm{A}\alpha}}$ and $N^{(1)}_{\overline{\mathrm{A}\beta\mathrm{A}\beta}}$ will be denoted as $x^{(1)}$ and $\hat{x}^{(1)}$, respectively; we shall also have to consider $y^{(1)} = N^{(1)}_{\overline{\mathrm{A}\alpha\mathrm{B}\alpha}}$ and $\hat{y}^{(1)} = N^{(1)}_{\overline{\mathrm{A}\beta\mathrm{B}\beta}}$. The slip of a second dislocation is going to restore the initial sites. In the case of n slips, it will be necessary to distinguish between even values ($n = 2\,p$) and odd values ($n = 2\,p + 1$) of n. Generalizing the former notations the following relations may be written:

$$x^{(2p+1)} = N^{(2p+1)}_{\overline{\mathrm{A}\alpha\mathrm{A}\alpha}}\,; \quad x^{(2p)} = N^{(2p)}_{\overline{\mathrm{A}\alpha\mathrm{A}\beta}},$$

$$y^{(2p+1)} = N^{(2p+1)}_{\overline{\mathrm{A}\alpha\mathrm{B}\alpha}}\,; \quad y^{(2p)} = N^{(2p)}_{\overline{\mathrm{A}\alpha\mathrm{B}\beta}},$$

$$\hat{x}^{(2p+1)} = N^{(2p+1)}_{\overline{\mathrm{A}\beta\mathrm{A}\beta}}\,; \quad \hat{x}^{(2p)} = N^{(2p)}_{\overline{\mathrm{A}\beta\mathrm{A}\alpha}},$$

$$\hat{y}^{(2p+1)} = N^{(2p+1)}_{\overline{\mathrm{A}\beta\mathrm{B}\beta}}\,; \quad \hat{y}^{(2p)} = N^{(2p)}_{\overline{\mathrm{A}\beta\mathrm{B}\alpha}}.$$

The exponents $2p$ and $(2p+1)$ mean we deal with $2p$ or $(2p+1)$ slips. It is now convenient to consider the probabilities

$$p_1 = P_{\overline{\mathrm{A}\beta\mathrm{A}\alpha}} = P_{\overline{\mathrm{B}\alpha\mathrm{B}\beta}} = \frac{1}{2}\frac{1-\sigma}{1+S},$$

$$p_2 = P_{\overline{\mathrm{A}\alpha\mathrm{B}\beta}} = P_{\overline{\mathrm{B}\beta\mathrm{A}\alpha}} = \frac{1}{2}\frac{1+\sigma+2S}{1+S},$$

$$p_3 = 1 - p_1 = P_{\overline{\mathrm{A}\beta\mathrm{B}\alpha}} = P_{\overline{\mathrm{B}\alpha\mathrm{A}\beta}},$$

$$p_4 = 1 - p_2 = P_{\overline{\mathrm{A}\alpha\mathrm{A}\beta}} = P_{\overline{\mathrm{B}\beta\mathrm{B}\alpha}}.$$

We define the notations for the first probability (for the others they are analogous): $P_{\overline{\mathrm{A}\beta\mathrm{A}\alpha}}$ is the probability that the nearest neighbour in the slip direction of an A atom on a β-site is an A atom on an α-site. The calculations of $x^{(n)}$, $y^{(n)}$, $\hat{x}^{(n)}$, $\hat{y}^{(n)}$ may then be performed. After one slip the following relations are obtained:

$$\begin{cases} x^{(1)} = x\,p_1 + y\,p_2, \\ y^{(1)} = x\,p_3 + y\,p_4, \end{cases}$$

or, using matrices,

$$\begin{pmatrix} x^{(1)} \\ y^{(1)} \end{pmatrix} = \begin{pmatrix} p_1 & p_2 \\ p_3 & p_4 \end{pmatrix} \begin{pmatrix} x \\ y \end{pmatrix}.$$

After two slips we have

$$\begin{cases} x^{(2)} = p_4\,x^{(1)} + p_3\,y^{(1)}, \\ y^{(2)} = p_2\,x^{(1)} + p_1\,y^{(1)}, \end{cases}$$

or, using matrices

$$\begin{pmatrix} x^{(2)} \\ y^{(2)} \end{pmatrix} = \begin{pmatrix} p_4 & p_3 \\ p_2 & p_1 \end{pmatrix} \begin{pmatrix} x^{(1)} \\ y^{(1)} \end{pmatrix} = \begin{pmatrix} p_4 & p_3 \\ p_2 & p_1 \end{pmatrix} \begin{pmatrix} p_1 & p_2 \\ p_3 & p_4 \end{pmatrix} \begin{pmatrix} x \\ y \end{pmatrix}.$$

In what follows

$$\begin{pmatrix} p_4 & p_3 \\ p_2 & p_1 \end{pmatrix} \begin{pmatrix} p_1 & p_2 \\ p_3 & p_4 \end{pmatrix}$$

will be denoted as M. In the case of n slips we are led to

$$\begin{pmatrix} x^{(2p)} \\ y^{(2p)} \end{pmatrix} = M^p \begin{pmatrix} x \\ y \end{pmatrix} \qquad \text{for} \quad n = 2p,$$

$$\begin{pmatrix} x^{(2p+1)} \\ y^{(2p+1)} \end{pmatrix} = \begin{pmatrix} p_1 & p_2 \\ p_3 & p_4 \end{pmatrix} M^p \begin{pmatrix} x \\ y \end{pmatrix} \qquad \text{for} \quad n = 2p+1.$$

The following relations may also be written

$$\begin{pmatrix}\hat{x}^{(2p)}\\ \hat{y}^{(2p)}\end{pmatrix} = \hat{M}^p \begin{pmatrix}\hat{x}\\ \hat{y}\end{pmatrix} \qquad \text{for} \quad n = 2\,p\,,$$

$$\begin{pmatrix}\hat{x}^{(2p+1)}\\ \hat{y}^{(2p+1)}\end{pmatrix} = \begin{pmatrix}p_4 & p_3\\ p_2 & p_1\end{pmatrix} \hat{M}^p \begin{pmatrix}\hat{x}\\ \hat{y}\end{pmatrix} \qquad \text{for} \quad n = 2\,p + 1\,,$$

where

$$\hat{M} = \begin{pmatrix}p_1 & p_2\\ p_3 & p_4\end{pmatrix}\begin{pmatrix}p_4 & p_3\\ p_2 & p_1\end{pmatrix}.$$

The diagonalization of the matrices M and $\hat{M}$ allows to calculate easily M^p and $\hat{M}^p$; we know effectively that, if D is a diagonal matrix similar to M, a matrix P may be found so that $M^p = P\,D^p\,P^{-1}$. The matrices M and $\hat{M}$ have the same eigenvalues: 1 and $(p_1 - p_2)^2$; the calculations lead to the following results:

$$M^p = \frac{1}{1 - (p_1 - p_2)^2}\begin{pmatrix}(p_2^2 - p_1\,p_2 + p_1 - 1) & 1\\ (p_1^2 - p_1\,p_2 - p_1) & -1\end{pmatrix}\begin{pmatrix}1 & 0\\ 0 & (p_1 - p_2)^{2p}\end{pmatrix} \times$$
$$\times \begin{pmatrix}-1 & -1\\ (-p_1^2 + p_1\,p_2 + p_1) & (p_2^2 - p_1\,p_2 + p_1 - 1)\end{pmatrix},$$

$$\hat{M}^p = \frac{1}{(p_1 - p_2)^2 - 1}\begin{pmatrix}(p_1^2 - p_1\,p_2 - p_1) & -1\\ (p_2^2 - p_1\,p_2 + p_1) & -1\end{pmatrix}\begin{pmatrix}1 & 0\\ 0 & (p_1 - p_2)^{2p}\end{pmatrix} \times$$
$$\times \begin{pmatrix}1 & 1\\ (-p_2^2 + p_1\,p_2 - p_1 + 1) & (p_1^2 - p_1\,p_2 - p_1)\end{pmatrix}.$$

$\Delta N^{(n)}_{\overline{\mathrm{AA}}}$ may then be written

$$\Delta N^{(2p)}_{\overline{\mathrm{AA}}} = N^{(2p)}_{\overline{\mathrm{AA}}} - N_{\overline{\mathrm{AA}}} = x^{(2p)} + \hat{x}^{(2p)} - (x + \hat{x}) =$$
$$= \frac{N\,Z}{4}\,[1 - (p_1 - p_2)^{2p}]\,(\sigma - S^2)\,,$$

$$\Delta N^{(2p+1)}_{\overline{\mathrm{AA}}} = N^{(2p+1)}_{\overline{\mathrm{AA}}} - N_{\overline{\mathrm{AA}}} = x^{(2p+1)} + \hat{x}^{(2p+1)} - (x + \hat{x}) =$$
$$= \frac{N\,Z}{4}\,[(\sigma + S^2) - (\sigma - S^2)\,(p_1 - p_2)^{2p+1}]\,.$$

At last, taking into account the relation $\Gamma^{(n)}_{(hkl)} = 2\,V\,\Delta N^{(n)}_{\overline{\mathrm{AA}}}$, the energy $\Gamma^{(n)}_{(hkl)}$ may be expressed as

$$\Gamma^{(2p)}_{(hkl)} = \frac{N\,Z\,v}{2}\,[1 - (p_1 - p_2)^{2p}]\,(\sigma - S^2) \qquad \text{for} \quad n = 2\,p,$$

$$\Gamma^{(2p+1)}_{(hkl)} = \frac{N\,Z\,v}{2}\,[(\sigma + S^2) - (\sigma - S^2)\,(p_1 - p_2)^{2p+1}] \quad \text{for} \quad n = 2\,p + 1$$

with

$$p_1 - p_2 = \frac{S^2 - \sigma}{1 - S^2}.$$

For the first values of n, the former expressions give

$$\Gamma^{(1)}_{(hkl)} = \frac{N\,Z\,v}{2}\,\frac{S^2 + \sigma^2 - 3\,S^2\,\sigma + \sigma}{1 - S^2} \qquad \text{for} \quad n = 1\,,$$

$$\Gamma^{(2)}_{(hkl)} = \frac{N\,Z\,v}{2}\,\frac{(\sigma - S^2)\,(1 - \sigma)\,(1 + \sigma - 2\,S^2)}{(1 - S^2)^2} \qquad \text{for} \quad n = 2\,,$$

$$\Gamma^{(3)}_{(hkl)} = \frac{N\,Z\,v}{2} \times$$

$$\times \frac{\sigma + S^2 + \sigma^4 - 2\sigma S^2 - S^4 - 3\,S^2\sigma + 2\sigma^2 S^2 - 3\sigma^3 S^2 + 3\,S^4\sigma + 3\sigma^2 S^4 + S^6 - 3\,S^6\sigma}{(1 - S^2)^3}$$

$$\text{for} \quad n = 3\,.$$

3. Conclusions

3.1 S different from 0

When S tends to unity, the parameter σ tends to S^2; the limits of $\Gamma^{(2p)}_{(hkl)}$ and $\Gamma^{(2p)}_{(hkl)}$ do not depend upon p and are, respectively, 0 and $2\,S^2\,N\,Z\,v/2$. It may then be inferred that the slip of an even number of dislocations does not alter the alloy whereas the glide of an odd number gives rise to a fault whose energy is proportional to S^2; the dislocations will then tend to associate in pairs. There we meet again known results. In fact, the approximation $\sigma = S^2$ being valid for $0.7 \leqq S \leqq 1$, the former results remain true in this range of values of S.

If the former approximation is no longer valid, i.e. for $0 < S < 0.7$, the energy $\Gamma^{(2p)}_{(hkl)}$ is no longer zero and takes increasing values when S tends to 0; for its part $\Gamma^{(2p+1)}_{(hkl)}$ decreases with S. The numerical values of the function $\Gamma^{(n)}_{(hkl)} = f(n)$ have been estimated for the alloy Fe–Co (50–50%) in the case of (110) slip planes which implies $Z = 2$ and $N = \sqrt{2}/a^2$ (a is the lattice parameter); the values of S and σ have been deducted from Bethe's theory [3]; the results are presented in Fig. 1 for $0 \leqq S \leqq 0.2$. The oscillations of $\Gamma^{(n)}_{(110)}$ show that the dislocations still tend to associate in pairs. The fact that $\Gamma^{(1)}_{(110)}$ is higher than $\Gamma^{(3)}_{(110)}$ and $\Gamma^{(2p+1)}_{(110)}$ for $p > 1$ implies, on the one hand that the first dislocation pair must have a spacing slightly smaller than the others, and on the other hand that a stress sufficiently large to move the lead dislocation will move the following ones. Thus we find again the notion of "dislocation group" that Cohen and Fine [6] had introduced in the case of short-range order. These results are obviously

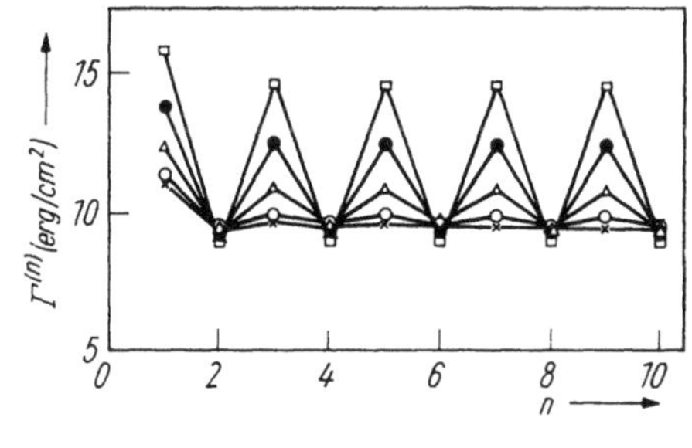

Fig. 1. Energy of antiphase area as a function of the number of slipping dislocations. × $S = 0$, $\sigma = 0.14$; ○ $S = 0.05$, $\sigma = 0.145$; △ $S = 0.1$, $\sigma = 0.15$; ● $S = 0.15$, $\sigma = 0.16$; □ $S = 0.2$; $\sigma = 0.174$

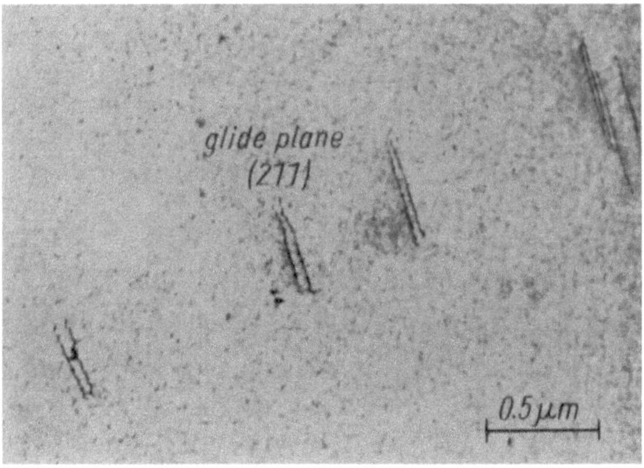

Fig. 2. Paired dislocations in Fe–Co–V alloy whose long-range order parameter is between 0 and 0.2. (111) face; $\boldsymbol{g}$ = (1̄10)

valid for any slip plane. The glide of paired dislocations is illustrated on Fig. 2 in the case of Fe–Co–V alloy whose parameter S ranges between 0 and 0.2.

3.2 $S = 0$

Now we deal with an alloy having only short-range order. The expressions of $\Gamma_{(hkl)}^{(2p)}$ and $\Gamma_{(hkl)}^{(2p+1)}$ are then

$$\Gamma_{(hkl)}^{(2p)} = \frac{N Z v}{2} (\sigma - \sigma^{2p+1}) ,$$

$$\Gamma_{(hkl)}^{(2p+1)} = \frac{N Z v}{2} (\sigma + \sigma^{2(p+1)}) .$$

When p tends to ∞, the terms between parentheses tend to σ by higher and lower values, respectively. If we consider the slip of n dislocations, some oscillations of the fault energy occur and then decrease very quickly because the value of σ is low. On the other hand, $\Gamma_{(hkl)}^{(1)}$ is still higher than $\Gamma_{(hkl)}^{(n)}$ for $n > 1$ which leads once more to the notion of "dislocation group" of Cohen and Fine [6].

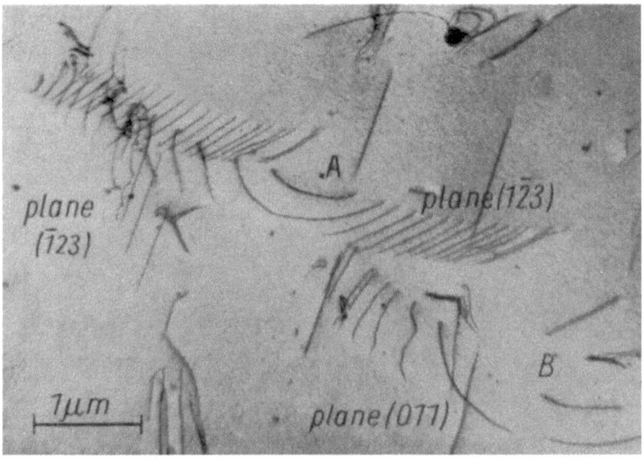

Fig. 3. Slip planes in short-range ordered Fe–Co–V. (113) face; $\boldsymbol{g}$ = (1̄10)

The variations in $\Gamma_{(110)}^{(n)}$ as a function of n for $S = 0$ are presented on Fig. 1 for the case of the alloy Fe–Co (50–50%).

The former considerations show that it is possible to find one superdislocation (or two at the most) at the head of a dislocation slip, the following dislocations being unpaired. Such configurations may be observed only after slight strains. For our part, we have noticed them in Fe–Co–V alloy quenched from 884 °C, i.e. in the range of short-range order; the dislocations have been probably generated during quenching. Fig. 3 shows dislocations generated by sources situated at A and B for instance and forming pile-ups; the slip planes are of (110) and (123) type; paired dislocations may be noticed at the head of the pile-ups and unpaired afterwards. Analogous observations have already been reported by Calvayrac and Fayard [7] concerning short-range ordered Ni_3Fe.

These results are in agreement with the calculations of Cohen and Fine [6] who had predicted the destruction of short-range order by the dislocations. In the case of our computations, the low amplitude of the energy oscillations theoretically predicted do not integrally interpret the observed phenomenons but gives a qualitative aspect of them; we have also to remind that we have to do with pile-ups observed on thin foils by electron microscopy. At last, we may think that the study of the problem using only one parameter in the case of short-range order is not sufficient.

References

[1] P. Moine and P. Grosbras, C.R. Acad. Sci. (France) **268**, 759 (1969).
[2] W. L. Bragg and E. J. Williams, Proc. Roy. Soc. **A151**, 540 (1935).
[3] A. H. Bethe, Proc. Roy. Soc. **A150**, 552 (1935).
[4] K. Sumino, Sci. Rep. Res. Inst. Tôhoku Univ., Ser. A **10**, 283 (1958).
[5] P. S. Rudman, Acta metall. **10**, 253 (1962).
[6] J. B. Cohen and M. E. Fine, J. Phys. Radium **23**, 749 (1962).
[7] Y. Calvayrac and M. Fayard, C.R. Acad. Sci. (France) **258**, 4531 (1964).

(Received June 23, 1972)

phys. stat. sol. (a) **14**, 439 (1972)

Subject classification: 13.4 and 14.3; 22.2.3

A. F. Ioffe Physico-Technical Institute, Academy of Sciences of the USSR, Leningrad

Ion Pairing in p-InSb Induced by X-Rays

By

T. V. Mashovets, G. A. Vikhlii, and N. A. Vitovskii

A decrease in the donor and acceptor concentrations has been observed in p-InSb under X-irradiation at low temperatures. Radiation-induced migration of Ge atoms and donor–acceptor ion pairing have been suggested.

Сообщается об изменениях концентраций акцепторов и доноров в антимониде индия p-типа, происходящих в результате низкотемпературного облучения рентгеновскими лучами. Делается предположение о протекании процесса радиационно-стимулированной миграции примеси (германий) и объединении доноров и акцепторов в электрически неактивные комплексы.

1. Introduction

Bulk properties of p-InSb have been observed to change [1 to 3] under low-temperature X-irradiation or visible light illumination and to recover at about 100 °K. These changes were shown in [3] not to be due to electron trapping in the oxide surface layer (such a process was investigated in [4]). The effect was discussed in [1 to 3] in terms of non-impact defect creation.

The experiments described in this paper show that the hole concentration increase [1 to 3] under irradiation happens to be only the most obvious result of the inner modifications in the spatial arrangement and electrical activity of the impurities. The rearrangement seems to be possible due to the radiation-enhanced migration [5, 6] of the impurity atoms and it is not necessary to admit Frenkel defects to be generated.

2. Experimental Methods and Results

250 keV X-irradiation of the Ge-doped p-InSb was performed at 77 °K. The choice of the quantum energy takes into account the following two purposes:

1. It provides the homogeneous energy absorption and diminishes the role of the surface processes under irradiation.

2. It eliminates the defect generation by means of the elastic impact mechanism.

For the first purpose the highest energy was desirable, for the second it had to be limited by the threshold energy (268 keV [7]).

The temperature dependences of Hall coefficient R and conductivity σ were investigated, as well as the isochronal annealing process. On the basis of the usual statistical analysis [8] of the experimental data the following quantities were determined: hole concentration p, donor and acceptor concentrations N_d and N_a, the compensation ratio $K = N_d/N_a$ and the acceptor activation energy ε_1. Fig. 1 and 2 and Table 1 illustrate the experimental results. The post-irradiation data correspond to the "saturating" integral dose, so that an additional irradiation did not change any parameters.

Table 1

sample No.	before irradiation $(N_a - N_d)$ $(10^{-14}\,cm^{-3})$	N_a $(10^{-15}\,cm^{-3})$	N_d $(10^{-15}\,cm^{-3})$	K (%)	ε_1 (meV)	after irradiation $(N_a - N_d)$ $(10^{-14}\,cm^{-3})$	N_a $(10^{-15}\,cm^{-3})$	N_d $(10^{-15}\,cm^{3})$	K (%)	ε_1 (meV)	ΔN_a $(10^{-15}\,cm^{-3})$	ΔN_d $(10^{-15}\,cm^{-3})$	Δp
1	1.6	1.6	1.4	90	8.6	2.8	0.58	0.30	52	9.2	1.0	1.1	12
2	0.1	10²*	10²*	99.99*	3.7	1.3	1.2	1.1	91	8.8	10²*	10²*	12
3	1.1	7.0*	6.9*	98.5*	6.9	2.2	1.9	1.7	89	8.1	5.1*	5.2*	11
4	0.1	40*	40*	99.97*	5.0	1.6	3.0*	2.8*	93*	6.4	37*	37*	15

The parameters mentioned above change under irradiation in the following way:

1. The concentrations N_a and N_d decrease approximately to the same limiting value of about 10^{15} cm^{-3}.

2. The values of ε_1, measured in non-irradiated samples are less than 9 to 9.5 meV, known for the isolated acceptor [9]. Some attempts were made to estimate ε_1 as a function of N_a (e.g. [10]), but no theory describes this dependence satisfactorily up to date.[1]) The decrease of ε_1 due to increase of N_a is established experimentally [11]. As it can be seen from the Table 1, ε_1 increases under irradiation and reaches the value $\varepsilon_1 \approx 9$ meV. The maximum in R, which must be attributed for the beginning of the hopping conductivity shifts to lower temperatures. All this can result from a decrease of N_a.

3. The Hall mobility $R\sigma$ increases under irradiation in the impurity scattering region (Fig. 2). A very sharp decrease in $R\sigma$ at low temperatures obviously results from the hopping conductivity [12]. This process can be observed at lower temperatures in irradiated samples.

4. The ratio $K = N_d/N_a$ decreases.

5. The hole concentration p_{77} at 77 °K increases. The changes are approximately the same for all samples and are equal to about 1.2×10^{14} cm^{-3}.

6. The decrease in $N_d(\Delta N_d)$ overtops that of ΔN_a by a relatively small amount equal to the increase in $p_{77}(\Delta p)$.

3. Discussion and Conclusions

All these effects seem to be well explicable on the base of the assumption that a radiation-enhanced migration of the impurity atoms takes place. This migration can result in:

1. Creation of the electrically inactive Ge_2 complexes.

2. Creation of the complexes of the donor impurity with some unknown defect (e.g. Sb-vacancy).

[1]) For this reason the calculated values of N_a, N_d and K seem to be rather doubtful for $\varepsilon_1 < 8$ meV. Nevertheless, these values are put in Table 1 (marked by *).

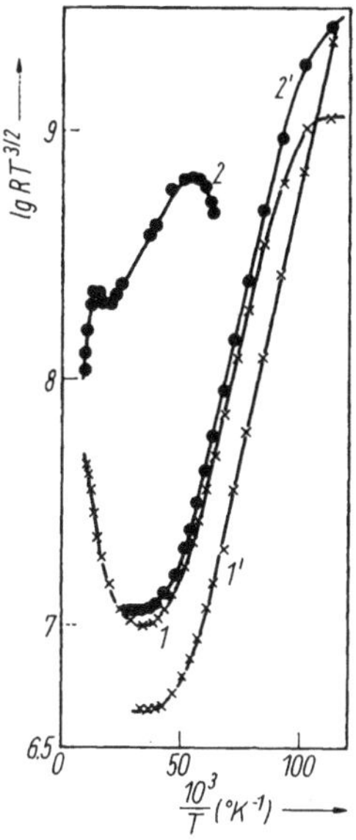

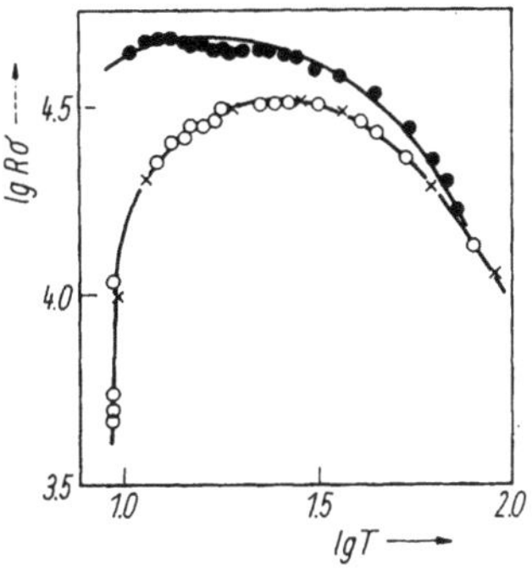

Fig. 2. The temperature dependences of lg $(R\sigma)$ for the sample 1. ○: Before irradiation; ●: after irradiation; ×: after irradiation and annealing up to 120 °K

Fig. 1. The temperature dependences of the value lg $(R\,T^{3/2})$ for two samples (1 and 2). 1, 2 Before irradiation; 1′, 2′ after irradiation

1. It is a fact that Ge atoms act as donors if they substitute In atoms (Ge_{In}) and as acceptors if they substitute Sb atoms (Ge_{Sb}). Let us suppose, that the Ge atoms can migrate (it seems possible that only one short of them — Ge_{In} or Ge_{Sb} — is mobile). As the migration occurs in the Coloumb field of the ionized donors and acceptors, ion pairing can take place, ($Ge_{In}^+ + Ge_{Sb}^- \rightarrow Ge_2$) each pair occupying two adjacent lattice sites in the two sublattices. The result is the formation of the electrically inactive complex Ge_2.

This process leads to the decrease of N_d and N_a (equally) and to the increase of the carrier mobility and the activation energy ε_1.

The maximum of R and the temperature range of sharp $R\sigma$ decrease connected with the hopping conductivity shift to low temperatures. After N_a and N_d become small enough the reaction $Ge_{In}^+ + Ge_{Sb}^- \rightarrow Ge_2$ ceases which accounts for the equality of the residial concentrations N_a and N_d ($\approx 10^{15}$ cm^{-3}). The annealing at about 100 °K results in the disintegration of Ge_2 complexes and recovery of N_a and N_d.

Unfortunately the overlapping of the impurity band with the conduction band prevents the observation of this effect in n-InSb by the method used.

2. The increase of hole concentration p_{77} can be accounted for by the trapping of migrating donors by some genetic defects provided for this trapping makes donors electrically inactive.[2]) The limiting concentration Δp in such a case is determined by the concentration of the traps (e.g. Sb vacancies) in the non-irradiated material. As it has been shown in [2], the rate of this process depends on the Fermi level μ in the non-irradiated samples. This means that the charge state of the trapping centres is important. The increase of p under irradiation takes place only if μ is lower than $E_v + 0.09$ eV at 77 °K. It seems reasonable,

[2]) The formation of the donor–vacancy complexes in Si under irradiation has been investigated in [13], the decrease in V-group donor state concentration in Ge under irradiation has been observed in [14].

that the generation of the vacancy–donor (Sb vacancy + Ge atom in the In site) complex is possible unless the vacancy state (E_v + 0.09 eV) is occupied by the electron. The disappearance of the abundant donor leads us to the assumption, that the Ge atoms migrate mainly into the In sublattice. As it has been shown in [2], no changes in the parameters measured at 77 °K take place if the irradiation is performed at 4.2 °K. It is an important argument for the diffusion character of the investigated process. The activation energy E for the hole generation process was estimated from the temperature dependence of the stationary hole concentrations (p_{77}) after visible light illumination (xenon lamp spectrum, $\approx 10^{22}$ quanta/cm^2). This estimation gives the value $E = 0.24 \pm 0.08$ eV. If the proposed model is valid, this value can be regarded as the donor migration energy under illumination.

The assumption of two types of complexes generated accounts for the whole number of observed effects.

Acknowledgement

The authors would like to thank Prof. S. M. Ryvkin for his interest in this work.

References

[1] G. W. Arnold and F. L. Vook, Phys. Rev. **137**, A1839 (1965).
[2] N. A. Vitovskii, G. A. Vikhlii, V. V. Galavanov, T. V. Mashovets, and R. Yu. Khansevarov, Fiz. Tekh. Poluprov. **3**, 132 (1969).
[3] N. A. Vitovskii, G. A. Vikhlii, V. S. Ivleva, and T. V. Mashovets, Fiz. Tekh. Poluprov. **5**, 1943 (1971).
[4] E. W. Krentz, Z. angew. Phys. **27**, 244 (1969).
[5] A. E. Kiv and F. T. Umarova, Fiz. tverd. Tela **9**, 2225 (1967).
[6] J. C. Bourgoin and J. W. Corbett, Phys. Letters (Netherland) **38A**, 135 (1972).
[7] F. H. Eisen, Phys. Rev. **135**, A1394 (1964).
[8] J. S. Blakemore, Semiconductor Statistics, Pergamon Press, 1962.
[9] M. Ya. Dashevskii, V. S. Ivleva, L. Ya. Krol, I. N. Kurilenko, L. B. Litvak-Gorskaya, R. S. Mitrofanova, and E. Yu. Fridlyand, Fiz. Tekh. Poluprov. **5**, 858 (1971).
[10] G. F. Neumark, Phys. Rev. **5**, 408 (1972).
[11] E. A. Davis and W. D. Compton, Phys. Rev. **140**, A2183 (1965).
[12] M. Cuevas, Phys. Rev. **164**, 1021 (1967).
[13] G. D. Watkins and J. W. Corbett, Phys. Rev. **134**, A1359 (1964).
[14] V. V. Emtsev, T. V. Mashovets, M. Maksimov, and N. A. Vitovskii, in: Radiation Effects in Semiconductors, Ed. J. W. Corbett and G. D. Watkins, Gordon & Breach, New York/London/Paris 1971.

(Received June 26, 1972)

phys. stat. sol. (a) **14**, 443 (1972)

Subject classification: 10.2; 22.9

Institut für Kristallographie der Universität zu Köln

Elastische Energie und Vorzugsrichtungen geradliniger Versetzungen in aus der Lösung gewachsenen organischen Kristallen

II. Thioharnstoff

Von

H. KLAPPER

Die elastische Energie und die Vorzugsrichtungen der Versetzungen des Thioharnstoffs mit den Burgersvektoren [100], [010], [001] und [101] werden berechnet. Infolge der hohen elastischen Anisotropie variiert die Energie der Versetzungen mit $\boldsymbol{b}$ = [010] mit der Richtung um den Faktor sechs. Die Übereinstimmung zwischen den beobachteten und den berechneten Vorzugsrichtungen ist gut. Die Rangfolge in der „Leichtigkeit" der Gleitung in den verschiedenen möglichen Gleitsystemen nach einem Kriterium von Eshelby wird berechnet. Das einzige und häufig beobachtete System (100) [001] steht hier nicht an erster Stelle. Aus den inversen Wulff-Diagrammen ist ersichtlich, daß vier Bereiche instabiler Richtungen existieren. Diese sind röntgentopographisch nicht eindeutig zu erkennen.

The elastic energy and the preferred directions of dislocations in thiourea with Burgers vectors [100], [010], [001], and [101] are calculated. As a consequence of the high elastic anisotropy the energy of dislocations with $\boldsymbol{b}$ = [010] varies by a factor of six. The agreement of observed and calculated preferred directions is satisfactory. The sequence in the ease of gliding of several possible glide systems as defined by a criterion of Eshelby is calculated. In this sequence the only and frequently observed system (100) [001] is not found to be in the first place. From the inverse Wulff plots is deduced that there are four ranges of instabilities. They do not appear distinctly on the X-ray topographs.

1. Einleitung

Im ersten Teil dieser Arbeit wurde über den Zusammenhang zwischen der elastischen Energie und den Vorzugsrichtungen der eingewachsenen geradlinigen Versetzungen in Einkristallen des Benzils berichtet [1]. Im folgenden werden die Ergebnisse einer entsprechenden Untersuchung über die Versetzungen des Thioharnstoffs, $(NH_2)_2CS$ (Raumgruppe Pnma, Gitterkonstanten: $a = 7{,}655$ Å, $b = 8{,}537$ Å, $c = 5{,}52$ Å [2]) mitgeteilt. Die theoretischen Grundlagen und die Bezeichnungen der Parameter sind dieselben wie in [1].

Die Kristalle des Thioharnstoffs zeigen eine extreme elastische Anisotropie, welche die Berechnung der Richtungsabhängigkeit der Versetzungsenergien besonders interessant erscheinen läßt. Die elastischen Konstanten sind für 20 °C (Einheit: 10^{11} dyn cm^{-2}) (Haussühl [3]):

$$\begin{array}{lll} c_{11} = 1{,}024 & c_{12} = 0{,}227 & c_{44} = 0{,}222 \\ c_{22} = 2{,}595 & c_{13} = 0{,}443 & c_{55} = 0{,}608 \\ c_{33} = 1{,}503 & c_{23} = 0{,}567 & c_{66} = 0{,}055\,. \end{array}$$

Hier fällt der ungewöhnlich kleine Scherwiderstand c_{66} auf. Er ist etwa 11mal kleiner als der maximale Schwerwiderstand c_{55}.

Über die Ergebnisse einer röntgentopographischen Untersuchung der Defektstrukturen in aus der Lösung gezüchteten Einkristallen des Thioharnstoffs wird in [4] berichtet. Es wurden geradlinige Versetzungen mit den Burgersvektoren (BV) [100], [010], [001] und [101] (bzw. [10$\bar{1}$]) beobachtet. Diese Versetzungen besitzen ausgeprägte Vorzugsrichtungen, die für ihren BV und für den Wachstumssektor, in dem sie verlaufen, charakteristisch sind.

a
b
c

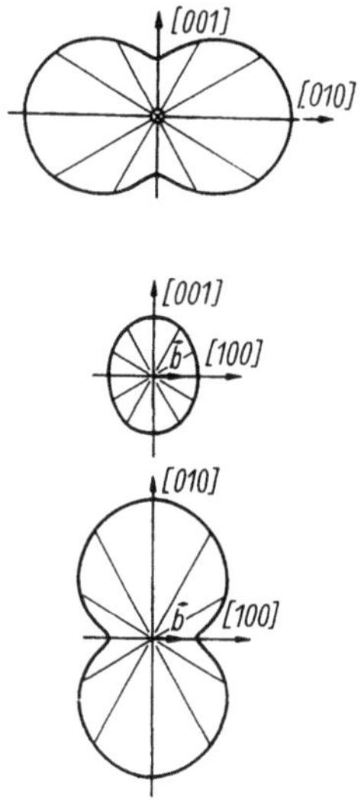

Fig. 1

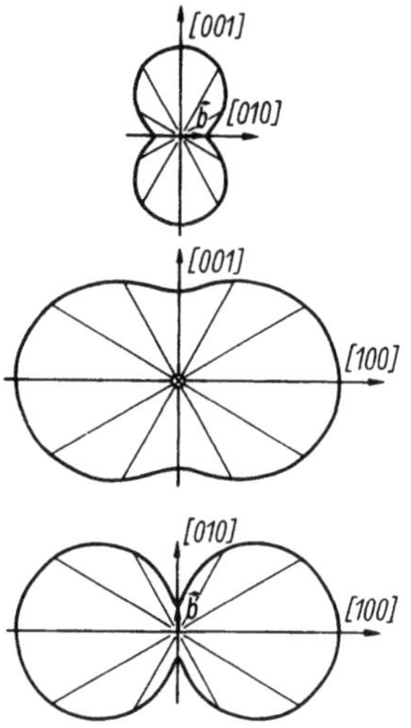

Fig. 2

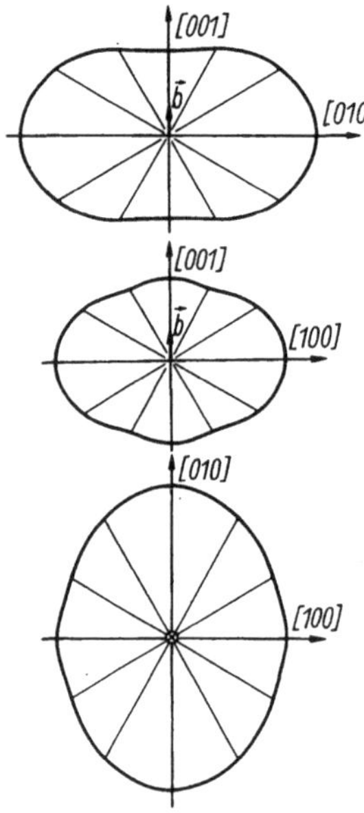

Fig. 3

a
b
c

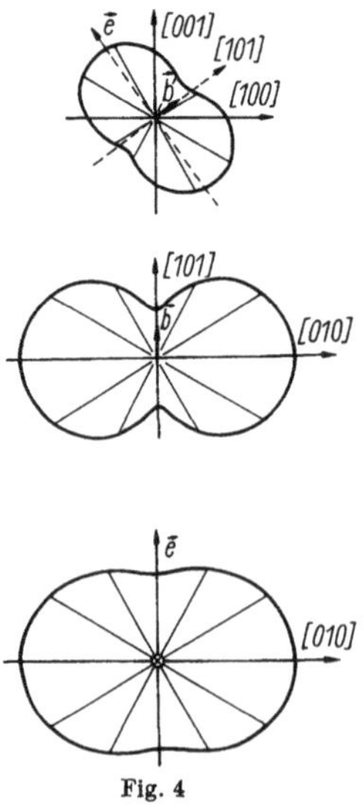

Fig. 4

Fig. 1. Richtungsabhängigkeit des Energiefaktors K (Polarkoordinaten) für $\boldsymbol{b}$ = [100] in den Ebenen a) (100); b) (010); c) (001)

Fig. 2. Richtungsabhängigkeit des Energiefaktors K (Polarkoordinaten) für $\boldsymbol{b}$ = [010] in den Ebenen a) (100); b) (010); c) (001)

Fig. 3. Richtungsabhängigkeit des Energiefaktors K (Polarkoordinaten) für $\boldsymbol{b}$ = [001] in den Ebenen a) (100); b) (010); c) (001)

Fig. 4. Richtungsabhängigkeit des Energiefaktors K (Polarkoordinaten) für $\boldsymbol{b}$ = [101] in den Ebenen a) (010); b) (10$\bar{1}$); c) senkrecht zu $\boldsymbol{b}$ = [101]

2. Ergebnisse und Diskussion

2.1 Die Richtungsabhängigkeit des Energiefaktors

Die Fig. 1 bis 4 zeigen die berechnete Richtungsabhängigkeit des Energiefaktors K für die vier beobachteten Versetzungstypen in den drei Koordinatenebenen bzw. — in Fig. 4 — in drei zueinander senkrechten Ebenen parallel und senkrecht zum BV $\boldsymbol{b} = [101]$. Tabelle 1 enthält die numerischen Werte von K und $K\,b^2$ (diese Größe ist proportional zur Energie E pro Längeneinheit der Versetzung) für die Richtungen $\boldsymbol{l}$ der Versetzungslinien parallel zu den drei zweizähligen Achsen des Kristalls.

Tabelle 1

Energiefaktor K (I) (Einheit: 10^{11} dyn cm^{-2}) und $K\,b^2$ (II) (Einheit: 10^{-5} dyn) für Versetzungen mit verschiedenen Burgersvektoren $\boldsymbol{b}$ und Richtungen $\boldsymbol{l}$ parallel zu den zweizähligen Achsen des Kristalls

$\boldsymbol{b}$	$\boldsymbol{l}$					
	[100]		[010]		[001]	
	I	II	I	II	I	II
[100]	0,183	10,72	0,543	31,80	0,228	13,38
[010]	0,680	49,55	0,110	8,04	0,363	26,49
[001]	0,517	15,77	0,657	20,03	0,367	11,19
[101]	0,297	26,46	0,582	51,86	0,276	24,57

In den Fig. 1 bis 4 fallen folgende Besonderheiten auf:

Eine besonders starke Richtungsabhängigkeit der Energie E bzw. des Energiefaktors K besitzen die Versetzungen mit $\boldsymbol{b} = [010]$ in der Ebene (001) (Fig. 2c). Hier ist der Energiefaktor K_e der reinen Stufenversetzung etwa 6 mal so groß wie derjenige der reinen Schraubenversetzung K_s. Im Falle elastischer Isotropie ist stets $K_e/K_s < 2$.

Die Energie der reinen Schraubenversetzung mit $\boldsymbol{b} = [001]$ (Fig. 3a, b) ist zwar kleiner als die der reinen Stufenversetzungen (Fig. 3c), das Minimum der Energie besitzt jedoch die gemischte Versetzung, deren Richtung $\boldsymbol{l}$ in der Ebene (010) um 19° gegen [001] geneigt ist (Fig. 3b).

In den Fig. 1a, 2b, 3c und 4c besitzen die Versetzungen in allen Orientierungen reinen Stufencharakter. Im Falle elastischer Isotropie würden sich hier Kreise ergeben. Besonders ausgeprägt ist die Richtungsabhängigkeit von K für die Stufenversetzungen mit $\boldsymbol{b} = [100]$ (Fig. 1a). Hier ändert sich K um den Faktor 2,4.

Liegen die BVs der Versetzungen parallel zu einer der zweizähligen Achsen des Kristalls (Fig. 1, 2, 3), so nehmen die zugehörigen Richtungsabhängigkeiten von K die volle Symmetrie der Punktgruppe mmm des Kristalls an. Für Versetzungen mit $\boldsymbol{b} = [101]$ zeigt K die Punktsymmetrie 2/m, wobei die zweizählige Achse parallel [010] liegt (Fig. 4). In diesem Fall besitzt diejenige gemischte Versetzung das Minimum des Energiefaktors $K_{\min} = 0.182$ (Einheit hier und im folgenden: 10^{11} dyn cm^{-2}), deren Richtung $\boldsymbol{l}$ in der Ebene (010) um 11° gegen den BV $\boldsymbol{b} = [101]$ ($K_s = 0.193$) geneigt ist. Ebenso bildet in derselben Ebene die Richtung der reinen Stufenversetzung ($K_e = 0.350$) mit der Richtung von $K_{\max} = 0.360$ einen Winkel von 16°.

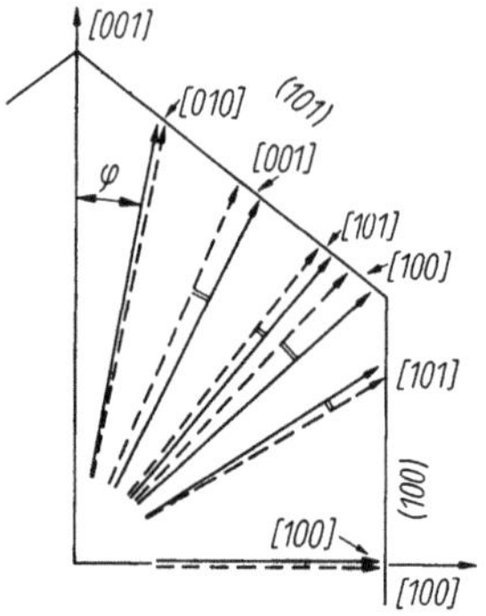

Fig. 5. Berechnete (——→) und beobachtete (- - -→) Vorzugsrichtungen l_0 in den Sektoren {101} und {100}

2.2 *Vorzugsrichtungen der eingewachsenen Versetzungen*

Wie in [1] näher begründet wurde, sind die Vorzugsrichtungen der eingewachsenen Versetzungen in einem Wachstumssektor durch das Minimum der elastischen Energie W pro Längeneinheit in Wachstumsrichtung ausgezeichnet. W ist proportional zu $K/\cos(\mathbf{n}, \mathbf{l})$, wobei $(\mathbf{n}, \mathbf{l})$ der Winkel zwischen der Richtung $\mathbf{l}$ der Versetzungslinie und der Wachstumsrichtung $\mathbf{n}$ ist. Die so definierten Vorzugsrichtungen wurden für verschiedene Typen von Versetzungen in den Sektoren der drei Wachstumsflächen {100}, {101} und {010} berechnet.

Die in den Sektoren {101} und {100} beobachteten Richtungen der eingewachsenen Versetzungen liegen parallel zur Ebene (010). Auslenkungen (bis maximal 5°) aus (010) wurden nur vereinzelt beobachtet. Die Versetzungen besitzen innerhalb dieser Ebene ausgeprägte Vorzugsrichtungen mit einer Streuung von weniger als 3°. In der Tabelle 2 sowie in Fig. 5 (die zugleich die Morphologie des Thioharnstoffs erkennen läßt) sind die beobachteten und berechneten Richtungen einander gegenübergestellt. Die Übereinstimmung ist gut. Die Abweichungen betragen in keinem Fall mehr als 5°.

Tabelle 2

Beobachtete und berechnete Vorzugsrichtungen l_0 in den Sektoren {101} und {100}. φ ist der Winkel zwischen l_0 und der Achse [001]. Die Burgersvektoren der Versetzungen mit den eingeklammerten Richtungen konnten röntgentopographisch nicht eindeutig bestimmt werden (siehe Text)

Sektor	$\mathbf{b}$	beob.	ber.
{101}	[100]	44°	49°
	[010]	12°	11°
	[001]	24°	28°
	[101]	39°	41°
{100}	[101]	60°	58°
	[100]	(90°)	90° (A)
	[001]	(74°)	79° (B)

Die in der Tabelle 2 aufgeführten Versetzungen mit den eingeklammerten Richtungen konnten aus den röntgentopographischen Aufnahmen nicht eindeutig charakterisiert werden. Im Falle (A) handelt es sich um reine Stufenversetzungen mit einer Vorzugsrichtung parallel zur Wachstumsrichtung [100] und BVs parallel [010] oder [001]. Die berechnete Vorzugsrichtung für die Ver-

Fig. 6. Stufenversetzungen ($\boldsymbol{b}$ = [100] oder [001]) im Sektor {010}. Wachstumsrichtung [010] vertikal. Platte 1,2 mm dick, parallel (001). Ausschnitt 4,6 × 4,6 mm². Reflex (200), MoKα-Strahlung

setzungen mit $\boldsymbol{b}$ = [010] liegt parallel zur beobachteten Richtung [100]. Dagegen nimmt für Versetzungen mit $\boldsymbol{b}$ = [001] die Energie W pro Wachstumslänge in der Ebene (010) bei Auslenkung aus der Wachstumsrichtung [100] zunächst geringfügig (um weniger als 1‰) auf ein flaches Minimum bei $\varphi = 79°$ ab und steigt dann an.

Im Fall (B) handelt es sich um eine einmalige Beobachtung eines Bündels von etwa 10 geradlinigen Versetzungen mit $\varphi = 74°$. Einige dieser Versetzungen erfahren beim Eindringen in einen {101}-Sektor eine Richtungsänderung in die Vorzugsrichtung der [001]-Versetzungen in diesem Sektor ($\varphi = 24°$). Die Annahme, daß es Versetzungen mit $\boldsymbol{b}$ = [001] sind, steht im Einklang mit der Rechnung, nach der ein Minimum von W bei $\varphi = 79°$ liegt. Allerdings ist die Streuung der Richtungen geringer, als auf Grund des flachen Verlaufs der Energiekurve zu erwarten ist.

In den Sektoren {010} werden zwei Typen von Versetzungen beobachtet [4]:

a) Reine Schraubenversetzungen ($\boldsymbol{b}$ = [010]) mit einer scharfen Vorzugsrichtung parallel zur Wachstumsrichtung [010], die sich auch aus der Rechnung ergibt.

b) Versetzungen mit reinem oder überwiegendem Stufencharakter. Ihre BVs ([100] oder [001]) ließen sich nicht eindeutig ermitteln. Die mittlere Richtung dieser Versetzungslinien ist die Wachstumsnormale [010]. Die Streuung der Richtungen ist groß: Abweichungen von [010] (in jeder Richtung) bis zu 30° wurden oft beobachtet. Dabei sind diese Versetzungen nicht streng geradlinig: sie haben einen welligen Verlauf. Überdies zeigen sie eine feine, zick-zackähnliche Strukturierung, die infolge der beschränkten Auflösung der röntgentopographischen Abbildung gerade noch erkennbar ist. Fig. 6 zeigt eine topographische Aufnahme der Versetzungen dieses Typs.

Zur Erklärung des beobachteten Verlaufs dieser Stufenversetzungen sind drei Gesichtspunkte zu diskutieren:

1. Die Rechnung zeigt, daß die Energie W pro Wachstumslänge für die Versetzungen mit $\boldsymbol{b}$ = [100] oder [001] in den Ebenen (100) und (001) ein flaches Minimum parallel der Wachstumsnormalen [010] besitzt oder sogar um etwa 0,5% auf ein Minimum bei ungefähr $\pm 10°$ gegen [010] abnimmt. Es existiert also um [010] ein kegelförmiger Bereich, in dem keine Richtung durch ein ausgeprägtes Minimum von W ausgezeichnet ist.

2. Die {010}-Sektoren der von uns untersuchten Kristalle sind durch den Einbau von Fremdsubstanz stark gestört [4]. Trübungen durch mikroskopisch

kleine Einschlüsse und Wachstumsschichten mit starkem topographischen Kontrast wurden fast immer beobachtet. Die beschriebenen Stufenversetzungen sind teilweise und in unterschiedlichem Maße mit Fremdteilchen dekoriert. Insbesondere traten deutliche Richtungsänderungen der Versetzungslinien beim Durchdringen von Wachstumsschichten mit starkem Kontrast auf. Offensichtlich werden die Richtungen der eingewachsenen Versetzungen durch diese ebenfalls beim Wachstum entstandenen Störungen beeinflußt.

3. Die Richtungen der Stufenversetzungen mit $\boldsymbol{b} = [001]$ sind in der Umgebung der Achse [010] energetisch instabil (siehe Abschnitt 2.4).

In [1] wurden drei Regeln über die Vorzugsrichtungen geradliniger Versetzungen abgeleitet. Diese sind für die im Thioharnstoff beobachteten Versetzungen, mit Ausnahme der Versetzungen mit $\boldsymbol{b} = [001]$ im Sektor $\{100\}$ und $\boldsymbol{b} = [010]$ in $\{101\}$ erfüllt. Im ersten Fall nimmt der Energiefaktor K der Versetzungslinie bei einer Auslenkung aus der Wachstumsrichtung stark ab. Dabei kommt es nicht zu einer Kompensation durch die Längenzunahme der Versetzungslinie, wie bei der Ableitung der betroffenen Regel angenommen wird. Dies tritt erst bei größeren Auslenkungen ein. Im zweiten Fall besitzt die Versetzung mit $\boldsymbol{b} = [010]$ parallel zur Wachstumsrichtung $\boldsymbol{n}$ ($\varphi = 36°$) zwar reinen Stufencharakter, jedoch kein Maximum von K. K nimmt bei Änderung der Richtung $\boldsymbol{l}$ der Versetzungslinie von $\boldsymbol{n}$ nach [001] auf ein Minimum bei [001] ab (Fig. 2b). Das Minimum von $W \sim K/\cos(\boldsymbol{n}, \boldsymbol{l})$ liegt daher zwischen $\boldsymbol{n}$ und [001] ($\varphi = 12°$).

2.3 Gleitversetzungen

Im Thioharnstoff wurden zahlreiche Gleitversetzungen mit der Gleitebene (100) (zugleich Spaltebene des Kristalls) und Gleitrichtung [001] (kürzeste Gittertranslation der Struktur) beobachtet [4]. Diese Versetzungen, die meist in Gestalt von Halbschleifen von den Oberflächen in das Innere des Kristalls hineingewandert sind (Fig. 7), wurden durch plastische Verformung infolge mechanischer und thermischer Spannungen während der Präparation hervorgerufen. Bei vorläufigen Verformungsversuchen wurden nur diese Versetzungen und keine anderen erzeugt. Somit kann man annehmen, daß in der Rangfolge der im Thioharnstoff möglichen Gleitsysteme das beobachtete System (100) [001] an erster Stelle steht.

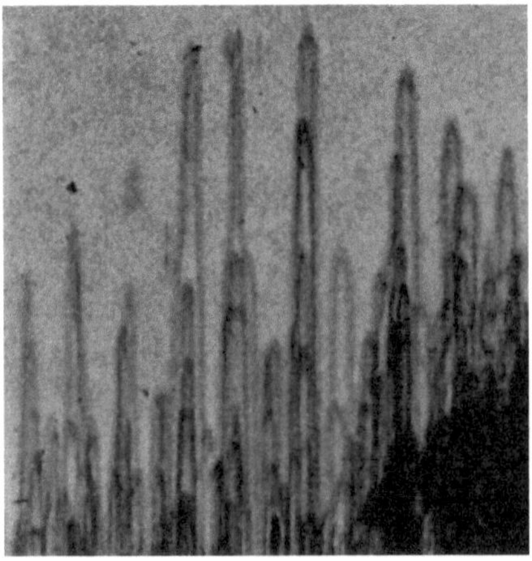

Fig. 7. Gleitversetzungen, Gleitrichtung [001] vertikal. Platte 1 mm dick, parallel (100). Ausschnitt 2,6 × × 2,6 mm². Reflex (002), MoKα-Strahlung

Eshelby [5] hat 1949 als Maß für die „Leichtigkeit“ einer Gleitung eine Größe ζ/b („ease of glide“) eingeführt. Es bedeuten

$$\zeta = \frac{1}{2} d\, K\, s$$

die Versetzungsweite in der Gleitebene [5, 6], d Abstand der Gleitebenen, K Energiefaktor, s elastischer Scherkoeffizient für eine Scherung auf der Gleitebene in der Richtung des Burgersvektors. In der Tabelle 3 sind die Größen ζ und ζ/b für Versetzungen mit den Burgersvektoren $\boldsymbol{b}$ und den Richtungen $\boldsymbol{l}$ parallel zu den Achsen des Kristalls für verschiedene Gleitebenen angegeben. Die Werte für das beobachtete Gleitsystem sind unterstrichen. Man erkennt, daß in der Rangfolge nach dem Kristerium von Eshelby das beobachtete System (100) [001] an 8. und 9. Stelle steht. Es ist daher anzunehmen, daß dieses Kriterium für Molekülkristalle mit stark asymmetrischen Bausteinen nicht gültig ist. Bemerkenswert ist auch, daß sich hier für die reine Schraubenversetzung mit $\boldsymbol{b} = [100]$ in der Ebene (001) eine Weite ζ ergibt, die mit 0,83 Å wesentlich kleiner als der Durchmesser eines Moleküls ist. Sie ist um den Faktor 17 kleiner als die Weite derselben Versetzung in der Gleitebene (010).

Tabelle 3

Versetzungsweite ζ (Einheit: Å), ζ/b und Nummer N in der Rangfolge der Gleitungen nach dem Kriterium von Eshelby für Versetzungen des Thioharnstoffs in verschiedenen Gleitebenen (hkl)

$\boldsymbol{b}$	$\boldsymbol{l}$											
	[100]				[010]				[001]			
	(hkl)	ζ	ζ/b	N	(hkl)	ζ	ζ/b	N	(hkl)	ζ	ζ/b	N
[100]	(010)	14,22	1,86	3	(001)	2,46	0,32	10	(010)	17,72	2,31	2
	(001)	0,83	0,11	12								
[010]	(001)	8,45	0,99	6	(100)	7,68	0,90	7	(100)	25,30	2,96	1
					(001)	1,37	0,16	11				
[001]	(010)	9,95	1,80	4	(100)	4,14	0,75	8	(100)	2,31	0,42	9
									(010)	7,06	1,28	5

2.4 Instabile Richtungen

Die hohe elastische Anisotropie des Thioharnstoffs läßt Winkelintervalle erwarten, in denen die Richtungen der Versetzungen energetisch instabil sind. Richtungsinstabilitäten treten in solchen Bereichen auf, in denen das inverse Wulff-Diagramm (Frank [7]) der Energie E bzw. des Energiefaktors K (d. i. die Richtungsabhängigkeit von E^{-1} bzw. K^{-1} in Polarkoordinatendarstellung) einen konkaven Verlauf besitzt (Head [8]).

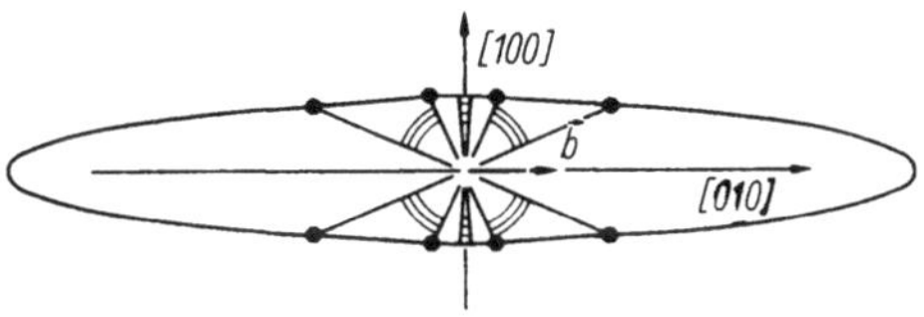

Fig. 8. Inverses Wulff-Diagramm für Versetzungen mit $\boldsymbol{b} = [010]$ in der Ebene (001)

Die in den inversen Wulff-Diagrammen der Versetzungen des Thioharnstoffs auftretenden konkaven Bereiche wurden berechnet. Fig. 8 zeigt ein solches Diagramm für Versetzungen mit $\boldsymbol{b}$ = [010] in der Ebene (001). Insgesamt wurden 4 schwach konkave Bereiche (symmetrieäquivalente Bereiche nicht gezählt) festgestellt. Sie sind in der Tabelle 4 angegeben. Die Bereiche A und B sind in Fig. 8 durch Kreisbögen gekennzeichnet.

Tabelle 4

Instabilitätsbereiche der Versetzungen im Thioharnstoff. (*hkl*) ist die Ebene, in der das instabile Richtungsintervall angegeben wird, [*uvw*] die Achse, gegen die der Winkel θ gemessen wird

	$\boldsymbol{b}$	(*hkl*)	θ	[*uvw*]
A	[010]	(001)	23°–66°	[100]
B	[010]	(001)	± 4°	[100]
C	[001]	(010)	±11°	[100]
		(001)	± 5°	
D	[001]	(100)	±14°	[010]
		(001)	± 8°	

Wie bei Head [8] näher erläutert wird, strebt eine Versetzung in einem Bereich instabiler Richtungen einen zick-zack-ähnlichen Verlauf an, dessen Linienelemente zu den Begrenzungsrichtungen des Instabilitätsintervalls parallel sind. Bei den in den Bereichen B und C befindlichen Versetzungen konnte ein solcher Verlauf nicht beobachtet werden. Das Instabilitätsintervall B ist zu klein. Es ist auch nicht sicher, ob diese Versetzungen den für den größeren Bereich C richtigen BV [001] besitzen. Dagegen zeigen die Stufenversetzungen im Bereich D eine zick-zack-ähnliche Strukturierung (Abschnitt 2.1, Fig. 6). Ob diese durch die berechnete Richtungsinstabilität erzeugt wurde, konnte nicht entschieden werden. Im Bereich A wurden keine Versetzungen beobachtet.

3. Schlußbemerkungen

Die am Benzil [1] und am Thioharnstoff durchgeführten Untersuchungen haben bestätigt, daß die bevorzugten Richtungen der geradlinigen Versetzungen in aus der Lösung gewachsenen Kristallen durch ein Minimum der elastischen Energie pro Längeneinheit in Wachstumsrichtung ausgezeichnet sind. Demnach sollte die wahrscheinlichste Erscheinungsform der eingewachsenen Versetzungen in diesen Kristallen die gerade Linie mit einer charakteristischen Richtung sein, die im allgemeinen nur wenig von der Wachstumsrichtung des betreffenden Wachstumssektors abweicht. Diese Geometrie der Versetzungen wird tatsächlich in den meisten aus der Lösung gezüchteten Kristallen beobachtet (siehe z. B. [9, 10]).

In vielen aus der Lösung gewachsenen Kristallen treten jedoch auch eingewachsene Versetzungen mit einem krummlinigen Verlauf auf. Hier kann man oft feststellen, daß diese Versetzungen in Kristallbereichen liegen, in denen andere, auch während des Wachstums entstandene Defekte (z. B. Einschlüsse, Wachstumsschichtungen) weitläufige Deformationsfelder hervorrufen. Das geradlinige Wachstum der Versetzungen wird dadurch gestört.

Der krummlinige Verlauf kann aber auch durch eine Bewegung der eingewachsenen Versetzungen nach Beendigung des Kristallwachstums bedingt sein. Dieses konnte in den Kristallen des Lithiumformiat-Hydrats [11], des Ammoniumhydrogenoxalat-Hemihydrats [12] und des Triglycinsulfats [13] nachgewiesen werden.

Der Autor dankt Herrn Prof. Dr. S. Haussühl für die Überlassung der Kristalle und für anregende Diskussionen. Die Rechnungen wurden am Rechenzentrum der Universität Köln durchgeführt.

Literatur

[1] H. Klapper, phys. stat. sol. (a) **14**, 99 (1972).
[2] M. M. Elcombe und J. C. Taylor, Acta cryst. **A24**, 410 (1968).
[3] S. Haussühl, Die elastischen Eigenschaften des Thioharnstoffs, in Vorbereitung.
[4] H. Klapper, J. Crystal Growth **15**, 281 (1972).
[5] J. D. Eshelby, Phil. Mag. **40**, 903 (1949).
[6] A. Seeger, in: Hdb. Phys., Bd. VII/1, Ed. S. Flügge, Springer-Verlag, 1955 (S. 577).
[7] F. C. Frank, Metal Surfaces, ASM, Metals Park (Ohio) 1963.
[8] A. K. Head, phys. stat. sol. **19**, 185 (1967).
[9] H. Klapper, J. Crystal Growth **10**, 13 (1971).
[10] V. F. Myuskov, V. P. Konstantinova und A. I. Gusev, Soviet Phys. — Cryst. **13**, 791 (1966).
[11] H. Klapper, in Vorbereitung.
[12] H. Klapper und H. Küppers, in Vorbereitung.
[13] A. Izrael, J. F. Petroff, Z. Malek und A. Authier, J. Crystal Growth **16**, 131 (1972).

(Received August 3, 1972)

phys. stat. sol. (a) **14**, 453 (1972)

Subject classification: 6; 13.4; 14.3; 22.1.1; 22.2.1

Physik-Department der Technischen Universität München

Shallow Donors in He-Cooled Semiconductors Used for Detection and Free Path Determination of Single Phonons

By

N. RIEHL, G. BAUR[1]), A. MÜLLER, and U. PUCHNER

Trapped charge carriers in phosphors can be used for detection and free path determination of single phonons at low temperatures [2]. It is shown that shallow donor or acceptor levels in semiconductors can be used for the same purpose and for other measurements on single phonons.

Getrapte Ladungsträger in Phosphoren können als Detektoren für einzelne Phononen bei der Bestimmung von Phononen-Weglängen bei tiefen Temperaturen dienen [2]. Es wird gezeigt, daß auch flache Donatoren oder Akzeptoren in Halbleitern für den gleichen Zweck sowie für andere Messungen an einzelnen Phononen dienen können.

1. Introduction

As reported in previous publications some phosphors contain very shallow traps ($E_{th} \approx 0.01$ eV) [1] which can be emptied by a single phonon and therefore can be used as detectors for single phonons at low temperatures [2]. A special technique avoiding an emptying by undesired outer IR is necessary for such experiments [1]. If a limited number of single phonons is generated (e.g. by bombardment with gas atoms of room temperature or in another manner) in a He-cooled and previously excited phosphor crystal and if its temperature remains unchanged, the phonons have long free paths thus being able without degradation to reach the trapped charges distributed in the crystal and to release them. The release is announced by a light spike from recombination of the released charges with luminous centres. (The effect is not caused by a simple heating of the crystal by the bombardment.) From the distance in which the phonons loose their ability to release charge carriers, the range of undissipated phonons can be determined. The use of traps with different depths enables a phonon spectroscopy and an investigation of the phonon degradation rate. This idea could be confirmed on several phosphors [2].

This principle of phonon detection can be applied also to semiconductors with shallow donors or acceptors. The observable response is in this case a current spike.

2. Experimental

2.1 Experiments on shallow donors

A typical experiment of this kind may be described. A He-cooled n-Ge crystal with 3×10^{15} Sb atoms per cm^3 and with low compensation (donor depth $E_{th} = 0.0096$ eV; spec. resistance at room temperature: 10 Ωcm; thickness 3 mm)

[1]) Now at Institut für angewandte Festkörperphysik der Fraunhofer-Gesellschaft, Freiburg/Br.

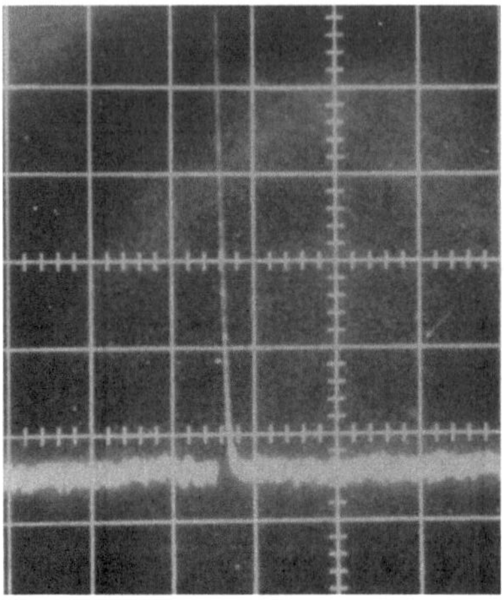

Fig. 1. Current spike caused by bombardment of a He-cooled Ge(Sb) crystal with a short pulse of Ne atoms. One small unit on the x-axis equals 10^{-3} s

has been bombarded with a short pulse of Ne atoms (total flux about 10^{16} atoms/cm^2 s during 10^{-4} s). This caused a spike of conductivity (Fig. 1). The rise time of the spike equals that of the atom pulse which corresponds to the rapidity of the used gas valve ($<10^{-6}$ s). (The used bombardment technique is described further down.)

The question arises whether the effect is actually due to electrons from the 0.0096 eV donors or to any other unknown levels. This was checked by measurements of the temperature dependence of the spike magnitude. The population density of donors (as a function of T) is proportional to $1/[1+A \exp(-E/2KT)]$ where E is the donor depth and A a constant. The experimental curve for the spike magnitude as a function of T agrees very well with this relation and yields a value of 0.0098 eV for E which agrees with the depth 0.0096 eV of the Sb donors within limits of error. Therefore we can be sure that the phonons coming from the bombarded crystal surface release the electrons actually from the Sb donor levels. The phonons, responsible for the release of electrons from the 0.0096 eV Sb donor levels in Ge, must be LA phonons (max. energy 0.0287 eV [3]).

It is easy to rule out the possibility that the effect is due to a simple heating up of the crystal by the bombardment. If considered as a trivial heat transfer phenomenon the possible temperature enhancement ΔT can be estimated from the energy delivered to the crystal per time unit by the hot gas atoms and the energy removed by heat conductivity through the crystal to the metallic holder. It results a value of ΔT less than 0.001 K, i.e. much too small for any observable thermal electron release in the usual sense. A similar result is obtainable from the minimal initial slope dn/dt of our experimental spike (n number of free carriers) and the value dn/dT from the temperature dependence of the conductivity. From $dn/dt = dn/dT \cdot dT/dt$ we can see which dT/dt would be necessary to obtain the observed dn/dt. The result is that the energy delivered by the hot gas atoms is thousandfold too small to reach the required dT/dt. Thus the spike effect is due only to the action of undissipated phonons and not to any heat in thermal equilibrium.

2.2 Experiments on shallow traps

Another typical experiment on the action of single phonons in semiconductors concerns the release of charges from traps (instead from donors). A semi-insulating GaAs crystal doped with oxygen and chromium (thickness 1 mm) has been studied. After its excitation with IR at He temperature and subsequent heating a TSC curve ('thermal stimulated conductivity') shows current maxima at 7, 13, and 20 K which are caused by charges released from three kinds of shallow traps. If at He temperature after excitation bombarded with Ne atoms, the crystal shows a high current spike. It can be shown that it is due to charges released from the shallowest of the three traps. In a TSC curve taken after the bombardment the TSC maximum at 7 K has almost vanished, but the other two maxima remain unchanged (Fig. 2). The deficiency of the charges stored in the 7 K traps is just equal to the charges released during the bombardment. The almost complete diminution of the 7 K maximum shows that the free paths of the phonons generated at the crystal surface reach throughout the whole crystal thickness, i.e. the free path is at least 1 mm. This experiment is analogous to the earlier experiments on phosphors with the difference that the release of charge carriers from traps is signaled by a current spike and not by light emission. The release from traps in GaAs is to be attributed to LA phonons of maximal energy 0.03 eV [3].

2.3 Experiments on acceptors

We have also tested the possibility of phonon detection by release of holes in p-type semiconductors. A p-Ge crystal with 2.5×10^{15} atoms indium per cm^3 has been used (E_{th} of the acceptor levels 0.0112 eV; specific resistance of the sample at room temperature 15 Ωcm). At low flux densities of the phonons the result was the same as in the n-type semiconductors: a conductivity spike appeared as response to the bombardment with Ne atoms. But if the phonon flux density was drastically enhanced (by using higher Ne atom flux densities), a reversed behaviour namely a negative spike, i.e. a drop of the conductivity was observed. A complete explanation of this effect cannot be given as yet.

3. Methodical Improvements and Future Applications

In our first experiments we used only crystals with a very low conductivity hoping to observe the spikes easier if the background conductivity is weak. Therefore the first experiments were restricted to low-doped crystals. But in such crystals some unknown impurities and levels can play a relatively high disturbing role, so it seemed desirable to find a way for an extension of the experiments to higher-doped crystals with greater conductivity. For this purpose we

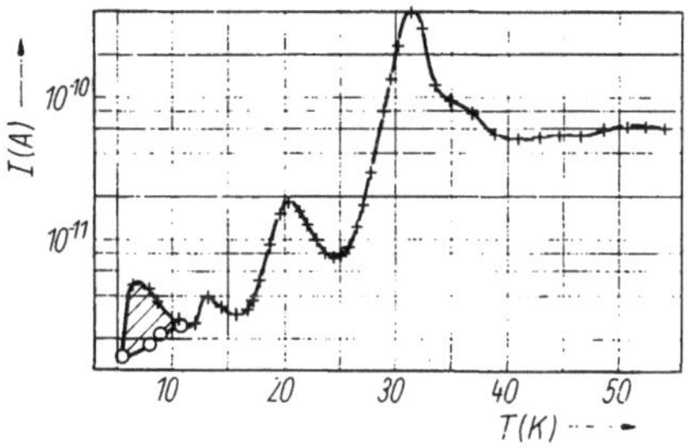

Fig. 2. TSC curve of a semi-insulating GaAs crystal before and after bombardment

made use of the fact that the observed current displays a steep spike with a very short rise time so that it can be easily amplified and thus separated from the background dc conductivity. With such a differential method we are able to get spike effects at semiconductors of very low resistance, e.g. 1 Ωcm at 300 K.

Another methodical improvement deals with the bombardment technique. Whereas in the first experiments we used room temperated Ne or He atoms with roughly Maxwellian distribution of velocities we now use gas atoms of higher and much more homogeneous velocity produced with the so-called Skimmer technique [4] which allows to obtain gas atom energies up to 3 to 4 kT with a relatively narrow Gaussian distribution. (Dr. Gspann from the Kernforschungszentrum Karlsruhe kindly suggested the use of a special nozzle designed by Dr. Hagena). Although the total flux of the produced beam was nearly the same as in the first experiments, the rise time of the current (or light) spike was faster by a factor of 100.

Using crystals which are doped (and therefore sensitive for phonons) only in a thin layer at the surface, we are able to measure the transit time for phonons of different energies. Measurements of this kind can lead to a simple determination of phonon dispersion curves. Such experiments as well as experiments with small band gap semiconductors, e.g. $Cd_xHg_{1-x}Te$ are in progress. The principal difficulty caused by the high conductivity of the latter ones (near 1 Ωcm at room temperature) is overcome by the described differential technique of spike measuring.

Acknowledgements

The authors are indepted to Prof. W. Heywang, Dr. L. Mader, H. Reinke, and J. Weidlich from the Siemens-Laboratory for semiconductor samples and to W. H. Koschel (now at Institut für angewandte Festkörperphysik der Fraunhofer-Gesellschaft, Freiburg/Br.) for suggestions concerning GaAs.

References

[1] N. RIEHL, Proc. Internat. Conf. Luminescence, Budapest 1966 (p. 974); Festkörperprobleme **8**, 232 (1968).
G. BAUR, L. MADER, and N. RIEHL, Z. Naturf. **24a**, 1296 (1969); J. Luminescence **1**, 1 (1970).
[2] N. RIEHL, Verh. DPG **H9**, 679 (1971).
N. RIEHL and A. MÜLLER, Phys. Letters (Netherlands) **36A**, 487 (1971).
[3] B. N. BROCKHOUSE, Phys. Rev. Letters **2**, 256 (1959).
J. L. T. WAUGH and G. DOLLING, Phys. Rev. **132**, 2410 (1963).
S. M. SZE, Physics of Semiconductor Devices, Wiley-Interscience, New York 1969.
[4] A. KANTROWITZ and J. GREY, Rev. sci. Instrum. **22**, 328 (1951).
U. BOSSEL, Thesis, Univ. of California, Berkeley 1968, Rep. No. AS -68-6.
U. BOSSEL, Entropie **30**, 11 (1969).

(Received September 22, 1972)

phys. stat. sol. (a) **14**, 457 (1972)

Subject classification: 14.3.3; 22.2.2; 22.2.4

VEB Werk für Fernseheelektronik, Berlin

On the Potential Barrier of Schottky Junctions

By

H. Klose [1])

It is shown on the basis of the thermionic emission model, how the effective potential barrier height φ_{Bn0} at 0 °K, its mean temperature coefficient $\bar{\alpha}_T$, and its temperature coefficient α_T for different temperatures can be determined from the forward I–U-characteristics of Schottky junctions. The derived equations were examined with data from literature and yield values of the temperature coefficient of 0.18 mV/deg, 0.20 mV/deg, and 0.22 to 0.24 mV/deg for Cr–n-GaP, Cr-n-GaAsP, and Au-n-GaP, respectively. Under the assumption that for high densities of surface states the potential barrier height φ_{Bn0} and the band gap ΔE_0 are directly proportional to each other a new procedure for the determination of the parameter of composition x of ternary semiconductor compounds is proposed.

Auf der Grundlage des Modells der Thermoemission wird gezeigt, wie aus den Flußkennlinien von Schottky-Übergängen die effektive Potentialbarriere φ_{Bn0} bei 0 °K, ihr mittlerer Temperaturkoeffizient $\bar{\alpha}_T$ und ihr Temperaturkoeffizient α_T für unterschiedliche Temperaturen bestimmt werden können. Die abgeleiteten Gleichungen werden mit den Daten aus der Literatur geprüft und ergeben für Cr–n-GaP, Cr–n-GaAsP bzw. Au–n-GaP Werte der Temperaturkoeffizienten von 0,18 mV/grd, 0,20 mV/grd bzw. von 0,22 bis 0,24 mV/grd. Unter der Annahme, daß für große Oberflächenzustandsdichten die Potentialbarriere φ_{Bn0} und der Bandabstand ΔE_0 einander direkt proportional sind, wird ein neues Verfahren zur Bestimmung des Stöchiometrieparameters x ternärer Halbleiterverbindungen vorgeschlagen.

1. Introduction

The potential barrier φ_{Bn} of a semiconductor–metal contact (Schottky junction) is the determining parameter of the I–U-characteristic. It is determined by the properties of the metal, the interface, and the semiconductor.

In many studies the potential barriers for different metals on the semiconductors Ge, Si, GaAs, GaP, and CdS were investigated [1]. Often, however, the temperature dependence of the potential barrier φ_{Bn} is neglected. Only in [2 to 4] values of the temperature coefficient α_T of the potential barrier are given, which is defined by

$$\alpha_T = \frac{\mathrm{d}}{\mathrm{d}T} (\varphi^*_{Bn} - \Delta\varphi) = \frac{\mathrm{d}}{\mathrm{d}T} \varphi_{Bn} , \tag{1}$$

where the effective potential φ_{Bn} includes the image force $\Delta\varphi$ ($\varphi_{Bn} = \varphi^*_{Bn} - \Delta\varphi$). The temperature coefficient determined in [2 to 4] was about 0.20 mV/deg for Au–n-Si, Ag–n-Si, Cu–n-Si, Au–n-GaAs, and Au–n-GaP.

---

[1]) Present address: Institute of Semiconductor Physics, Academy of Science of USSR, Siberian Branch, Novosibirsk 90.

Sometimes the determination of the temperature dependence of the potential barrier φ_{Bn} is difficult, because the C–U or the I_{photo}–$h\nu$ characteristics have to be taken at different temperatures.

In this work, we want to show how it is possible to determine the temperature coefficient α_T or $\bar{\alpha}_T$ and the effective potential $\varphi_{\mathrm{Bn}0}$ ($T = 0$ °K) from the I–U characteristic.

In this case we suppose that the current transport is accomplished by the thermal excitation of charge carriers. Thus the current density in forward direction is given by

$$j_{\mathrm{ft}} = I_{\mathrm{s}} \left(\exp \left\{ \frac{q\,U_{\mathrm{f}}}{n\,kT} \right\} - 1 \right) = A^*\,T^2 \exp \left\{ -\frac{q\,\varphi_{\mathrm{Bn}}}{kT} \right\} \left(\exp \left\{ \frac{q\,U_{\mathrm{f}}}{n\,kT} \right\} - 1 \right), \tag{2}$$

where A^* is the effective Richardson constant ($A^* = 4\pi q\,m_{\mathrm{n}}^*\,k^2/h^3$) and n is the diode value conditioned by the interface.

In this study we only want to deal the thermionic emission model, because up to now the current transport by carrier diffusion (Schottky model) has been observed seldom, but the results obtained are similar.

2. Basic Equations

For $q\,U_{\mathrm{f}} \geqq 3\,kT$, it follows from equation (2)

$$\frac{U_{\mathrm{f}}}{n} = \varphi_{\mathrm{Bn}} - \frac{kT}{q} \ln \left\{ \frac{A^*\,T^2}{j_{\mathrm{ft}}} \right\}. \tag{3}$$

The derivation of equation (3) with respect to T gives, assuming $\mathrm{d}m_{\mathrm{n}}^*/\mathrm{d}T = \mathrm{d}j_{\mathrm{ft}}/\mathrm{d}T \equiv 0$ (m_{n}^* is the effective mass), the temperature coefficient γ_T of the forward voltage under constant current conditions:

$$\gamma_T = \left.\frac{\mathrm{d}U_{\mathrm{f}}}{\mathrm{d}T}\right|_{j_{\mathrm{ft}}=\mathrm{const}} = -\frac{n\,k}{q} \left(2 + \left[1 + \frac{T}{n} \frac{\mathrm{d}n}{\mathrm{d}T} \right] \ln \left\{ \frac{A^*\,T^2}{j_{\mathrm{ft}}} \right\} \right) + \varphi_{\mathrm{Bn}} \frac{\mathrm{d}n}{\mathrm{d}T} + n\,\alpha_T\,. \tag{4}$$

Since today with a careful surface preparation Schottky junctions with $n \approx 1$ and thus with $\mathrm{d}n/\mathrm{d}T \leqq 10^{-5}\ \mathrm{deg}^{-1}$ can be produced, equation (4) can be simplified to

$$\gamma_T = -\frac{n\,k}{q} \left(2 + \ln \left\{ \frac{A^*\,T^2}{j_{\mathrm{ft}}} \right\} \right) + n\,\alpha_T\,. \tag{5}$$

Under the presumption of a linear temperature dependence the effective barrier height φ_{Bn} changes according to

$$\varphi_{\mathrm{Bn}} = \varphi_{\mathrm{Bn}0} - \bar{\alpha}_T\,T\,, \tag{6}$$

where $\bar{\alpha}_T$ is the mean temperature coefficient. With equation (6) we find for the so-called saturation current I_{s} according to equation (2)

$$\ln \left\{ \frac{I_{\mathrm{s}}}{T^2} \right\} = -\frac{q}{kT}\,\varphi_{\mathrm{Bn}0} + \frac{q}{k}\,\bar{\alpha}_T + \ln A^*\,. \tag{7}$$

Analogously for the forward current density from equations (2) and (6) one obtains for $q\,U_f \geqq 3\,kT$

$$\ln\left\{\frac{j_{ft}}{T^2}\right\} = \frac{q\,U_f}{n\,kT} - \frac{q}{kT}\,\varphi_{Bn\,0} + \frac{q}{k}\,\bar{\alpha}_T + \ln A^*\,. \tag{8}$$

Equations (3) to (5), (7) and (8) allow to determine both the barrier height $\varphi_{Bn\,0}$ and the mean temperature coefficient $\bar{\alpha}_T$ or α_T according to equations (4) and (5) for a certain temperature.

3. Discussion

The plot of the forward voltage U_f/n versus $T \ln (A^*\,T^2/j_{ft})$ under constant forward current conditions, gives the value φ_{Bn0} by extrapolation of T to zero along the Y-axis. From the slope of the straight line one obtains the value of k/q which gives information about the exactness of the experimental data. The parameter n strongly dependent on temperature must be known at each temperature used for the evaluation.

We have evaluated the I–U characteristics for Cr–n-GaP according to [5] and for Cr–n-GaAsP according to [6] in agreement with equation (3). We used $A^*_{GaAs} = 7.4$ [7], $A^*_{GaP} = 35$ [8], and therefore for $A^*_{GaAsP} = 20$ corresponding to the effective masses.

In Fig. 1 and 2 we show the results. We measured the diode value n for each temperature [9]. n is not constant over the total current density range at a given temperature, as especially at high current densities temperature variations occur. From the barrier height φ_{Bn0} determined in this way and that indicated by the authors at room temperature it is possible to evaluate $\bar{\alpha}_T$. The results are included in Table 1.

From the current density dependence of the temperature coefficient $\gamma_T = f(\ln 1/j_{ft})$ it is possible to determine the parameters dn/dT and α_T according to equation (4), if n and φ_{Bn} are known. If the temperature dependence of n is negligible, equation (5) allows to evaluate the coefficient α_T without knowing φ_{Bn}. In [9] we give a more detailed discussion of equations (4) and (5).

Now we will consider equation (7) and (8). In the literature they were occasionally used for the determination of the effective Richardson constant and the effective mass. Neglecting the coefficient $\bar{\alpha}_T$ however, often too high values for A^* or m_n^* are obtained (see [10] as an example). In Fig. 3 one can see the evalua-

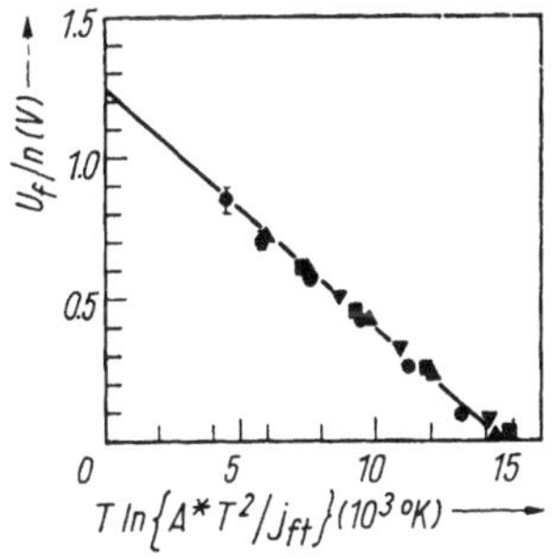

Fig. 1. Forward voltage–temperature characteristic for Cr-n-GaP. $j_{ft} = 10^0$ (●); 10^{-2} (▲); 10^{-4} (■); 10^{-6} (▼) A cm^{-2}

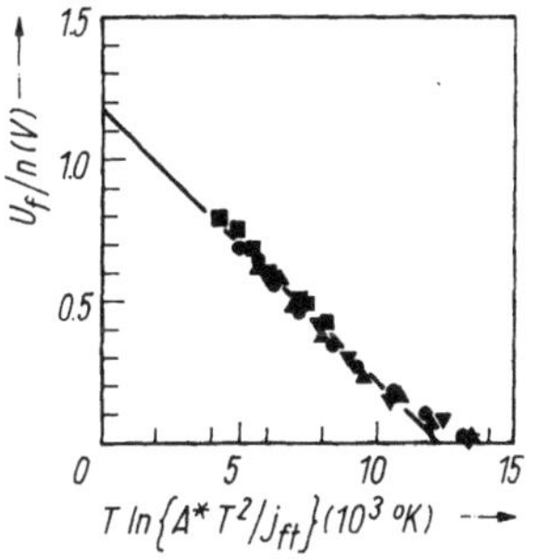

Fig. 2. Forward voltage–temperature characteristic for Cr-n-GaAsP. $j_{ft} = 10^0$ (■); 10^{-1} (●); 10^{-2} (▲); 10^{-3} (▼) A cm^{-2}

Table 1

Results for φ_{Bn}, φ_{Bn0}, and $\bar{\alpha}_T$

junction	Cr–n-GaP	Cr–n-GaAsP	Au–n-GaAs	Au–n-GaP
φ_{Bn}(V) (T = 300 °K)	[5] 1.18	[6] 1.12	[11] 0.89	[4, 14] 1.18
φ_{Bn0}(V) equation (3) or $U_f = f(T)$	1.235 1.24 ($\bar{n}$ = 1.12)	1.18 —	— $\bar{n}$ = 0.95	— —
φ_{Bn0}(V) equations (7), (8)	1.25	—	—	1.25
$\bar{\alpha}_T$ (mV/deg) equations (7), (8)	0.18	0.17	—	0.24
$\bar{\alpha}_T = \frac{\varphi_{Bn} - \varphi_{Bn0}}{\Delta T}$	0.20	0.20	≈0.1	0.22

tion of the I–U characteristics of [4] and [5] for Au–n-GaP ($\ln[j_{ft}/T^2] = f(T^{-1})$, $U_f = 0.73$ V, ■), and for Cr–n-GaP ($\ln[j_{ft}/T^2] = f(T^{-1})$, $U_f = 0.2$ V, ● and $U_f = 0.4$ V, ▼ respectively; $\ln[I_s/T^2] = f(T^{-1})$, ▲). In Table 1 the values for $\bar{\alpha}_T$ and φ_{Bn0} calculated from Fig. 3 are given. The variations for Au–n-GaP versus the value of $\bar{\alpha}_T = 0.20$ mV/deg according to [4] can probably be explained by our inaccurate knowledge of the area of the junction (we assumed a value of 1.5×10^{-2} cm^2).

Since in equation (3) the temperature dependence of U_f via $\ln T^2$ is only weak, it is easier at very small dn/dT to evaluate the I–U characteristics by plotting U_f versus T. From Fig. 4 which according to [5] with the mean value $\bar{n} = 1.12$ gives almost the same value for φ_{Bn0} as equation (3) (see Table 1) one can notice that the error is very small.

For comparison in Fig. 4 also the data for a Au–n-GaAs junction according to [11] are involved. The value resulting from it for $\bar{n}\,\varphi_{Bn0} = 0.95$ V however,

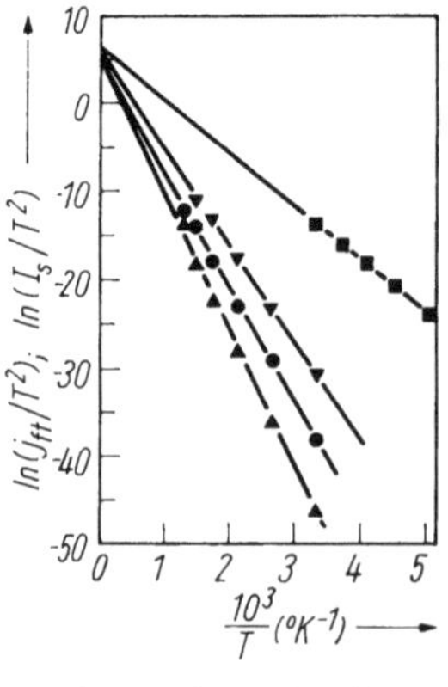

Fig. 3. Forward current or saturation current as functions of inverse temperature. ■: $\ln[j_{ft}/T^2]$ for Au–n-GaP (U_f = 0.73 V); ●: $\ln[j_{ft}/T^2]$ (U_f = 0.2 V); ▼: $\ln[j_{ft}/T^2]$ (U_f = 0.4 V), and ▲: $\ln[I_s/T^2]$ for Cr–n-GaP

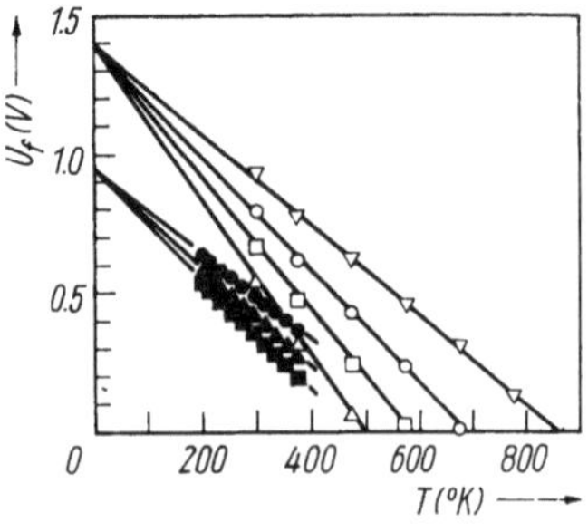

Fig. 4. Forward voltage–temperature characteristics for Cr–n-GaP (open symbols) and Au–n-GaAs (black symbols). $j_{ft} = 10^0$ (▽; ●); 10^{-1} (▲); 10^{-2} (○; ■); 10^{-4} (□); and 10^{-6} (△) A cm^{-2}

seems to be too small. The reason for this might be that the I–U characteristics are measured at lower temperatures.

With a small dn/dT and $n \approx 1$, which is possible with present preparation techniques one can obtain the barrier height φ_{Bn0} from $U_f = f(T)$ in a relatively short time. Since with a high density of surface states the barrier height and the band gap ΔE are connected by

$$\varphi_{Bn} \approx \frac{2}{3} \Delta E , \tag{9}$$

the parameter x can be obtained by means of Schottky diodes on an angle-lapped sample of $Ga_xAl_{1-x}As$ or $GaAs_xP_{1-x}$ from the location dependence of φ_{Bn0}. If equation (9) applies, as recently was confirmed, e.g. for $Ga_xAl_{1-x}As$ by [12], φ_{Bn0} and ΔE_0 always differ by a constant value. Since $\Delta E_0 = f(x)$ is known from literature a determination of x might be possible from a temperature dependent voltage measurement with a constant forward current. This method is partly easier than optical methods [2, 13], because one can operate with relatively high current densities and thus with large signals.

4. Conclusions

From our considerations and the results obtained (Table 1) we can draw the following conclusions:

1. From the temperature dependence of the forward voltage with a constant forward current the effective barrier height φ_{Bn0} can be determined, analogously as in [15] the band gap ΔE_0 was evaluated from p–n junctions under high level injection conditions.
2. The current density dependence of the temperature coefficient of the forward voltage γ_T renders it possible to determine the temperature dependence of the temperature coefficient of the effective potential barrier α_T.
3. For the determination of the mean temperature coefficient of the potential barrier $\bar{\alpha}_T$ the temperature dependence of the forward current or of the saturation current can be used.
4. If equation (9) applies to a number of Schottky junctions on an angle-lapped sample produced under the same conditions it is possible to determine the parameter x of ternary semiconductor compounds from the temperature dependence of the forward voltage under constant forward current conditions.

Acknowledgements

I would like to thank Prof. Dr. K. Thiessen and Dipl.-Phys. P. Hottewitzsch for the critical reading of the manuscript.

References

[1] C. A. Mead, Solid State Electronics **9**, 1023 (1966).
K. Heime, Solid State Electronics **13**, 1505 (1970).
[2] C. R. Crowell, S. M. Sze, and W. G. Spitzer, Appl. Phys. Letters **4**, 91 (1964).
[3] F. A. Padovani, J. appl. Phys. **37**, 921 (1966).
T. Arizumi and M. Hirose, Japan J. appl. Phys. **8**, 749 (1969).
[4] B. V. Tsarenkov, Yu. A. Goldberg, A. P. Izergin, E. A. Posse, V. N. Ravich, T. Yu. Rafiev, and N. F. Silvestrova, Fiz. Tekh. Poluprov. **6**, 710 (1972).
[5] Y. Nannichi and G. L. Pearson, Solid State Electronics **12**, 341 (1969).

[6] D. A. NEAMEN and W. W. GRANNEMANN, Solid State Electronics **14**, 1319 (1971).
[7] S. M. SZE, Physics of Semiconductor Devices, Wiley, New York 1969.
[8] S. A. ABAGYAN, G. A. IVANOV, A. P. IZERGIN, and YU. E. SCHANURIN, Fiz. Tekh. Poluprov. **6**, 1130 (1972).
[9] H. A. KLOSE and YU. E. MARONTCHUK, to be published.
[10] C. F. GENZABELLA and C. M. HONWELL, in: Gallium Arsenide, Proc. Internat. Symp., Reading 1966 (p. 131).
[11] F. A. PADOVANI and G. G. SUMMER, J. appl. Phys. **36**, 3744 (1965).
[12] YU. A. GOLDBERG, T. YU. RAFIEV, B. V. TSARENKOV, and YU. A. YAKOLEV, Fiz. Tekh. Poluprov. **6**, 462 (1972).
[13] K. GRAFF and H. PIEPER, Solid State Electronics **15**, 831 (1972).
[14] YU. A. GOLDBERG, E. A. POSSE, and B. V. TSARENKOV, Electronics Letters **7**, 601 (1971).
[15] N. SCLAR and D. P. POLLOCK, Solid State Electronics **15**, 473 (1972).

(Received August 29, 1972)

phys. stat. sol. (a) 14, 463 (1972)

Subject classification: 10.1; 18.2; 21.1; 21.1.1

Research Institute for Iron, Steel, and Other Metals, Tôhoku University, Sendai

Plastic Deformation of Iron–Nickel Invar Alloys

By

J. ECHIGOYA, S. HAYASHI, and M. YAMAMOTO

Plastic deformation of Fe–Ni invar alloys which have peculiar magnetic properties is studied by means of the tensile and stress relaxation tests. The plastic properties of these alloys in comparison with other face centred cubic alloys are characterized by the large temperature dependence of the flow stress, the straight dislocation configuration, and not having clearly defined three-stage hardening. These results reveal a contribution to the resistance to dislocation motion from the drag due to making the new region of the magnetization decrease around a dislocation.

Die plastische Deformation der Eisen–Nickel-Legierungen, die besonderes magnetisches Verhalten zeigen, wird mittels Zug- und Spannungsrelaxation untersucht. Das plastische Verhalten dieser Legierungen wird im Vergleich mit anderen kubisch-flächenzentrierten Legierungen durch eine große Temperaturabhängigkeit der Fließspannung, eine Verfestigungskurve ohne die drei klar definierten Stufen und eine gerade Versetzungsanordnung charakterisiert. Diese Ergebnisse weisen auf einen Hinderungsanteil zur Versetzungsbewegung hin, die durch den neuen Bereich der Magnetisierungsabnahme um eine Versetzung hervorgerufen wird.

1. Introduction

Fe–Ni invar alloys consisting of iron and about 36 wt% nickel, have abnormally small thermal expansion coefficients at room temperature. This peculiar property has so far been explained phenomenologically by a large positive volume magnetostriction. Also, other physical properties such as a rapid decrease of the saturation magnetization and of the Curie temperature in the invar region were studied by many workers [1]. Further, it is necessary to notice that the invar alloy has the property of a large magnetic volume effect, which is a mutual dependence between magnetic properties and volume changes.

Plastic properties of invar alloys have so far been discussed mainly in their relation with the martensitic transformation [2]. Recently Bolling and Richman [3] studied experimentally the invar-type Fe–Ni–C alloys and found that the plastic properties of the invar-type alloys have the following characteristics: (i) a large temperature dependence of the initial flow stress, (ii) the existence of a flow stress threshold near the Curie temperature, and (iii) inhibition of martensite nucleation by the invar properties. Kray [4] tried to interpret these plastic properties on the basis of micromagnetics, but he could not explain the large temperature dependence of the flow stress.

In order to investigate plastic properties and dislocation motions in invar alloys which have the peculiar magnetic properties stated above, we have studied systematically the temperature dependence of flow stress and activation volume, which are obtained by tensile and stress relaxation tests, and the dislocation configuration in samples deformed at temperatures above and below the Curie temperature for several iron–nickel alloys. Further, we shall discuss these plastic properties on the basis of the magnetic frictional stress due to peculiar magnetic properties of invar alloys.

2. Experimental Procedure

Iron–nickel alloy polycrystalline specimens were prepared by vacuum melting 99.9% purity electrolytic iron and the same purity nickel together, casting, and then drawing into wire of 1.5 mm diameter. There were seven charges with 31 to 45 wt% nickel. Also, single crystals of 1.5 mm diameter were grown by the Bridgman method. Specimens used for tensile tests, both single crystals and polycrystals, were annealed at about 1200 °C for a week in vacuum. The tensile tests were performed in an argon atmosphere, using an Instron-type testing machine, at the strain rate of $5.5 \times 10^{-4}\ s^{-1}$.

Our experimental apparatus for the temperature change during tensile tests is schematically shown in Fig. 1. The range of the temperature change during tests was from 20 to 320 °C, using a bifilarly wound electric furnace. The measurements of the temperature dependence of the flow stress and the activation volume were covered for one specimen over the whole temperature range, since their results were consistent with the results of the initial deformation of each one of many specimens. The total strain when all tests for one specimen were finished was below 3% and the influence of the work-hardening was corrected simply. The flow stress was determined from the load at 0.1% strain from the yield at constant temperature. Then, the cross-head motion was stopped and the stress relaxation test was carried out. The concentration of the specimens was determined by chemical analysis after all tests had been finished.

3. Experimental Results

3.1 Stress–strain curves

Shear stress–shear strain curves of Fe–31, 34, and 41 wt% Ni and Co–60% Ni [5] single crystals were shown in Fig. 2. It is seen that the stress–strain curves of invar alloy single crystals in comparison with that of the Co–60% Ni single

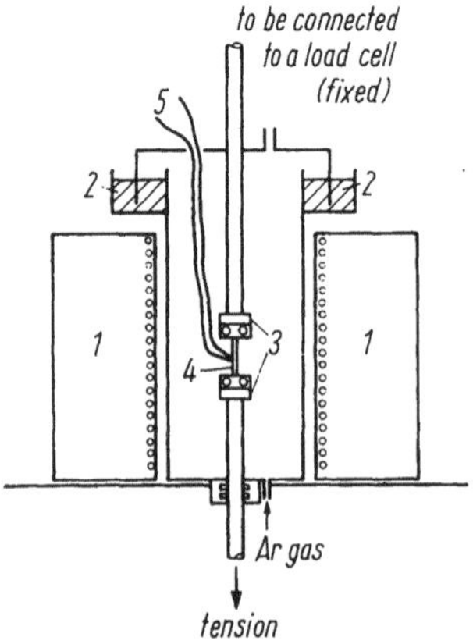

Fig. 1. Scheme of the apparatus used for the high-temperature deformation. 1 electric furnace, 2 silicon oil, 3 specimen chuck, 4 specimen, 5 alumel–chromel thermocouple

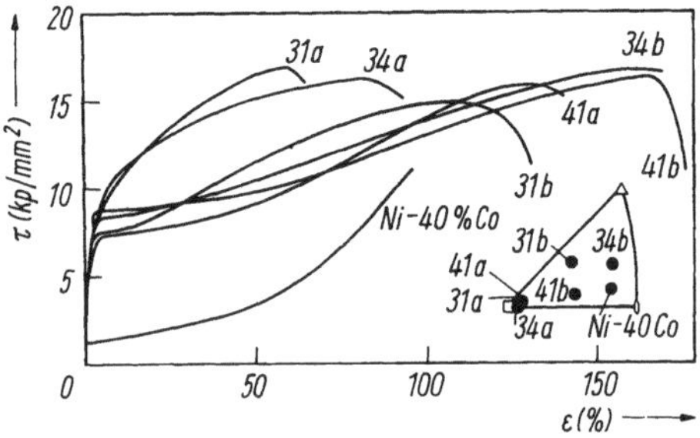

Fig. 2. Stress–strain curves of Fe–31, 34, and 41% Ni and of Co–60% Ni single crystals. Crystal orientations are shown in the stereographic triangle

crystal are as follows: a) The absolute value of the flow stress at room temperature is very much larger than in the latter alloys. The yield stresses of the former alloy single crystals are 4 to 5 times larger than that of the latter ones. b) The work-hardening rate in stage II of the former alloys is smaller than that of the latter one. But, in stage I, the work-hardening rate is of the same order of $10^{-4}\,\mu$ as other f.c.c. metals and alloys, where μ is the shear modulus, while, the work-hardening rate in stage II for invar alloys is 2 to 3 times smaller than the rate of normal f.c.c. alloys, which are of the order of $\mu/300$. This value for invar alloys is of the same order as that of b.c.c. alloys, for example, Fe–3% Si [6] as shown in Table 1. c) The stress–strain curves of Fe–Ni invar alloys do

Table 1

Comparison of the hardening rates at stage I and II of various f.c.c. and b.c.c. alloy single crystals

alloy	hardening rate in stage I ($10^{-3}\,\mu$)	hardening rate in stage II ($10^{-3}\,\mu$)
Fe–31% Ni	0.43	1.3
Fe–34% Ni	0.6	1.1
Fe–40% Ni	0.48	1.38
Fe–42% Ni	0.5	2.0
Co–50% Ni	0.38	2.4
Fe–3% Si		1.5

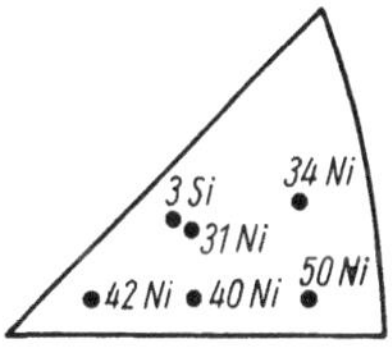

not show three clear-cut stages of work-hardening as other f.c.c. alloys like Co–Ni single crystals. In spite of the middle orientation on the stereographic triangle, the specimen of 31b in Fig. 2, for example, shows neither a clear stage I nor the transition from stage II to stage III. Thus, it is found that the stress–strain curves of Fe–Ni invar alloys resemble those of b.c.c. single crystals at low temperature rather than those of other normal f.c.c. metals.

3.2 Temperature dependence of the flow stress

The results of the temperature dependence of the flow stress are given for single crystals of the invar, Co–Ni [7] and Cu–Ni [8] alloys in Fig. 3. These curves indicate that the temperature dependence of the flow stress of the invar alloys is very much larger than that of Co–Ni and Cu–Ni alloys in the vicinity of room temperature. Similar results for polycrystalline Fe–Ni alloys are given in Fig. 4. In this case the Fe–45% Ni alloy, which does not have peculiar magnetic properties like invar, indicates a smaller temperature dependence of the flow stress than that of invar alloys. Thus, it may be said that the strong temperature dependence of the flow stress is a peculiar plastic property of invar alloys. The flow stress versus temperature curve for the invar region (Fe–30 to 37% Ni) can be seen to be separated into three parts, a) a part of steep and linear temperature dependence of the flow stress, b) a normal temperature dependence which can be seen for the Fe–45% Ni alloy, and c) a plateau part in which the

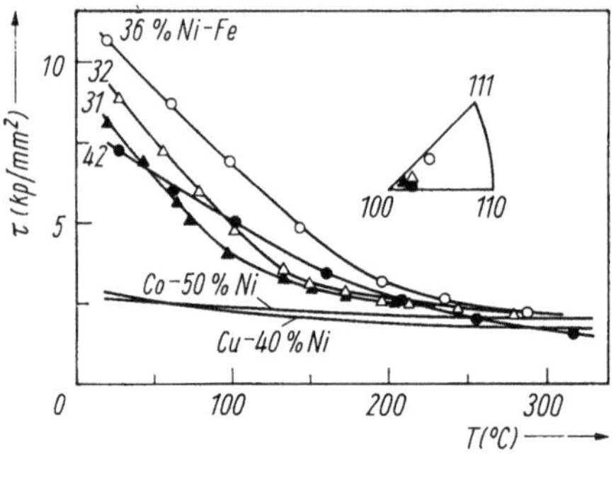

Fig. 3. The temperature dependence of the flow stress for Fe–Ni invar alloy single crystals as compared with Co–50% Ni and Cu–40% Ni alloy single crystals. Orientations of the single crystals are shown in the triangle

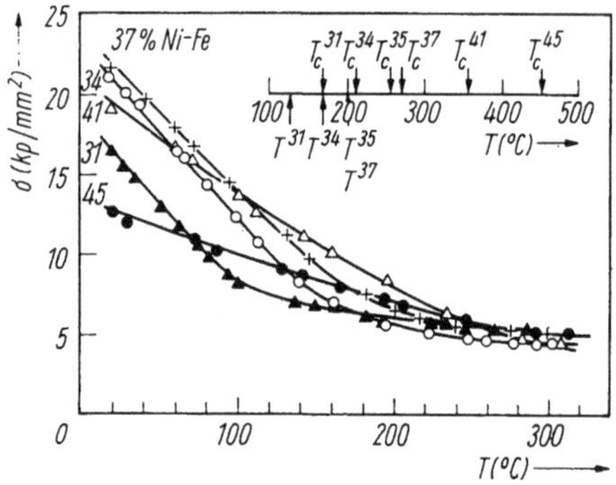

Fig. 4. The temperature dependence of the flow stress of polycrystalline Fe–Ni alloys. The inset shows the comparison of the threshold temperature with the Curie temperature of the specimens

flow stress is almost independent of temperature. For non-invar alloys, the a) part does not appear and the flow stress depends weakly on temperature as for normal f.c.c. alloys near room temperature, and shown for the Fe–45% Ni alloy in Fig. 4. These differences between invar and non-invar alloys suggest that the invar properties affect the plastic property of the steep and linear temperature dependence of the flow stress. Further, the extrapolated intersection of the steep and linear part and the plateau part corresponds to the temperature region of rapid decrease of the saturation magnetization rather than the Curie temperature. The relations between the intersection temperatures and the Curie temperatures are shown in the inset of Fig. 4, where T indicates the intersection temperature of the flow stress and T_c the Curie temperature. The correspondence of the intersection temperature of the flow stress and the temperature of the rapid decrease of magnetization is good for Fe–30 to 35% Ni alloys and gradually deviates with increasing nickel concentrations. These correspondences may show that the large temperature dependence of the flow stress of the invar alloys is caused by their peculiar magnetic properties.

The nickel concentration dependence of the flow stress at room temperature is shown in Fig. 5. This figure shows that the flow stress has a peak at the Fe–37% Ni alloy and this tendency does not resemble that of normal solid solution hardening such as that of the Cu–Ni alloy system [8] which has a peak near 50% con-

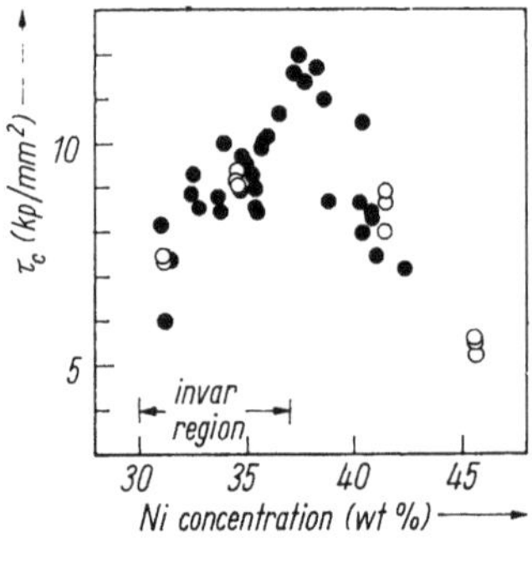

Fig. 5. The concentration dependence of the critical shear stress τ_c at room temperature. The open circles are the reduced values from polycrystalline data for the comparison with single crystal data (full circles)

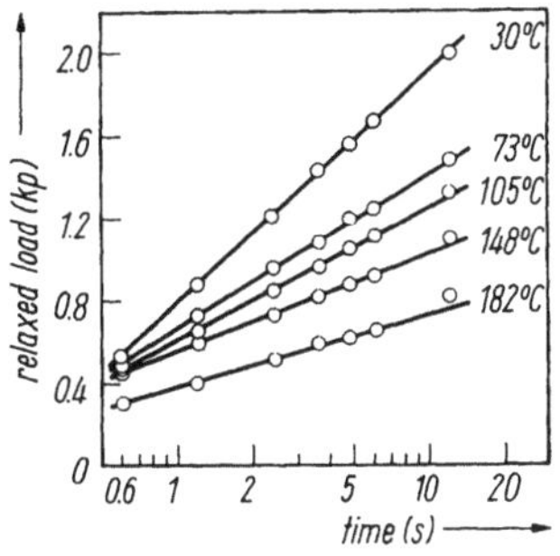

Fig. 6. Logarithmic representation of the stress relaxation curve at various temperatures for the 31% Ni–Fe alloy

centration. This experimental result may be interpreted by the higher Curie temperature of alloys with higher nickel content.

3.3 Activation volume

The activation volume is determined from experiments on stress relaxation. The stress-relaxation curve is described empirically by the following equation:

$$\tau_0 - \tau = \lambda \lg (t + c) - \lambda \lg c \, , \tag{1}$$

where τ_0 and τ are the stress values at the times $t = 0$ and $t = t$ at which the cross head stops, and λ and c are constants. This empirical equation holds in the case of invar alloys also, as shown for the Fe–31% Ni alloy in Fig. 6. From the experimental condition for the stress relaxation, we have

$$\dot{\varepsilon}_{\text{p}} = - M \dot{\tau} \, , \tag{2}$$

where $\dot{\varepsilon}_{\text{p}}$, M, and $\dot{\tau}$ are the rate of plastic deformation, a constant, and the rate of stress relaxation, respectively.

We assume only one thermally activated process, so the process represents an Arrhenius type equation,

$$\dot{\varepsilon}_{\text{p}} = \varphi \, b \, A \, N \, \nu_0 \exp\left(- \frac{U(\tau)}{kT} \right), \tag{3}$$

where φ is a geometrical factor, b the magnitude of a Burgers vector, A the area swept out after activation, N the number of activation sites, ν_0 the frequency factor which depends on the nature of obstacles and the way of overcoming them, $U(\tau)$ the activation energy for the activation event, k the Boltzmann constant, and T the absolute temperature.

Further, the activation energy for the activation event may be expressed as

$$U(\tau) = U^* - \tau^* \, v \, , \tag{4}$$

where U^* is the activation energy which is less than the total energy of an obstacle by applying the effective stress, τ^* the effective stress at the obstacle, and v the activation volume. If U^* has little flow stress dependence, i.e., U^* equals U_0 of the total energy, from equations (1) to (4) the activation volume, v, is given by

$$v = 2.3 \frac{kT}{\lambda} . \tag{5}$$

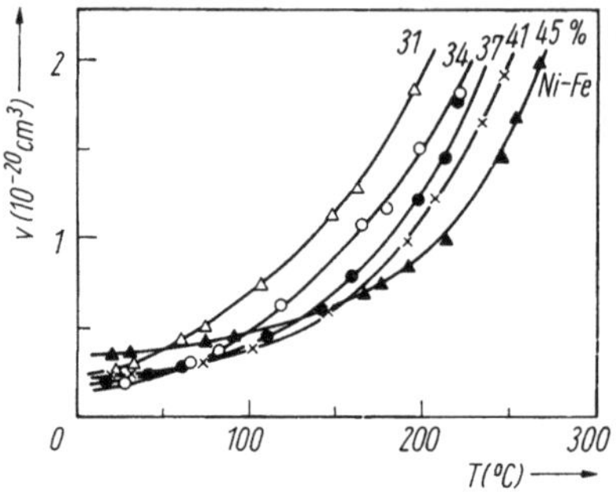

Fig. 7. The temperature dependence of the activation volume for 31, 34, 37, 41, and 45% Ni–Fe alloys

Thus, it is seen that the activation volume is obtained experimentally from the gradient, λ, of the $(\tau_0 - \tau)$ versus $\lg t$ curve. The temperature dependence of the activation volume determined by equation (5) is shown in Fig. 7. The value of the activation volume for invar alloys is slightly smaller than that of Co–Ni alloys, but for the Fe–Ni alloy system it has the same value, irrespective of them being invar and non-invar alloys at room temperature. The magnitude of the activation volume for iron–nickel alloys at room temperature is of the order of $100\ b^3$. The rapid increase of the activation volume at high temperature may be caused by the inadequacy of the assumption of one thermally activated process and the temperature at which the activation volume increases rapidly does not correspond to a magnetic transition point. Thus, the properties peculiar to invar are not shown up on the activation volume.

4. Discussion

One of the most important plastic properties of invar alloys is the strong temperature dependence of the flow stress and its disappearance near the Curie temperature. At first, we discuss this on the basis of a magnetic frictional stress due to the peculiar magnetic properties, because it cannot be explained by the magneto-elastic interaction between dislocations [4].

Now, it has already been mentioned that the invar alloys have a large volume-magneto effect, which means magnetic properties are very sensitive to the volume change. Actually, a strong decrease of saturation magnetization and Curie temperature on applying a hydrostatic pressure have been observed and this effect for invar alloys is 100 times larger than that of normal ferromagnetic metals [9]. We notice the experimental results of the decrease of saturation magnetization at plastic deformation for invar alloys at room temperature [10]. According to Evert and Kussmann [9], the relation between the change of hydrostatic pressure, ΔP, and the magnetization decrease, ΔI, is given by

$$\Delta I = \alpha\ I\ \Delta P\ , \tag{6}$$

where α is a constant expressing the magnetization sensitivity for the hydrostatic pressure. The experimental results are explained by the hydrostatic pressure around an edge dislocation increased by the plastic deformation.

Now, we show that a moving edge dislocation gets the resistance by producing a new region of magnetization decrease around it. Assuming that E is the energy change when a dislocation moves by a distance r, the friction stress for the motion of a dislocation of unit length, τ, is written as

$$\tau\, b\, r = E\,. \tag{7}$$

E is calculated from the magnetization decrease by the pressure around an edge dislocation and is given as follows using the simple molecular field approximation:

$$E = -\varrho\, r\, d \int\limits_{I}^{I-\Delta I_d} w\, I\, dI\,, \tag{8}$$

where ϱ, d, w, and ΔI_d are, respectively, the density, the size which indicates the region of magnetization decrease by the hydrostatic pressure around an edge dislocation and is about 100 b for invar alloys, the molecular field constant, and the quantity of time-dependent magnetization decrease for the pressure change. E in equation (8) takes a non-zero value when time delay of magnetization decrease exists for the pressure change by dislocation motion, because the difference between time delayed and balanced magnetization decrease under hydrostatic pressure contributes to the energy change for the dislocation motion. According to experimental results of Schlossser et al. [11], it is expected that a very large relaxation of magnetization decrease exists for the pressure change, and the relaxation of the magnetization decrease is very large in comparison with d/v, where v is the dislocation velocity. Thus,

$$\Delta I - \Delta I_d \approx 0\,. \tag{9}$$

From equations (6) to (9), τ of equation (1) is given by

$$\tau \approx \frac{w}{b}\, I^2\, (\alpha\, \Delta P)\, \varrho\, d\,. \tag{10}$$

In equation (10) τ is 7, 10, and 1 kp/mm² for 30, 35, and 45% Ni–Fe alloys at room temperature, respectively. These values are in good agreement with experimental results. Further, the temperature dependence of τ is also in agreement with experimental results, as shown in Fig. 8, assuming that only the magnetization changes with temperature. So it may be concluded that the strong temperature dependence and the high value of the flow stress at room temperature are caused by the magnetic friction for the dislocation motion.

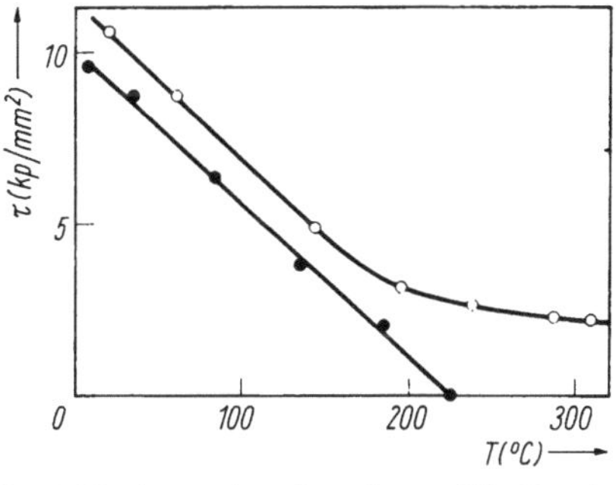

Fig. 8. The temperature dependence of the flow stress for the 35% Ni–Fe alloy calculated from equation (10) (full circles). The experimental results are shown by open circles

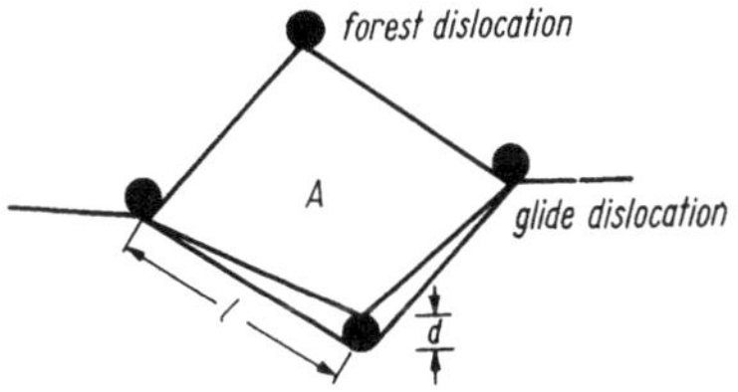

Fig. 9. The process of dislocation intersection with forest dislocations

The mechanisms of the magnetic friction mentioned above are connected with the edge dislocations. In the early stage of deformation, the flow stress is determined by the process of dislocation intersection with forest dislocations. Thus, a screw dislocation is bent as shown in Fig. 9 during the dislocation intersection process, and this bent dislocation comes to have an edge component which has a hydrostatic pressure. Thus, in this way the moving dislocation experience a magnetic friction stress.

These considerations have been confirmed by the electron microscopic observation of a polycrystalline Fe–31% Ni alloy. The samples used for the observation were strained by about 3% in tension at room temperature and above the Curie temperature. Then they were thinned by chemical polishing in a mixture of fluoric acid, perhydrooxide, water, and nitric acid and by electropolishing in perchloric acid and acetic acid and then the observation was carried out. The configuration of dislocations for an invar alloy deformed at room temperature, as shown in Fig. 10, is very different from other normal f.c.c. metals, for example, Co–Ni alloys [5]. The dislocation configuration for invar alloys is characterized by their uniformity and by consisting of very straight and squarely bent dislocations. This may be understood from the fact that the dislocation in invar alloys has a tendency to decrease the edge component. On the other hand, on the dislocation structure of the 3% strained specimen at about 200 °C (above the Curie temperature) for the Fe–31% Ni invar alloy, wavy and tangle dislocations are observed which are similar to those of other normal f.c.c. metals such as Co–Ni and non-invar Fe–45% Ni alloys deformed at room temperature. This configuration is caused by the alloy not having invar properties as a result of the deformation above the Curie temperature. Thus, a small work-hardening rate in stage II of the stress–strain curve for invar alloys at room temperature is

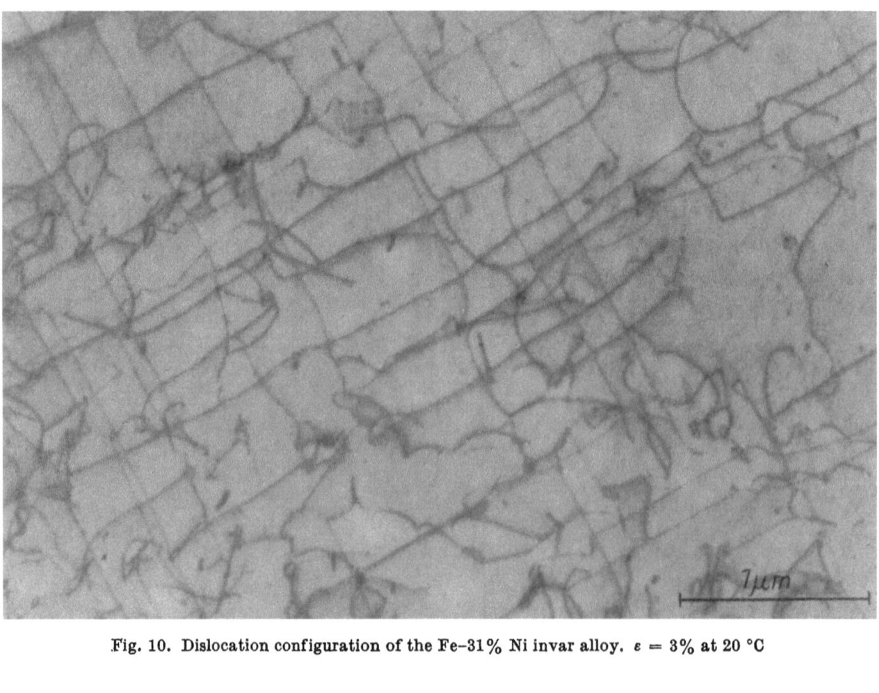

Fig. 10. Dislocation configuration of the Fe–31% Ni invar alloy. $\varepsilon = 3\%$ at 20 °C

phenomenologically understood from an uniform dislocation distribution which scarcely has dislocation tangles because of invar properties. We report the results of the detailed investigation of dislocations in a subsequent paper.

Acknowledgements

The authors thank to Dr. T. Nakamichi for his interest and for discussing the results and M. Suenaga and M. Simizu for their invaluable assistance.

References

[1] S. Chikazumi, T. Mizoguchi, N. Yamaguchi, and P. Beckwith, J. appl. Phys. **39**, 939 (1968).
[2] A. W. McReynolds, J. appl. Phys. **20**, 896 (1949).
[3] G. F. Bolling and R. H. Richman, Phil. Mag. **19**, 247 (1969).
[4] U. Kray, Phil. Mag. **20**, 1295 (1969).
[5] S. Mader, Electron Microscopy and Strength of Crystals, John Wiley and Sons, 1963 (p. 183).
[6] A. S. Keh and S. Weissmann, ibid. (p. 231).
[7] F. Pfaff, Z. Metallk. **53**, 441 (1962).
[8] H. Suzuki, Dislocations and Mechanical Properties of Crystals, John Wiley and Sons, 1957 (p. 361).
[9] H. Evert and A. Kussmann, Z. Phys. **38**, 437 (1937).
[10] J. Echigoya, S. Hayashi, T. Nakamichi, and M. Yamamoto, J. Phys. Soc. Japan **30**, 289 (1971).
[11] W. F. Schlosser, P. P. M. Meincke, E. Vittoratos, and G. M. Graham, J. Phys. Soc. Japan **27**, 1696 (1969).

(Received August 15, 1972)

phys. stat. sol. (a) 14, 473 (1972)

Subject classification: 1.5 and 3; 22.9

Department of Chemistry and Electron Microscope Centre,
University of Queensland, St. Lucia

The Structure of Vacuum Deposited Thin Films of Tetracene

By

W. RIEMER and J. V. HARDY

The structure of vacuum deposited thin films of tetracene was investigated for a range of deposition conditions (film thickness 34 to 500 nm, deposition rate 0.2 to 2.5 nm s^{-1}, on substrates of glass, glass coated with aluminium or tin oxide, and collodion) by replica and direct transmission electron microscopy. The films were found to be polycrystalline with the crystal size decreasing slightly with deposition rate, but increasing with film thickness. In view of their physical properties, the detection of whiskers was of special significance.

Durch Vakuumsublimation erhaltene Tetrazenschichten wurden elektronenmikroskopisch durch Abdruck und Durchstrahlung untersucht. Dabei wurde die Abhängigkeit der Struktur von den Aufdampfparametern (Schichtdicke 34 bis 500 nm; Aufdampfgeschwindigkeit 0,2 bis 2,5 nm s^{-1}; Unterlagen aus Glas, Aluminium- oder Zinnoxyd-bedampftem Glas, und Kollodium) berücksichtigt. Die Schichten erwiesen sich als polykristallin, wobei die Kristallgröße mit der Schichtdicke zunahm, jedoch bei größerer Aufdampfgeschwindigkeit abnahm. Im Hinblick auf deren besonderen Eigenschaften ist die Beobachtung von Whiskern von Bedeutung.

1. Introduction

Molecular crystals of aromatic hydrocarbons, such as tetracene, can be regarded as electrical insulators. The carrier generation, transport, and trapping processes which give rise to the electronic conduction properties [1, 2] of these materials are determined by the distribution of energy levels in the crystal. It has been suggested that in the absence of impurities the energy spectrum of localized states in molecular crystals can be related to random static displacements of molecules [3], i.e. to structural defects. Because of the large bandgap of tetracene, the observed conductivity must result from carrier injection at the surface. One mechanism of photoinjection depends on the generation and dissociation of excitons near the surface of the crystal [4], a process that involves surface states.

A number of experiments on various aspects of electronic processes in thin films of tetracene have been reported: photoemission [5], contact potential difference [6], photoconductance [7], and the photovoltaic effect [8, 9] were investigated. The thin films used in these experiments were prepared by vacuum deposition and were thought to be microcrystalline.

A different mechanism that may lead to greatly enhanced electronic conductivity is associated with the formation of conductive filaments in an amorphous thin film. Observations of this phenomenon have recently been reported for thin films of tetracene that were shown, by X-ray diffraction, to be essentially amorphous [10, 11].

In view of the importance of structure for electronic properties, but also because, in multilayer devices, the thin film surface structure propagates and affects subsequent layers [8], an electron microscopy study was undertaken to investigate the structure of thin films of tetracene for a range of deposition conditions.

2. Experimental

The thin films of tetracene were deposited by evaporation from a tantalum source in an oil diffusion pump system at a pressure not exceeding 5×10^{-6} Torr. The film thickness and deposition rate were monitored by a quartz crystal oscillator deposition meter which had been calibrated using interferometric techniques. The substrate was maintained at room temperature during deposition. For replica work Corning 7059 glass, bare or freshly coated with 4N aluminium or tin oxide, was used. For direct observation the tetracene was deposited on electron microscope specimen support grids which had been filmed with collodion.

The thin films were investigated both by direct observation and by replica techniques in a Siemens Elmiskop I electron microscope. For direct observation it was found to be necessary to keep the electron beam intensity at a minimum to avoid evaporating the tetracene.

Electron diffraction patterns were obtained either by the selected-area microdiffraction technique (in which the pattern is obtained from a small area a few microns across) or by using the microscope as a diffraction camera, switching off the image-forming lenses (when the pattern is obtained from a much larger area). In either case, evaporated aluminium was used to calibrate the camera-constant. Usually 60 kV electrons were used; sometimes when thicker specimens were examined, it was found advantageous to increase the voltage to 100 kV.

A replica-technique was used to examine tetracene on glass, aluminium, or tin oxide. The specimen was placed in a vacuum evaporation unit and shadowed with platinum/palladium. It was then coated with evaporated carbon. The specimen was rotated during the deposition of carbon in order to obtain an even, continuous film. The replica film was then scribed into small squares and then released from the specimen by immersing it in a chloroform bath which dissolved the tetracene and allowed the carbon film with shadowing metal to float free. After washing with chloroform, the squares were mounted on support grids and examined in the microscope. Electron micrographs of shadowed replicas were printed in reversed-contrast in order to obtain dark shadows and hence a more "realistic" appearance.

3. Results

For the range of deposition conditions investigated all thin films were found to be polycrystalline. Films were deposited with a thickness between 34 and 500 nm, at deposition rates from 0.2 to 2.5 nm s^{-1}, on substrates of glass, glass coated with aluminium or tin oxide, and collodion. The diffraction pattern for a film of tetracene 34 nm thick on collodion is shown in Fig. 1a. The radii of the rings are in agreement with the structure expected for tetracene [12, 13]. Although the diffraction pattern acquired increased detail with film thickness, the energy dissipation associated with the electron beam striking the film caused the film to sublimate and the diffraction pattern to diffuse before it could be photographed. The diffraction pattern of Fig. 1b for a film 90 nm thick, indicates a significant increase in crystallite size.

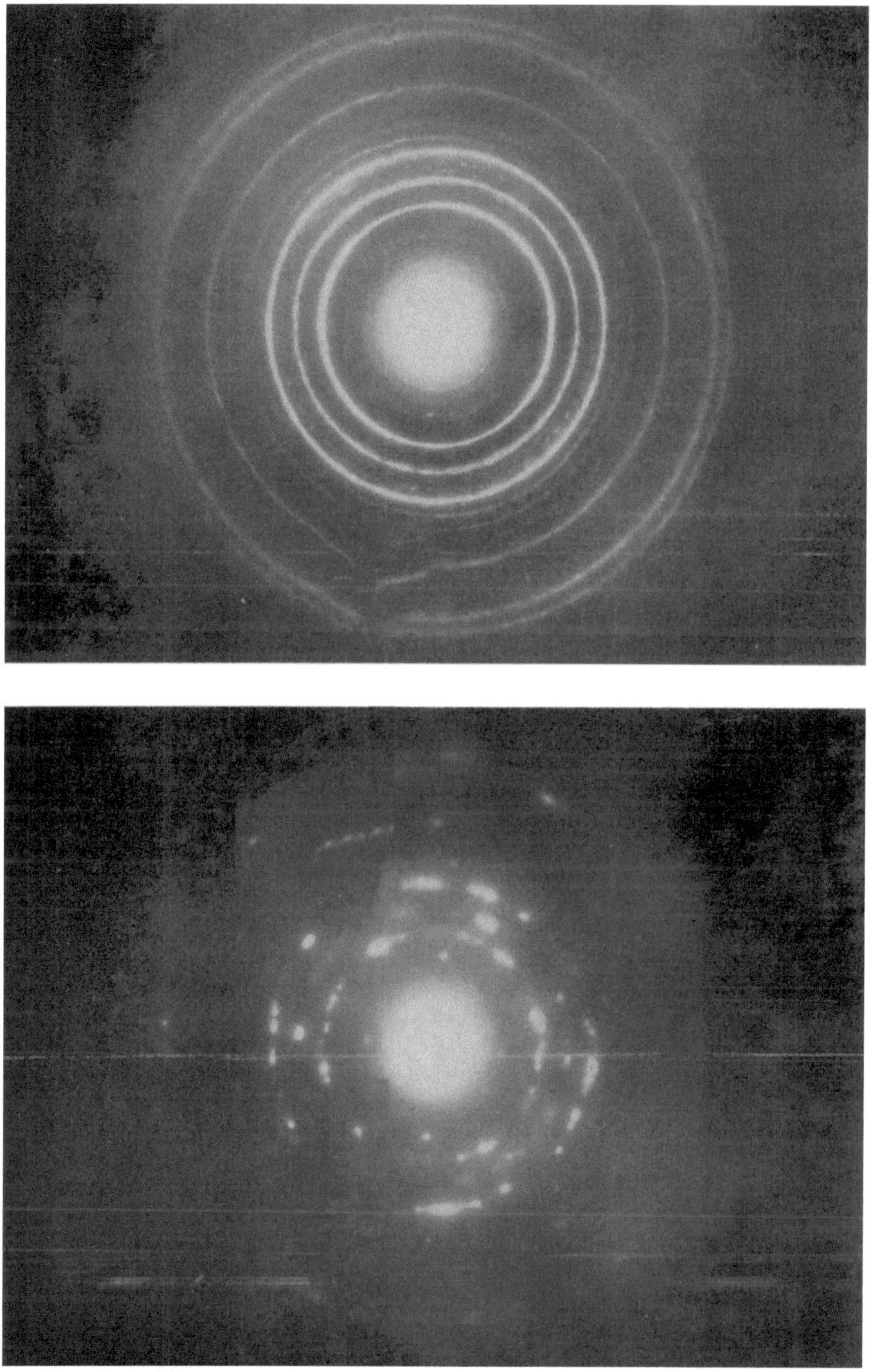

a

b

Fig. 1. Selected-area diffraction pattern for a thin film of tetracene deposited on collodion. a) Film thickness 34 nm, b) film thickness 90 nm

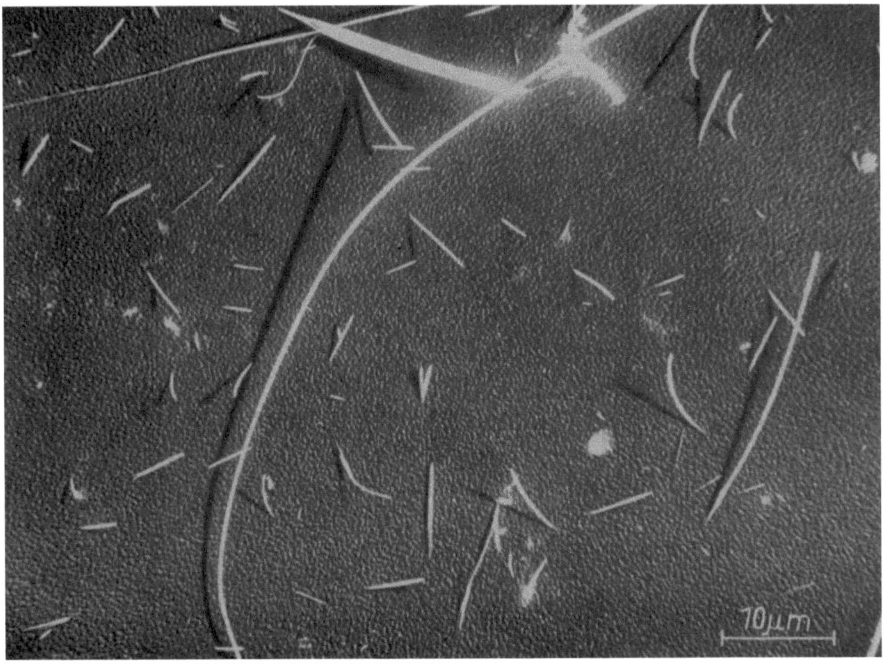

Fig. 2. Replica of thin film of tetracene 450 nm thick

In order to determine the effect of deposition rate and film tickness on crystallite size, a series of tetracene films were deposited on collodion and examined by direct observation. It was found that for a constant film thickness of 90 nm the crystallite size decreased slightly with increasing (0.19 to 1.7 nm s^{-1}) deposition rate. For a constant deposition rate of 0.45 nm s^{-1}, the crystallite size increased with film thickness (34 to 115 nm), most significantly so at the low values of this range.

Fig. 2 illustrates the surface structure of a tetracene film 450 nm thick as represented by a replica. A noteworthy feature encountered is the presence of whiskers, crystal filaments with parallel sides and tapered tips. The whiskers appear in a distribution of lengths and orientations. The developed planes of the base of a whisker are shown in Fig. 3a. In Fig. 3b, besides a whisker, a number of coalesced grains separated by narrow channels can be recognized. Disc-like platforms sometimes project from the grains. Other grains have fine, parallel striations which are probably evidence of repeated twinning.

Although, for similar deposition conditions, the grain size on tin-oxide coated glass was somewhat smaller than that on glass or aluminized glass, the substrate did not significantly affect the grain size. Whiskers were observed on all substrates, but were most prevalent at low deposition rates (less than 1 nm s^{-1}).

Even for very thin films, whiskers would attain remarkable lengths. Fig. 4a, the micrograph of a tetracene film 63 nm thick deposited at a rate of 0.45 nm s^{-1} on collodion, shows a number of whiskers. It was observed that the whiskers were more resistent to sublimation by the electron beam than was the polycrystalline film. Under electron beam illumination whiskers could be deflected

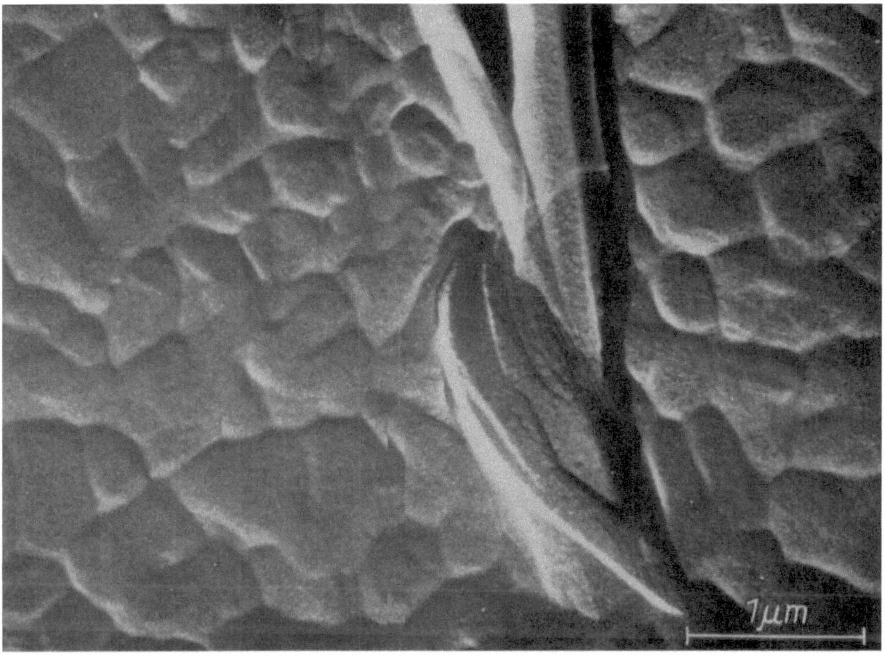

a

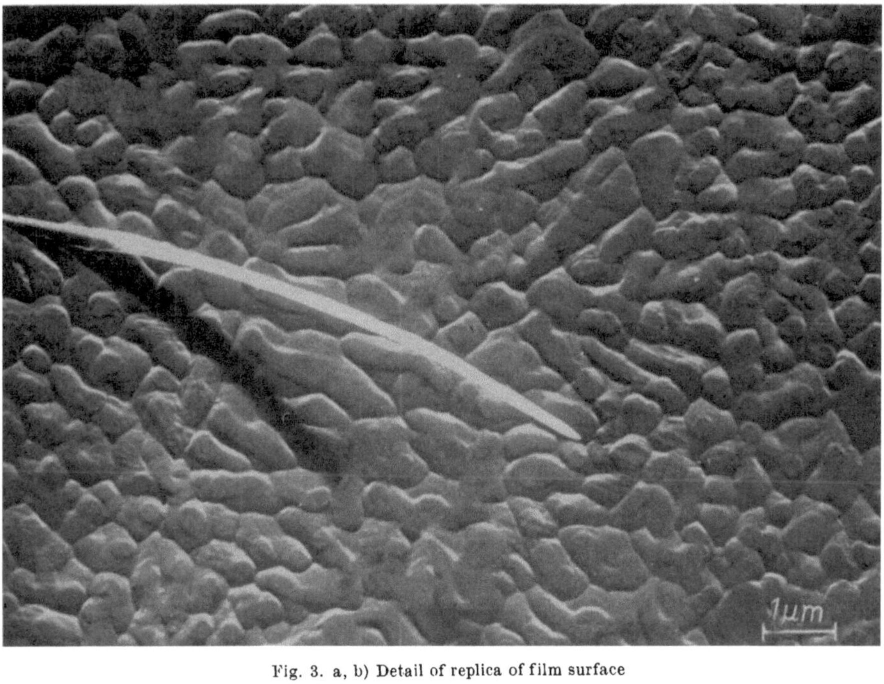

b

Fig. 3. a, b) Detail of replica of film surface

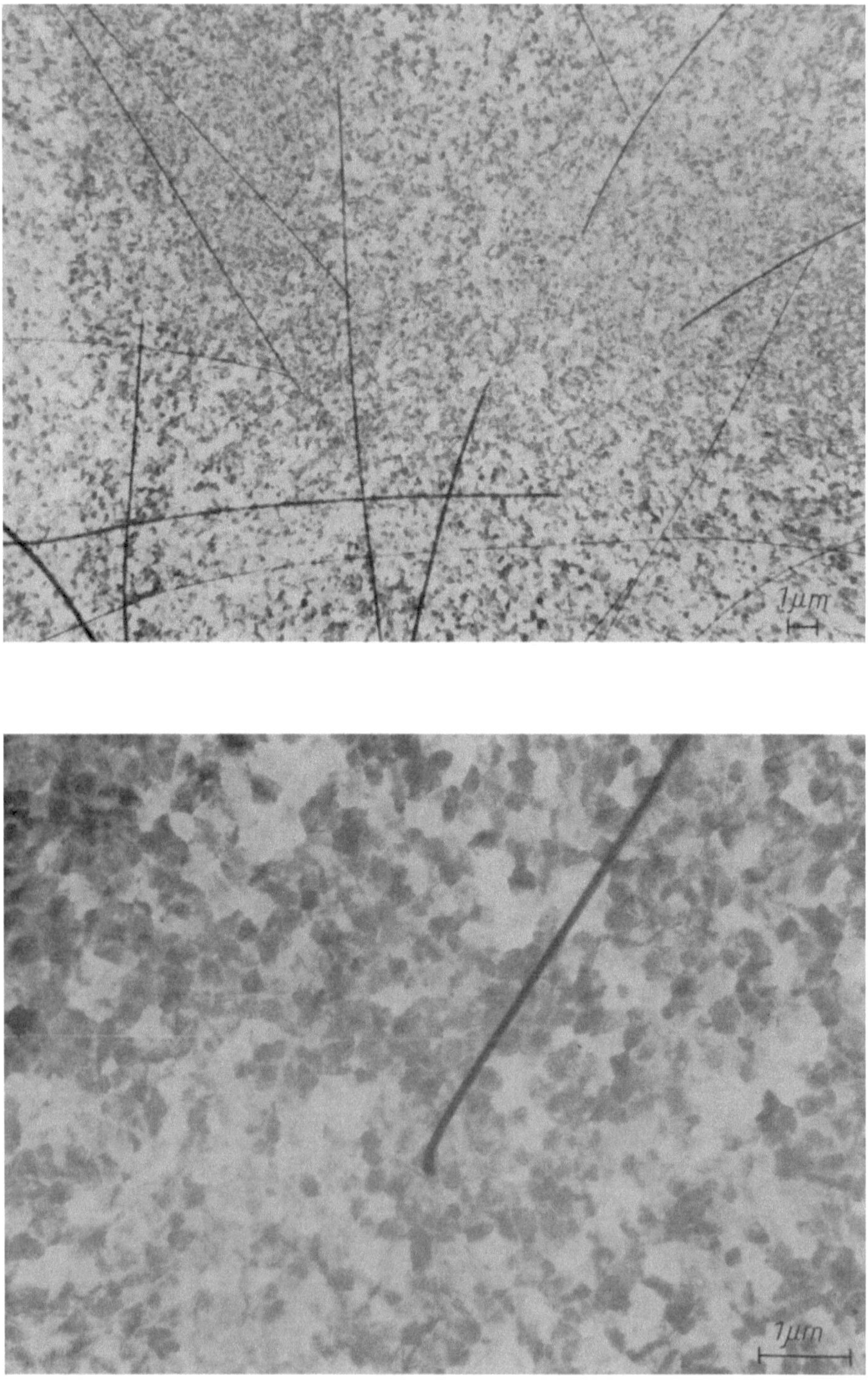

a

b

Fig. 4. a, b) Micrograph of tetracene film 63 nm thick deposited at a rate of 0.45 nm s^{-1} on collodion

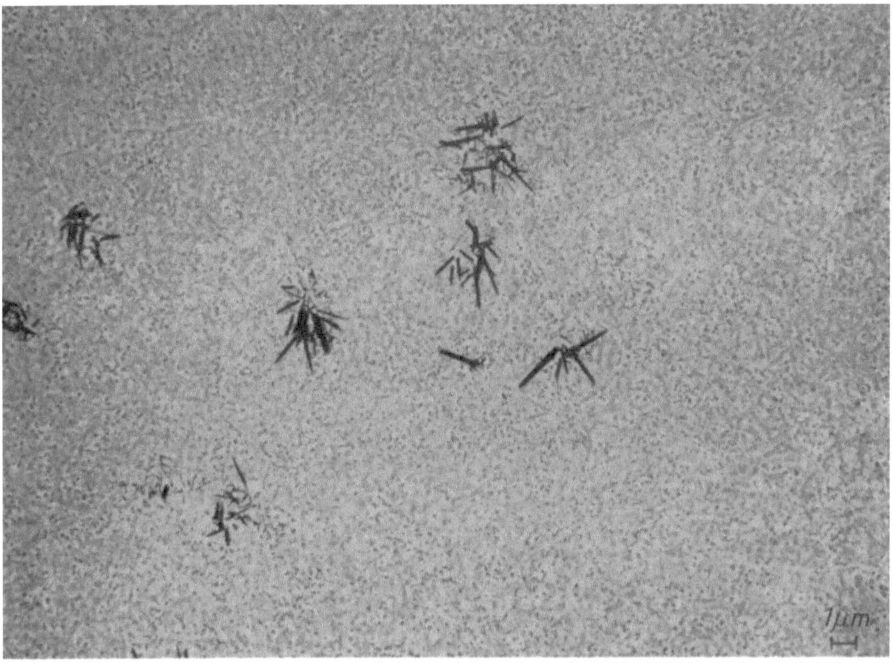

a

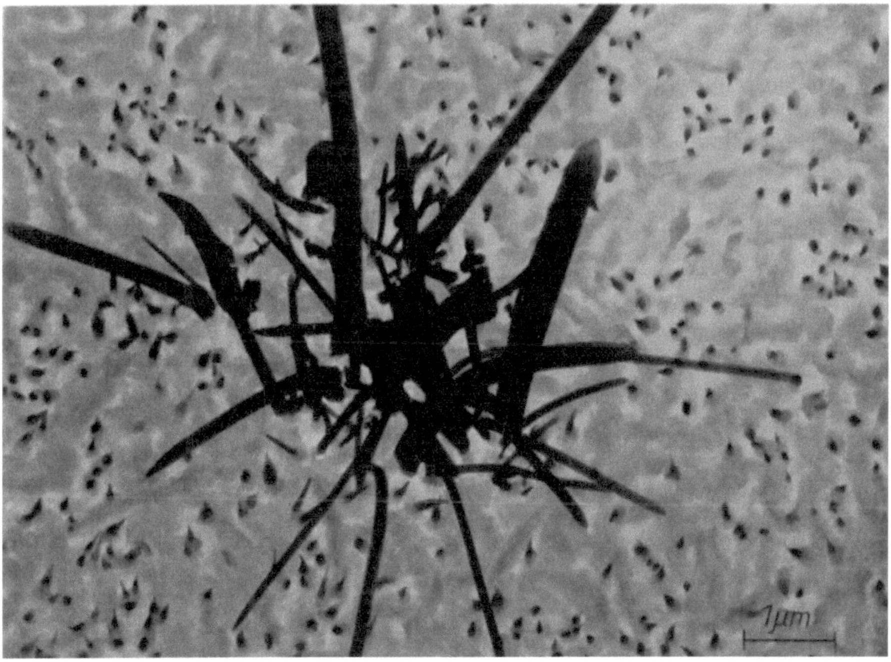

b

Fig. 5. a) Micrograph of tetracene film 34 nm thick deposited on collodion. Clusters of whiskers. b) Detail of cluster of whiskers

through large angles due to their acquisition of an electrical charge which made them subject to electrostatic forces. The whisker of Fig. 4b could thus be rotated about its base without permanent deformation.

On very thin films (34 nm) of tetracene on collodion clusters of whiskers, as in Fig. 5a, were observed. The micrograph of the details of the cluster, Fig. 5b, shows whiskers at various stages of growth, and the germinal pyramids can be recognized. The deflection of the whiskers as well as the diffuseness of the background are caused by the electron beam.

4. Discussion

In a recent paper it was suggested that the permeation of oxygen through thin films of tetracene was ascribable to grain boundaries rather than gross imperfections such as pinholes [14]. The results derived from the electron microscopy study of the structure support this interpretation: the thin films were found to be polycrystalline and free of gross imperfections for the range of deposition conditions investigated. Pinholes can arise from the contamination of the substrate by dust particles. In the preparation procedure used this was minimized by enclosing the deposition apparatus in a filtered laminar flow.

The structure of thin films, as determined by the condensation processes, has been reviewed by Chopra [15]. Briefly, the impinging vapor molecule interacts with the substrate surface, diffuses and, if it can nucleate before re-evaporation, contributes to the film growth. With continued exposure to the vapor the nuclei grow until they coalesce, giving rise to the polycrystalline texture of the film. Two important factors which influence the nucleation and growth processes are the vapor flux density and the surface mobility of the molecule. Aromatic hydrocarbons, such as tetracene, form molecular crystals in which the intermolecular bonding is by weak van der Waals forces. The expectation that tetracene molecules, therefore, have a large surface mobility is borne out by the observations: the films are polycrystalline (rather than amorphous) for room temperature substrates, the crystallites are separated by relatively little amorphous material, and the crystal size increases significantly with film thickness. Only a small decrease in crystal size with increasing deposition rate was observed suggesting that surface diffusion predominated and few nucleation centres were formed. From these observations it would appear that amorphous films of tetracene can only be prepared by reducing the substrate temperature (reducing the surface mobility) and greatly increasing the deposition rate (thus increasing the number of nucleating centers).

The nucleation of vapor molecules involves the interaction of the molecules with the surface environment, and is therefore very sensitive to the presence of surface imperfections and also to the degree of supersaturation [16]. The degree of supersaturation required for lattice plane nucleation by the two-dimensional island mechanism is much larger than that observed experimentally [17]. Frank [18] proposed that screw dislocations emerging at the surface provide the preferred sites and the continuous growth spiral for nucleation. This growth mechanism characterizes the surface with growth cones or pyramids and whiskers. Whiskers, which grow by adsorption from the vapor in a nutrient area followed by diffusion along the whisker sides to the tip and accretion of molecules into the spiral dislocation at the tip, have few imperfections and physical properties quite different from the crystallites in the film. The growth of whiskers has been reviewed by Nabarro and Jackson [19]. The micrograph obtained for a very thin film of tetracene (Fig. 5b) shows whiskers at all stages of growth. The disc-like projections observed in thick films (Fig. 3b) may well be artefacts of growth spirals.

It has been shown that a correlation between charge carrier trap density and the density of dislocations [20] exists, thus affecting the injection mechanisms. For thin films of tetracene, which on a non-orienting substrate are dominated by anisotropic growth effects, reproducibility and interpretation of characteristics due to electronic processes at the contact-film surface interface would be difficult unless deposition parameters are regulated. The propagation of surface features, such as whiskers, to successive layers in multilayer devices may contribute to device failure.

It has not been possible to establish the conditions reponsible for the initiation of whisker growth at particular sites, to explain why in certain instances whiskers grow in clusters, or to determine what causes the filament growth to terminate. By depositing the tetracene films in the electron microscope, it would be possible to continuously monitor the growth. It is anticipated, however, that this would be difficult to accomplish since a beam intensity adequate for observing the deposition would cause the film to sublimate.

5. Conclusions

The experiments have demonstrated that the structure of thin films of tetracene is determined primarily by those deposition parameters that affect the degree of supersaturation. For the range of deposition conditions investigated (representative of the conditions in device preparation) it was found that the films were polycrystalline, with a variety of crystal habits. Since the physical properties of these differ and are anisotropic, it is necessary that the deposition parameters be regulated if a reproducible surface structure is required.

Acknowledgements

The authors express their thanks to Prof. L. E. Lyons for stimulating discussions and financial support through an A.R.G.C. Fellowship to W.R. It is a pleasure to acknowledge programming assistance from J. Bonham.

References

[1] F. Gutmann and L. E. Lyons, Organic Semiconductors, Wiley, New York 1967.
[2] M. Pope and H. Kallmann, Disc. Faraday Soc. **51**, 7 (1971).
[3] E. A. Silinsh, phys. stat. sol. (a) **3**, 817 (1970).
[4] O. H. LeBlanc, Phyics and Chemistry of the Organic Solid State, Vol. 3, Ed. D. Fox, M. M. Labes, and A. Weissberger, Interscience Publishers, New York 1967 (p. 133).
[5] Y. Harada and H. Inokuchi, Bull. Chem. Soc. Japan **39**, 1443 (1966).
[6] M. Kotani and H. Akamatu, Disc. Faraday Soc. **51**, 94 (1971).
[7] E. L. Frankevich, Disc. Faraday Soc. **51**, 37 (1971).
[8] A. Golubovic, Proc. O.A.R. Res. Appl. Conf., Washington 1967 (p. 211).
[9] L. E. Lyons and O. M. G. Newman, Austral. J. Chem. **24**, 13 (1971).
[10] A. Szymanski, D. C. Larson, and M. M. Labes, Appl. Phys. Letters **14**, 88 (1969).
[11] J. Kevorkian, M. M. Labes, D. C. Larson, and D. C. Wu, Disc. Faraday Soc. **51**, 139 (1971).

[12] J. M. Robertson, V. C. Sinclair, and J. Trotter, Acta cryst. **14**, 697 (1961).
[13] R. B. Campbell, J. M. Robertson, and J. Trotter. Acta cryst. **15**, 289 (1962).
[14] W. Riemer, Thin Solid Films accepted for publication.
[15] K. L. Chopra, Thin Film Phenomena, McGraw-Hill Publ. Co., New York 1969.
[16] D. R. Frankl and J. A. Venables, Adv. Phys. **19**, 409 (1970).
[17] W. D. Lawson and S. Nielsen, Preparation of Single Crystals, Butterworths, London 1958.
[18] F. C. Frank, Disc. Faraday Soc. **5**, 48 (1949).
[19] F. R. N. Nabarro and P. J. Jackson, in: Growth and Perfection of Crystals, Ed. R. H. Doremus, B. W. Roberts, and T. Turnbull, Wiley, New York 1958 (p. 13).
[20] J. M. Thomas and J. O. Williams, in: Organic Solid State Chemistry, Ed. G. Adler, Gordon & Breach, New York 1969 (p. 59).

(Received September 13, 1972)

phys. stat. sol. (a) **14**, 483 (1972)
Subject classification: 18.2; 1; 13; 21.1; 21.4

Laboratoire de Magnétisme, C.N.R.S. (a) and
Laboratoire de Diffraction Neutronique, C.E.N.-G., Grenoble (b)

Magnetic Structure of the TmNi Compound Crystal Field Effect

By
D. GIGNOUX (a), J. ROSSAT-MIGNOD (b), and F. TCHEOU (b)

The TmNi compound crystallizes in the FeB-type structure (space group Pnma) where the Tm^{3+} ions occupy a very low symmetry site (m). Below its Curie temperature (7 K), TmNi shows a ferromagnetic behaviour. From neutron diffraction studies at 1.3 K a non-collinear magnetic structure of this compound is deduced. The exchange energy is small with respect to the total crystal field splitting of the ground multiplet J. With a point charge model the crystal field parameters are calculated and the direction and magnitude of the magnetic moments of the Tm^{3+} ions in each site are determined. The good agreement between experimental results and calculations confirms the predominance of the crystal field in determining the magnetic stucture of this compound.

Le composé TmNi cristallise dans la structure Pnma, isotype de FeB, dans laquelle les ions Tm^{3+} sont dans un site de très basse symétrie (m). Au-dessous de sa température de Curie (7 K) TmNi présente un comportement ferromagnétique. A partir d'un diagramme de diffraction neutronique à 1,3 K nous avons déterminé la structure magnétique non colinéaire du composé. L'énergie d'échange est beaucoup plus faible que la décomposition totale du multiplet fondamental J par le champ cristallin. A l'aide d'un modèle de charges ponctuelles nous calculons les paramètres de champ cristallin et nous déterminons la direction et la valeur des moments magnétiques des ions Tm^{3+} sur chaque site. Le bon accord entre les résultats observés et calculés confirme le rôle prépondérant du champ cristallin dans la stabilité de la structure magnétique.

1. Introduction

In the equiatomic rare earth–nickel compound RNi the exchange interactions are produced through conduction electrons. Furthermore the exchange interaction is of the same order of magnitude as that of the crystal field. In some compounds, the total splitting of the ground state multiplet J by crystal field is preponderant. This happens for the TmNi compound. Using a point charge model, we deduce the direction and magnitude of the Tm^{3+} ion magnetic moment and finally we compare these results to those obtained by neutron diffraction.

In 1964, Abrahams et al. [1] and also Walline and Wallace [2] studied the crystallographic structures and magnetic properties of the RNi compounds. The first authors found all RNi alloys isomorphous to the CrB orthorhombic compound, whereas Walline and Wallace specified that compounds from dysprosium to thulium crystallize in the FeB-type structure. More recently, Paccard [3] and Lemaire and Paccard [4] have re-examined the RNi compounds and made a precise crystal structure determination of TbNi.

The FeB-type crystallographic structure of TmNi belongs to the Pnma (D_{2h}^{16}) space group and can be described by trigonal prisms formed of rare earth atoms at the corners and one nickel atom in the centre. Thulium and nickel atoms are in the same (4c) site with m(Cs) symmetry whose atomic positions are: (1) in $(x, \frac{1}{4}, z)$, (2) in $(-x, \frac{3}{4}, -z)$, (3) in $(\frac{1}{2} - x, \frac{3}{4}, \frac{1}{2} + z)$, and (4) in $(\frac{1}{2} + x, \frac{1}{4}, \frac{1}{2} - z)$.

2. Experimental

The rare-earth elements used were 99.9% pure, the nickel was 99.99% pure. The intermetallic compound was prepared by induction melting of the constituents at 1100 °C in an alumina crucible, under helium atmosphere of 3 bar to reduce the evaporation of Tm. No attack reaction between the crucible and the constituents was observed. The purity of the sample was checked by X-ray analysis. Magnetic measurements were made at the "Laboratoire de Magnétisme du Centre National de la Recherche Scientifique, Grenoble" in fields up to 24 kOe produced by an electromagnet. The magnetic structure was studied at the "Laboratoire de Diffraction Neutronique du Centre d'Etudes Nucléaires de Grenoble". Wavelength of the neutrons was 1.14 Å.

3. Magnetic Measurements

Below its Curie temperature, TmNi exhibits a ferromagnetic behaviour [1, 2]. On a polycrystalline sample, we studied the magnetization versus applied field variation at 1.8 K and the magnetization versus temperature variation in a 100 Oe external field (Fig. 1). At 1.8 K in an applied field of 24 kOe, the magnetization does not attain saturation; it is 4.1 μ_B per thulium atom. The ferromagnetic Curie temperature is 7 K. These results are in rather good agreement with those obtained by the previous authors [1, 2]. In the paramagnetic region, the observed deviation from a Curie-Weiss law confirms the importance of crystal field effects on the magnetic behaviour of TmNi.

4. Neutron Diffraction Results

To resolve the magnetic structure of this compound, neutron diffraction patterns have been made at 1.3 and 77 K (Fig. 2). The nuclear Bragg peaks observed at 77 K can be indexed in the orthorhombic cell with the parameters:

$$a = 6.96 \text{ Å}, \quad b = 4.11 \text{ Å}, \quad c = 5.40 \text{ Å}.$$

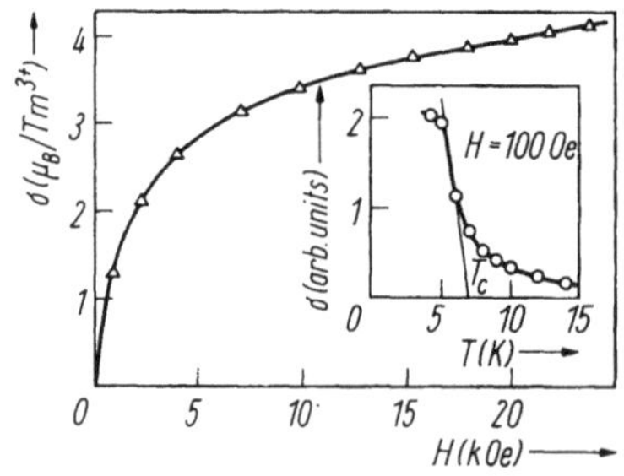

Fig. 1. Variation of magnetization as a function of the applied field at 1.8 K. The insert shows the variation of magnetization as function of temperature under 100 Oe

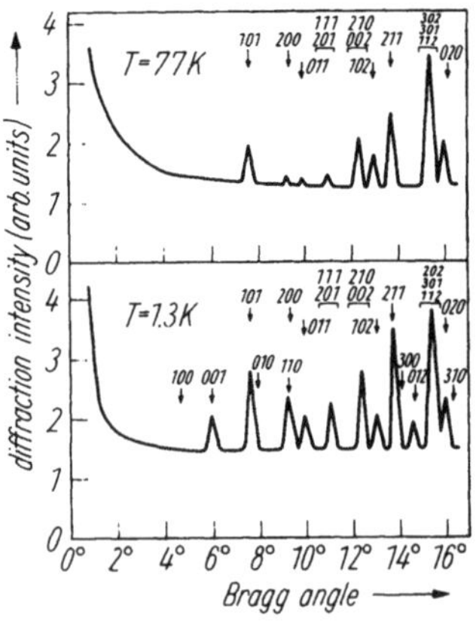

Fig. 2. Neutron diffraction patterns at 77 and 1.3 K; $\lambda = 1.14$ Å

Table 1
Observed and calculated neutron diffraction intensities at 77 K ($R = 4.2\%$)

h k l	θ	I_{cal}		I_{obs}
1 0 1	7.68	17.3		16.5
2 0 0	9.43	10.3		9.1
0 1 1	10.04	7.4		8.7
1 1 1	11.11	3.0	3.1	3.0
2 0 1	11.25	0.1		
0 0 2	12.20	0.5	55.8	53.7
2 1 0	12.39	55.3		
1 0 2	13.10	35.3		42.3
2 1 1	13.84	104.4		102.4
1 1 2	15.40	193.9	242.2	245.6
2 0 2	15.51	48.3		
3 0 1	15.52	0.0		
0 2 0	16.10	78.9		73.4

With these data we have refined the atomic position parameters of thulium and nickel; these are for thulium, $x = 0.177$ and $z = 0.134$ and for nickel, $x = 0.040$ and $z = 0.630$. The calculated nuclear intensities are compared with the observed ones in Table 1, where the discrepancy factor,

$$R = \frac{\sum |I_{obs} - I_{cal}|}{\sum I_{obs}},$$

is 4.2%. At 1.3 K the neutron diffraction pattern (Fig. 2) exhibits an increase in intensity of most of the nuclear reflections and moreover superstructure peaks which can be indexed with the same orthorhombic cell. The extinction conditions limiting the possible reflections of the Pnma space group are no longer valid. The superstructure lines are characteristic of an antiferromagnetic arrangement $\boldsymbol{C}_a$ of the thulium magnetic component parallel to $\boldsymbol{a}$ (Fig. 3). The increase of nuclear peaks is due to a ferromagnetic arrangement $\boldsymbol{F}_c$ of the thulium magnetic component parallel to $\boldsymbol{c}$. This results in a non-collinear structure with the moments lying in the *ac* plane. We obtain the best agreement with no magnetic moment on nickel and a moment of $(5.9 \pm 0.5)\,\mu_B$ on thulium which makes an angle of $(38 \pm 2)°$ with *c*-axis. The discrepancy factor is 7.2% (Table 2). On Fig. 3 we represent the projec-

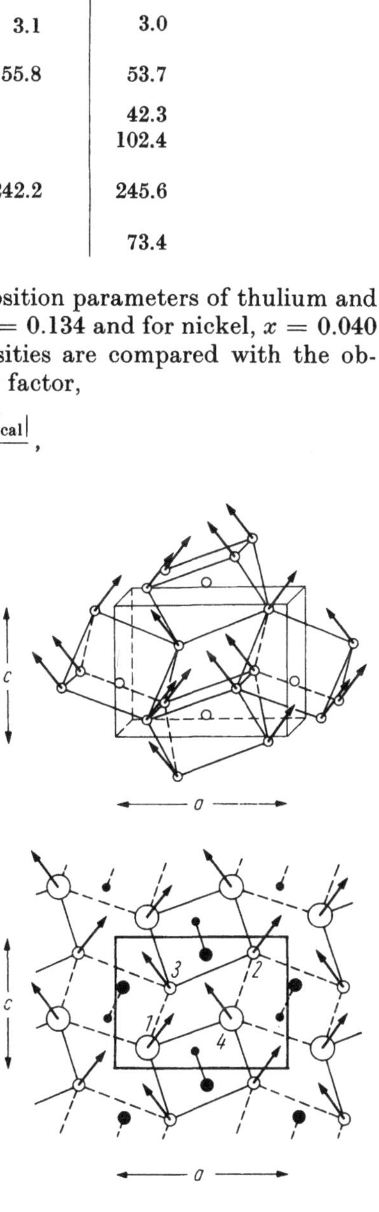

Fig. 3. Magnetic structure of TmNi defined with the basic vectors
$\boldsymbol{C}_a = \boldsymbol{M}_{1a} + \boldsymbol{M}_{2a} - \boldsymbol{M}_{3a} - \boldsymbol{M}_{4a}$,
$\boldsymbol{F}_c = \boldsymbol{M}_{1c} + \boldsymbol{M}_{2c} + \boldsymbol{M}_{3c} + \boldsymbol{M}_{4c}$

Table 2
Observed and calculated neutron diffraction intensities at 1.3 K ($R = 7.2\%$)

$h\,k\,l$	θ	$I_{M_{\text{cal}}}$		$I_{M_{\text{cal}}} + I_{N_{\text{cal}}}$	I_{obs}
1 0 0	4.70	0.0		0.0	n.o.
0 0 1	6.06	12.4		12.4	11.8
1 0 1	7.68	30.8		48.1	40.3
0 1 0	7.97	0.0		0.0	n.o.
1 1 0	9.27	28.6	} 43.8	54.1	48.0
2 0 0	9.43	15.2			
0 1 1	10.04	27.2		34.6	26.8
1 1 1	11.11	36.5	} 44.2	47.3	46.9
2 0 1	11.25	7.7			
0 0 2	12.20	0.0	} 44.9	100.7	96.5
2 1 0	12.39	44.9			
1 0 2	13.10	6.2		41.5	42.6
2 1 1	13.84	47.2		151.6	155.5
3 0 0	14.23	0.0		0.0	n.o.
0 1 2	14.64	35.9		35.9	45.7
1 1 2	15.40	10.1			
2 0 2	15.51	15.1	} 50.6	292.8	311.1
3 0 1	15.52	25.4			
0 2 0	16.10	28.6	} 28.8	107.7	100.7
3 1 0	16.39	0.2			

tion of the magnetic structure on the *ac* plane. This structure is isomorphous to that of ErNi [5] and belongs to the magnetic space group Pn′m′a.

5. Crystal Field Calculation and Interpretation

In the orthorhombic cell the four Tm^{3+} ions are in the same (4c) site with very low symmetry C_s(m). For such a symmetry the crystal field splits totally the multiplet ground state $^3H_6$ ($J = 6$). For TmNi $T_c = 7$ K, thus the exchange energy between the thulium ions is small with respect to the total splitting of the free ion in the crystal field (a few hundred K). In these conditions only the low-lying states are relevant for the magnetic properties.

If the singlet ground state $|a\rangle$ is well separated from the first excited level (with respect to the exchange energy), there will be no possible magnetic order down to 0 K. Thus, for the compound to be ordered it is necessary that the two low-lying states $|a\rangle$ and $|b\rangle$ are separated by an energy Δ of only a few K. We can obtain a cooperative phenomenon only if the exchange energy in the case when the two singlets have the same energy ($\Delta = 0$), is greater than $\Delta/2$ [6 to 8].

The magnetic moment is proportional to matrix elements of the form $\langle a|\,\boldsymbol{J}\,|b\rangle$. The condition for a non-zero matrix element implies that the representation $\Gamma^{(J)}$, according to which the components of J transform exists in the decomposition of $\Gamma^{(a)} \times \Gamma^{(b)}$ representation. If $\Gamma^{(a)}$ and $\Gamma^{(b)}$ are one-dimensional, $\Gamma^{(J)}$ is also one-dimensional. Thus the magnetic moment is fixed in a direction which may be determined by the product $\Gamma^{(a)} \times \Gamma^{(b)}$. We obtain an Ising-like model.

Table 3

Irreducible representations of $C_s(m)$ and $C_{2v}(mm2)$ point groups. Components of $\boldsymbol{J}$ associated with these representations

$C_s(m)$		$C_{2v}(mm2)$	
$^{(1)}\Gamma_1$	J_z	$^{(1)}\Gamma_1$	
$^{(1)}\Gamma_2$	J_x or J_y	$^{(1)}\Gamma_2$	J_y
		$^{(1)}\Gamma_3$	J_z
		$^{(1)}\Gamma_4$	J_x

In the $C_s(m)$ group there are only two irreducible first-order representations, Γ_1 and Γ_2 (Table 3). If the two singlets belong to the same representation the moments will be in the direction (0z) perpendicular to the mirror plane. If they belong to two different representations they will be fixed in a direction of the mirror plane which cannot be determined because J_x and J_y transform in the same representation Γ_2 ($\Gamma_1 \times \Gamma_2 = \Gamma_2$). On the other hand in the $C_{2v}(mm2)$ group (Table 3), where J_x, J_y, and J_z transform in different representations, the magnetic moments are in the direction of a symmetry axis.

The crystal field Hamiltonian in the C_s symmetry can be written by using the Stevens equivalent operators [9]:

$$H_c = \alpha\,(V_2^0\,O_2^0 + V_2^{-2}\,O_2^{-2} + V_2^2\,O_2^2) + H_4 + H_6\,.$$

The quantization axis 0z (b-axis) is chosen perpendicular to the mirror plane; 0x and 0y are respectively taken to be parallel to the c- and a-axes (Fig. 4a). The $C_s(m)$ group differs from the C_{2v} group by the presence of imaginary terms like $V_2^{-2}\,O_2^{-2}$, $V_4^{-4}\,O_4^{-4}$, etc. in the Hamiltonian H_c with C_s symmetry. With the help of a point charge model we have calculated the V_l^m parameters for the (1) Tm^{3+} ion. We have taken a charge $+3\,e$ for Tm^{3+} ions and we have assumed a zero charge on nickel atoms. This supposition is justified since YNi exhibits only a Pauli paramagnetism [2]. Moreover we have neglected the conduction electron contribution to the crystal field. The calculation shows that the second-order terms play a preponderant part in H_c. Under these conditions Rossat-

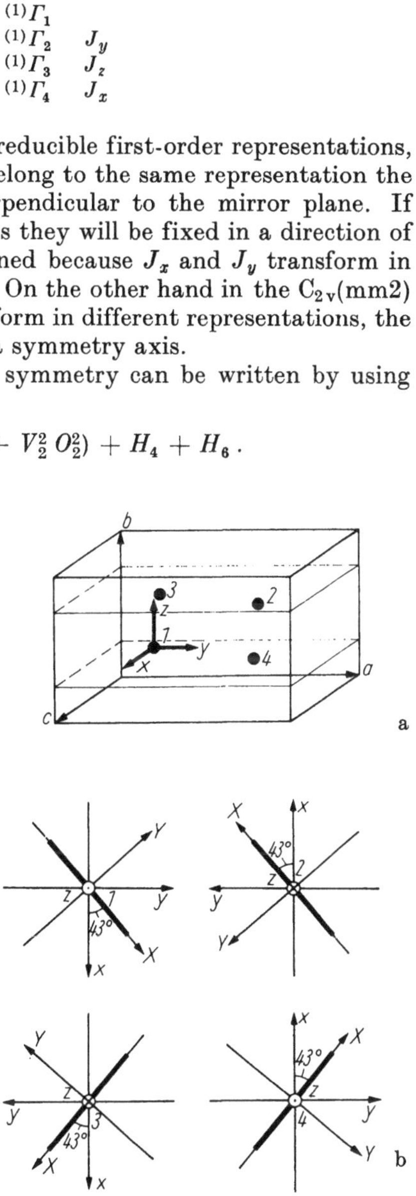

Fig. 4. a) Reference axes and atomic positions in the unit cell chosen for the calculations. b) Reference axes before and after rotation in each site deduced by symmetry operation

Mignod and Tcheou [10] have shown that by a rotation around the 0z-axis with angle φ such that tg $2\varphi = V_2^{-2}/V_2^2$, it is possible to get rid of the second-order term $V_2^{-2} O_2^{-2}$ and thus to obtain a crystal field Hamiltonian having a C_{2v}-like symmetry. This calculation gives $\varphi = +43°$. The diagonalization of H_c written in the $0XYz$-axes deduced from the $0xyz$ ones by the rotation around the 0z-axis ($\varphi = 43°$) shows that the ground state is composed of two closely spaced singlets belonging to the Γ_4 and Γ_1 representations of the C_{2v} group. The first excited state is at about 50 K and the highest state is at about 250 K. Thus the magnetic moments must be fixed in the $0X$-direction making an angle of 43° with the $0x$ direction, i.e. with the c-axis (Fig. 4b). Taking into account the above approximations these results are in good agreement with the angle of $(38 \pm 2)°$ found experimentally by neutron diffraction. The calculated eigen functions give a moment of 6.8 μ_B for Tm^{3+} assuming the same energy for the two singlets states. This value is very close to that of the free Tm^{3+} ion (7 μ_B). The discrepancy between the experimental value of the magnetic moment (5.9 μ_B) and the calculated one arises probably from the fact that the two singlets are separated by an energy gap Δ of a few K.

The direction of magnetic moments of the other ions in the unit cell is deduced from that of the ion (1) by symmetry operations: the inversion $\bar{1}|(1) \to (2), (3) \to (4)|$ and the two-fold rotation $2_x|(1) \to (3), (2) \to (4)|$. Fig. 4b shows the direction of the magnetic moment of these four ions. The result is in very good agreement with the structure determined by neutron diffraction.

6. Conclusion

Although the point charge model is over-simplified, it allows to interpret correctly the observed magnetic structure. The determination of the direction of magnetic moments depends only on the ratio V_2^{-2}/V_2^2 which eliminates the screening effect and the value of $\langle r^2 \rangle$. In the TmNi compound the ferromagnetic structure is essentially determined by the crystal field. The ground state is well described by a two singlet states model with $|M_J = \pm 6\rangle$ (the quantization axis being parallel to the easy magnetization direction) separated by an energy gap Δ of a few K.

Acknowledgement

We wish to thank R. Lemaire for many fruitful discussions.

References

[1] S. C. ABRAHAMS, J. L. BERNSTEIN, R. C. SHERWOOD, J. H. WERNICK, and H. J. WILLIAMS, J. Phys. Chem. Solids **25**, 1069 (1964).
[2] R. E. WALLINE and W. E. WALLACE, J. chem. Phys. **41**, 1587 (1964).
[3] D. PACCARD, Thèse, Université de Grenoble, 1969; No. 1478, A.O. — C.N.R.S.
[4] R. LEMAIRE and D. PACCARD, J. less-common Metals **21**, 403 (1970).
[5] R. LEMAIRE and D. PACCARD, Les éléments des terres rares, Vol. II, 1970 (p. 231).
[6] J. S. GRIFFITH, Phys. Rev. **132**, 316 (1963).
[7] B. BLEANEY, Proc. Roy. Soc. **276A**, 19 (1963).
[8] Y. L. WONG and B. R. COOPER, Phys. Rev. **172**, 539 (1968).
[9] M. T. HUTCHINGS, Solid State Phys. **16**, 227 (1964).
[10] J. ROSSAT-MIGNOD and F. TCHEOU, J. Physique **33**, 423 (1972).

(Received October 2, 1972)

phys. stat. sol. (a) **14**, 489 (1972)

Subject classification: 10.2 and 12.2; 22.1.1

Research Institute for Iron, Steel, and Other Metals, Tôhoku University, Sendai

Internal Friction in Deformed Germanium Crystals at Low Temperatures

By

K. OHORI and K. SUMINO

The internal friction in deformed germanium single crystals is investigated in the temperature range from 4.2 to 150 K at resonance frequencies about 70 kHz. Two relaxation peaks, denoted by P_3 and P_4, with activation energies of (0.08 ± 0.01) and (0.13 ± 0.01) eV, respectively, are observed to develop upon deformation. Peaks P_3 and P_4 are interpreted to be associated with the stress-induced motion of geometrical kinks on screw dislocations and 60° dislocations, respectively, passing over the Peierls potential barrier of the second kind. The results are discussed on the basis of the kink–chain model of abrupt kinks. It is shown that the kink–chain model leads to expressions for the relaxation strength and the relaxation time which are similar to those derived on the extensible string model by Southgate and Attard.

Die innere Reibung verformter Germaniumeinkristalle wird im Temperaturbereich 4,2 bis 150 K bei Frequenzen um 70 kHz untersucht. Zwei Relaxationsmaxima der inneren Reibung, P_3 und P_4, mit Aktivierungsenergien von (0,08 ± 0,01) bzw. (0,13 ± 0,01) eV werden nach der Verformung beobachtet. Die Maxima P_3 und P_4 werden der spannungsbedingten Wanderung der geometrischen Kinken längs Schraubenversetzungen bzw. 60°-Versetzungen, die die Peierlspotentialbarriere von zweiter Art überschreiten, zugeschrieben. Die Ergebnisse werden mit Hilfe des Kinkenkettenmodells der abrupten Kinken diskutiert. Es wird gezeigt, daß das Kinkenkettenmodell zu ähnlichen Ausdrücken für die Relaxationsstärke und die Relaxationszeit führt wie die, die vom Saitenmodell von Southgate und Attard abgeleitet wurden.

1. Introduction

Dislocations in covalent crystals are thought to have an extremely high Peierls barrier. The rate-controlling process in the dislocation motion over a long distance in such crystals is now believed to be the double-kink nucleation. Theoretical estimates by Suzuki [1] and Labusch [2] have shown the energy of a kink pair in germanium to be 1.67 and 1.75 eV, respectively. Because of such a high energy the internal friction peak associated with the double-kink nucleation (so-called Bordoni peak) is absent in germanium in the temperature range below the melting point. Possible causes for the relaxation process in dislocation damping in covalent crystals are thought to be the thermally activated motion of geometrical kinks past over some dragging points and that over the Peierls potential barrier of the second kind (kink potential barrier).

In a previous paper [3] the present authors have measured the internal friction of deformed germanium single crystals in the temperature range from 77 to 503 K at about 1200 Hz and found two relaxation peaks, termed as P_1 and P_2, with activation energies of 0.62 and 0.85 eV, respectively. Peak P_1 was interpreted to be associated with the kink–jog interaction while peak P_2 was attributed to the single-kink nucleation on the dislocations terminating at the crystal surface.

A theoretical estimate by Labusch [2] has shown the kink potential in germanium to be 0.04 eV. Thus, we can expect the internal friction peak associated with the stress-induced diffusive motion of geometrical kinks in the Peierls potential field of the second kind to be observed in the low-temperature range. An internal friction peak which is thought to be associated with such a motion of geometrical kinks has recently been reported for silicon at about 105 K at 84 Hz by Khiznichenko et al. [4]. As to germanium Kayano et al. [5] reported two internal friction peaks at about 107 and 154 K at 25 MHz the nature of which was not investigated in detail. The motion of geometrical kinks can also be detected by means of the modulus defect measurements. However, no measurement of the elastic modulus in covalent crystals has yet been made for such a purpose in the low-temperature region.

The purpose of the present investigation is to study the behaviour of the dislocation damping and the resonance frequency in deformed germanium single crystals in the temperature range down to liquid helium temperature and to get information on the stress-induced diffusive motion of geometrical kinks.

2. Experimental Procedure

Single crystals of germanium were grown by the Czochralski method starting with zone-refined ingots of purity 99.9999999%. Specimens were cut from single crystals by means of a diamond wheel so that they had a square cross-section of approximately 4×4 mm^2 and were 43 mm in length. The long axis of the specimens was parallel to either the [123] or the [112] direction. The side surface of the specimens was parallel to $(\bar{1}\bar{1}1)$. Details of the surface preparation were described in a previous paper.

Specimens were subjected to tensile deformation at 600 °C and a strain rate of 1.1×10^{-4} s^{-1} in a high-purity argon gas atmosphere with the use of an Instron type machine. Details concerning the procedure for high-temperature deformation have been given elsewhere [6]. The density of grown-in dislocations was 10^3 to 10^4 cm^{-2}. Specimens were deformed to various amounts of strain in order to introduce different densities and arrangements of dislocations. After the intended strain was reached, the specimen was cooled without removing the load. The deformed specimens were cut and polished to the shape of rods having a square cross-section 3.5×3.5 mm^2 and a length of approximately 37 mm which is a half of the wavelength of the longitudinal vibration in germanium at a frequency of about 70 kHz.

The internal friction and the resonance frequency were measured in the temperature range between 4.2 and 150 K at frequencies near 70 kHz by using a composite piezo-electric oscillator system originally developed by Marx [7]. During the measurements specimens were kept in a helium gas atmosphere of 10^{-2} Torr which ensured a good thermal contact between the specimen and its surroundings as well as a low background. The temperature was measured by means of a thermocouple located close to the specimen. The heating rate of the specimens was approximately 0.3 to 0.5 deg/min.

In order to determine the activation energy of the relaxation process from the peak shift measurements, the pulse echo method was also employed. Longitudinal ultrasonic attenuation was measured at frequencies between 15 and 35 MHz with the use of a Sperry attenuation comparator.

3. Experimental Results

3.1 General features of the internal friction peaks

Fig. 1 shows the temperature dependence of the internal friction and the resonance frequency in a deformed [112] specimen at about 70 kHz. Two broad peaks, denoted by P_3 and P_4 from the low-temperature side, are observed at 64 and 97 K, respectively. Both peaks are thought to be of the relaxation type since they are accompanied by the modulus defect effect and show no dependence on the strain amplitude. The widths of both peaks are larger than those for the single relaxation processes having the same activation energies. The width of peak P_4 is larger than that of peak P_3. It is worth noting that the magnitudes of both internal friction and resonance frequency in the deformed specimen are constant against temperature in the neighbourhood of liquid helium temperature.

Fig. 2 shows the results obtained by measuring the ultrasonic attenuation at 15 MHz. Three peaks are observed at about 55, 105, and 150 K. The peaks at about 105 and 150 K are identified as to be P_3 and P_4 in Fig. 1 from the shift of the peak temperature due to the frequency change. On the other hand, the position of the peak at 55 K is observed to be independent of frequency. It is also found that the height of this peak increases with increasing frequency. Thus, it may be concluded that this peak originates from the interaction between conduction electrons and sound waves as proposed by Weinreich et al. [8].

The activation energies of relaxation processes responsible for peaks P_3 and P_4 were determined from the relationship peak temperature versus resonance frequency. The logarithm of the resonance frequency is plotted against the inverse of the peak temperature in Fig. 3. The activation energy E and the frequency factor ν_0 determined for peaks P_3 and P_4 are as follows:

$$E_3 = 0.08 \pm 0.01 \text{ eV} ; \qquad \nu_{0,3} = 5.4 \times 10^{10} \text{ s}^{-1} ,$$
$$E_4 = 0.13 \pm 0.01 \text{ eV}; \qquad \nu_{0,4} = 2.5 \times 10^{11} \text{ s}^{-1} .$$

3.2 Effect of plastic deformation on the internal friction peaks

The internal friction in [123] specimens at about 70 kHz after various amounts of deformation is shown in Fig. 4a together with that in an undeformed specimen.

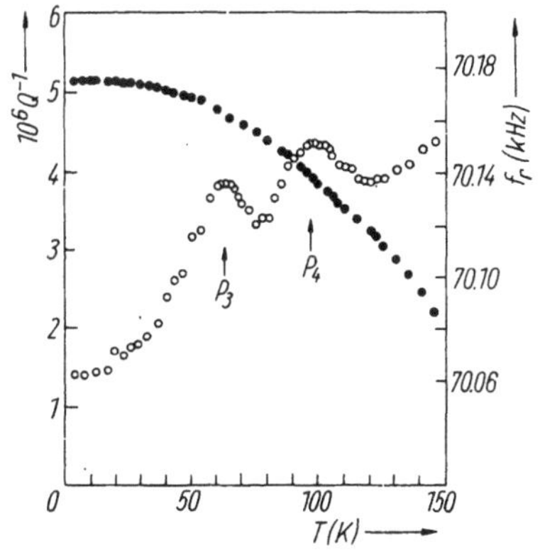

Fig. 1. Temperature dependence of the internal friction Q^{-1} and the resonance frequency f_r in a [112] specimen deformed to stage I

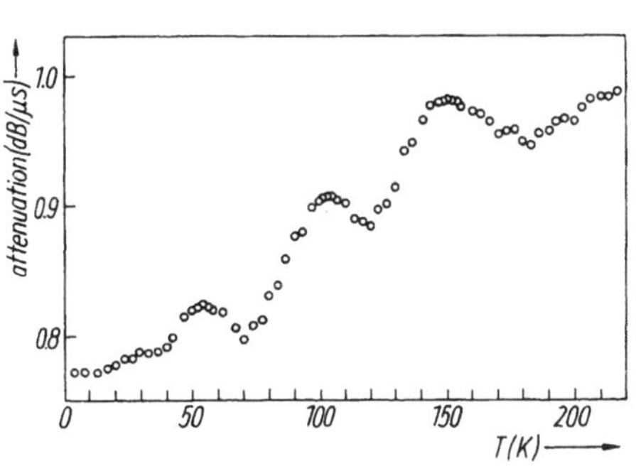

Fig. 2. Temperature dependence of the ultrasonic attenuation at 15 MHz in a [111] specimen deformed to the lower-yield point

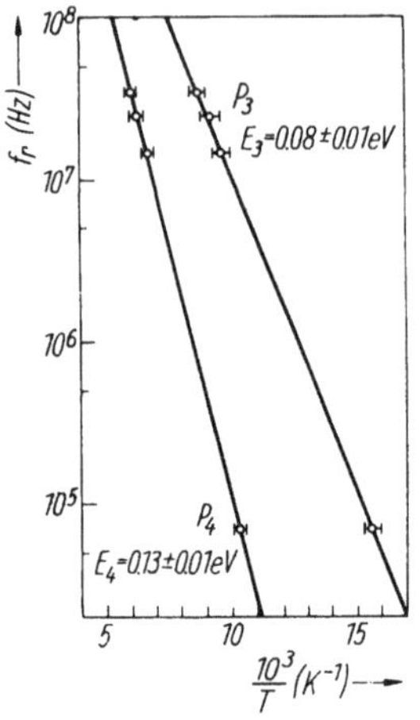

Fig. 3. Resonance frequencies f_r plotted against the inverse of the peak temperature T for peaks P_3 and P_4

The stress–strain curves of the specimens are shown in Fig. 4 b. Single slip takes place in specimens of this orientation. Neither peak P_3 nor peak P_4 are found in undeformed specimens. They are developed by plastic deformation. The height of peak P_3 grows with increasing strain while that of peak P_4 shows a maximum on deformation to stage I. In any case it may be concluded that the origins of both peaks P_3 and P_4 are associated with dislocations introduced by deformation. It would be reasonable to think these peaks to be due to the stress-induced diffusive motion of geometrical kinks along dislocation lines from their low values of the activation energy.

Fig. 5a shows the result on [112] specimens. The stress–strain curves of these specimens are shown in Fig. 5b. Double slip takes place from the early stage of deformation in specimens of this orientation. The magnitude of the work-hardening rate in stage I of these specimens is almost the same as that in stage II of [123] specimens [9]. The heights of both peaks P_3 and P_4 in these specimens are largest after deformation to stage I.

3.3 Effect of annealing on the internal friction peaks

The effect of annealing on peaks P_3 and P_4 in [123] specimens deformed to 12.4% shear strain (stage I) was investigated for various annealing temperatures above 400 °C. No appreciable change in the peak height was found upon annealing at temperatures below 500 °C for both peaks P_3 and P_4. Appreciable decreases in the peak heights were first observed upon annealing at 500 °C for 1 h. The effect of annealing at 500 and 600 °C is shown in Fig. 6. It should be noted that the decay of the peak height is more noticeable in peak P_3 than in peak P_4. The observed decreases in the peak heights may be ascribed to the decrease in the kink density as well as to a decrease in the dislocation density due to annealing.

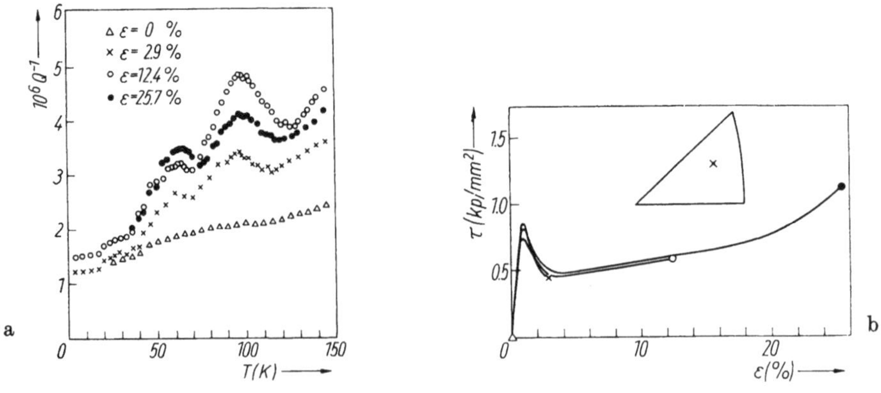

Fig. 4. a) Internal friction Q^{-1} plotted against temperature T; b) stress–strain curves in [123] specimens

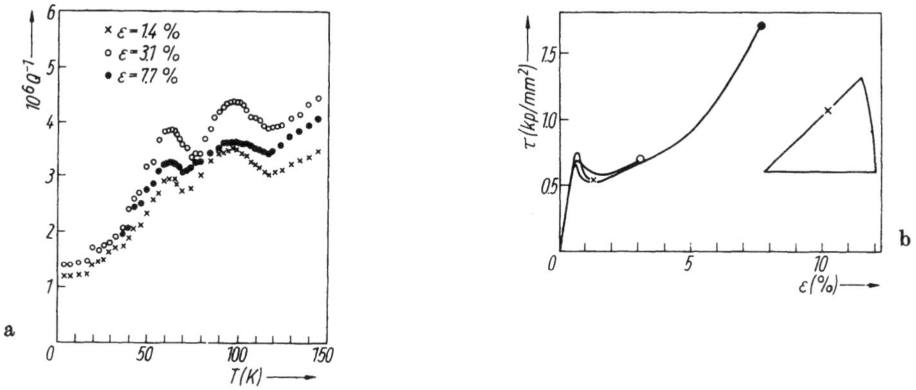

Fig. 5. a) Internal friction Q^{-1} plotted against temperature T; b) stress–strain curves in [112] specimens

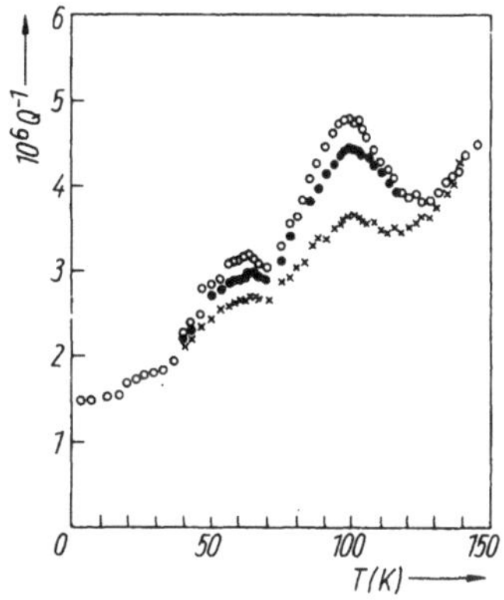

Fig. 6. Effect of annealing on the internal friction Q^{-1} in a [123] specimen deformed to 12.4% shear strain. ○ before annealing; ● after 1 h annealing at 500 °C; □ after 1 h annealing at 600 °C

It was also observed that the value of the dynamic modulus at temperatures around liquid helium temperature did not change upon annealing. This suggests that the motion of geometrical kinks is absent in this temperature region. This is contrary to the results on f.c.c. metals, b.c.c. metals, and ionic crystals in which the value of the elastic modulus in cold-worked materials in this temperature region has been observed to increase upon annealing (see for example [10]).

4. Discussion

Because of their low values of the activation energy, it seems reasonable to think that peaks P_3 and P_4 observed in the present investigation are due to the stress-induced diffusive motion of geometrical kinks along dislocation lines lying in the Peierls valleys. We now consider the cause for the existence of two relaxation peaks associated with the motion of kinks.

There are two kinds of dislocations lying in the Peierls valleys of a germanium crystal, namely, 60° dislocations and screw dislocations. If kinks on these two kinds of dislocations are of different mobility, then it is natural that two relaxation peaks are to be observed in the internal friction measurements. Fig. 7a and b show atomic configurations at a kink on a 60° dislocation; the distance between dangling bonds located at the ends of these two kinks is different in both cases. It should be noted that both types of kinks possess no dangling bonds. On the other hand, only one type of kink is available on a screw

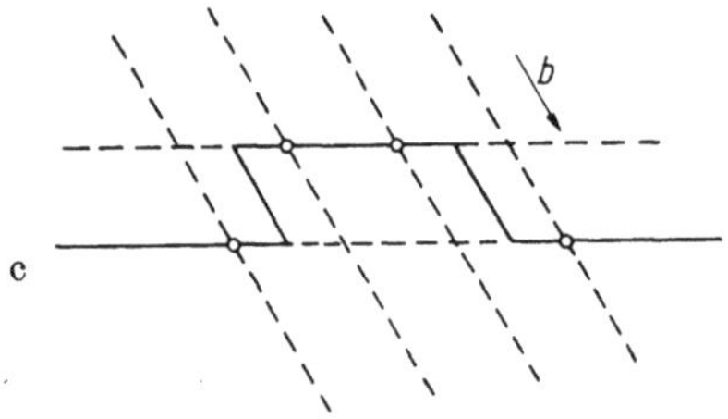

Fig. 7. a) and b) atomic configurations at two types of kinks on a 60° dislocation in the diamond lattice. The broken lines represent the centres of dislocation lines. c) Top view showing the difference in the separation of dangling bonds. The circles and broken lines represent dangling bonds and the Peierls valleys, respectively

dislocation. Such a kink is accompanied by a dangling bond as shown in Fig. 8. It may be possible that a kink having a dangling bond has a different mobility from those having no dangling bond as it is the case for dislocation lines [11].

We think the existence of two internal friction peaks to be due to the difference between the mobilities of kinks along 60° and screw dislocation lines for the following reasons. It has been reported that most of the dislocations remaining in germanium crystals with single glide orientations deformed to stage I are of edge character [6, 12 to 14]. The density of such edge dislocations increases monotonically with increasing strain from the lower yield point through stage I. The predominance of edge dislocations becomes less noticeable after stage II deformation. Thus, we may think the density of 60° dislocations to be much higher than that of screw dislocations in stage I. On the other hand, the theory based on the abrupt kink model [15, 16] predicts that the height of the relaxation peak Q^{-1}_{max} associated with kink motion depends on the density N of dislocations on which kinks occur and the free length L of dislocations as $Q^{-1}_{max} \sim N L^2$. Since the values of L on 60° dislocations and on screw dislocations would not be very different, the behaviour of the peak height against strain shown in Fig. 4 is reasonably interpreted by peak P_4 to be due to the kink motion on 60° dislocations. Furthermore, screw dislocations are expected to be annihilated more quickly than 60° dislocations upon annealing because of their

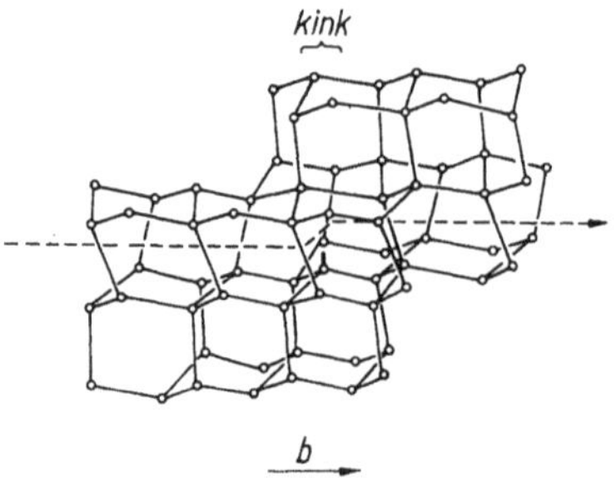

Fig. 8. The atomic configuration at a kink on a screw dislocation. The broken line represents the centre of a dislocation line

lower activation energy of motion [11]. It is seen in Fig. 6 that peak P_3 decays more rapidly than peak P_4 upon annealing. Thus, we may conclude that peak P_3 is associated with the motion of kinks on screw dislocations and peak P_4 with that on 60° dislocations.

Since the background component of the internal friction measured in the present investigation is relatively low, it is reasonable to compare the absolute magnitude of the observed relaxation strength with that derived directly from theory. Brailsford [15] first developed an abrupt kink model. Southgate and Attard [16] discussed the relaxation strength in the internal friction due to the thermally activated motion of geometrical kinks on the basis of the extensible string model of dislocations. In their theory the force F acting on a kink consists of two components, namely, the externally applied force and the force due to the line tension of the dislocation. Assuming the x-axis in the direction of the Peierls valley and the y-axis in the direction of the dislocation motion, F is expressed by

$$F = \sigma\, a\, b + E_0\, a \frac{\partial^2 y}{\partial x^2}, \tag{1}$$

where σ is the applied stress, a the period of the Peierls potential, b the magnitude of Burgers vector, and E_0 the line energy of the dislocation. The fundamental equation for the dislocation motion in their theory is

$$\frac{1}{\mu\, a^2\, p_0} \frac{\partial y}{\partial t} - E \frac{\partial^2 y}{\partial x^2} = \sigma\, b\,, \tag{2}$$

where μ is the kink mobility and p_0 the kink density along the dislocation line. E is the effective line tension of the kinked dislocation and is given by

$$E = E_0 + \frac{kT}{a^2\, p_0}\,. \tag{3}$$

The second term of the right-hand side of equation (3) is called "kink gas pressure" term. By neglecting the contribution from this second term compared with that from the first term the relaxation strength Δ_M is obtained

$$\Delta_M = \frac{8\, \Omega\, G\, b^2\, N\, L^2}{\pi^4\, E_0} \tag{4}$$

and the relaxation time τ

$$\tau = \frac{L^2}{p_0\, a^2\, \pi^2\, E_0\, \mu_0} \exp\left(\frac{U}{kT}\right), \tag{5}$$

where $\mu = \mu_0 \exp(-U/kT)$, Ω is the orientation factor, G the shear modulus, and U the activation energy for the motion of a kink.

The second term on the right-hand side of equation (1) originates from the string model of a dislocation. However, it should be noticed that such a macroscopic concept of line tension is inadequate to apply directly to the motion of abrupt kinks. The bowing out of a dislocation line between two pinning points due to the sidewise motion of abrupt kinks does not result in a change of the total length of the dislocation line though it brings about an apparent change in the curvature of the dislocation line. Correctly, the force acting on kinks which is associated with the bowing out of the dislocation line should be treated on the basis of the kink–kink interaction. This has been taken into account in the Southgate-Attard theory as the kink gas pressure. However, in their theory this term does not play any essential role in determining the relaxation strength.

The relaxation peak in the internal friction caused by the stress-induced diffusive motion of geometrical kinks is discussed on the basis of the kink–chain model in the Appendix. The expressions for Δ_M and τ derived on such a model are given by (A15) and (A16), respectively. They are slightly different from (4) and (5). Thus, the height of peak P_4 measured in the present investigation is compared with that given by (A15). The value of L is obtained by using the result of Kromer and Khiznichenko [17] in which the frequency factor ν_0 is given approximately by

$$\nu_0 = \left(\frac{2\,b}{L}\right)^2 \nu_D\,, \tag{6}$$

where ν_D is the Debye frequency. With $\nu_0 = 2.5 \times 10^{11}\ \mathrm{s}^{-1}$, $b = 4.0 \times 10^{-8}$ cm, $\nu_D = 10^{13}\ \mathrm{s}^{-1}$, L^2 turns out to be $2.6 \times 10^{-13}\ \mathrm{cm}^2$. Thus, substituting $\Omega = 0.2$, $N = 10^8\ \mathrm{cm}^{-2}$, and $L^2 = 2.6 \times 10^{-13}\ \mathrm{cm}^2$ into (A15) we obtain $Q^{-1}_{\max} = 3.0 \times 10^{-6}$. This value is in quantitative agreement with the measured heights of peak P_4 in the present experiment.

As mentioned earlier, the widths of both peaks P_3 and P_4 are larger than those of single relaxation processes. This could be attributed to the existence of spectra in the magnitude of L, p_0, and U. Such spectra lead to a spectrum in the relaxation time as seen from (A16) and hence, to the broadening of the observed peaks. It is natural to suppose that the values of L and p_0 have some fluctuation in any deformed crystal. The existence of two types of kinks on 60° dislocations seen in Fig. 7a and b could also be a cause for the spectrum in the value of U which would account for the larger width of peak P_4 compared with peak P_3.

Acknowledgements

Authors thank to Dr. K. Kamigaki and Dr. T. Fukase for advise on the technique in ultrasonic attenuation measurements.

Appendix

We derive here expressions for the relaxation strength and the relaxation time for the internal friction peak caused by the stress-assisted diffusive motion of geometrical kinks on the basis of the kink–chain model.

The kink–chain model was first discussed by Seeger and Schiller [18] and, then, developed by Suzuki and Elbaum [19] and Alefeld [20] in detail. Using this model a resonance type loss resulting from the sidewise motion of kinks has so far been discussed.

We assume here abrupt kinks undergoing diffusive motion by virtue of thermal fluctuations. In the presence of an applied stress σ they would undergo a drift motion obeying the Einstein relation

$$v = \sigma b a \mu \tag{A1}$$

with

$$\mu = \frac{D}{kT}, \tag{A2}$$

where v is the kink velocity along the dislocation line, b the magnitude of Burgers vector, a the distance between adjacent Peierls valleys, μ the kink mobility, and D the diffusion coefficient of a kink. A kink is thought to behave like an overdamped oscillator for which the viscous term is far greater than the inertia term. If we take account of the interaction between neighbouring kinks only, the equation of motion of the j-th kink may be given by the following equation

$$\frac{1}{\mu} \frac{\mathrm{d}u_j}{\mathrm{d}t} = \sigma b a + C (u_{j-1} - 2 u_j + u_{j+1}), \tag{A3}$$

where u_j is the displacement of the j-th kink from its equilibrium position. C is the interaction coefficient between kinks given by

$$C = \frac{G b^2 a^2}{4 \pi d_0^3} \beta, \tag{A4}$$

where

$$\beta = \left(\frac{1 + \nu}{1 - \nu}\right) \cos^2 \varphi + \left(\frac{1 - 2 \nu}{1 - \nu}\right) \sin^2 \varphi .$$

d_0 in the above equation is the distance between kinks when the external stress is absent, and G, ν, and φ are the shear modulus, the Poisson ratio, and the angle between Burgers vector and the direction of the Peierls valley, respectively [18, 21, 22]. Assuming

$$\sigma = \sigma_0 \exp (i \omega t) \tag{A5}$$

and the number of kinks on a free dislocation segment of length L to be n, we have a solution of the simultaneous differential equations (A3) as follows [19]:

$$u_j = \sum_{m=0}^{\infty} A_m \sin \left[\frac{\pi j d_0 (2 m + 1)}{L}\right] \exp (i \omega t),$$

$$j = 1, 2, 3, \ldots, n, \qquad \text{and} \qquad L = (n + 1) d_0 . \tag{A6}$$

A_m is given by

$$A_m = \frac{4 \sigma_0 b a}{\pi (2 m + 1) \{4 C \sin^2 [\pi d_0 (2 m + 1)/2 L] + (i \omega/\mu)\}} . \tag{A7}$$

The strain of the crystal with dislocation density N due to the kink motion is given by

$$\varepsilon_k = \frac{N}{L} b\, a \sum_{j=1}^{n} u_j\,. \tag{A8}$$

Since the term $m = 0$ in (A6) is shown to give the dominant contribution to the internal friction, terms other than $m = 0$ are neglected. Substitution of (A6) into (A8) gives

$$\varepsilon_k = \frac{b\, a\, N\, u_0}{L} \cot\left(\frac{\pi\, d_0}{2\, L}\right) \frac{\exp\,(i\,\omega\, t)}{i\,\omega\,\tau + 1}\,, \tag{A9}$$

$$u_0 = \frac{\sigma_0\, b\, a}{\pi\, C \sin^2\,(\pi\, d_0/2\, L)}\,, \tag{A10}$$

$$\tau = \left(4\,\mu\, C \sin^2 \frac{\pi\, d_0}{2\, L}\right)^{-1}. \tag{A11}$$

The internal friction Q^{-1} and the modulus defect $\Delta G/G$ are easily derived from (A9) and are given by

$$Q^{-1} = \frac{\omega\,\tau}{1 + \omega^2\,\tau^2}\, \Delta_{\mathrm{M}}\,, \tag{A12}$$

$$\frac{\Delta G}{G} = \frac{\Delta_{\mathrm{M}}}{1 + \omega^2\,\tau^2}\,, \tag{A13}$$

$$\Delta_{\mathrm{M}} = \frac{G\, a\, b\, u_0\, N}{\sigma_0\, L} \cot \frac{\pi\, d_0}{2\, L}\,. \tag{A14}$$

The relaxation time τ shows an exponential dependence on the temperature through the kink mobility μ. Thus, our model leads to an internal friction peak of the relaxation type.

In the case of high kink density ($L \gg d_0$), the relaxation strength Δ_{M} and the relaxation time τ reduce to

$$\Delta_{\mathrm{M}} = \frac{8\,\Omega\, G\, b^2\, N\, L^2}{\pi^4\, E'}\,, \tag{A15}$$

$$\tau = \frac{L^2}{\pi^2\, a^2\, p_0\, E'\,\mu}\,, \tag{A16}$$

where $E' = G\, b^2\, \beta/4\,\pi$, Ω is the orientation factor, and p_0 the kink density. These equations are identical with Southgate-Attard's relations of equation (4) and (5) except that E' takes the place of E_0 in the latter.

References

[1] H. SUZUKI, J. Phys. Soc. Japan 18, (Suppl. I), 182 (1963).
[2] R. LABUSCH, phys. stat. sol. **10**, 645 (1965).
[3] K. OHORI and K. SUMIMO, phys. stat. sol. (a) **9**, 151 (1972).
[4] L. P. KHIZNICHENKO, P. F. KROMER, D. K. KAIPNAZAROV, E. OTENYAZOV, D. YASUPOVA, and L. G. ZOTOVA, Soviet Phys. — Solid State 8, 1521 (1966); phys. stat. sol. **21**, 805 (1967).

[5] H. Kayano, K. Kamigaki, and S. Koda, J. Phys. Soc. Japan **23**, 1189 (1967).
[6] K. Kojima and K. Sumino, J. Phys. Soc. Japan **26**, 1213 (1969).
[7] J. Marx, Rev. sci. Instrum. **22**, 503 (1951).
[8] G. Weinreich, T. M. Sanders, and H. G. White, Phys. Rev. **114**, 33 (1959).
[9] M. Kozima and K. Sumino, unpublished.
[10] G. Alefeld, J. Fillox, and H. Harper, Dislocation Dynamics, McGraw-Hill Publ. Co., 1970 (p. 191).
[11] H. Schaumburg, phys. stat. sol. **40**, K1 (1970).
[12] H. Alexander and S. Mader, Acta metall. **10**, 887 (1962).
[13] H. Alexander, phys. stat. sol. **27**, 391 (1968).
[14] K. Sumino and K. Kojima, Crystal Lattice Defects **2**, 159 (1971).
[15] A. D. Brailsford, Phys. Rev. **122**, 778 (1961).
[16] P. D. Southgate and A. E. Attard, J. appl. Phys. **34**, 855 (1963).
[17] P. F. Kromer and L. P. Khiznichenko, phys. stat. sol. **21**, 811 (1967).
[18] A. Seeger and P. Schiller, Acta metall. **10**, 348 (1962).
[19] T. Suzuki and C. Elbaum, J. appl. Phys. **35**, 1539 (1964).
[20] G. Alefeld, J. appl. Phys. **36**, 2642 (1965).
[21] F. Kroupa and L. M. Brown, Phil. Mag. **6**, 1267 (1961).
[22] A. D. Brailsford, Phys. Rev. **128**, 1033 (1962).

(Received September 4, 1972)

phys. stat. sol. (a) **14**, 501 (1972)

Subject classification: 14.4.1 and 20; 22.8.1

II. Physikalisches Institut der Universität zu Köln

Optical Second Harmonic Generation during Ferroelectric Polarization Reversal

By

H. VOGT and D. WEINMANN

Optical second harmonic generation is strongly influenced by ferroelectric domain structures. The second harmonic intensity $I^{(2)}$ has been measured during ferroelectric switching in $NaNO_2$. $I^{(2)}$ proves to be proportional to the number N of domains, so that the time dependence of $I^{(2)}$ is determined by that of N. $I^{(2)}(t)$ is compared with the switching pulse $i(t)$. From both some information about the behaviour of the mean domain thickness during switching is obtained.

Ferroelektrische Domänen haben einen starken Einfluß auf die Erzeugung von Oberwellen im optischen Spektralbereich. Es wurde untersucht, wie sich die Intensität $I^{(2)}$ der in einem $NaNO_2$ Kristall erzeugten zweiten Harmonischen während der Umkehr der spontanen Polarisation verhält. $I^{(2)}$ erweist sich zur Anzahl N der Domänen proportional, so daß die Zeitabhängigkeit von $I^{(2)}$ durch die von N bestimmt wird. $I^{(2)}(t)$ wird mit dem Stromimpuls $i(t)$ verglichen, der während der Polarisationsumkehr entsteht. Hieraus ergeben sich Aussagen über das Verhalten der mittleren Domänendicke im Verlaufe der Polarisationsumkehr.

1. Introduction

As was first shown by Miller [1], optical second harmonic generation (SHG) in ferroelectric crystals is strongly influenced by the presence of ferroelectric domains. This effect can be explained by spatial variations of the optical nonlinear susceptibility χ^{NL} [2, 3]. If referred to a laboratory system the tensor χ^{NL} differs from domain to domain and is hence modulated spatially by the domain structure. In general the optical linear susceptibility χ^{L} is modulated, too, but in a different way, because χ^{NL} and χ^{L} are tensors of different rank.

In the case of 180°-domains, e.g., χ^{NL} reverses its sign from domain to domain, whereas χ^{L} remains the same in the whole crystal. Therefore, 180°-domains modify the nonlinear, but not the linear optical behaviour of ferroelectric crystals and can be investigated by SHG, although they are not directly observable by linear optical means.

Previous studies on ferroelectric domain structure by SHG [2, 3] have mainly been concentrated on static properties of domains and their temperature dependence in the vicinity of the Curie temperature.

The present paper deals with SHG during ferroelectric switching. We have measured the change of the second harmonic intensity $I^{(2)}$ when reversing the spontaneous polarization by an external electric field. As ferroelectric samples we used single crystals of $NaNO_2$. This substance proved to be appropriate for our study because of its long switching time [4] and its relatively simple 180°-domain structure [5].

After presenting our experimental results we compare the time dependence of $I^{(2)}$ with that of the electric switching current. From both we get some information about the number N of domains and the average domain thickness $\bar{a}$ as functions of time t during the switching process. $\bar{a}(t)$ is further discussed in terms of a simple rate equation.

2. Experimental Details

Our experimental equipment was similar to that described in [3]. A pulsed Nd^{3+} glass laser emitted the fundamental wave at 1.06 μm wavelength. After being split by a partially reflecting mirror the laser beam generated the second harmonic intensities I_s and I_R in the sample under investigation and in a reference crystal, respectively. The ratio I_s/I_R, averaged over the laser pulse, was a reproducible measure for the efficiency of SHG [3]. It is henceforth called I_A.

We observed SHG on rectangular crystal specimens of $NaNO_2$. Their edges were parallel to the crystallographic axes a, b, and c and had lengths of about 7,7, and 5 mm, respectively.

The wave vectors $\boldsymbol{k}^{(1)}$ and $\boldsymbol{k}^{(2)}$ of the fundamental and the harmonic wave pointed into the direction of the a-axis. They were normal to the domain walls spreading along (100). For ferroelectric switching we applied an electric field E_s along the b-axis via silver electrodes evaporated on the (010) surfaces. As function of time E_s had the form of a rectangular step rising from -2.5 kV/cm to $+2.5$ kV/cm. Therefore, our results refer to a transition from a one-domain state with negative spontaneous polarization P_S to a one-domain state with positive P_S.

The laser pulse followed the reversal of E_s after a delay time t, which could be varied between 1 and 360 ms. We measured I_A as a function of t in steps Δt of about 1 ms, thereby switching the crystal and operating the laser for each value of t. Δt was nearly equal to the duration of the laser pulse, which determined the time resolution for measuring $I_A(t)$.

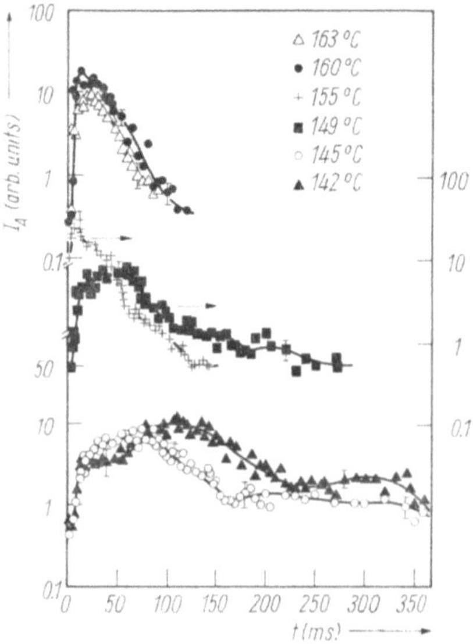

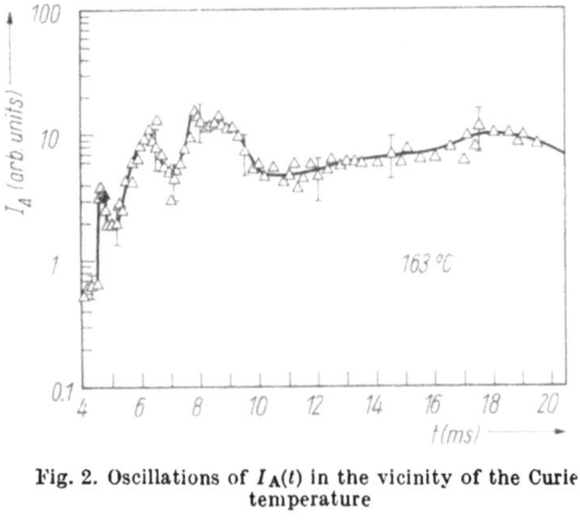

Fig. 2. Oscillations of $I_A(t)$ in the vicinity of the Curie temperature

Fig. 1. The second harmonic intensity during polarization reversal in $NaNO_2$

3. Results and Discussion

Our experimental results are presented in Fig. 1 to 3. Fig. 1 shows the time dependence of I_A at six temperatures between 142 and 163 °C. 142 °C was the lowest temperature at which thespontaneous polarization could be completely reversed by an electric field of about $\pm$ 2.5 kV/cm within 360 ms, whereas 163 °C was the highest temperature which we could maintain in our samples below their transition into the paraelectric phase.

In the range between 160 and 163 °C we observed oscillations of $I_A(t)$ at the beginning of the polarization reversal. Such oscillations can be seen in Fig. 2.

At first sight the curves in Fig. 1 look similar to the corresponding switching pulses $i(t)$, i.e. the time dependence of the displacement current. However the width of $i(t)$ differs remarkably from that of $I_A(t)$. This can bee seen in Fig. 3 where the switching time t_s is plotted as function of temperature:

The width $t_s^{(i)}$ of $i(t)$ is indicated by dots. As usual [6] $t_s^{(i)}$ is defined as the time necessary for $i(t)$ to drop to one tenth of its maximum value. In our case $t_s^{(i)}$ is nearly equal to the time, during which more then 98% of the spontaneous polarization P_S is reversed into the opposite direction. This can be seen from $P_S(t)$ obtained by integration of the measured switching pulse $i(t)$. If delay effects between the electrode surfaces and the crystal bulk are absent, $P_S(t)$ coincides with $M_S(t)/V$, where $M_S(t)$ is the spontaneous dipole moment of the whole crystal volume V.

The vertical bars in Fig. 3 indicate the width $t_s^{(I_A)}$ of $I_A(t)$. $t_s^{(I_A)}$ is much larger than $t_s^{(i)}$. This leads to the conclusion that some kind of domain structure still exists when P_S is almost completely reversed. One may assume the laser radiation to penetrate negative domains, which are so small, that they do not contribute to P_S considerably, although they have an influence on SHG.

Our experimental results can be interpreted in more detail, if the relations between I_A and characteristic parameters of the domain structure are known. As explained in [3] the influence of the domain structure on χ^{NL} and hence I_A can be best described by a modulation function $M(x)$ which takes the following values:

$$M(x) = \begin{cases} 0 \text{ outside the crystal,} \\ +1 \text{ in positive domains,} \\ -1 \text{ in negative domains,} \\ 0 \text{ in domain walls.} \end{cases}$$

The x-direction is assumed to be parallel to the wave vectors $\boldsymbol{k}^{(1)}$ and $\boldsymbol{k}^{(2)}$ of the fundamental and the harmonic wave. In defining $M(x)$ for domain walls too, we take into account recent results of Suzuki and Tagaki [7]. These authors have shown that the thickness of the domain walls in $NaNO_2$ can be about

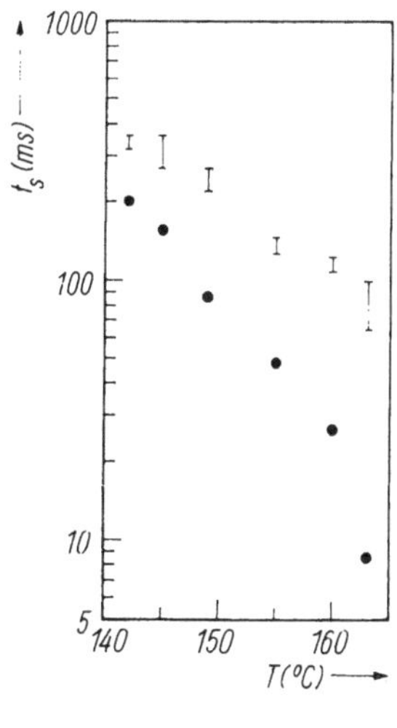

Fig. 3. The temperature dependence of switching time t_s. ● t_s taken from $i(t)$; I t_s taken from $I_A(t)$

1 μm and thus cannot be regarded as negligibly small. Because the domain walls have a disordered crystal structure, their optical nonlinear susceptibility and hence $M(x)$ is approximately zero as it is in the paraelectric phase.

I_A is determined by the $(k^{(2)} - 2\,k^{(1)})$ Fourier component of $M(x)$ [3]:

$$I_A \sim |M(k)|^2 \quad \text{with} \quad k = k^{(2)} - 2\,k^{(1)}\,.$$

Instead of $k = (k^{(2)} - 2\,k^{(1)})$ we may equivalently use the coherence length $l_c = \pi/(n^{(2)} - n^{(1)})$ for characterizing the phase mismatch between the second harmonic and the nonlinear polarization, induced by the fundamental wave. The value of l_c depends on the radiation geometry. In our experiments the radiation of the laser was unpolarized and propagated along the crystallographic a-axis. Therefore the components χ_{22}, χ_{23}, and χ_{34} of $\boldsymbol{\chi}^{NL}$ are involved in I_A. The contribution of χ_{22} to I_A, however, is negligibly small in comparison with that of χ_{23} and χ_{34}. This follows from [3] and from refractive index data measured by Hirotsu and Iio [8]. The coherence lengths corresponding to χ_{23} and χ_{34} are

$$l_{23} = \frac{\lambda}{4\,|n_b^{(2)} - n_c^{(1)}|} \approx 1.3\ \mu\text{m} \quad \text{and} \quad l_{34} = \frac{\lambda}{4\,|n_c^{(2)} - \frac{1}{2}\,(n_c^{(1)} + n_b^{(1)})|} \approx 1.7\ \mu\text{m}\,,$$

where λ is the fundamental wavelength and $n^{(2)}$, $n^{(1)}$ indicate principal refractive indices at harmonic and fundamental frequency, respectively. The harmonic waves due to χ_{23} and χ_{34} are polarized orthogonally to each other. Therefore, they do not interfere and one may write

$$I_A \sim \left|M\left(\frac{\pi}{l_{23}}\right)\right|^2 + \left|M\left(\frac{\pi}{l_{34}}\right)\right|^2 \approx 2\left|M\left(\frac{\pi}{l_c}\right)\right|^2. \tag{1}$$

with $l_c \approx 1.5\ \mu$m.

$M(\pi/l_c)$ can be best evaluated via the autocorrelation function

$$S(\xi) = \int_{-\infty}^{\infty} M(x)\,M(x + \xi)\,\mathrm{d}x\,. \tag{2}$$

$S(\xi)$ is related to $M(\pi/l_c)$ by the Fourier transformation

$$\left|M\left(\frac{\pi}{l_c}\right)\right|^2 = \frac{1}{4\,\pi^2}\int_{-\infty}^{\infty} S(\xi)\,\mathrm{e}^{-i\pi\frac{\xi}{l_c}}\,\mathrm{d}\xi\,. \tag{3}$$

The calculation of $S(\xi)$ becomes more convenient, if $M(x)$ is expressed by

$$M(x) = R(x)\,M_\infty(x) = R(x)\,[2\,m_+(x) + m_0(x) - 1]\,. \tag{4}$$

$M_\infty(x)$ is the modulation function of an infinite domain structure, while $R(x)$ takes into account the finite crystal thickness d.

$R(x) = 1$ for $0 < x < d$ and 0 outside the crystal. The rectangular function $m_+(x)$ only refers to positive domains. It has the value 1 within positive domains, but is equal to 0 otherwise. An analogous definition holds for $m_0(x)$ with respect to the domain walls.

Substitution of (4) into (2) leads to

$$S(\xi) = T(\xi)\,S_\infty(\xi)\,, \tag{5}$$

with

$$T(\xi) = d - |\xi| \quad \text{for} \quad -d < \xi < d \quad \text{and} \quad 0 \text{ otherwise [3]}\,,$$

and

$$S_\infty(\xi) = \lim_{L\to\infty} \frac{1}{2L} \int_{-L}^{L} M_\infty(x)\, M_\infty\,(x+\xi)\, \mathrm{d}x =$$

$$= 1 - 4\,W_+ - 2\,W_0 + 4\,S_\infty^{(+)}(\xi) + S_\infty^{(0)}(\xi) + 4\,S_\infty^{(0+)}(\xi)\,. \tag{6}$$

The quantities in the last equation have the following meaning:

$$W_+ = \lim_{L\to\infty} \frac{1}{2L} \int_{-L}^{L} m_+(x)\, \mathrm{d}x$$

represents the probability for the presence of a positive domain.

$$S_\infty^{(+)} = \lim_{L\to\infty} \frac{1}{2L} \int_{-L}^{L} m_+(x)\, m_+\,(x+\xi)\, \mathrm{d}x$$

is the autocorrelation function of $m_+(x)$. W_0 and $S_\infty^{(0)}(\xi)$ are defined in analogy to W_+ and $S_\infty^{(+)}(\xi)$, respectively.

$S_\infty^{(0+)}(\xi)$ describes the cross-correlation between $m_0(x)$ and $m_+(x)$, and has the form:

$$S_\infty^{(0+)}(\xi) = \lim_{L\to\infty} \frac{1}{2L} \int_{-L}^{L} m_+(x)\, m_0\,(x+\xi)\, \mathrm{d}x\,.$$

For further discussion of the correlation functions statistical concepts are needed. We assume that at each moment during the polarization reversal the thicknesses a and a' of the positive domains and the domain walls vary according to distribution functions $w_+(a)$ and $w_0(a')$, respectively. $w_+(a)$ and $w_0(a')$ are only well defined in the presence of many domains within the crystal and probably lose their meaning at the very beginning and end of the switching process.

After introducing $w_+(a)$ and $w_0(a')$ we can derive expressions for the correlation functions. These will at first be written down and then explained in detail. $S_\infty^{(+)}(\xi)$ and $S_\infty^{(0+)}(\xi)$ can be approximated by

$$S_\infty^{(+)}(\xi) = \frac{W_+}{\bar{a}_+} \int_{\xi}^{\infty} (a-\xi)\, w_+(a)\, \mathrm{d}a \tag{7}$$

and

$$S_\infty^{(0+)}(\xi) = S_\infty^{(0+)}\,(\xi;\, a' < a) + S_\infty^{(0+)}\,(\xi;\, a' > a)\,, \tag{8}$$

with

$$S_\infty^{(0+)}\,(\xi;\, a' < a) = \frac{W_+}{\bar{a}_+} \left\{ \int_{\xi/2}^{\xi} \mathrm{d}a \int_{\xi-a}^{a} \mathrm{d}a'\, (a + a' - \xi)\, w_+(a)\, w_0(a') + \right.$$

$$\left. + \int_{\xi}^{\infty} \mathrm{d}a \int_{0}^{\xi} \mathrm{d}a'\, a'\, w_+(a)\, w_0(a') + \int_{\xi}^{\infty} \mathrm{d}a \int_{\xi}^{a} \mathrm{d}a'\, \xi\, w_+(a)\, w_0(a') \right\}. \tag{9}$$

$S_\infty^{(0+)}(\xi; a' > a)$ is obtained from $S_\infty^{(0+)}(\xi; a' < a)$ by commutation of a and a'.

In (7) to (9) ξ has been confined to $\xi > 0$. This is sufficient, because all correlation functions are even in ξ.

The approximation (7) only takes into account the correlation between the values of $m_+(x)$ within the same domain. Thus each positive domain, which is larger than ξ, contributes by $(a - \xi)$ to $S_\infty^{(+)}(\xi)$. The constant before the integral follows from $S_\infty^{(+)}(0) = W_+$ and from the integral

$$\bar{a}_+ = \int_0^\infty a\, w_+(a)\, da\,,$$

which defines the mean thickness of the positive domains. The form of the autocorrelation function $S_\infty^{(0)}(\xi)$ is analogous to (7).

The expressions (8) and (9) describe the overlap between the positive domains and the neighbouring domain walls, if shifted by ξ.

The distributions of a' and a are assumed to be independent of each other.

In evaluating $S_\infty^{(0+)}(\xi; a' < a)$ according to (9) on has to distinguish four regions in the $(a'\ a)$-plane. These are

$$(a + a') \leqq \xi\,; \qquad a \leqq \xi \leqq (a + a')\,; \qquad a' \leqq \xi \leqq a\,; \qquad 0 \leqq \xi \leqq a'.$$

The first region does not contribute to $S_\infty^{(0+)}(\xi; a' < a)$, because there is no overlap between $m_+(x)$ and $m_0(x + \xi)$. The contributions of the remaining regions are given by the three integrals in (9).

In order to get the Fourier transform $S(k)$ of $S(\xi)$ and hence $|M(k)|^2$ one has to convolve the Fourier transform $S_\infty(k)$ of $S_\infty(\xi)$ with the Fourier transform $T(k)$ of the triangular function $T(\xi)$, defined by (5). The calculation of $S_\infty(k)$ becomes more convenient, if it is not directly performed via the integral expressions (7) and (9), but via the derivatives of $S_\infty(\xi)$. These are related to $S_\infty(k)$ by

$$S_\infty(k) = \frac{1}{2\pi}\int_{-\infty}^{\infty} S_\infty(\xi)\, e^{-ik\xi}\, d\xi = \frac{1}{\pi}\int_0^\infty S_\infty(\xi)\cos k\xi\, d\xi =$$

$$= -\frac{1}{\pi k^2}\left(\frac{dS_\infty(\xi)}{d\xi}\right)_{\xi=+0} - \frac{1}{\pi k^2}\int_0^\infty \left(\frac{d^2 S_\infty(\xi)}{d\xi^2}\right)\cos k\xi\, d\xi\,. \tag{10}$$

From (6) to (9) one gets

$$\frac{d^2 S_\infty(\xi)}{d\xi^2} = \frac{W_+}{\bar{a}_+}\left\{-2\, w_0(\xi) + 4\int_0^\xi w_0(\eta)\, w_+(\xi - \eta)\, d\eta\right\}. \tag{11}$$

In deriving formula (11) it is assumed that twice as many domain walls as positive domains are present and thus $W_0/\bar{a}_0 = 2\, W_+/\bar{a}_+$.

According to (6) one has

$$\left(\frac{dS_\infty(\xi)}{d\xi}\right)_{\xi=+0} =$$

$$= 4\left(\frac{dS_\infty^{(+)}(\xi)}{d\xi}\right)_{\xi=+0} + \left(\frac{dS_\infty^{(0)}(\xi)}{d\xi}\right)_{\xi=+0} + 4\left(\frac{dS_\infty^{(0+)}(\xi)}{d\xi}\right)_{\xi=+0} = -2\,\frac{W_+}{\bar{a}_+}\,. \tag{12}$$

Substitution of (12) and (11) into (10) leads to

$$S_\infty(k) = \frac{2\,W_+}{\pi\,k^2\,\bar{a}_+}\left[1 + \pi\,w_0^c(k) + 2\,\pi^2\left(w_0^s(k)\,w_+^s(k) - w_0^c(k)\,w_+^c(k)\right)\right], \quad (13)$$

where $w_0^c(k)$, $w_+^c(k)$ and $w_0^s(k)$, $w_+^s(k)$ are the cosine- and the sine-Fourier transforms of $w_0(a')$ and $w_+(a)$, respectively, e.g.

$$w_+^c(k) = \frac{1}{\pi}\int_0^\infty w_+(a)\cos ka\,\mathrm{d}a \quad \text{and} \quad w_+^s(k) = \frac{1}{\pi}\int_0^\infty w_+(a)\sin ka\,\mathrm{d}a\,.$$

As already mentioned $S_\infty(k)$ is related to $S(k)$ by the convolution

$$S(k) = \int_{-\infty}^{\infty} S_\infty(k')\,T\,(k - k')\,\mathrm{d}k'\,,$$

with

$$T\,(k - k') = \frac{2}{\pi\,(k - k')^2}\sin^2\frac{(k - k')}{2}\,d\,.$$

The width of $T\,(k - k')$, however, is much smaller than that of $S_\infty(k')$ [3]. Therefore $T\,(k - k')$ may be approximated by $d\,\delta\,(k - k')$, so that $S(k) \approx \approx d\,S_\infty(k)$. After combining (3) and (13) one finally arrives at

$$I_A \sim \left|M\left(\frac{\pi}{l_c}\right)\right|^2 = \frac{N_+\,l_c^2}{\pi^4}\left[1 + \pi\,w_0^c\left(\frac{\pi}{l_c}\right) + \right.$$
$$\left. + 2\,\pi^2\left(w_0^s\left(\frac{\pi}{l_c}\right)w_+^s\left(\frac{\pi}{l_c}\right) - w_0^c\left(\frac{\pi}{l_c}\right)w_+^c\left(\frac{\pi}{l_c}\right)\right)\right], \quad (14)$$

where $N_+ = d\,\frac{W_+}{\bar{a}_+}$ is the number of positive domains in the crystal. The expression (14) may be regarded as generalization of the formulae (7) and (15) in reference [3], which are only valid for negligibly small domain walls and a special distribution function $w_+(a)$.

The Fourier transforms in (14) are mainly determined by the ratios $\Delta a_+/l_c$ and $\Delta a_0/l_c$, where Δa_+ and Δa_0 indicate the width of $w_+(a)$ and $w_0(a')$, respectively.

If $\Delta a_+ \gg l_c$ and $\Delta a_0 \gg l_c$, many periods of $\cos\,(\pi/l_c)\,a$ or $\sin\,(\pi/l_c)\,a$ fall into the range of $w_+(a)$ and $w_0(a)$. Therefore, all Fourier transforms in (14) become rather small, so that I_A is proportional to N_+. In this case the irregular domain structure destroys the coherence of the second harmonic wavelets generated in different domains. Then I_A is equal to the sum of the wavelet intensities and hence becomes proportional to the total number N of domains. However, $N \approx \frac{1}{2}\,N_+$, because the numbers N_+ and N_- of positive and negative domains only differ by one.

The Fourier transforms in (14) are significant, if $\Delta a_+ \lesssim l_c$ and $\Delta a_0 \lesssim l_c$. We may assume that this can only happen when the mean values $\bar{a}_+$ and $\bar{a}_0$ are in the same order of magnitude as $l_c \approx 1.5\,\mu$m. In this case the terms in the square bracket of (14) become rather sensitive to a change of the domain structure. Therefore rapid variations of I_A are to be expected. These, however, are only observed between 160 and 163 °C at the beginning of the polarization

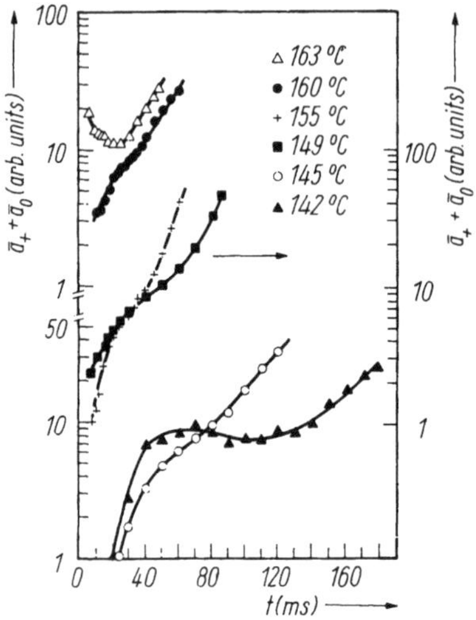

Fig. 4. The mean domain thickness during polarization reversal

reversal. Thus we conclude that the Fourier transforms of $w_+(a)$ and $w_0(a')$ are only to be taken into account for this range. It may be supposed that the domain wall distribution $w_0(a')$ and consequently $w_0^c(\pi/l_c)$ and $w_0^s(\pi/l_c)$ are mainly determined by the temperature and do not vary remarkably during the switching process. According to (14) the oscillations in Fig. 2 are then due to the behaviour of $w_+^c(\pi/l_c)$ and $w_+^s(\pi/l_c)$, indicating that the mean thickness of the positive domains has the order of magnitude of l_c.

On the other hand the curves $I_A(t)$ showing no oscillations mainly represent the time dependence of N_+. In their case $w_0^c(\pi/l_c)$ and $w_0^s(\pi/l_c)$ may be considered as almost constant, whereas $w_+^s(\pi/l_c)$ and $w_+^c(\pi/l_c)$ are rather small because of $\Delta a_+ \gg l_c$.

We are now able to combine $I_A(t)$ and $i(t)$. From $I_A(t)$ we get $N_+(t) = \frac{1}{2} N(t)$ and by integration of $i(t)$ we get $P_S(t)$.

$N(t)$ and $P_S(t)$ are related by

$$\frac{1}{2} N(t) \, [\bar{a}_+(t) - \bar{a}_-(t)] = d \frac{P_S(t)}{P_S(\infty)} . \tag{15}$$

$\bar{a}_+(t)$ and $\bar{a}_-(t)$ represent the mean thickness of the positive and negative domains. $P_S(\infty)$ is the saturation value of the spontaneous polarization $P_S(t)$. In (15) any delay between the switching processes at the electrode surfaces and those in the crystal bulk is neglected.

From (15) and from

$$\frac{1}{2} N(t) \, [\bar{a}_+(t) + \bar{a}_-(t) + 2 \, \bar{a}_0(t)] = d$$

we obtain

$$\bar{a}_+(t) + \bar{a}_0(t) = \frac{d}{N(t)} \left(1 + \frac{P_S(t)}{P_S(\infty)} \right). \tag{16}$$

In Fig. 4 $(\bar{a}_+ + \bar{a}_0)$ is plotted as function of time according to this formula. The unit of $(\bar{a}_+ + \bar{a}_0)$ is arbitrary, because we measured only relative values of $I_A(t)$ and hence $N(t)$. We have confined the application of (16) to the range of t, in which $I_A(t) \gtrsim 10 \, I_A(0)$. Thus we are sure that N is large and our statistical arguments are valid.

With the exception of the curve for 163 °C all curves in Fig. 4 show the same shape. At the beginning of the polarization reversal $(\bar{a}_+ + \bar{a}_0)$ depends linearly on t, but after a minimum of slope there is an exponential increase according to $(\bar{a}_+ + \bar{a}_0) \sim 1/N$.

For a better understanding of this behaviour $\bar{a}_+(t)$ is considered in more detail, whereas $\bar{a}_0$ is assumed to be almost constant during switching.

We take into account one dimension only, viz. the direction perpendicular to the domain walls. Any change of $\bar{a}_+(t)$ is due to one of the following three processes:

a) *nucleation of domains*: The nucleation of a new positive domain reduces $\bar{a}_+$ by the amount

$$-\Delta\bar{a}_+ = \bar{a}_+ - \frac{N_+\bar{a}_+ + r_n}{N_+ + 1} \approx \frac{\bar{a}_+ - r_n}{N_+}\,.$$

r_n describes the dimension of the domain nucleus. The total change of $\bar{a}_+$ due to $\dot{N}_n$ nucleations per second is

$$\left(\frac{d\bar{a}_+}{dt}\right)_n = -\left(\frac{\bar{a}_+ - r_n}{N_+}\right)\dot{N}_n\,. \tag{17a}$$

b) *sideways expansion of existing domains*: If all domain walls move with the same velocity v, this process contributes to (da_+/dt) by

$$\left(\frac{da_+}{dt}\right)_s = 2\,v\,. \tag{17b}$$

c) *coalescence of domains*: In analogy to (17a) we may write

$$\left(\frac{d\bar{a}_+}{dt}\right)_c = \left(\frac{\bar{a}_+ + r_c}{N_+}\right)\dot{N}_c\,. \tag{17c}$$

$\dot{N}_c$ is the coalescence rate, whereas r_c indicates the size of the region which disappears during coalescence. By summing up formulae (17a) to (17c) we get the rate equation

$$\dot{\bar{a}}_+ + \frac{\dot{N}_+}{N_+}\bar{a}_+ = 2\,v + \frac{1}{N_+}(r_n\,\dot{N}_n + r_c\,\dot{N}_c)\,, \tag{18}$$

where the dots above $\bar{a}_+$, N_+, N_n, and N_c denote differentiation with respect to time and $(\dot{N}_n - \dot{N}_c)$ is replaced by $\dot{N}_+$.

If r_n, r_c, and v do not vary in time, the solution of (18) takes the form

$$\bar{a}_+(t) = r_n + (r_n + r_c)\frac{N_c}{N_+} + \frac{2\,v}{N_+}\int_0^t N_+(\tau)\,d\tau\,. \tag{19}$$

At the beginning of the polarization reversal N_+ is nearly equal to the number N_n of domain nuclei, whereas the number N_c of coalescences may be neglected. Therefore, $\bar{a}_+(t)$ starts as

$$\bar{a}_+(t) = r_n + 2\,v\frac{1}{N_n}\int_0^t N_n(\tau)\,d\tau\,. \tag{19a}$$

The observed linear increase of $\bar{a}_+(t)$ is in accordance with this equation, if N_n varies like t^α with $\alpha \geqq 0$. Such a variation of N_n is expected in any case and is not a characteristic of a specific nucleation model.

At the end of the polarization reversal N_n remains constant. With $N_c = N_n - N_+$ we obtain

$$\bar{a}_+(t) = -r_c + \frac{1}{N_+}\left[(r_c + r_n) N_n + 2v\int_0^{t_s} N_+ \,d\tau\right], \tag{19b}$$

where the upper limit of integration t is already replaced by t_s. Thus $\bar{a}_+(t)$ increases like $1/N_+$. As this increase is due to the coalescence of positive domains and hence to the disappearance of rather small negative ones, it does not contribute to $P_S(t)$ and $i(t)$ considerably, although it mainly determines the decrease of $I_A(t)$. This may explain the discrepancy between the switching times $t_s^{(i)}$ and $t_s^{(I_A)}$ defined above.

At inflection points of $\bar{a}_+(t)$ the increase due to sideways expansion and coalescence is compensated by the decrease due to nucleation. A negative slope of $\bar{a}_+(t)$ is also allowed by (18) or (19). According to (17a) it is only possible, if $(d\bar{a}_+/dt)_n$ and hence the nucleation rate $\dot{N}_n$ are large and $\bar{a}_+ > r_n$.

Therefore, the exceptional behaviour of $\bar{a}_+(t)$ at 163 °C may indicate, that at this temperature many domain nuclei are present. These, however, expand and coalesce immediately after their formation, so that $\bar{a}_+$ becomes larger than r_n at once.

Acknowledgements

The authors wish to thank Prof. Dr. H. Happ for his interest in this work. The support of the Deutsche Forschungsgemeinschaft is gratefully acknowledged.

References

[1] R. C. MILLER, Phys. Rev. **134**, A1 313 (1964).
[2] G. DOLINO, J. LAJZEROWICZ, and M. VALLADE, Phys. Rev. **B2**, 2194 (1970).
[3] H. VOGT, H. HAPP, and H. G. HÄFELE, phys. stat. sol. (a) **1**, 439 (1970).
H. VOGT and H. HAPP, phys. stat. sol. (b) **44**, 207 (1971).
[4] I. HATTA, S. SAWADA, Y. ASAO, and T. YANAGI, J. Phys. Soc. Japan **18**, 1229 (1963).
I. HATTA and S. SAWADA, Japan. J. appl. Phys. **4**, 389 (1965).
[5] S. NOMURA, Y. ASAO, and S. SAWADA, J. Phys. Soc. Japan **16**, 917 (1961).
[6] E. FATUZZO and W. J. MERZ, Ferroelectricity, North-Holland Publ. Co., Amsterdam 1967 (p. 199).
[7] S. SUZUKI and M. TAKAGI, J. Phys. Soc. Japan **30**, 188 (1971); **32**, 1302 (1972).
[8] S. HIROTSU, private communication.
K. IIO, J. Phys. Soc. Japan, to be published.

(Received September 11, 1972)

phys. stat. sol. (a) 14, 511 (1972)

Subject classification: 18.3 and 19; 22.8.2

Zentralinstitut für Festkörperphysik und Werkstofforschung der Akademie der Wissenschaften der DDR, Institutsteil für magnetische Werkstoffe, Jena

Relaxation Behaviour of Co^{2+} Ions in Mg Ferrite Single Crystals with Different Inversion Degrees

By

P. Görnert

Results of parallel pumping experiments at 9.3 GHz and measurements of the FMR-linewidth at various microwave frequencies are reported for $Mg_{1-y}Co_yFe_2O_{4+\gamma'}$ single crystal spheres and disks with $y = 0; 0.005; 0.010; 0.015; 0.022; 0.050$, and different inversion degrees. The parallel pumping experiments showed that both the critical microwave field strength h_c and the spin wave ellipticity e_k depend on inversion degree, but the low wave number spin wave linewidth $\Delta H_{k\to 0}$ is independent of the cation distribution. $\Delta H_{k\to 0}$ increases isotropically by approximately 10 Oe for each percent of cobalt. FMR linewidth measurements at $Mg_{0.95}Co_{0.05}Fe_2O_{4+\gamma'}$ samples showed that the linewidth ΔH varies only slowly with the shape of samples but ΔH increases linearly with the frequency. These experimental results cannot be explained by the magnon–magnon scattering model suggested by Haas and Callen. The frequency dependence indicates a direct spin–lattice relaxation process.

Es werden Parallelfeld-Pumpexperimente bei 9,3 GHz and Messungen der FMR-Linienbreite bei verschiedenen Mikrowellenfrequenzen an $Mg_{1-y}Co_yFe_2O_{4+\gamma'}$-Einkristallkugeln und -scheiben mit $y = 0; 0{,}005; 0{,}010; 0{,}015; 0{,}022; 0{,}050$ sowie unterschiedlichen Inversionsgraden angegeben. Die Parallelfeld-Pumpexperimente an Kugeln ergaben, daß sowohl die kritische Mikrowellenfeldstärke h_c als auch die Spinwellenelliptizität e_k vom Inversionsgrad abhängen, dagegen die Spinwellenlinienbreite $\Delta H_{k\to 0}$ für Spinwellen mit kleinen Wellenzahlen unabhängig von der Kationenverteilung ist. Die isotrope Vergrößerung der Spinwellenlinienbreite für jedes Prozent Kobalt beträgt etwa 10 Oe. FMR-Linienbreitenmessungen an $Mg_{0,95}Co_{0,05}Fe_2O_{4+\gamma'}$-Proben zeigten, daß die Linienbreite ΔH kaum von der Probengeometrie abhängt, aber linear mit der Frequenz anwächst. Diese experimentellen Ergebnisse können nicht mit dem von Haas und Callen vorgeschlagenen Magnon–Magnon-Streuprozeß erklärt werden. Dagegen weist vor allem die Frequenzabhängigkeit auf einen direkten Spin–Gitter-Relaxationsprozeß hin.

1. Introduction

It has been known since long that in ferrite and garnet single crystals small admixtures of impurity ions with degenerate orbital ground state increase the linewidth of ferromagnetic resonance drastically. The linewidths in rare-earth garnets have generally been interpreted in terms of the direct spin–lattice relaxation theories.

White [1] has suggested that certain ions of the iron transition series may play a part in relaxation processes in ferrites similar to that played by the rare-earth ions. The Co^{2+} ions when located on an octahedral sublattice is one of these ions. Haas and Callen [2], on the other hand, have predicted that Co^{2+} ions may contribute significantly to relaxation by a magnon–magnon scattering mechanism (2 M process). Measurements by Bunina et al. [3] are in qualitative agreement with the 2 M model. But from experiments of many authors [4 to 12] one may conclude that such processes do not play an important role.

This paper reports parallel pumping measurements at 9.3 GHz and ferromagnetic resonance measurements at different frequencies at room temperature for Co^{2+} substituted Mg ferrite single crystal spheres and disks with different inversion degrees. Our results indicate that the Haas-Callen mechanism of line broadening [2] is unimportant.

The preparation and characterization of $Mg_{1-y}Co_yFe_2O_{4+\gamma'}$ crystal specimens used with $y = 0$; 0.005; 0.010; 0.015; 0.022; 0.050 and the effect of cobalt doping upon the magnetocrystalline anisotropy have been described in [13]. Our first results of parallel pumping measurements at Co^{2+} substituted Mg ferrite single crystals with different inversion degrees were published earlier [14].

2. Experimental Results

2.1 Parallel pumping experiments at 9.3 GHz

From measurements with two pulse durations, $\tau_1 = 1.9$ μs and $\tau_2 = 5$ μs, it is possible to calculate the true critical microwave field strength h_c [15]:

$$h_c = 2\,\Delta H_k \min_{\omega_p = 2\omega_k} \left(\frac{1}{e_k - \frac{1}{e_k}} \right), \tag{1}$$

where ΔH_k is the spin wave linewidth of a spin wave with wave number k, $\omega_p = 2\pi$ 9.3 GHz the pump frequency, ω_k the frequency of the spin wave, and e_k the ellipticity of the spin wave [16, 17]. The experimental methods for the determination of h_c have been described in detail in [18].

Curves of h_c corresponding to the "butterfly" minimum for Co^{2+} substituted Mg ferrite single crystal spheres as a function of the cation distribution parameter x

$$\underbrace{(Mg_xFe_{1-x})}_{\text{tetrahedral (A) sites}} \qquad \underbrace{[Mg_{1-x}Fe_{1+x}]\,O_4}_{\text{octahedral (B) sites}} \tag{2}$$

are presented in Fig. 1 in the range from $x \approx 0.16$ to $x = 0.275$. Different x-values were obtained by heat treatment in air between 600 and 1000 °C and quenching in water [13]. Quench temperatures above 1000 °C were not convenient because the samples were strained strongly.

It has been shown experimentally [19][1]) that the transient growth of spin waves excited by parallel pumping technique in Mg ferrite single crystals depends very strongly upon the inversion degree, the larger x the better the typical microwave pulse deformation for $h > h_c$. Therefore h_c measurements with pure or Co^{2+} substituted Mg ferrites for $x \lesssim 0.16$ were hampered.

In Fig. 2, the spin wave ellipticity e_k has been plotted against the cation distribution parameter x for $Mg_{1-y}Co_yFc_2O_{4+\gamma'}$ spheres with $y = 0$ and $y = 0.022$ in ⟨100⟩, ⟨111⟩, and ⟨110⟩ directions. One can easily see from Fig. 1 and Fig. 2 that h_c as well as e_k depend on inversion degree.

Low wave number spin wave linewidths $\Delta H_{k\to 0}$ were determined with the help of equation (1). The results are shown in Fig. 3. We found that the contri-

---

[1]) Due to a production error the symbols of the Bild 2 on page 300 of the "Nachrichtentechnik" were puzzled. The right symbols are $x = 0.163$ (●); 0.180 (◐); 0.208 (○); 0.245 (×), and 0.268 (△) but not $x = 0.163$ (○); 0.180 (×); 0.208 (●); 0.245 (◐), and 0.268 (△).

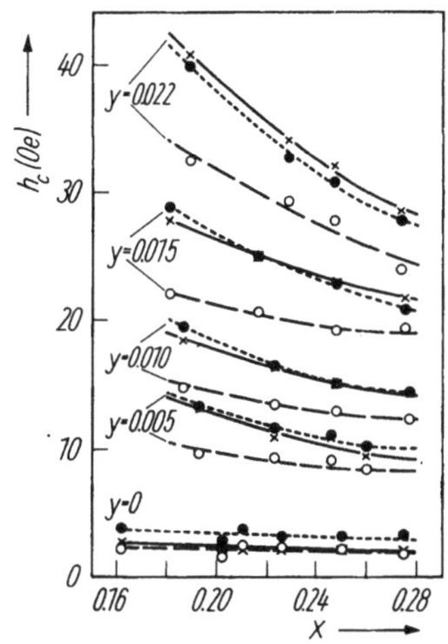

Fig. 1. The critical microwave field strength h_c corresponding to the "butterfly" minimum vs. cation distribution parameter x for $Mg_{1-y}Co_yFe_2O_{4+\gamma'}$ single crystal spheres with the static magnetizing field oriented along ⟨100⟩ (×), ⟨110⟩ (○), and ⟨111⟩ (●) directions

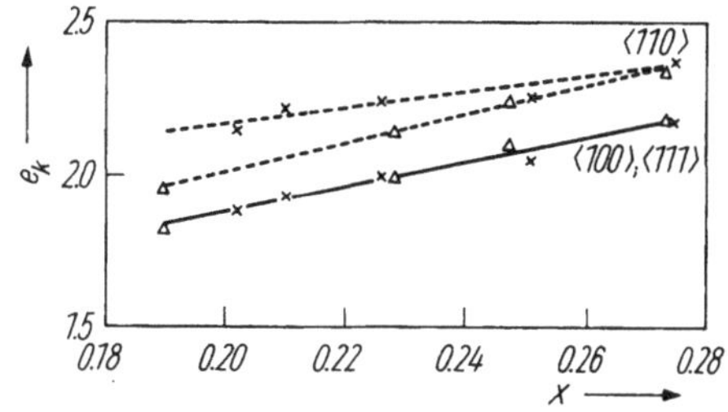

Fig. 2. The spin wave ellipticity e_k as a function of cation distribution parameter x for $Mg_{1-y}Co_yFe_2O_{4+\gamma'}$ spheres with $y = 0$ (×) and $y = 0.022$ (△); dc magnetic field parallel to ⟨100⟩, ⟨111⟩, and ⟨110⟩ crystallographic directions

bution of Co^{2+} ions to $\Delta H_{k\to 0}$ is isotropic within 10% independent of the cation distribution. It exists a linear dependence between $\Delta H_{k\to 0}$ and the Co^{2+} concentration y, as presented in Fig. 4. The spin wave linewidth $\Delta H_{k\to 0}$ increases by approximately 10 Oe for each per cent of cobalt.

2.2 Ferromagnetic resonance measurements at different frequencies

The experimental apparatus was described in [20]. In contrast to the parallel pumping measurements of Section 2.1 it is difficult to estimate the effect of Co^{2+} ions on FMR linewidth ΔH because of the large scatter in the measured

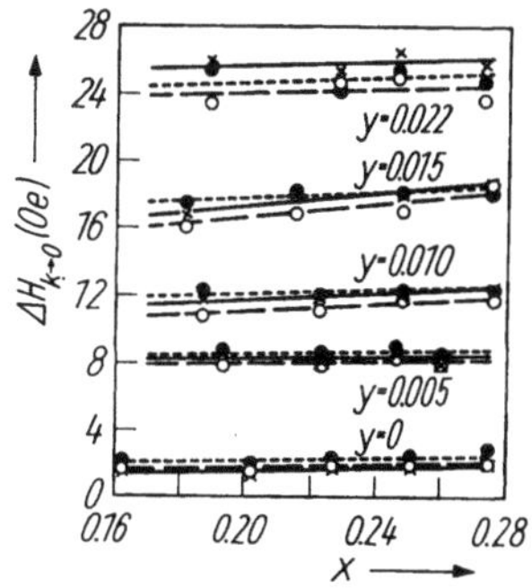

Fig. 3. $\Delta H_{k\to 0}$ vs. x for $Mg_{1-y}Co_yFe_2O_{4+\gamma'}$ spheres with the static magnetic field oriented along ⟨100⟩ (×), ⟨110⟩ (○), and ⟨111⟩ (●) crystallographic directions

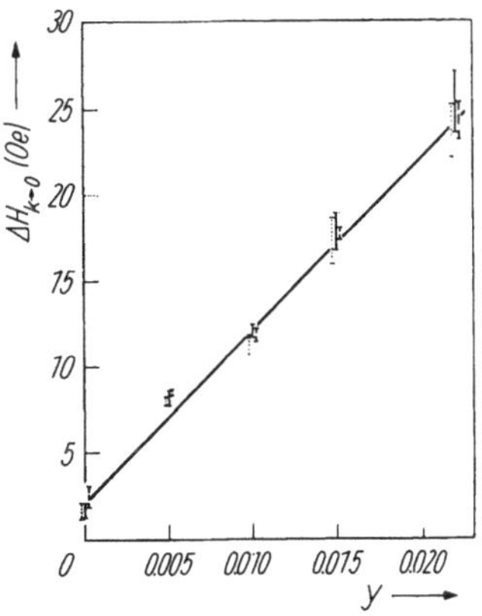

Fig. 4. $\Delta H_{k\to 0}$ as a function of y for $Mg_{1-y}Co_yFe_2O_{4+\gamma'}$ spheres along the ⟨111⟩ (— — —), ⟨110⟩ (····), and ⟨100⟩ (———) directions in the range $0.16 \lesssim x \lesssim 0.28$

Table 1

Dependence of FMR linewidth on the transverse demagnetizing coefficients N_t of samples which are an ellipsoid of revolution with the axes $a = b \geqq c$ for two $Mg_{0.95}Co_{0.05}Fe_2O_{4+\gamma'}$ spheres ($\Delta H^{\langle 110 \rangle}$) and two disks ($\Delta H_{\perp}^{\langle 110 \rangle}$) at 9.48 GHz, $x = 0.151$,
$\Delta H_{\perp}^{\langle 110 \rangle}$ for $MgFe_2O_{4+\gamma'}$ disks $\approx$ 15 Oe,
$\Delta H^{\langle 110 \rangle}$ for $MgFe_2O_{4+\gamma'}$ spheres 15 up to 30 Oe

a (mm)	b (mm)	c (mm)	a/c	$\frac{N_t}{4\pi}$	$\Delta H^{\langle 110 \rangle}$ (Oe)
1.25	1.25	1.25	1	0.333	188
1.35	1.35	1.35	1	0.333	178
					$\Delta H_{\perp}^{\langle 110 \rangle}$ (Oe)
3	3	0.50	6.0	0.108	122
3	3	0.40	7.5	0.089	130
3	3	0.28	10.7	0.065	117
3	3	0.25	12.0	0.059	117
3	3	0.18	16.7	0.043	117
3	3	0.14	21.4	0.034	117
3	3	0.50	6.0	0.108	146
3	3	0.40	7.5	0.089	139
3	3	0.27	11.1	0.063	137
3	3	0.23	13.0	0.055	130
3	3	0.17	17.6	0.041	132

linewidths for small Co^{2+} concentrations y. This scattering may be caused by 2 M processes induced by surface and volume non-uniformities. Therefore, we carried out measurements of the linewidth ΔH as a function of the shape of the sample (Table 1) and as a function of frequency (Fig. 5) on $Mg_{1-y}Co_yFe_2O_{4+\gamma'}$ samples with $y = 0.050$. Here the ΔH broadening as a result of the large Co^{2+} concentration dominates.

In Table 1 results of linewidth measurements of two disks (dc magnetic field $\boldsymbol{H}$ perpendicular to the plane) and two spheres are summarized. The transverse demagnetizing coefficients N_t were calculated after Osborn [21].

Fig. 5 shows that the contribution of Co^{2+} ions to the linewidths ΔH and

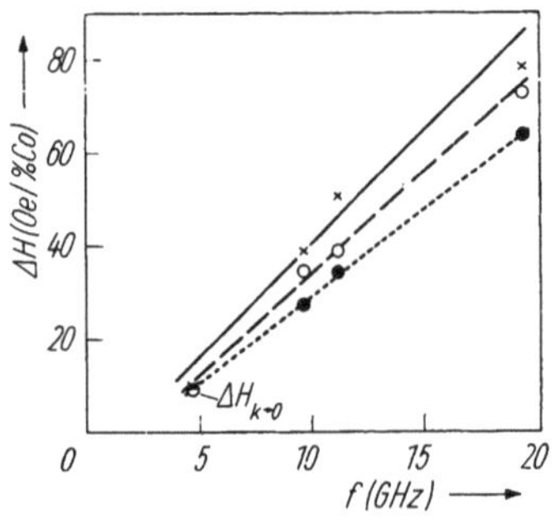

Fig. 5. Frequency dependence of the linewidths ΔH and $\Delta H_{k \to 0}$ per % Co with the static magnetic field oriented along ⟨111⟩ (●), ⟨110⟩ (○), and ⟨100⟩ (×) crystallographic directions. For $\Delta H_{k \to 0}$ we have to take $f = \omega_p/(2 \times 2\pi)$

$\Delta H_{k\to 0}$ is a linear function of frequency. This experimental fact is in agreement with a prediction from Vrehen et al. [8].

3. Discussion

In this section, the experimental results will be discussed in terms of the magnon–magnon scattering model [2]. In this model the contribution to the linewidth arising from small amounts of Co^{2+} ions on octahedral sites is given by

$$\Delta H \approx \frac{1}{4}\left(\frac{\alpha\,\lambda}{\gamma\,\hbar}\right)^2\left(\frac{4\,\pi\,M}{H_{ex}^3}\right)^{1/2}(3\,I_1 + 2\,I_2)\,c\,W(x) \qquad (3)$$

with $|\alpha\,\lambda|$ spin–orbit coupling constant of a Co^{2+} ion [22] in a trigonal crystal field, γ gyromagnetic ratio, M magnetization of the host ferrite, H_{ex} exchange field, I_1, I_2 dimensionsless functions depending only on sample shape and $\omega/4\,\pi\,\gamma\,M$ where ω is the resonance frequency of the $k = 0$ mode [23], c fraction of octahedral sites occupied by Co^{2+} ions, $W(x)$ probability for effectiveness of one Co^{2+} ion.

For a quantitative comparison of the results of parallel pumping measurements of Section 2.1 with equation (3) we have taken the parameters given in Table 2. One gets for the cation distribution parameter $x = 0.18$ a linewidth

$$\Delta H \approx 1.7\times 10^3\,c\,W(x)\ \text{Oe}$$

and for $x = 0.275$ a linewidth

$$\Delta H \approx 2.6\times 10^3\,c\,W(x)\ \text{Oe}\,.$$

Taking $y = 0.01$ or $c = 0.005$ and the probabilities [13]

$$W(x) = \frac{1.75 - 5.1\,x}{9.62}$$

in the range $0.11 \lessapprox x \lessapprox 0.28$ we obtain for $x = 0.18$ $\Delta H \approx 0.7$ Oe and for $x = 0.275$ $\Delta H \approx 0.5$ Oe. The experimental value found by parallel pumping measurements was 10 Oe for $y = 0.01$ independent of inversion degree. Hence the calculated contribution to linewidth is several times smaller than the experimental value.

Disks with the transverse demagnetizing coefficients N_t presented in Table 1 should give ΔH roughly one-twentieth of ΔH for a sphere if 2M scattering is dominant [2, 23]. But experimentally we found only a very much smaller reduction of linewidth (see Table 1). This result is in qualitative agreement with Teale [4]. The 2M linewidth of equation (3) should be nearly independent of the frequency. On the other hand Fig. 5 shows that the magnetic losses due to octahedral Co^{2+} ions increase strongly with frequency.

Table 2

Parameters of the equation (3) for the cation distributions $x = 0.180$ and $x = 0.275$

	$x = 0.180$	$x = 0.275$
$4\,\pi\,M$ (G)	≈ 2000	≈ 2900
effective Landé factor g_{eff}	≈ 2	≈ 2
$(3\,I_1 + 2\,I_2)$ for a sphere [23]	≈ 0.043	≈ 0.053
H_{ex} (Oe)	$\approx 10^6$	$\approx 10^6$
$\lvert\alpha\,\lambda\rvert$ (erg)	3.51×10^{-14}	3.51×10^{-14}

The conclusion of the present work is that the effect of Co^{2+} doping upon the magnetic losses cannot be explained by the 2M model. The linear frequency dependence, shown in Fig. 5, indicates a direct spin–lattice relaxation process.

Acknowledgements

The author would like to thank Prof. Dr. F. Voigt and Dr. E. Glauche for numerous helpful discussions on all aspects of this problem. Thanks are due to Miss A. Schmitz for helping in the measurements.

References

[1] R. L. WHITE, Phys. Rev. Letters **2**, 465 (1959).
[2] C. W. HAAS and H. B. CALLEN, Phys. Rev. **122**, 59 (1961).
[3] M. BUNINA, E. V. LEBEDEVA, A. I. PILCHIKOV, and I. I. SILVESTROVICH, Fiz. tverd. Tela **12**, 433 (1970).
[4] R. W. TEALE, J. appl. Phys. **33**, 1295 (1962).
[5] G. A. PETRAKOVSKII, I. K. PUCHOV, V. N. SELEZNEV, and A. I. DROKIN, Fiz. tverd. Tela **10**, 3186 (1968).
[6] T. OKADA, H. SEKIZAWA, and S. IIDA, J. Phys. Soc. Japan **18**, 981 (1963).
[7] M. D. STURGE, E. M. GYORGY, R. C. LECRAW, and J. P. REMEIKA, Phys. Rev. **180**, 413 (1969).
[8] Q. H. F. VREHEN, A. BROESE VAN GROENOU, and J. G. M. DE LAU, Phys. Rev. **B1**, 2332 (1970).
[9] I. K. PUCHOV, G. A. PETRAKOVSKII, V. N. SELEZNEV, and A. D. PROCHOROV, Izv. Akad. Nauk SSSR, Ser. fiz. **34**, 991 (1970).
[10] F. VOIGT, M. MANZEL, and A. SCHMITZ, phys. stat. sol. **35**, 329 (1969).
[11] F. VOIGT, XIV. Internat. Wiss. Coll., Vol. 3, TH Ilmenau 1969 (p. 71).
[12] M. ROSENBLOOM and R. W. TEALE, J. Phys. C **4**, 537 (1971).
[13] P. GÖRNERT, phys. stat. sol. (a) **13**, 101 (1972).
[14] P. GÖRNERT, XIV. Internat. Wiss. Coll., Vol. 3, TH Ilmenau 1969 (p. 81).
[15] E. SCHLÖMANN, J. appl. Phys. **34**, 1998 (1963).
[16] W. HAUBENREISSER, Z. angew. Phys. **18**, 432 (1965).
[17] F. VOIGT, phys. stat. sol. **31**, 621 (1969).
[18] D. LINZEN and F. VOIGT, Exper. Tech. Phys. **16**, 271 (1968).
[19] P. GÖRNERT, Promotion A, Techn. Univ. Dresden, 1971; Nachrichtentechnik **22**, 299 (1972).
[20] E. GLAUCHE and F. VOIGT, Exper. Tech. Phys. **11**, 196 (1963).
[21] J. A. OSBORN, Phys. Rev. **67**, 351 (1945).
[22] J. C. SLONCZEWSKI, Phys. Rev. **110**, 1341 (1958).
[23] H. B. CALLEN and F. PITTELLI, Phys. Rev. **119**, 1523 (1960).

(Received October 3, 1972)

phys. stat. sol. (a) 14, 517 (1972)

Subject classification: 14; 15; 22.8

Institute for Low Temperature and Structure Research, Polish Academy of Sciences, Wrocław

Electrical Resistivity of UAsY Single Crystals (Y=S, Se, Te)

By

A. Wojakowski, Z. Henkie, and Z. Kletowski

The results of electrical resistivity measurements on single crystals of UAsY type compounds in the temperature range from 4.2 to 650 K and in an electric field perpendicular to the [001] axis are presented and discussed.

Измерено электрическое сопротивление монокристаллов соединений типа UAsY в направлении перпендикулярном оси [001] в пределе температур от 4,2 до 650 K и проведено обсуждение полученных результатов.

1. Introduction

The UAsY compounds crystallize in the tetragonal system either with PbFCl (UAsS, UAsSe) or UGeTe (UAsTe) type of structure [1, 2]. Of a similar structure type are also the UX_2 (X pnigogen) and UOY (Y chalcogen) compounds which are antiferromagnetic at low temperatures. The compounds UAsY display a relatively strong anisotropy of magnetic properties [3]. They undergo a transition to magnetic ordering at 123, 113, and 66 K for UAsS, UAsSe, and UAsTe, respectively. According to the authors [3] the magnetic moments of uranium are aligned parallel to the [001] axis. Contrary to the UOY compounds, all the UAsY and UX_2 [4] compounds exhibit metallic conductivity and a relatively small coefficient of the thermo-electric power [1]. The electrical properties of the UAsY compounds have not been investigated as yet, except for a brief information given in [1].

2. Experimental Part

Ternary UAsY compounds in powder form were prepared by direct reaction of stoichiometric quantities of the components sealed in quartz capsules under vacuum. The components were gradually heated to 800 °C and held at this temperature during 3 d. Single crystals were obtained by the chemical transport method using iodine or bromine as carrier gas (about 5 mg/cm^3). For the temperature gradient 1000 → 950 °C the best conditions of the single crystal growth were obtained. Single crystals obtained in this way did have the shape of thin plates with maximum dimensions $15\times10\times0.5$ mm^3. X-ray examinations revealed that the c-axis was always perpendicular to the plate surface. Hence, the resistivity measurements were carried out in the direction perpendicular to the [001] axis. The measurements were performed with the conventional dc method on samples polished to dimensions of about $8\times1.5\times0.5$ mm^3, in the temperature range 4.2 to 650 K. The current and voltage leads were soldered

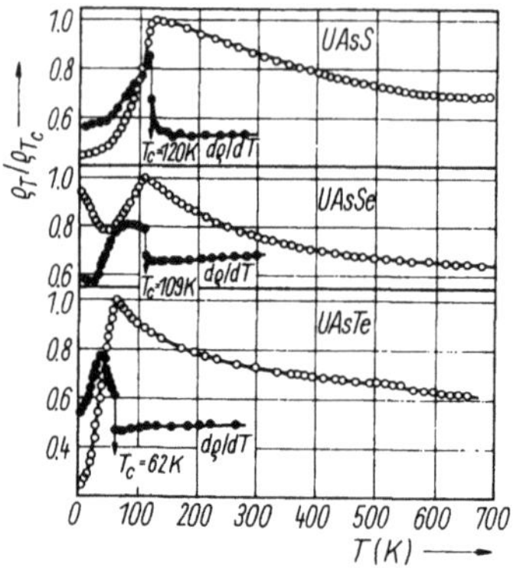

Fig. 1. The relative electrical resistivity ϱ_T/ϱ_{T_c} vs. T (open circles) and its temperature coefficient $d\varrho/dT$ (full circles) of UAsY compounds

to the sample by means of the Ga–Sn–Cu alloy for $T > 300$ K or the Viking alloy for $T <$ < 300 K.

3. Results and Discussion

The results of the resistivity measurements on UAsS, UAsSe, and UAsTe in the direction perpendicular to the [001] axis are given in Fig. 1 in the form of the $\varrho_T/\varrho_{T_c} = f(T)$ dependence. Absolute values of the transversal resistivity and thermo-electric power at room temperature are presented in Table 1. It is worth noting that the thermo-electric power values $\alpha \parallel [001]$ and $\alpha \perp [001]$ given in the table considerably differ from each other. This indicates a large anisotropy of the transport properties in these compounds.

Table 1

		UAsS	UAsSe	UAsTe
$\varrho_{300} \perp [001]$	($\mu\Omega$cm)	270 ± 20	280 ± 20	270 ± 20
$\alpha \perp [001]$	(μV/K)	+37	+17	+10
$\alpha \parallel [001]$	(μV/K)	+5	+4	+3
T_c (K)		120	109	62

The course of the temperature dependence of the resistivity indicates that the mechanism of current carrier scattering is rather complicated. It is known that the magnetic part of the electrical resistivity depends on both the type of the magnetic structure and its variation with temperature [5, 6], and on the domain structure [7] as well. These factors affect the transport properties due to changes in the shape of the Fermi surface, and the magnitude of this effect is anisotropic with respect to the magnetic reciprocal lattice vector $\boldsymbol{K}$. The lack of knowledge of both the shape of the Fermi surface and the magnetic structure in detail does not allow to discuss the obtained results quantitatively. However, it is worthwhile to compare the temperature dependence of the transversal resistivity of the UAsY and UX_2 compounds because in the direction perpendicular to the vector $\boldsymbol{K}$ this dependence exhibits quite the same anomaly in the vicinity of the critical temperature for both ferromagnets and antiferromagnets [8].

It is well known [9] that for magnetic materials displaying metallic conductivity the resistivity which arises from the scattering of carriers on spin waves follows at low temperatures ($T < (1/3)\ T_c$) either a $\varrho \sim T^2$ or a $\varrho \sim T^2 \exp(-\Delta/kT)$ law. In the antiferromagnetic UX_2 compounds, which have a similar distribution of atoms in the lattice as the UAsY compounds and a uniaxial character of magnetic ordering, the transversal resistivity follows either the theoretical T^2-dependence (in a wider temperature range than one would expect) or the $T^{5/2}$ dependence which has not been theoretically predicted so far. Similar

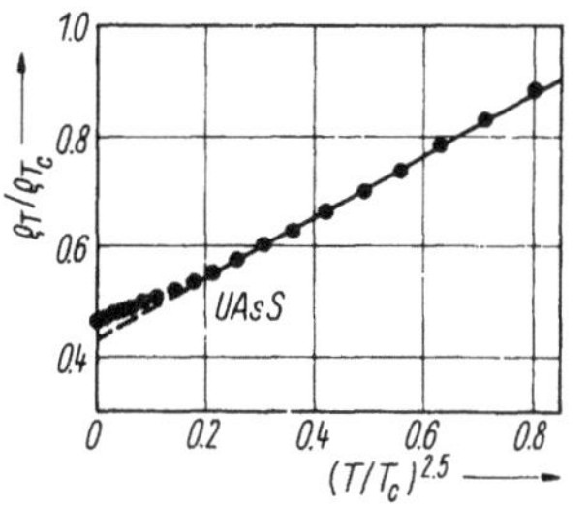

Fig. 2. The relative electrical resistivity ϱ_T/ϱ_{T_c} vs. $(T/T_c)^{2.5}$ for UAsS

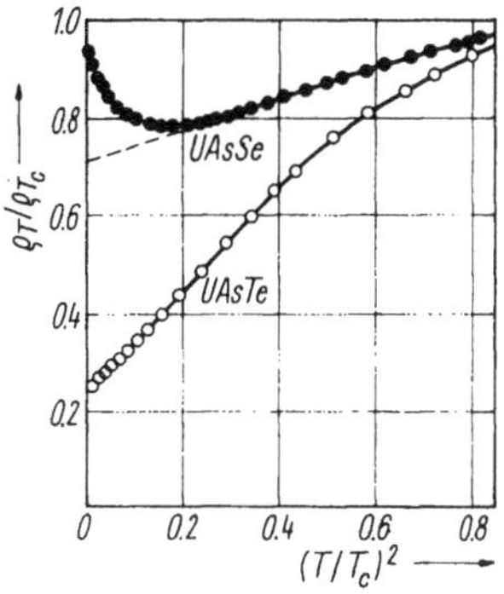

Fig. 3. The relative electrical resistivity ϱ_T/ϱ_{T_c} vs. $(T/T_c)^2$ for UAsSe and UAsTe

dependences have been found in the UAsY compounds, i.e., a quadratic power for UAsSe and UAsTe and a 5/2 power for UAsS, with the exception of the lowest temperatures (Fig. 2 and 3). In contrast to the UX_2 compounds, however, the three compounds examined display considerably lower values of the slope of the function $\varrho \sim A\ T^{\alpha}$, with the cofficients A being about one order of magnitude smaller. It should be noted that at very low temperatures some deviations of the resistivity curves in Fig. 2 and 3 from linearity are observed. The difference between the measured and extrapolated values of the residual resistivity is the lowest for UAsTe and the highest for UAsSe. In contrast to UAsS and UAsTe, the latter displays a negative temperature coefficient of the resistivity below 45 K. It has already been reported [3] that the magnetization of UAsS and UAsSe increases abruptly above a certain critical value H_{cr} of the magnetic field ($H_{cr} \approx 4$ kOe at $T = 4.2$ K). Bazan and Zygmunt [3] suggest that such a variation of the magnetization with field strength is due to a sudden transition from the compensated 180° domain structure into a one-domain structure. It can therefore be concluded that the high values of the residual resistivity and the anomaly in the $\varrho(T)$ function being observed for UAsY compounds have the same origin.

At the temperatures of the magnetic phase transition there appear knee-like maxima on the $\varrho(T)$ curves (Fig. 1). It seems that the occurrence of these maxima does not arise from the critical fluctuations in the nearest vicinity of T_c only. It was found that even at the highest temperatures measured ($\approx 10\ T_c$) the temperature coefficient of the resistivity remains still negative.

The temperature variation of the derivative $d\varrho/dT$ indicates a systematic change in the scattering character when passing from UAsS to UAsTe (Fig. 1). However, it seems that the peculiarities of the $d\varrho/dT$ curves found in the neighbourhood of T_c result mainly from the critical effects. Considering qualitatively the $d\varrho/dT = f(T)$ functions for $T > T_c$ one might conclude that in case of the scattering of current carriers in UAsS the short-range part of the spin fluctuations gives rise to a maximum of $d\varrho/dT$ for T_c^+. Such a behaviour of the $d\varrho/dT$ curves was also found in the UX_2 compounds [4]. On the other hand, the contribution of long-range spin fluctuations [11] increases when passing from UAsS to UAsTe, which manifests itself in the derivative $d\varrho/dT$ for T_c^+ being minimum in the case of the latter compound. A similar gradual variation of the character of the spin fluctuations has been observed for the Laves phases of rare-earth compounds and for dilute solutions of Fe in Pd [12].

The transition temperatures determined from the jump of the derivative $d\varrho/dT$ are 120, 109, and 62 K for UAsS, UAsSe, and UAsTe, respectively. There temperatures slightly deviate from those determined from magnetic measurements.

Acknowledgement

The authors are greatly indebted to Prof. W. Trzebiatowski for his stimulating interest in this work.

References

[1] F. HULLIGER, J. less-common Metals **16**, 113 (1968).
[2] A. ZYGMUNT and M. DUCZMAL, phys. stat. sol. (a) **9**, 659 (1972).
[3] C. BAZAN and A. ZYGMUNT, phys. stat. sol. (a) **12**, 649 (1972).
[4] Z. HENKIE and Z. KLETOWSKI, Acta phys. Polon., in the press.
[5] R. J. ELLIOT and P. A. WEDGWOOD, Proc. Phys. Soc. **81**, 846 (1963).
[6] Z. HENKIE and C. BAZAN, phys. stat. sol. (a) **5**, 259 (1971).
[7] R. C. FIVAZ, Phys. Rev. Letters **30A**, 72 (1969).
[8] Y. SUEZAKI and H. MORI, Progr. theor. Phys. (Kyoto) **41**, 1177 (1969).
[9] G. T. MEADEN, Contemp. Phys. **12**, 313 (1971).
[10] M. E. FISCHER and J. S. LANGER, Phys. Rev. Letters **20**, 665 (1968).
[11] D. J. KIM, Progr. theor. Phys. (Kyoto) **31**, 921 (1964).
[12] M. P. KAWATRA and J. I. BUDNICK, Internat. J. Magnetism **1**, 61 (1970).

(Received September 14, 1972)

phys. stat. sol. (a) **14**, 521 (1972)

Subject classification: 9; 21.1.1

Natuurkundig Laboratorium der Universiteit van Amsterdam (a) and Instituut voor Aardwetenschappen, Vrije Universiteit, Amsterdam (b)

Diffusion of Si in Fe–Si Containing 8 to 11 at% Si

By

H. V. M. MIRANI (a) and P. MAASKANT (b)

By means of an electron microprobe the diffusion of silicon in iron–silicon was measured in the temperature range from 900 to 1100 °C, using diffusion couples containing 7.64 and 11.1 at% Si. The diffusion coefficients turned out to be concentration dependent. Activation energies and frequency factors deduced are given.

Die Diffusion von Silizium in Eisen–Silizium wurde mittels Mikrosondenanalyse im Temperaturbereich zwischen 900 und 1100 °C gemessen. Die verwendeten Proben enthielten 7,64 und 11,1 At% Si. Die Diffusionskoeffizienten erwiesen sich als konzentrationsabhängig. Die ermittelten Aktivierungsenergien und Frequenzfaktoren werden angegeben.

1. Introduction

The temperature dependence of the diffusion coefficient D is usually given by the Arrhenius relationship

$$D = D_0 \, e^{-Q/RT} \,, \tag{1}$$

where D_0 and Q are constants called frequency factor and activation energy, respectively. T is the absolute temperature and R the gas constant.

For silicon diffusion in iron–silicon alloys only three measurements of D_0 and Q have – to the authors' knowledge – been reported in the literature [1 to 3]. These measurements cover various ranges of temperature and concentration of Si and yield different values for D_0 and Q. Moreover, measurements of

Table 1

Previous measurements of the activation energy and frequency factor for silicon diffusion in iron–silicon alloys

	ref.	Q (kcal/mol)	D_0 (cm²/s)	temperature region (°C)	concentration range (at%)	method
Batz et al.	[1]	48 ± 6	0.4	1095 to 1347	4.5 to 7.2	diffusion
Borg and Lai	[2]	52.5 ± 1.0	0.7 to 1.2	900 to 1450	0 to 4.7	diffusion
Mitani et al.	[3]	44	0.16	1130 to 1220	7	diffusion
Boesono et al.	[4]	69 ± 2	625	500 to 600	7.64	elastic Zener relaxation
Brommer and 't Hooft	[5]	68.5 ± 2	500	400 to 600	3 to 12	magnetic Zener relaxation

elastic [4] and magnetic [5] Zener relaxation, performed at temperatures between 400 and 600 °C, yield much higher values (see Table 1).

The aim of this investigation is to determine the activation energy and frequency factor at concentrations between 8 and 11 at% Si and to verify the difference between the relaxation data and the diffusion measurements.

The first author is responsible for the physical experiment and for the text of the paper; the microprobe analyses were carried out by the second author.

2. Experimental Procedure

With the aid of a spark erosion machine discs were cut from two monocrystalline iron–silicon bars containing 7.64 and 11.1 at% Si, respectively. These concentrations were checked by means of X-ray fluorescence as well as by electron microprobe analysis. After careful polishing the discs were annealed at 1100 °C during one hour. Three diffusion couples were made by pressing the polished surfaces of discs of both concentrations against each other during 15 min at a temperature of 950 °C and under a pressure of 0.25 kp/mm². After this treatment no recrystallization was observed.

The couples were spark-cut into eight samples, which were diffusion-annealed in a vacuum resistance furnace during times long enough to render the time of welding negligible. Measurement and control of the temperature of the furnace were estimated to be within 1 deg. After the diffusion anneal a layer of 1 mm thickness was cut from a surface perpendicular to the interface, to eliminate effects of surface diffusion. Then, after polishing this surface, concentration profiles were made by use of a Cambridge electron microprobe (type Geoscan), operated at an acceleration potential of 20 kV and with an electron beam diameter of about 1 μm. The intensity of the SiKα radiation was detected by two spectrometers, both equipped with KAP crystals, at take-off angles of 75°. On each sample scans were made on three separate lines, perpendicular

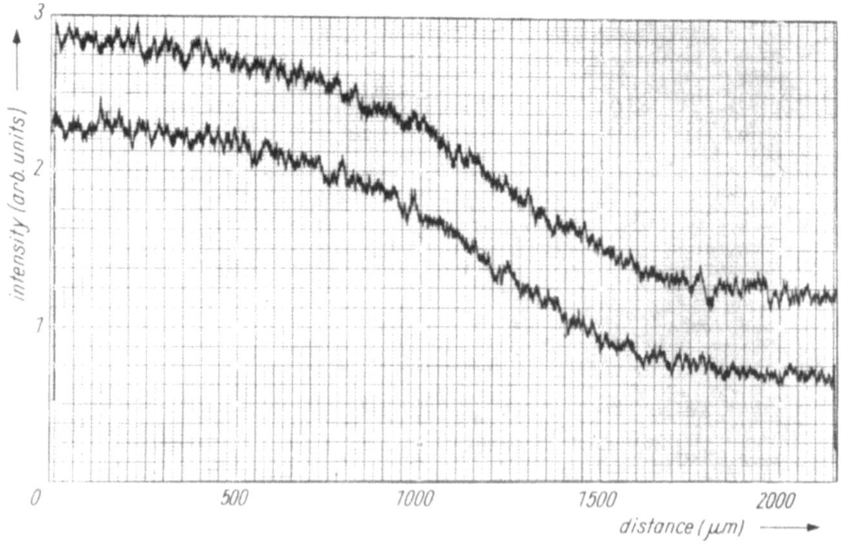

Fig. 1. SiKα intensities, obtained by the two spectrometers, for one of the scans

to the interface. The samples were moved with a speed of 50 μm/min, while the integrated number of counts was printed out every 10 s. The counting rate at 7.64% Si was about 4000 counts per 10 s, about 5% of this number being due to background radiation.

As a check the intensities were also continuously plotted on a pen-recorder. One of these plots is shown in Fig. 1.

The above-mentioned profiles were obtained by point analyses. Besides, for each sample also one scan has been made with the electron beam oscillating parallel to the interface, at a frequency of 400 s^{-1} and in a 20 μm wide interval, resulting in a larger volume analysed.

During the measurements the variations in specimen current were less than 0.5%.

For the calculation of the diffusion coefficients in each of the samples the numbers of counts per 10 s were used of every scan (if not deviating from the mean by more than 5 σ), as initial calibration measurements showed a linear dependence of concentration on X-ray intensity.

3. Results

The data were analysed by an electronic computer using a modified method of Matano, due to den Broeder [6]. His equation for the interdiffusion coefficient is

$$\tilde{D}_{c^*} = \frac{-1}{2\,t\left(\dfrac{\partial c}{\partial x}\right)_{x^*}} \left\{(1 - y_{c^*}) \int_{-\infty}^{x^*} (c - c_1)\,\mathrm{d}x + y_{c^*} \int_{x^*}^{\infty} (c_2 - c_1)\,\mathrm{d}x\right\}, \tag{2}$$

where c_1 and c_2 represent the initial concentration, c^* is the concentration, and x^* the corresponding x-coordinate at which D is to be calculated, y_{c^*} is a parameter given by

$$y_{c^*} = \frac{c^* - c_1}{c_2 - c_1}. \tag{3}$$

Darken's equation, written for this case as

$$\tilde{D} = \left(c_{\mathrm{Fe}}\, D_{\mathrm{Si}}^{\mathrm{FeSi}} + c_{\mathrm{Si}}\, D_{\mathrm{Fe}}^{\mathrm{FeSi}}\right)\left\{1 + \frac{\partial \ln \gamma_{\mathrm{Fe}}}{\partial \ln c_{\mathrm{Fe}}}\right\}, \tag{4}$$

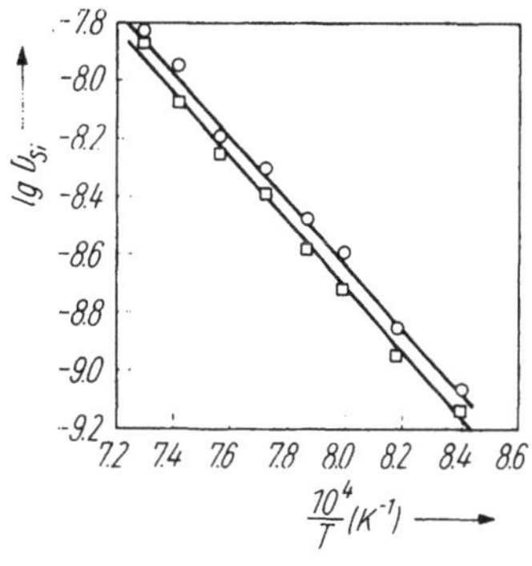

Fig. 2. Plot of lg D_{Si} vs. reciprocal temperature, calculated at concentrations of 8.35 (□) and 10.41% Si (○), respectively

gives a relation between the interdiffusion coefficient and the selfdiffusion coefficients of iron and silicon in the alloy, γ_{Fe} being the activity coefficient of iron. Measurements of the activity of iron in iron–silicon alloys have been performed by Vecher et al. [7]. They find a linear dependence of the activity on the atom fraction of iron in the region between 88 and 94.5% iron. This means that we can consider γ to be constant at our concentrations. Further using $D_{Si} \approx 5\, D_{Fe}$ as an approximate relation [2], silicon diffusion coefficients were calculated for seven concentrations in the interval between 7.64 and 11.1% Si. All samples showed about linearly increasing diffusion coefficients with growing silicon concentration. For clearness only the values at 8.35 and 10.41% are

Table 2

Our results, as calculated by least squares fit of the data. At 9.38 at% the data appear to have the best "correlation" (expressed by the correlation coefficient r)

c_{Si} (at%)	Q (kcal/mol)	D_0 (cm²/s)	r (%)	ΔQ (kcal/mol)	ΔD_0 (cm²/s)
8.35	51.43	1.82	99.79	1.4	+1.2 −0.7
8.69	51.37	1.87	99.87	1.1	+0.9 −0.5
9.04	51.12	1.77	99.90	0.9	+0.8 −0.5
9.38	50.82	1.62	99.90	0.9	+0.7 −0.5
9.73	50.58	1.52	99.88	1.0	+0.7 −0.5
10.07	50.55	1.55	99.84	1.2	+0.8 −0.5
10.41	50.66	1.66	99.79	1.3	+1.1 −0.6

given in Fig. 2. At each concentration D_0 and Q were calculated by least squares analysis. Their values have been listed in Table 2, together with the correlation coefficient r and the respective standard errors.

4. Discussion

As can be seen, comparing Table 2 with Table 1, the obtained values for the activation energy and frequency factor are not much different from the results of Borg and Lai [2]. However, our values tend to increase with decreasing silicon content. (The standard errors given in Table 2 are, of course, less representative in case of dependence of Q and D_0 on concentration.) Noticing this tendency it seems reasonable to attribute Borg and Lai's higher value of Q to the lower concentrations of their samples.

The results of Mitani et al. [3] are not confirmed by this investigation. The difference between the diffusion experiments and the relaxation measurements remains remarkable. Unfortunately no silicon diffusion measurements can be made at low temperatures, lacking a suitable silicon radioisotope.

Acknowledgements

The authors would like to thank Mr. J. M. E. de Lange for his aid in the microprobe measurements and Prof. Dr. G. de Vries for critically reading the manuscript.

Furthermore, the first author is indebted to Mr. R. Harthoorn for the computer programming and Dr. H. Bakker for helpful discussions.

References

[1] W. Batz, H. W. Mead, and C. E. Birchenall, J. Metals **4**, 1070 (1952).
[2] R. J. Borg and D. Y. F. Lai, J. appl. Phys. **41**, 5193 (1970).
[3] H. Mitani, M. Omishi, and T. Shikano, J. Japan Inst. Metals **30**, 56 (1966).
[4] Ir. Boesono, G. J. Ernst, M. C. Lemmens, M. J. van Langen, and G. de Vries, phys. stat. sol. **19**, 107 (1967).
[5] P. E. Brommer and H. A. 't Hooft, Phys. Letters (Netherlands) **26A**, 52 (1967).
[6] F. J. A. den Broeder, Scripta metall. **3**, 321 (1969).
[7] R. A. Vecher, Ya. I. Gerasimov, and V. A. Geiderikh, Zh. fiz. Khim. **39**, 650 (1965).

(Received September 11, 1972)

phys. stat. sol. (a) **14**, 527 (1972)

Subject classification: 13.4 and 16; 22.4.2

Lund Institute of Technology, Department of Solid State Physics, Lund, Sweden (a), and Department of Physics, University of St. Andrews, Fife, Scotland (b)

Photocapacitance Measurements of the Energy Levels in ZnSe:Mn

By

S. Braun (a), H. G. Grimmeiss (a), and J. W. Allen (b)

Photocapacitance measurements, using two simultaneous sources of illumination, show that manganese introduces an energy level in ZnSe which is 2.0 eV below the conduction band and 0.65 eV above the valence band at room temperature. There is some evidence that when the level is filled the manganese is in the d^5 configuration.

Unter gleichzeitiger Verwendung zweier Lichtquellen wird mit Hilfe von Photokapazitätsmessungen gezeigt, daß Mangan in ZnSe ein Energieniveau hervorruft, das bei Zimmertemperatur 2,0 eV unterhalb des Leitungsbandes und 0,65 eV oberhalb des Valenzbandes liegt. Es besteht Grund zur Annahme, daß das Mangan sich in der d^5-Konfiguration befindet, wenn das Niveau besetzt ist.

1. Introduction

Measurements of photoconductivity can give a great deal of information concerning the properties of impurities in semiconductors. If the semiconductor is highly conducting then the method is often not applicable, because the photosignal may be much less than the current noise. One way of overcoming this limitation is to remove free carriers from some region of the material by forming a rectifying junction, for then the photoresponse of the junction may be greater than the junction dark current. For example, Grimmeiss and Ottosson [1] investigated the properties of GaP:Cu in this way using an alloyed diode and Björklund and Grimmeiss [2] investigated the oxygen level in epitaxial GaP junctions. One may measure the effect of light on either the charge in the junction or the rate of flow of charge across the junction. The former is a photocapacitance measurement, the latter a photocurrent, or equivalently a photovoltage, measurement. Variations of these methods can give information on photoionization cross section spectra [2, 3], impurity concentration [2, 4], thermal emission rates [5], and so on. In general, photocurrent or photovoltage measurements give more accurate data because one is usually measuring a large signal on a small background whereas in the photocapacitance methods one measures a small change on a large background. Photocapacitance measurements, however, can give valuable information on the sign of photo-induced charge changes.

If one wants information on impurities in bulk material, there is an advantage in using Schottky barrier diodes since the depletion region is then of the same composition as the bulk material, unaffected by alloying or diffusion. However, in this case an extra complication arises in measuring photocurrent or photovoltage if one works in a spectral region where photoemission from the metal to the semiconductor gives an important contribution to the current. We have shown, in work to be reported elsewhere, that this effect can be separated from the impurity effect by using time discrimination. Photocapacitance measurements are not seriously affected by photoemission.

The material investigated here is ZnSe:Mn. Although transition metal impurities are of considerable importance as radiative or non-radiative recombination centres in semiconductors and phosphors, there is no adequate theory of their energy levels with respect to the host crystal energy levels. Experimental evidence is therefore valuable in throwing light on the problem. In addition, ZnSe:Mn is a useful electroluminescent material [6] so knowledge of its energy levels is of practical value.

2. Experiment

In order to avoid complications due to unwanted impurities, the ZnSe used in these experiments was synthesized from the elements of better than 5N purity. The concentration of zinc vacancies and of fast-diffusing impurities was reduced by treating the crystals in molten zinc. For purposes of comparison, ZnSe ingots either with or without added manganese were grown from the vapour. Aluminium was added during growth to provide shallow donors. Dices, approximately $2 \times 2 \times 1$ mm^3, were cut from the ingots. Schottky diodes were made by alloying an indium ohmic contact and evaporating a gold rectifying contact. All experimental results obtained here were obtained at room temperature.

Fig. 1 shows the spectrum of the short-circuit photocurrent. As mentioned in the Introduction, photoemission from the gold into the ZnSe accounts for a large part of the current. The threshold is about 1.4 eV; this corresponds to the barrier height and is close to the value reported by Swank et al. [7]. Although the spectra for ZnSe with and without manganese differ somewhat in the region around 2.5 eV, no clear conclusion concerning manganese can be drawn from the simple photocurrent curves.

Photocapacitance spectra are shown in Fig. 2. There is a pronounced peak between 2.1 eV and the bandgap for material containing manganese. There is no corresponding strong feature in the material without manganese, although there is a weak feature with a threshold about 0.1 eV lower than that of the manganese peak. The sign in the change in capacity for ZnSe:Mn shows that the positive space-charge in the depletion region is being increased. It follows that the action of light is to remove electrons from the manganese to the con-

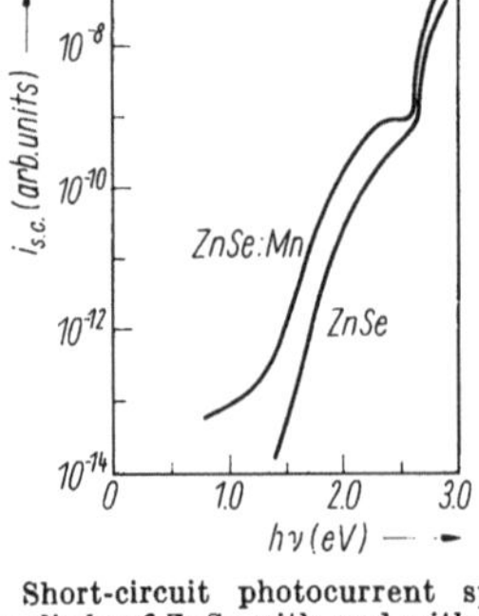

Fig. 1. Short-circuit photocurrent spectrum of a Schottky diode of ZnSe with and without manganese

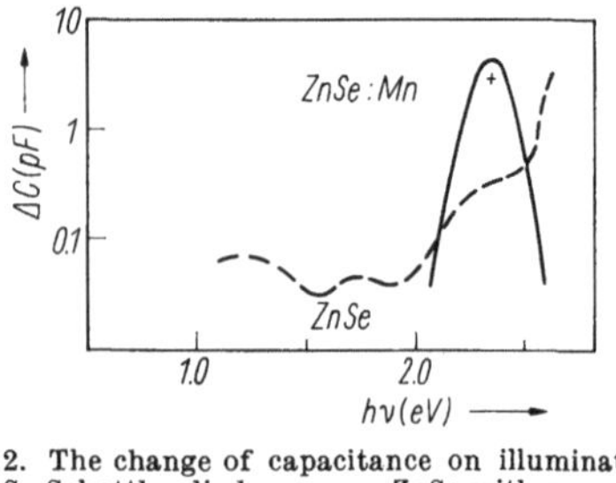

Fig. 2. The change of capacitance on illumination of a ZnSe Schottky diode. ——— ZnSe with manganese, – – – ZnSe without manganese

duction band. From the threshold of the excitation spectrum, we conclude that the filled manganese level lies roughly 2.1 eV below the conduction band.

Because the ZnSe is n-type, and because the Schottky barrier height is less than the depth of the manganese level, all the manganese levels are filled in the equilibrium state. Transitions from the valence band to the manganese levels are therefore not observed in the simple photocapacitance measurement. Suppose now that one illuminates the diode with a steady source of light of fixed wavelength, which we call the primary beam. If the primary photons have energy greater than the energy gap and can therefore produce holes, or if they have energy greater than the depth of the manganese level and can therefore photoionize the manganese, then some proportion f of the manganese levels will be emptied. The value of f can be obtained from the steady-state rate equations but since it depends on several unknowns, e.g., the various recombination cross sections, it is unnecessary for present purposes to write down an explicit expression for it. If a second beam of light, of variable wavelength, which we call the secondary beam, falls on the diode simultaneously with the primary beam then it can produce transitions from the valence band to the impurity, or from the impurity to the conduction band, depending on the wavelength. Fig. 3 is a plot of the photocapacitance spectra obtained in this way, in which the primary beam photons have energy 2.35 eV. In ZnSe:Mn there is a threshold at about 0.65 eV at which the secondary beam can excite electrons from the valence band to those manganese levels emptied by the primary beam. The sign of the photocapacitance change confirms that the transition is transferring electrons to the manganese. Let the photoionization cross section for transitions from the valence band to the impurity be $\sigma_v(h\nu)$, and let the concentration of impurity centres be N. Then if at some wavelength of the secondary beam the relationship

$$(1 - f)\, N\, \sigma_c(h\nu) = f\, N\, \sigma_v(h\nu) \tag{1}$$

can be satisfied, then the photocapacitance change produced by the secondary beam will go through zero. Obviously the photon energy at which this occurs is not the threshold energy for transitions to the conduction band but is that at which

$$\frac{\sigma_c}{\sigma_v} = f\,(1 - f)^{-1}\,. \tag{2}$$

Since the fraction of emptied centres, f, now depends on the intensity and wavelength of both the primary beam and the secondary beam, the position of the zero will also depend on these parameters. The zero must, however, lie to the high energy side of the threshold for transitions to the conduction band.

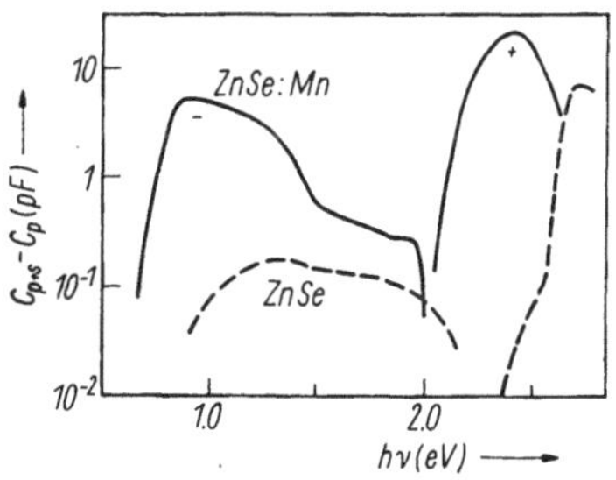

Fig. 3. The change of capacitance of a ZnSe Schottky diode produced by a secondary light source, with a constant primary light source of photon energy 2.35 eV. ——— ZnSe with manganese, – – – ZnSe without manganese

Also it is clear that if, in the vicinity of the zero, $\sigma_v f$ is small and only slowly changing while $\sigma_c (1 - f)$ is rapidly rising to a large value, then the zero will occur close to the threshold for $\sigma_c(h\nu)$. Fig. 3 shows this to be the case for ZnSe:Mn and we deduce that the manganese level is about 2.0 eV below the conduction band.

3. Discussion

The photocapacitance measurements described here show that in ZnSe:Mn there is an impurity level lying 2.0 eV below the conduction band and 0.65 eV above the valence band. Any possible ambiguity as to whether the level is 2.0 eV below the conduction band or 2.0 eV above the valence band has been resolved by observing the sign of the photocapacitance change. Zinc selenide prepared in a similar manner but not containing manganese gives a weaker photocapacitance signal, presumably due to stray impurities or lattice defects. We therefore conclude that the observed level is associated with manganese.

Optical absorption [8] and spin resonance [9, 10] measurements show that manganese in ZnSe is normally in the d^5 configuration. A detailed investigation of the optical absorption of ZnSe:Mn by Dr. E. M. Wray of the Department of Physics in the University of St. Andrews has shown that the manganese is on an isolated substitutional site and in the d^5 configuration, whether or not the material has been treated in molten zinc. No evidence for manganese in other states or associated in complexes has been found. There is therefore good reason for believing that the level 2.0 eV below the conduction band is produced by substitutional manganese atoms, and that when the level is filled the manganese is in the d^5 configuration. To make the identification certain, one should show that the density of levels and the density of Mn d^5 centres are equal. We have not yet attempted this.

It can be seen from Fig. 3 that the threshold energies can be determined to within $\pm$ 0.05 eV. In order to obtain more accurate information one may use photoconductance measurements. With more accurate data, one must then take phonon interactions into proper account in the interpretation. Work along these lines is in progress.

Acknowledgements

We thank Dr. E. M. Wray for letting us use his unpublished data on optical absorption. J. W. Allen is grateful to Tullis Russell Co. for a Fellowship.

References

[1] H. G. GRIMMEISS and M. O. OTTOSSON, phys. stat. sol. (a) **5**, 481 (1971).
[2] G. BJÖRKLUND and H. G. GRIMMEISS, Solid State Electronics **14**, 589 (1971).
[3] A. F. TASCH and C. T. SAH, Phys. Rev. **B1**, 800 (1970).
[4] C. T. SAH, L. L. ROSIER, and L. FORBES, Appl. Phys. Letters **15**, 161 (1969).
[5] C. T. SAH, L. FORBES, L. L. ROSIER, A. F. TASCH, and A. B. TOLE, Appl. Phys. Letters **15**, 145 (1969).
[6] J. W. ALLEN, A. W. LIVINGSTONE, and K. TURVEY, Solid State Electronics, to be published.
[7] R. K. SWANK, M. AVEN, and J. Z. DEVINE, J. appl. Phys. **40**, 89 (1969).
[8] D. W. LANGER and H. J. RICHTER, Phys. Rev. **146**, 554 (1966).
[9] R. S. TITLE, Phys. Rev. **131**, 2503 (1963).
[10] B. C. CAVENETT, Proc. Phys. Soc. **84**, 1 (1964).

(Received September 15, 1972)

phys. stat. sol. (a) **14**, 531 (1972)

Subject classification: 14.3; 13.1; 13.4; 22.2.4

Laboratoire de Physique des Solides, C.N.R.S., Meudon/Bellevue

Electrical Properties of S-Doped $Ga_xIn_{1-x}P$ Alloys

By

J. CHEVALLIER

Hall effect and resistivity measurements were performed on S-doped $Ga_xIn_{1-x}P$ alloys in the whole range of compositions x. In strictly direct band gap materials ($x < 0.57$), sulphur behaves like a hydrogenic impurity. Near the cross-over composition ($0.57 < x < 0.71$), a deep level due to sulphur appears; applying a two conduction band–two donor level model for the analysis, this level is shown to be tied to the X conduction band. Generally, another shallower level is present because of residual impurities. The depth of sulphur level is 170 meV under the X conduction band for $x = 0.65$ whereas it is 73 meV in GaP.

Des mesures électriques d'effet Hall et de résistivité ont été réalisées dans toute la gamme de composition des alliages $Ga_xIn_{1-x}P$ dopés au soufre. Dans les alliages à bande interdite purement directe ($x < 0,57$), le soufre se comporte comme une impureté hydrogénoïde. Au voisinage du point de croisement ($0,57 < x < 0,71$), il apparaît un niveau profond dû au soufre. Une analyse à l'aide d'un modèle à deux bandes de conduction et deux niveaux donneurs montre que ce niveau est lié à la bande de conduction X. Il apparaît aussi, en général, un niveau plus superficiel dû à la présence d'impuretés résiduelles. La profondeur du niveau donneur du soufre est de 170 meV pour $x = 0,65$ alors qu'elle est de 73 meV dans GaP.

1. Introduction

The choice of donor impurities in an electroluminescent junction depends upon some chemical and physical characteristics of the donor such as: solubility coefficient, electrical ionization energy, and carrier capture cross-section. The energy and the free carrier concentration are fundamental parameters in radiative recombination processes. Until now, no detailed information is available about electrical properties of undoped or doped $Ga_xIn_{1-x}P$ alloys. In this paper, we report Hall effect measurements on sulphur-doped $Ga_xIn_{1-x}P$ alloys in the whole range of compositions ($x = 0$ to 1) and temperatures (between 300 to 4 °K when possible). From these measurements, we obtain some results about the variation of the position of the sulphur level as a function of composition. It is well known that this serie of alloys consists of direct band gap materials in In-rich alloys, and indirect bandgap materials in Ga-rich alloys, due to a cross-over of Γ and X conduction bands.

2. Sample Preparation

Alloys were prepared by a solution growth method, as described elsewhere [1, 2]. Typically, the growth temperature is 980 °C. The composition of crystals is determined by an electron microprobe. The residual impurity content of non-intentionally doped samples will be indicated below. Roughly, the sign of the thermoelectric power for undoped crystals shows that they are n-type for

Table 1

band gap nature	No.	x	sulphur in bath (at. ppm)	n_H at 300 °K (cm^{-3})	μ_H at 300 °K (cm^2/Vs)	N_a (cm^{-3})	N_d (cm^{-3})	$N_{d'}$ (cm^{-3})	C_S *) (cm^{-3})	C_{Si} *) **) (cm^{-3})	E_d (meV)	$E_{d'}$ (meV)
direct gap	1	0.32	0	2.7×10^{16}	360				1.5×10^{16}	1.6×10^{16}		
	2	0.32	16	1.2×10^{17}	500							
	3	0.43	16	8×10^{16}	800							
cross-over region	4	0.586	16	8.8×10^{16}	390							
	5	0.60	50	5×10^{17}	110	8×10^{16}	2×10^{18}	2.3×10^{17}			45	≈0
	6	0.62	16	4×10^{16}	550	1.49×10^{17}	2×10^{17}	1.5×10^{17}	2.1×10^{17}	3×10^{16}	82	10
	7	0.63	16	4.3×10^{16}	100	1.49×10^{17}	2.7×10^{17}	1.5×10^{17}			90	22
	8	0.64	230	3.2×10^{17}	100	1.49×10^{17}	4.3×10^{18}	1.5×10^{17}	$\approx1.5\times10^{18}$	$\approx10^{17}$	95	≈0
–	9	0.675	70	1.3×10^{17}	90	9.9×10^{17}	3×10^{18}	10^{18}			156	58
	10	0.71	25	1.7×10^{16}	300	7.39×10^{17}	2.8×10^{18}	7.47×10^{17}			235	100
indirect gap	11	0.71	230	4×10^{17}	20	$>2\times10^{17}$	$>2.8\times10^{18}$		$\approx1.5\times10^{18}$	$\approx10^{17}$	136	
	12	0.83	16	7×10^{16}	120	1.7×10^{17}	3.2×10^{17}				99	
	13	1	16	10^{18}	63	1.3×10^{17}	3.5×10^{18}		7×10^{17}	2×10^{16}	73	

*) Sulphur and silicon concentrations were determined from semi-quantitative spectrographic mass analysis.
**) All other elements are less than 5×10^{16} cm^{-3} ($\approx$ 1 at. ppm) except for C ($\leqq6\times10^{17}$) and O ($\leqq3\times10^{17}$).

direct gap alloy compositions and p-type for other compositions. Crystals are doped by adding sulphur in the melt. Concentrations from 16 to 230 at. ppm were added. In few cases, the S (and also Si) content was known by mass spectrographic analysis. Table 1 presents the characteristics of sulphurdoped samples (No. 2 to 13) which will be used in this work.

Measurements were made on small parallelepipedic plates of $0.5 \times 1.5 \times 5$ mm³. In these conditions, the homogeneity $\Delta x/x$ is better than 1.5%. After polishing and etching in a hot 4 HNO_3/1 HCl mixture, five small-area contacts were made. Good ohmic contacts are obtained by fusing tin shot under hydrogen atmosphere. Samples, kept dark, were cooled in a liquid-helium cryostat by an exchange gas (helium).

3. Results

3.1 Direct band gap materials

Fig. 1 shows the variation of $n_H = 1/e\, R_H$ as a function of reciprocal temperature for three strictly direct gap crystals. On crystal with composition 0.32, we see that the introduction of sulphur (16 at. ppm in the indium bath) increases the number of carriers, particularly at low temperatures. Moreover, n_H is constant for all temperatures in doped samples. Sulphur introduces probably a very shallow level and for moderate and high dopings, this level merges in the conduction band. This behaviour is characteristic of a hydrogenic donor (a few meV) as in pure InP.

3.2 Cross-over region and indirect band gap materials

For $0.57 < x < 0.71$, a non-hydrogenic behaviour of the impurity appears as seen on Fig. 2. Curves n_H vs. $1/T$ have two slopes. Craford et al. [3] found exactly the same shape of curves in S-doped $GaAs_{1-x}P_x$ alloys. For $x > 0.71$, samples have an indirect band gap and the slope decreases. The curves become similar to those of S-doped GaP (Fig. 3).

3.3 Analysis of indirect band gap alloys

In a strictly indirect band gap semiconductor, characterized by M ellipsoidal valleys, the carrier concentration n_1 is given by

$$n_1 = r_1\, F_1 n_H$$

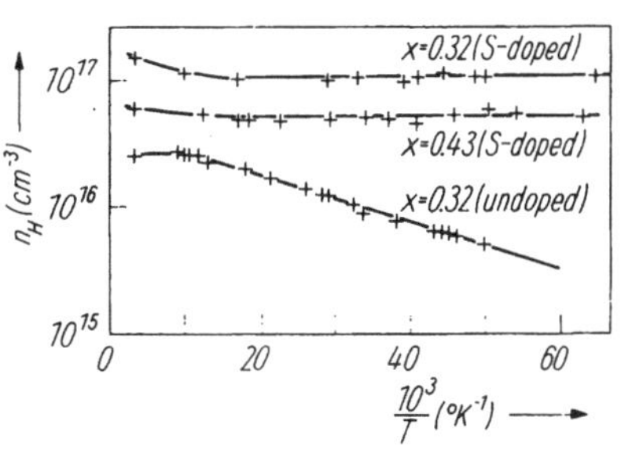

Fig. 1. Hall carrier concentration versus reciprocal temperature for S-doped and undoped direct band gap $Ga_xIn_{1-x}P$ alloys

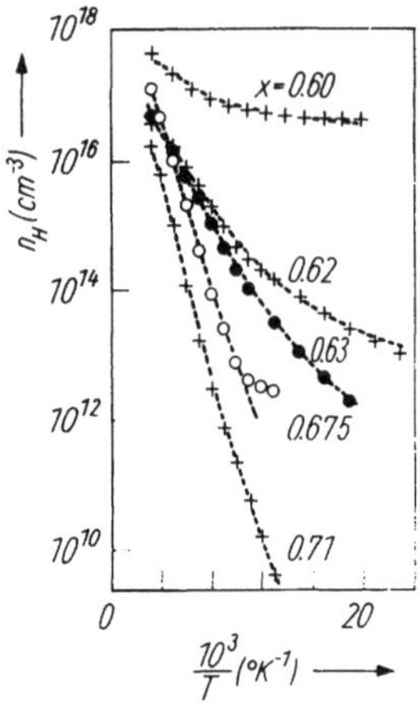

Fig. 2. Points show experimental Hall carrier concentration for compositions around the cross-over point. Dashed lines are calculated curves

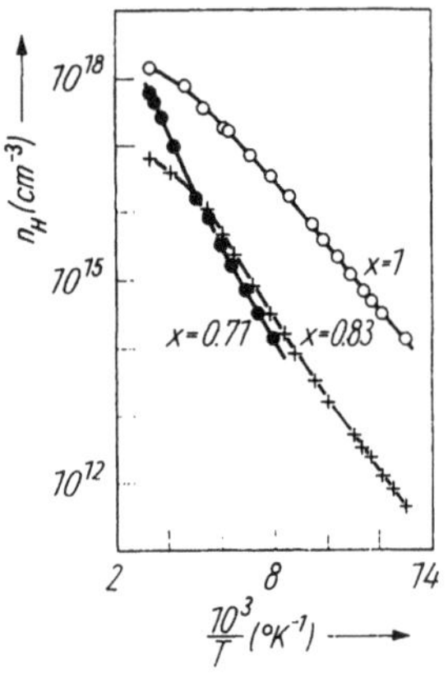

Fig. 3. Hall carrier concentration versus reciprocal temperature for S-doped indirect band gap $Ga_xIn_{1-x}P$ alloys

with

$$r_1 = \frac{\langle \tau^2 \rangle}{\langle \tau \rangle^2},$$

$$F_1 = \frac{3\,K\,(K+2)}{(2\,K+1)^2},$$

$$K = \frac{m_l^*}{m_t^*},$$

τ being the electron relaxation time and m_l^* and m_t^* the longitudinal and transverse electron effective masses, respectively.

Here, we shall always assume that r_i ($i = 1$ or 2) equals unity.

For a non-degenerate n-type semiconductor, with N_d donors and N_a acceptors, the donor ionization energy E_d is given by the well-known expression

$$\frac{n_1\,(n_1 + N_a)}{N_d - N_a - n_1} = \beta\,N_{c1} \exp\left(\frac{-E_d}{kT}\right),$$

where $1/\beta$ is the spin degeneracy of the donor (taken to be 2), N_{c1} the effective density of states in the conduction band. Assuming parabolic bands, we have

$$N_{c1} = 2\left(\frac{2\,\pi\,m_1^*\,kT}{h^2}\right)^{3/2},$$

where m_1^* is the density of states effective mass, given in a semiconductor with a silicon-type band structure, by

$$m_1^* = M^{2/3}\,(m_t^{*2}\,m_l^*)^{1/3}\,.$$

Young and Bass [4] developed recently some arguments showing that the conduction band in GaP is composed of three equivalent valleys. So, we shall take $M = 3$ for the X-band of all alloys. The values of m_l^* and m_t^* have been calculated by Onton and Taylor [5]. They find

$$m_l^* = 1.5\,m_0; \qquad m_t^* = 0.18\,m_0 .$$

We deduce

$$m_1^* = 0.76\,m_0; \qquad F_1 = 0.83 .$$

The energy of the X conduction band is not very sensitive to the composition x of alloys. So, we shall assume for all alloys the same parameter m_1^* as for GaP.

Results on strictly indirect band gap compounds are listed in the bottom of Table 1. In general, the ionization energy of a donor depends upon its concentration. Montgomery and Feldmann [6] gave this variation for sulphur doped GaP. For $N_d = 3.5\times10^{18}\ cm^{-3}$, our values (73 meV) well agrees with their results (76 meV) for $x = 1$. As seen on Table 1, the ionization energy of sulphur increases (from 73 to 136 meV) as x decreases from 1 to 0.71.

3.4 Analysis of alloys near the cross-over composition

Since we wish especially to interprete the samples 4 to 10, of composition near the Γ–X cross-over composition, we have to consider electrons located in two conduction bands. However this is strictly necessary only for samples near the cross-over composition (6 to 9). From simple Boltzmann factor arguments, it can be shown that, in our range of temperatures, only one band is populated in samples 5 and 10. The fact that these samples present two slopes in the R_H vs. $1/T$ curve suggest the presence of a second donor, besides S. This is the reason why we shall use a two-band, two-donor level model. This model can be defined using the following parameters:

$\Delta E = E_{c1} - E_{c2}$ the energy difference between the X and Γ conduction bands, with $i = 1$ for X-band, $i = 2$ for Γ-band, m_i^* the density-of-states effective mass in band i, N_{ci} effective density-of-states in band i, n_i electron concentration in band i, μ_i mobility of heavy ($i = 1$) and light ($i = 2$) electrons, $\sigma_i = n_i\, e\, \mu_i$ the contribution of each band to conductivity, R_2 Hall coefficient of light electrons (isotropic band), R_1 Hall coefficient of heavy electrons (anisotropic band).

In the limit of low magnetic field, Hall constant and resistivity of a semiconductor with two types of carriers are given by

$$R_H = \frac{R_1\,\sigma_1^2 + R_2\,\sigma_2^2}{(\sigma_1 + \sigma_2)^2}\,, \tag{1}$$

$$\varrho = \frac{1}{\sigma_1 + \sigma_2}\,.$$

Usually, the method for calculating R_1, σ_1, R_2, σ_2, consists of using expressions of R_H and ϱ at high magnetic field B and to measure their variations as a function of B. Plotting $\varrho_0/\Delta\varrho$ and $R_0/\Delta R$ versus B^{-2}, ϱ_0 and R_0 being zero-field values, one obtains two straight lines. Their slopes allow to calculate the four unknown parameters [7]. In our case, this method was not applied because of too low mobility values. Variations of only 2% for R and ϱ would be obtained for magnetic fields between 0 and 70 kG.

Replacing in (1) R_1, σ_1, R_2, σ_2 by the corresponding expressions, R_H and ϱ can be written

$$R_H = \frac{1}{e\,n_H} = \frac{1}{e\,n_2} \frac{1 + F_1 a^2 y}{(1 + a\,y)^2}, \tag{2}$$

$$\varrho = \frac{1}{n_2 e \mu_2 (1 + a\,y)}, \tag{3}$$

where

$$a = \frac{\mu_1}{\mu_2}, \qquad y = \frac{n_1}{n_2}. \tag{4}$$

The two carrier concentrations are known from general expressions:

$$n_1 = N_{c1}\,\mathcal{F}_{1/2}\left(\eta - \frac{\Delta E}{kT}\right),$$

$$n_2 = N_{c2}\,\mathcal{F}_{1/2}(\eta)$$

with

$$\eta = \frac{E_F - E_{c2}}{kT} \qquad (E_F \text{ Fermi level}) .$$

Hence, for a non-degenerate semiconductor, the ratio y is given by

$$y = \frac{N_{c1}}{N_{c2}} \exp\left(-\frac{\Delta E(T)}{kT}\right) = \left(\frac{m_1^*}{m_2^*}\right)^{3/2} \exp\left(-\frac{\Delta E(T)}{kT}\right). \tag{5}$$

From equations (2), (3), (4), (5), we can deduce, at all temperatures, the four values n_1, n_2, μ_1, μ_2 if the parameter a is known.

Experimental electron mobility values $\mu_H = R_H/\varrho$ are somewhat scattered (see Table 1). In strictly direct band gap polycrystals, the average mobility is about 570 cm²/Vs and in strictly indirect band gap materials it is about 65 cm²/Vs at room temperature. We shall take the average value 0.125 for a, so that

$$\frac{1}{a} = \frac{\mu_2}{\mu_1} = 8 .$$

As previously mentioned, m_1^* is taken to be constant throughout the whole composition range. On the contrary, m_2^* varies with composition. From recent electroreflectance measurements [8], this variation has been evaluated.

The determination of ΔE as a function of composition and temperature is more delicate. Γ conduction band energies, at 4.2 °K, were evaluated in the range $0 < x < 0.50$, from optical absorption measurements of Bachrach and Hakki [9]: these data are reported on Fig. 4. Recently, Onton et al. [10] determined the separation energy between Γ and X conduction bands in InP at 8 °K. Taking account of this energy ($\Delta E = 960$ meV) and photoluminescence results on indirect band gap materials [11] we can calculate the energy variation of the X conduction band vs. composition, assuming a quadratic law (Fig. 4). In these conditions, one finds the cross-over composition $x = 0.68$ at 4.2 °K.

Electroreflectance measurements [8] allow us to know accurately the energy variation of the Γ conduction band at 300 °K. Assuming the same temperature coefficient for the X conduction band in InP and in GaP, it is possible, from opti-

Fig. 4. Energy variation of different levels versus composition of S-doped $Ga_xIn_{1-x}P$ at 4.2 °K. Triangles (△) show S level variation (energy E_d). Crosses (×) correspond to the shallower level (energy E_d'), and □ to a deeper level found on sample 10. We have also plotted measurements (●) on Te-doped InP (Onton et al. [10])

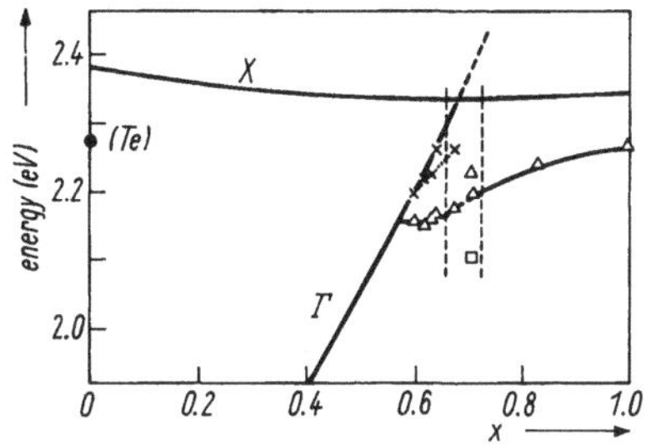

cal absorption measurements in indirect band gap alloys, to give the energy variation of the X conduction band. The cross-over composition is $x = 0.64$ at 300 °K.

Consequently, $\Delta E(x)$ will be taken to be constant from 80 to 4.2 °K and will be assumed to vary linearly with T, from 300 to 80 °K:

$$\Delta E(x, T) = \Delta E(x, 80\ °\mathrm{K}) - \alpha(x)\,(T - 80°)\,.$$

We now introduce the two donor levels, having ionization energies E_d and E_d' with respect to the lowest conduction band.

The neutrality equation can be written

$$N_d + N_d' - N_{dn} - N_{dn}' = n_1 + n_2 + N_a\,, \tag{6}$$

where N_a is the acceptor concentration, N_d and N_d', the donor concentrations. If $E_{c1} > E_{c2}$ (the opposite case would need some simple changes), the number of neutral donors $N_{dn}^{(i)}$ is given for each species by

$$N_{dn}^{(i)} = \frac{N_d^{(i)}}{1 + \beta \exp \dfrac{(E_{c2} - E_d^{(i)} - E_F)}{kT}}\,. \tag{7}$$

For a non-degenerate semiconductor:

$$\exp \frac{E_F - E_{c2}}{kT} = \frac{n_2}{N_{c2}} = \frac{n_1}{N_{c1}} \exp \frac{\Delta E}{kT}\,. \tag{8}$$

Using equations (2), (5), (6), (7), and (8), one expresses n_H as the solution of the equation

$$A\, n_H^3 + B\, n_H^2 + C\, n_H + D = 0\,,$$

where

$$A = \frac{(1+y)\,(1+F_1 a^2 y)^3}{\beta N_{c2}^2\,(1+a\,y)^6} \exp \frac{E_d + E_d'}{kT}\,,$$

$$B = \left\{N_a \exp \frac{E_d + E_d'}{kT} + \beta N_{c2}\,(1+y)\left(\exp \frac{E_d}{kT} + \exp \frac{E_d'}{kT}\right)\right\} \frac{(1+F_1 a^2 y)^2}{(1+a\,y)^4\, \beta^2 N_{c2}^2}\,,$$

$$C = \left\{\beta N_{c2}\,(1+y) + (N_a - N_d) \exp \frac{E_d'}{kT} + (N_a - N_d') \exp \frac{E_d}{kT}\right\} \frac{(1+F_1 a^2 y)}{(1+a\,y)^2\, \beta N_{c2}}\,,$$

$$D = N_a - N_d - N_d'\,.$$

4. Discussion

Inspection of Table 1 leads to some interesting remarks:

a) Let us compare samples 12 and 13. For an equivalent doping in the bath, we find, by mass spectrographic analysis, a larger sulphur concentration in sample 13 (GaP) than in sample 12 ($Ga_{0.83}In_{0.17}P$). For pure GaP, the relatively high sulphur content may be explained by a larger segregation coefficient of sulphur between gallium and GaP than between indium and $Ga_xIn_{1-x}P$.

b) Considering alloys with low doping level (<30 at. ppm), we find for N_d: 2.0×10^{17} to 3.2×10^{17} cm^{-3} (except sample 10: see below).

For doping levels in the range 30 to 70 at. ppm N_d increases to values of 10^{18} to 3×10^{18} cm^{-3}. At higher sulphur dopings (230 at. ppm), N_d increases up to 4.3×10^{18} cm^{-3}. So, donor concentrations N_d are well correlated to sulphur quantities which are introduced in the indium bath. They are also correlated to spectrographic analysis, although the latter tend to give a lower sulphur content than electrical measurements.

c) Sample 10 is exceptional in that its two donor levels seem to correspond to sulphur (100 meV) and to a *higher* ionization energy donor (235 meV) in higher concentration (2.8×10^{18} cm^{-3}).

d) For all other samples, the second donor is shallower than sulphur. Its origin is not clear, it may differ from sample to sample. In samples 5 and 8 it is particularly shallow (hydrogenic).

Considering samples 6, 7, and 9, we see that N'_d may be related to the sulphur concentration in the bath. So, we suppose that sulphur participates in the shallow donor although residual impurities (C, O, or Si) may be implied too.

The parameters R_d, E'_d, N_d, N'_d, and N_a are obtained from a least mean-square calculation. This two donor model gives good agreement between experimental and calculated values as seen on Fig. 2. Table 1 presents all results.

e) Fig. 4 shows energy variations of all relevant levels with alloy composition. Our experimental results are similar to those of Craford et al. for $GaAs_{1-x}P_x$ [3], but our interpretation is different. These authors neglect the existence of the shallower level (although they mention it for $x < 0.45$). Because of well-known compensation effects, they consequently find too low values of E_d in these alloys, which results in a sharp anomaly of $E_d(x)$. On the contrary we find that the distance between sulphur level and X band varies *uniformly* in the whole composition range $0.57 < x < 1$ (we recall that E_d is defined relatively to the lowest of the two conduction bands). Our value of 170 meV for $x = 0.65$ is in good agreement with the independent determination (by pressure experiments) of Craford (160 meV). Another possible comparison is with Onton et al. [10] who gives 106 meV for the distance between Te level and the X band in InP. We expect a somewhat larger ionization energy for sulphur; thence we conclude that the ionization energy of sulphur passes through a smooth maximum near the cross-over composition.

In our calculation, E_d and E'_d were assumed independent of temperature. In fact, if the S level is rigidly tied to the X conduction band, E_d varies with temperature, in the direct gap range, since ΔE is temperature dependent. This effect might be taken in account, leading to a small correction for the value of E_d.

5. Conclusions

In conclusion, we have performed Hall effect measurements, in the whole composition range, for S-doped $Ga_xIn_{1-x}P$ alloys. In direct gap materials, S introduces, as in InP, a shallow hydrogenic level. As x increases, the results were analysed with a model implying two conduction bands and two donor levels. We find that a level tied to the X conduction band appears for compositions $0.57 < x < 1$. This level is assigned to the S donor.

Its energy increases smoothly as x decreases. Another level appears in some crystals; its origin seems to be due to sulphur, carbon, oxygen, or silicon.

References

[1] J. Chevallier and H. Rodot, C.R. Acad. Sci. (France) **B271**, 1037 (1970).
[2] A. Laugier and J. Chevallier, phys. stat. sol. (a) **7**, 427 (1971).
[3] M. G. Craford, G. E. Stillman, J. A. Rossi, and N. Holonyak, Jr., Phys. Rev. **168**, 867 (1968).
[4] M. L. Young and S. J. Bass, J. Phys. D (Appl. Phys.) **4**, 995 (1971).
[5] A. Onton and R. C. Taylor, Phys. Rev. **B1**, 2587 (1970).
[6] H. C. Montgomery and W. L. Feldmann, J. appl. Phys. **36**, 3228 (1965).
[7] C. C. Y. Kwan, J. Basinski, and J. G. Woolley, phys. stat. sol. (b) **48**, 699 (1971).
[8] C. Alibert, G. Bordure, A. Laugier, and J. Chevallier, Phys. Rev. **B6**, 1301 (1972).
[9] R. Z. Bachrach and B. W. Hakki, J. appl. Phys. **42**, 5102 (1971).
[10] A. Onton, Y. Yacoby, and R. J. Chicotka, Phys. Rev. Letters **28**, 966 (1972).
[11] J. Chevallier and A. Laugier, phys. stat. sol. (a) **8**, 437 (1971).

(Received August 8, 1972)

phys. stat. sol. (a) **14**, 541 (1972)
Subject classification: 14.3 and 20.1; 11; 22.8

Institute of Physics, Bucharest (a), and
Helsinki University of Technology and Technical Research Centre of Finland, Otaniemi (b)

Reactor Irradiation of the Chalcopyrite and Sphalerite Forms of $ZnSnAs_2$

By
M. POPESCU (a), O. J. A. TIAINEN (b)[1]), and T. O. TUOMI (b)

The effect of reactor irradiation at 300 to 310 K on the resistivity and electroreflectance of chalcopyrite and sphalerite forms of $ZnSnAs_2$ was studied. It is found that the resistivity of the chalcopyrite $ZnSnAs_2$ increases with increasing fast neutron fluence while that of the sphalerite $ZnSnAs_2$ is rather insensitive to irradiation. The irradiation induced changes in the electroreflectance peaks in the region 2.0 to 2.8 eV attributed to N_{1v}–N_{1c} transitions in the chalcopyrite $ZnSnAs_2$ were measured. In the sphalerite $ZnSnAs_2$ a splitting of the broad E_1 peak at 2.1 eV is observed after irradiation.

Die Auswirkung der Bestrahlung im Reaktor bei 300 bis 310 K auf den spezifischen Widerstand und die Elektroreflexion des $ZnSnAs_2$ mit Chalcopyrit- und Sphaleritstruktur wurde untersucht. Der spezifische Widerstand des Chalkopyrit-$ZnSnAs_2$ nimmt mit zunehmender Dosis der schnellen Neutronen zu, während der spezifische Widerstand von Sphalerit-$ZnSnAs_2$ sich nicht merkbar verändert. Die Elektroreflexionsmaxima, die auf N_{1v}–N_{1c}-Übergänge von Chalkopyrit-$ZnSnAs_2$ im Bereich 2,0 bis 2,8 eV zurückzuführen sind, wurden vor und nach den Bestrahlungen gemessen. In Sphalerit-$ZnSnAs_2$ wurde eine Aufspaltung des breiten E_1-Maximums bei 2,1 eV beobachtet.

1. Introduction

The ternary $A^{II}B^{IV}C_2^{V}$ semiconducting compound $ZnSnAs_2$ studied in this work crystallizes in two forms: the sphalerite form which is of cubic structure where zinc and tin atoms can be thought to replace the gallium atoms in GaAs in a random manner and the chalcopyrite form which is tetragonal with $c/a = 2$, characterized by the ordered arrangement of zinc and tin atoms in the cation sub-lattice. The two forms of $ZnSnAs_2$ show important differences in their electrical and optical properties [1, 2]. Accordingly, it will be interesting to find out to what extent the effect of disorder produced by reactor irradiation could be seen, especially in the electroreflectance (ER) spectra of the two types of $ZnSnAs_2$.

2. Experimental Procedure

The samples were prepared by direct synthesis from stoichiometric quantities of the elements. The tetragonal form was obtained by slow cooling of the melt (about 100 deg/h) and the cubic form was prepared by water quenching. The samples obtained were polycrystalline with coarse grain size in the chalcopyrite form and finer grain size ($\approx$ 1 mm) in the sphalerite form.

The samples studied were irradiated in the core of the Triga Mark II reactor in Otaniemi. During irradiation the samples were cooled with a gaseous N_2

[1]) Present address: Helsinki Electricity Works.

stream through the irradiation chamber in order to keep the irradiation temperature at 300 to 310 K. Thermal neutrons were absorbed in a 0.4 mm thick cadmium shield surrounding the sample. The resistivity, thermoelectric power, and ER spectrum were measured before and after several irradiation periods. The ER measurements were carried out in the energy range of 1.3 to 2.8 eV using the electrolyte technique with KCl as the electrolyte. This derivative method gives a much sharper structure than the conventional reflectance measurement used in [3].

3. Experimental Results and Conclusions

$ZnSnAs_2$ exhibits p-type conductivity. The resistivity values, ϱ, at room temperature, after several fast neutron fluences are given in Table 1. In the chalcopyrite $ZnSnAs_2$ the hole concentration, p, decreased somewhat according

Table 1

The effect of reactor irradiation on the resistivity of chalcopyrite and sphalerite $ZnSnAs_2$

fluence (10^{17} neutrons/cm^2)*) (neutron energy >10 keV)	$ZnSnAs_2$ chalcopyrite-type (Ωcm)	$ZnSnAs_2$ sphalerite-type (Ωcm)
0	3.1×10^{-2}	1.9×10^{-3}
0.57	—	1.8×10^{-3}
1.1	5.0×10^{-2}	—
2.0	—	1.8×10^{-3}
2.3	1.1×10^{-1}	—
3.7	1.7×10^{-1}	—

*) The simultaneous gamma dose: 410 Mr per 10^{17} neutrons/cm^2.

to the thermoelectric power measurements (from p_0 before irradiation down to $\approx 0.5\,p_0$ after irradiation with 3.7×10^{17} neutrons/cm^2). The decrease in p was not sufficient to explain the increase in ϱ. Thus, after fast neutron irradiation, the hole mobility was decreased probably due to large scattering centres [4, 5]. The plot of the logarithm of the conductivity, σ, versus the fast neutron fluence $(nvt)_f$ in the chalcopyrite $ZnSnAs_2$ is approximately linear as shown in Fig. 1. This behaviour may also be interpreted as Crawford and Cleland did for silicon [6]. They suggested such a distribution of energy levels as to result in

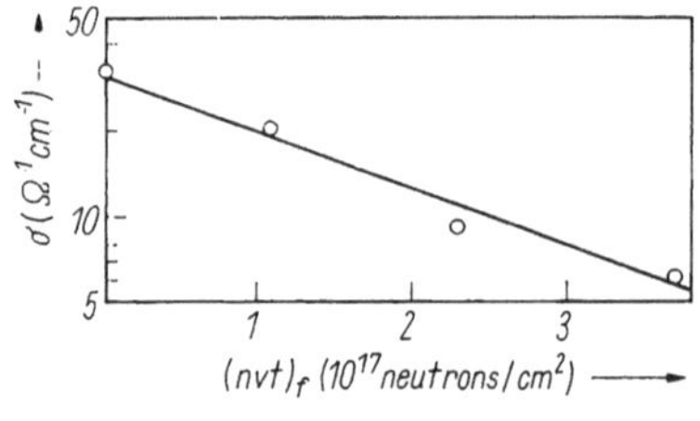

Fig. 1. The conductivity, σ, of chalcopyrite $ZnSnAs_2$ as a function of fast neutron fluence $(nvt)_f$ (neutron energy > 10 keV) measured outside the reactor at 300 K

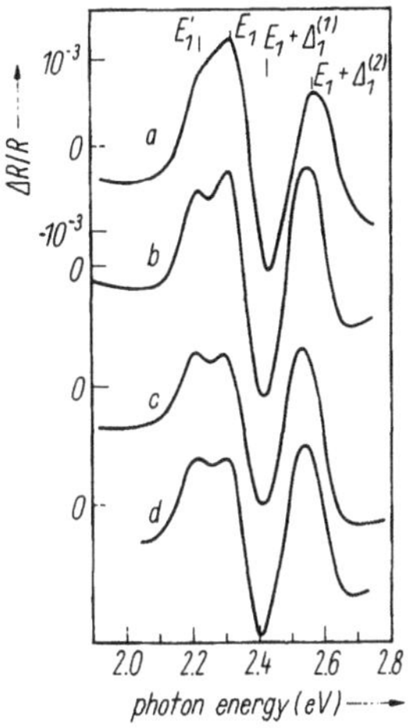

Fig. 2. The electroreflectance, $\Delta R/R$, of chalcopyrite $ZnSnAs_2$ at 300 K (dc voltage −1 V, ac voltage 1 V rms) before irradiation (a) and after fast neutron fluence, 1.6×10^{16} neutrons/cm² (b), 5.7×10^{16} neutrons/cm² (c), and 2.0×10^{17} neutrons/cm² (d)

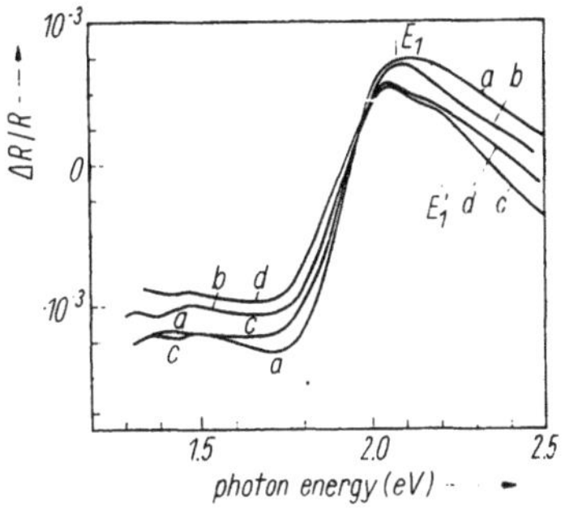

Fig. 3. The electroreflectance, $\Delta R/R$, of sphalerite $ZnSnAs_2$ at 300 K (dc voltage −1 V, ac voltage 1 V rms) before irradiation (a) and after fast neutron fluence, 1.6×10^{16} neutrons/cm² (b), 5.7×10^{16} neutrons/cm² (c), and 2.0×10^{17} neutrons/cm² (d)

a linear depression of the Fermi level towards the middle of the gap with neutron exposure. The resistivity of the sphalerite $ZnSnAs_2$ was rather insensitive to irradiation.

Fig. 2 and 3 show the ER spectra of the chalcopyrite and sphalerite forms of $ZnSnAs_2$ before irradiation and after several irradiation periods. The main radiation effects in the chalcopyrite $ZnSnAs_2$ are the non-linear shift of E_1 and $E_1 + \Delta_1$ towards lower photon energies and the changes in the relative intensities of the ER peaks. In Fig. 4 the energy values of the peaks are plotted as a function of the fast neutron fluence. The position of E_1' (2.21 eV) remains practically unchanged, the change being less than 0.005 eV after a fluence of 2×10^{17} neutrons/cm². The spin–orbit splitting Δ_1 was 0.25 eV before irradiation which is somewhat larger than that obtained by Kwan and Woolley [7]. The irradiation produced a small decrease in Δ_1 (in the order of 0.005 eV at $2 \times \times 10^{17}$ neutrons/cm²).

The ER peaks E_1', E_1, and $E_1 + \Delta_1$ can be attributed to the electronic transitions N_{1v}–N_{1c} derived from the transitions L_{3v}–L_{1c} in the sphalerite form [8]. All these transitions have a common final state, i.e. the lowest conduction

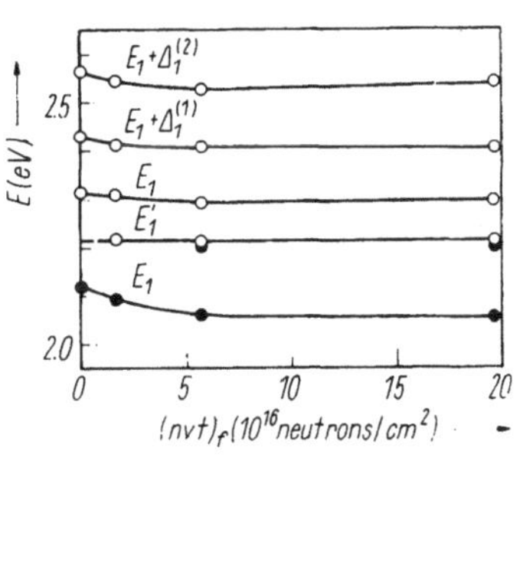

Fig. 4. The electroreflectance peak positions as a function of fast neutron fluence (peak denotions in Fig. 2 and 3). ○ chalcopyrite, ● sphalerite

band at the point N of the Brillouin zone. The different behaviour of E_1' from that of E_1 and $E_1 + \Delta_1$ means that irradiation has a stronger effect on the spin–orbit split valence bands than on the highest valence and the lowest conduction band at N. According to studies of the stress dependence of optical transitions in several semiconductors [9] the changes in the peak positions may be a result of an increase in the average lattice parameter or more likely due to internal stresses in the vicinity of the damaged regions.

The disorder induced by irradiation diminishes the intensity of the ER response as can be seen easily from Fig. 2. The ratio of the intensity of the peaks $E_1 + \Delta_1$ and E_1 increases with exposure due to different effects of the disorder on the two types of transitions.

In the ER spectrum of the sphalerite $ZnSnAs_2$ the E_1 peak shifts to lower energies with increasing fast neutron fluence. It must, however, be noticed that the photon energy values where $\Delta R/R = 0$ do not in practice change. This suggests that the shape of the reflectance maximum also changes when the sample is irradiated. The increased width of the ER signal may be explained by the effect of the imperfections and strains introduced in the material by water quenching. The irradiation tends to split the E_1 peak, the high energy split-off component of which is labeled as E_1'. Theoretical calculations [10] predict two peaks in the region investigated: at 2.0 eV corresponding to the Σ_2–Σ_1 transition and at 2.3 eV corresponding to the L_3–L_1 transition. The new peak E_1' at 2.22 eV may correspond to the theoretical peak at 2.3 eV. The appearing of the peak after irradiation may be explained assuming that the intensities of both peaks are affected by disordering in such a manner as to permit the observation of E_1'. Another explanation is also possible: splitting of the peak may be a result of local anisotropic stresses induced by irradiation.

Acknowledgement

The authors wish to thank Dr. R. Grigorovici for his interest in this work and valuable observations.

References

[1] K. MASUMOTO and S. ISOMURA, J. Phys. Chem. Solids **26**, 163 (1965).
[2] C. C. Y. KWAN and J. C. WOOLLEY, Canad. J. Phys. **48**, 2085 (1970).
[3] F. P. KESAMANLY, S. G. KROITORY, YU. V. RUD', V. V. SOBOLEV, and N. N. SYRBU, Dokl. Akad. Nauk SSSR **163**, 868 (1965).
[4] B. R. GOSSICK, J. appl. Phys. **30**, 1214 (1959).
[5] L. R. WEISBERG, J. appl. Phys. **33**, 1817 (1962).
[6] J. H. CRAWFORD, JR. and J. W. CLELAND, J. appl. Phys. **30**, 1204 (1959).
[7] C. C. Y. KWAN and J. C. WOOLLEY, phys. stat. sol. (b) **44**, K93 (1971).
[8] R. A. BENDORIUS, G. Z. KRIVAITE, A. YU. SHILEIKA, G. F. KARAVAEV, and V. A. CHALDYSHEV, Proc. 11th Internat. Conf. Phys. Semicond., Warsaw 1972, to be published.
[9] F. H. POLLAK and M. CARDONA, Phys. Rev. **172**, 816 (1968).
[10] V. V. SOBOLEV and V. I. DONETSKIKH, Izv. Akad. Nauk SSSR, Ser. neorganicheskie materialy **7**, 1141 (1971).

(Received September 15, 1972)

phys. stat. sol. (a) **14**, 545 (1972)
Subject classification: 10.1; 21.3; 21.6

System Development Division, International Business Machines Corporation, Kingston, New York

The Operative Slip Systems and Strain Hardening Behaviour of β'-AgMg Single Crystals

By
M. O. ABOELFOTOH

The operative slip systems, and the strain hardening behaviour in tension of single crystals of β'-AgMg (CsCl structure) has been investigated in the temperature range from 25 to 298 °K; a few compressive experiments are also reported. It is shown that β'-AgMg crystals deform by slip on the {112} ⟨111⟩, {110} ⟨111⟩, {123} ⟨111⟩, {120} ⟨001⟩, and {110} ⟨001⟩ systems. The activation of dislocations of $\boldsymbol{b} = \langle 111 \rangle$ and ⟨001⟩ depends on the crystal orientation, sense of the applied stress, and temperature. Comparison of the observed slip with those predicted from anisotropic elasticity theory suggests that factors other than dislocation energy and mobility must be considered in predicting the slip systems. The ordering energy and the separation width of $\frac{a}{2}$ ⟨111⟩ partials and the asymmetry of the Peierls stress on {112} planes appear to play an important role. Crystals which exhibit a linear hardening region (stage II) of relatively low rate of strain hardening slip by the movement of dislocations of $\boldsymbol{b} = \langle 111 \rangle$, while those exhibiting high rate of strain hardening slip by the movement of dislocations of $\boldsymbol{b} = \langle 001 \rangle$ which do not disorder the lattice.

Das wirkende Gleitsystem und das Druck–Verfestigungs-Verhalten von β'-AgMg (CsCl Struktur)-Einkristallen unter Spannung wurde im Temperaturbereich zwischen 25 und 298 °K untersucht. Zusätzlich wird über einige Kompressions-Experimente an diesen Einkristallen berichtet. Es zeigt sich, daß β'-AgMg-Kristalle sich durch Verschieben der {112} ⟨111⟩, {110} ⟨111⟩, {123} ⟨111⟩, {120} ⟨001⟩, und {110} ⟨001⟩ Systeme deformieren. Die Aktivierung der Versetzung $\boldsymbol{b} = \langle 111 \rangle$ und ⟨001⟩ hängt von der Kristallorientierung, der Spannungsrichtung und der Temperatur ab. Vergleiche zwischen den experimentell beobachteten und den aus der anisotropen Elastizitätstheorie abgeleiteten Versetzungen lassen vermuten, daß zusätzliche Faktoren zu der Versetzungsenergie und der Beweglichkeit in Betracht gezogen werden müssen, um die Verschiebung zu erklären. Die Anordnungsenergie und die Trennungsbreite von $\frac{a}{2}$ ⟨111⟩ Teilen und die Asymmetrie des Peierlsdrucks in der {112}-Ebene scheinen eine wichtige Rolle zu spielen. Kristalle mit einem linearem Härtebereich (Stufe II) mit relativ niedrigem Druckhärtungsanteil gleiten beim Versetzen der Verschiebung mit $\boldsymbol{b} = \langle 111 \rangle$, während die mit hohem Druckhärtungsanteil mit $\boldsymbol{b} = \langle 001 \rangle$ gleiten, was nicht zu einer Gitterstörung führt.

1. Introduction

The operative glide modes, the mechanisms of slip, and the work hardening characteristics of long-range ordered alloys and intermetallic compounds that crystallize in the B_2 lattice (CsCl type structure) are interesting not only from a practical viewpoint but equally from their theoretical implications. Whereas some B_2 alloys slip along the ⟨111⟩ vector, in a manner similar to that for b.c.c. metals and their random solid solution alloys, others slip along the ⟨001⟩ vector somewhat typical for ionic materials. The origin of these differences in terms of the antiphase boundary (APB) energies, σ, was first suggested by Rachinger and Cottrell [1].

Recently, Ball and Smallman [2] have shown that the Rachinger and Cottrell treatment does not satisfactorily predict the slip system in B_2 superlattices. They calculated dislocation energies and mobilities for the three compounds CuZn, NiAl, and CsBr, using anisotropic elasticity theory, and obtained excellent correlation between the observed and predicted slip systems.

The energies and mobilities of dislocations of $\boldsymbol{b} = \langle 001\rangle$, $\langle 011\rangle$, and $\langle 111\rangle$ have been calculated for the compound β'-AgMg and the results are summarized in Table 1. A unit $a\,\langle 111\rangle$ dislocation will exist in β'-AgMg as pairs of $\frac{a}{2}\,\langle 111\rangle$

Table 1

Dislocation parameters for β'-AgMg

compound	A	$\boldsymbol{b}$	plane	screw edge mixed	energy factor $10^{-1} K_{ij}[12]$ (dyn/cm²)	energy $10^1\,E$ (erg/cm)	$10^{-12}\,G$ (dyn/cm²)	ξ/b	S
β'-AgMg	3.48	[111]	($\bar{1}$10)	S	0.206	6.75	0.183	0.23	0.682
				E	0.504	16.516	0.183	0.562	0.207
				M (35°)	0.312	10.224	0.183	0.348	0.492
				M(55°)	0.406	13.304	0.183	0.453	0.331
		[111]	(1$\bar{1}$2)	S	0.206	6.75	0.183	0.133	0.725
				E	0.467	15.303	0.183	0.301	0.571
		[110]	($\bar{1}$10)	S	0.260	5.680	0.140	0.464	0.316
				E	0.366	7.996	0.140	0.654	0.135
		[110]	(001)	S	0.260	5.680	0.140	0.657	0.133
				E	0.493	10.770	0.140	1.245	0.006
				M(45°)	0.426	9.306	0.140	1.076	0.016
		[010]	($\bar{1}$01)	S	0.485	5.298	0.485	0.354	0.481
				E	0.416	4.544	0.485	0.303	0.568
		[100]	(010)	S	0.485	5.298	0.485	0.25	0.653
				E	0.366	3.998	0.485	0.189	0.724
				M(45°)	0.377	4.118	0.485	0.195	0.719

$A = 2\,C_{44}\,(C_{11} - C_{12})$ is the elastic anisotropy factor, where C_{ij} are the three independent elastic constants. $E = (K_{ij}\,b_i\,b_j)\,\frac{1}{4\pi}\ln\left(\frac{R}{r_0}\right)$ is the energy of dislocation per unit length. A value of 1.0 was taken for $\frac{1}{4\pi}\ln\left(\frac{R}{r_0}\right)$. ξ is the width of a dislocation, and S is a measure of its "relative mobility".

dislocations separated by a ribbon of antiphase boundary (APB). The energy of this extended dislocation is

$$E_e = E_1 + E_2 + E_{12} + \sigma\,w\,, \qquad (1)$$

where E_1 and E_2 are the energies of the superpartials bounding the APB,

$$E_{12} = \frac{K\,b_1\,b_2}{2\pi}\,\ln\left(\frac{R}{w}\right)$$

is the elastic interaction energy between the superpartials, and $\sigma\,w$ is the energy of the APB ribbon of width w and energy σ per unit area. The equilibrium separation of the superpartials, w, was calculated from the expression $w = K\, b_1\, b_2/2\,\pi\,\sigma$ using the value of the antiphase boundary energy estimated from the expression [3, 4].

$$\sigma = \frac{-4\, n^2\, h\, \varepsilon}{a^2\,(h^2 + K^2 + l^2)^{1/2}} \quad \text{for} \quad h \geqq K + l, \tag{2}$$

where

$$\varepsilon = \varepsilon_{AB} - \frac{1}{2}\,(\varepsilon_{AA} + \varepsilon_{BB}) = -\frac{K\,T_c}{4},$$

n is unity for stoichiometric compounds, (hkl) is the plane of APB, ε_{AB} is the energy of an A–B bond, ε_{AA} that of an A–A bond, ε_{BB} that of B–B bond, T_c is the critical disordering temperature and K is Boltzmann's constant. The compound β'-AgMg is ordered up to the melting point [5], in which case T_c was chosen to be the melting temperature. The calculated values of σ, w, and E_e are presented in Table 2.

Table 2

Calculated antiphase boundary widths, and superdislocation energies on various slip systems

compound	lattice parameter (Å)	bond ordering energy $10^{-14}\,E$ (erg)	slip system	antiphase boundary energy σ (erg cm^{-2})	antiphase boundary width w (Å)		energy $10^{-4}\,E_e$ (erg/cm)	
					screw	edge	screw	edge
β'-AgMg	3.305 [9]	3.77	{110}⟨111⟩	97.6	27.5	67.3	5.51	12.9
			{112}⟨111⟩	112.75	23.86	54.03	5.55	12.04

The results in Tables 1 and 2 show that the elastic energy of the dissociated $a\,\langle 111\rangle$ screw dislocation is about the same as that of $a\,\langle 010\rangle\,\{\bar{1}01\}$ dislocation in β'-AgMg. In addition, the elastic energy of the $a\,\langle 110\rangle\,\{110\}$ dislocation is approximately equal to that of the dissociated $a\,\langle 111\rangle$ dislocation. It can be seen then that in β'-AgMg the elastic energy of the $a\,\langle 010\rangle$, $a\,\langle 110\rangle$, and $a\,\langle 111\rangle$ dislocations are about the same and, hence, the glide parameter, S, may be the determining factor in predicting the slip mode.

The values of the relative mobility term in Table 1 suggest that slip in crystals stressed very close to $\langle 001\rangle$, where the resolved shear stress on dislocations of $\boldsymbol{b} = \langle 001\rangle$ is a very small fraction of the applied stress, should occur by the movement of dislocations of $\boldsymbol{b} = \langle 011\rangle$. The results reported in an earlier paper [6] and those of the present investigation show that crystals oriented near the $\langle 001\rangle$ pole deform in tension and compression by the operation of the $\{112\}\,\langle 11\bar{1}\rangle$ slip mode. Thus, even though $a\,\langle 110\rangle$ dislocations are expected to be present in β'-AgMg [the reaction $a\,\langle 100\rangle + a\,\langle 010\rangle \rightarrow a\,\langle 110\rangle$ to form $a\,\langle 110\rangle$ screw dislocation is energetically favorable (see Table 1)], these dislocations were not observed.

Recently, Potter [7] pointed out that the glide parameter, S, does not properly reflect the ease of movement of dislocations in CsCl-type intermetallic compounds. He suggested that the ordering energy plays a role, in addition to the dislocation mobility, since, for example, the ⟨110⟩ slip leads to forcing like atoms to become nearest neighbors at the half-slipped position. This would lead to the conclusion that dislocations of $\boldsymbol{b} = \langle 111 \rangle$ may be more mobile than those of $\boldsymbol{b} = \langle 110 \rangle$, and hence this would explain the experimental observations.

The results in Table 1 also show that the glide parameters of dislocations of $\boldsymbol{b} = \langle 111 \rangle$ and ⟨010⟩ are about the same. The dissociation of the $a\,\langle 111 \rangle$ dislocation into a superdislocation configuration will result in an increase in its mobility [8]. Thus, according to these results, the energies and mobilities of $a\,\langle 010 \rangle$ and $a\,\langle 111 \rangle$ dislocations are very close, and hence it is difficult to make predictions based on energy and mobility considerations alone. It will be shown that crystals with orientations near the ⟨110⟩ and ⟨111⟩ poles deform in tension by the operation of $\{hk0\}\,\langle 001 \rangle$ slip system and in compression by the operation of {112} ⟨111⟩ slip system, and that for crystals oriented near the ⟨001⟩ pole, the critical resolved shear stress in compression is higher than that in tension for slip on the {112} ⟨111⟩ system. This marked asymmetry in slip on the {112} ⟨111⟩ system will also be shown to arise from a tendency toward asymmetric splitting of the $\frac{a}{2}\,\langle 111 \rangle$ screw dislocation [9]. It is clear then that all the slip systems in β′-AgMg cannot be predicted from only a consideration of the energy and mobility (calculated in this way) of dislocations.

The study of strain hardening of long-range ordered alloys and intermetallic compounds that crystallize in the B_2 lattice (CsCl type structure) is still not frequent. Wood and Westbrook [10], and Robinson and Bever [11] have shown that the strain-hardening rate of stoichiometric polycrystalline β′-AgMg at ambient temperature is three to four times greater than that of nonstoichiometric alloys. These authors concluded that the two partials of $a\,\langle 111 \rangle$ superdislocations in the nonstoichiometric alloys are widely separated and move independently, creating ribbons of antiphase boundary (APB) and extensive strain-induced disorder. On the other hand, in the stoichiometric alloy, the superdislocation halves move as a unit with nonaligned jogs, hence, forming APB tubes that increase the strain hardening.

Flinn [3], Bragg [12], and Cupschalk and Brown [13] have pointed out that antiphase domains in the CsCl structure are unstable and hence a domain network would not be present in alloys of this structure annealed below the critical ordering temperature. Consequently, an increase in strain-hardening rate resulting from the cutting of domain walls by mobile superdislocations would not be expected in β′-AgMg. Furthermore, strain hardening arising from the cross-slip mechanism of Kear [14] or Davis and Stoloff [15] does not appear to be applicable to alloys having the CsCl lattice since this structure has no preferred low-energy orientation for APB corresponding to the cube planes in the $L1_2$ lattice [3, 13, 16]. It appears then that the high strain hardening in stoichiometric β′AgMg might arise from the Vidoz-Brown [17] mechanism of formation of APB tubes by motion of jogged superdislocations as suggested by Robinson and Bever [11], Terry and Smallman [18], and Marcinkowski and Chessin [19]. However, it will be shown that the high strain-hardening rate exhibited by β′-AgMg crystals is generally associated with the motion of dislocations of $\boldsymbol{b} = \langle 001 \rangle$ which does not disorder the lattice.

The present paper reports the results on the operative slip systems and strain hardening behaviour of β′-AgMg crystals deformed in the temperature range from 25 to 298 °K. The objective was elucidation of the principal mechanism of strain hardening.

Fig. 1. Crystallographic orientations of β'-AgMg single crystals

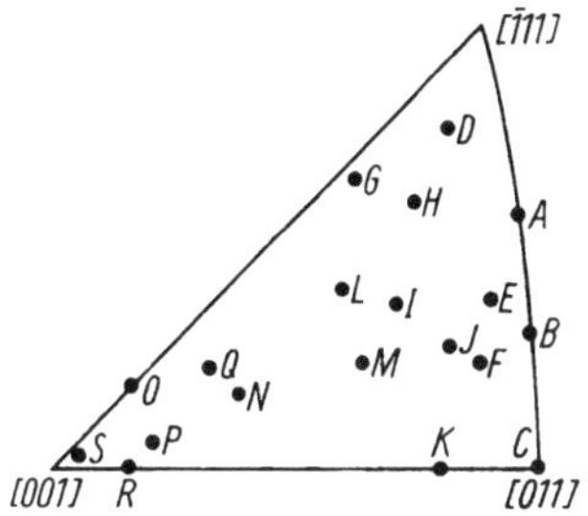

2. Experimental Procedure

The composition, method of preparation of β'-AgMg crystals, and mechanical testing of the crystals have been described in a previous paper [8]. Back-reflection Laue X-ray patterns were taken of all deformed crystals to determine the rotation of the tensile and compressive axes. Several crystals were X-rayed after various amount of tensile and compressive deformation to follow the rotation of the crystal during straining. The slip lines were observed using the Normarski interference contrast technique. Operative slip planes were determined by the two-surface trace analysis method. The slip-plane poles determined by the two-surface analyses were all within eight degrees of those reported as the operative slip planes. Stress and strain were resolved on the observed primary slip systems. The crystal orientations investigated are shown in Fig. 1.

3. Experimental Results

The orientation dependence of the strain hardening curves is shown in Fig. 2 for crystals tested in tension and compression at 237.5 °K. Comparison of the curves for crystals with tensile axis orientation near the [001] pole with those for orientations near the [011] and [$\bar{1}$11] poles shows a marked difference in deformation behaviour with orientation. Crystals with orientations S, R, P, N, and Q exhibited a region of decreasing hardening rate, stage O, followed by a region of easy glide, stage I, and then a linear region of relatively low rate of strain hardening, stage II. Tensile crystals with orientations A, B, C, D, E, F, H, and I exhibited parabolic stress–strain curves with higher rates of strain hardening. Crystals with orientations L and M exhibited a transitional deformation behaviour. These crystals displayed a short region of stage O and easy

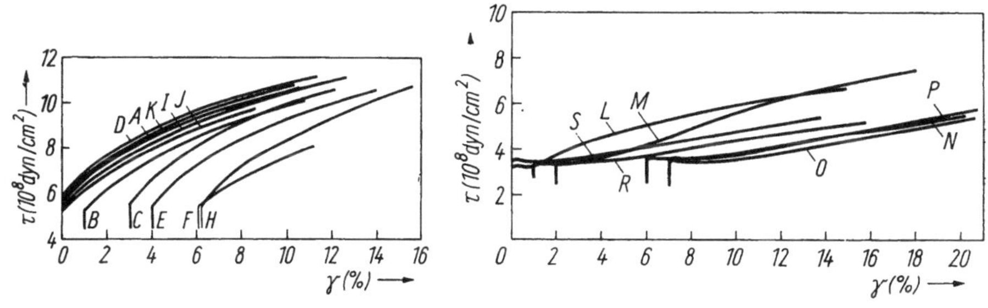

Fig. 2. Shear stress–shear strain curves of β'-AgMg crystals deformed in tension at 237.5 °K as a function of orientation. Strain rate $\gamma = 1.67 \times 10^{-5}\,s^{-1}$

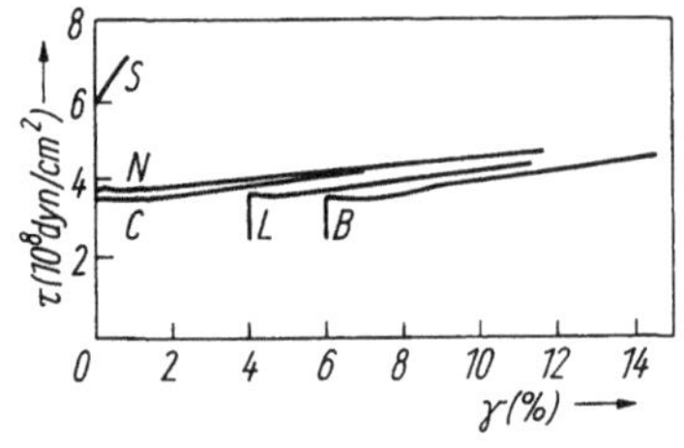

Fig. 3. Shear stress–shear strain curves of β'-AgMg crystals deformed in compression at 237.5 °K as a function of orientation. Strain rate $\dot{\gamma} = 1.67 \times 10^{-5}\ s^{-1}$

glide, similar to those crystals oriented near the [001] pole. With continued straining, the crystals developed parabolic stress–strain curves similar to tensile crystals with orientations near the [011] and [$\bar{1}$11] poles.

The orientation dependence in compression, Fig. 3, is generally the reverse of the orientation dependence in tension. Crystals oriented near the [011] and [$\bar{1}$11] poles exhibited a short region of stage O, easy glide, and linear strain hard-

Table 3

Observed slip systems in β'-AgMg single crystals stressed in tension and compression at 237.5 °K ($\dot{\gamma} = 1.67 \times 10^{-5}\ s^{-1}$)

orientation	tensile slip systems			compression slip system		
	primary	secondary	CRSS (10^8 dyn/cm²)	primary	secondary	CRSS ($10^8 \frac{dyn}{cm^2}$)
A	($\bar{1}$20) [001]		5.61			
B	($\bar{1}$20) [001]		5.21	($\bar{2}$11) [$\bar{1}\bar{1}\bar{1}$]		3.47
C	(010) [001]		5.5	($\bar{2}$11) [$\bar{1}\bar{1}\bar{1}$]		3.45
	(001) [010]			(211) [1$\bar{1}\bar{1}$]		
D	($\bar{1}$20) [001]	($\bar{2}$10) [001]	5.78	($\bar{2}$11) [$\bar{1}\bar{1}\bar{1}$]*		3.53*
E	($\bar{1}$20) [001]		5.45			
F	($\bar{1}$20) [001]		5.40			
H	($\bar{1}$20) [001]		5.48			
I	($\bar{1}$20) [001]		5.38			
J	($\bar{1}$20) [001]		5.35	($\bar{2}$11) [$\bar{1}\bar{1}\bar{1}$]	(211) [1$\bar{1}\bar{1}$]	3.69
K	(010) [001]*		5.45	($\bar{2}$11) [$\bar{1}\bar{1}\bar{1}$]		3.60
				(211) [1$\bar{1}\bar{1}$]		
L	(1$\bar{1}$2) [$\bar{1}$11]		3.21	($\bar{2}$11) [$\bar{1}\bar{1}\bar{1}$]		3.61
	($\bar{1}$20) [001]					
M	(1$\bar{1}$2) [$\bar{1}$11]		3.34			
	($\bar{1}$20) [001]					
N	(1$\bar{1}$2) [$\bar{1}$11]		3.55	($\bar{2}$11) [$\bar{1}\bar{1}\bar{1}$]		3.70
O	(1$\bar{1}$2) [$\bar{1}$11]*	(2$\bar{1}$3) [$\bar{1}$11]*	3.44			
		(1$\bar{2}$3) [$\bar{1}$11]*				
P	(1$\bar{1}$2) [$\bar{1}$11]	(2$\bar{1}$3) [$\bar{1}$11]	3.56			
Q	(1$\bar{1}$2) [$\bar{1}$11]	(2$\bar{1}$3) [$\bar{1}$11]	3.40			
R	(1$\bar{1}$2) [$\bar{1}$11]		3.38			
	($\bar{1}\bar{1}$2) [111]					
S	(1$\bar{1}$2) [$\bar{1}$11]	($\bar{1}\bar{1}$2) [111]	3.5	(1$\bar{1}$2) [$\bar{1}\bar{1}\bar{1}$]*	(211) [1$\bar{1}\bar{1}$]	6.0

* These results were reported in an earlier paper [6].

ening similar to the tensile stress–strain curves for crystals in orientations N through S. Those in orientations S near the [001] pole yielded parabolic stress–strain curves.

The operative slip systems for the various orientations stressed in tension and compression at 237.5 °K are presented along with the calculated values of the critical resolved shear stress (CRSS) on the primary slip planes in Table 3. It is evident from these results that β'-AgMg crystals deform by two different slip modes, depending on crystal orientation and sense of the applied stress.

It is also evident that there is a direct correlation between the stress–strain behaviour and the type of operative slip mode. Crystals exhibiting a short region of decreasing hardening rate (stage O), easy glide (stage I), and linear strain hardening (stage II) deformed by slip on the $\{112\}\langle 111\rangle$ primary system. Those exhibiting parabolic stress–strain curves generally deformed by slip on the $\{hk0\}\langle 001\rangle$ system.

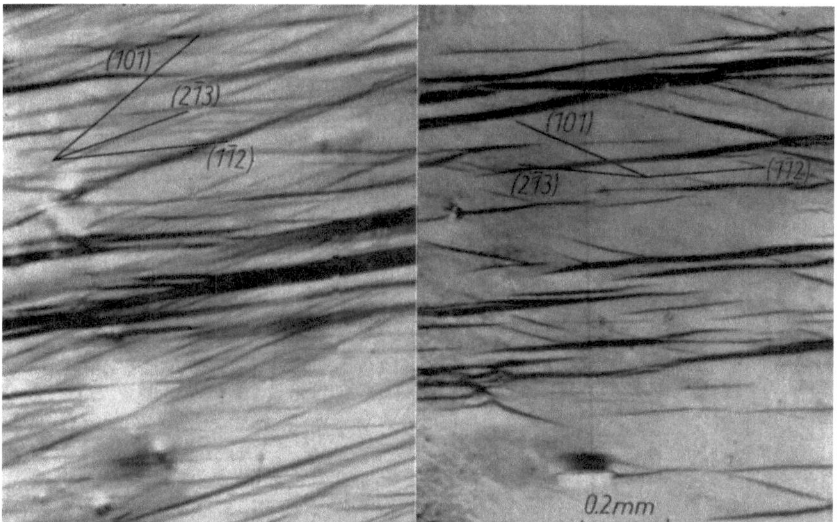

a b

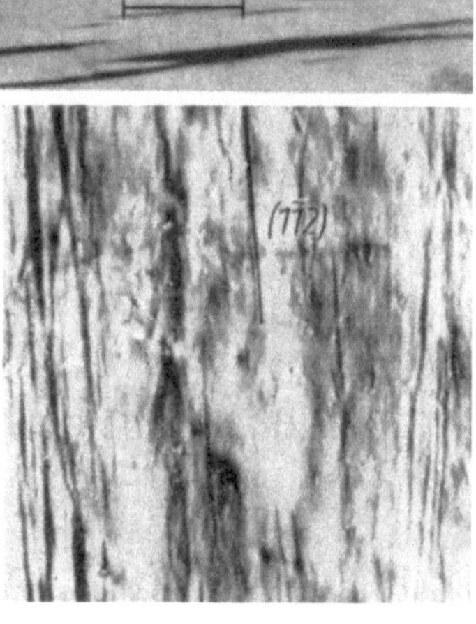

Fig. 4. Slip traces of crystals with orientation Q strained in tension at a) 237.5 °K, b) 158 °K, and c) 77° K. $\gamma = 10\%$

c

The slip lines corresponding to the primary $(1\bar{1}2)$ $[\bar{1}11]$ in Fig. 4 were initially straight, coarse, and inhomogeneous, separated by relatively wide undeformed regions. During easy glide (stage I) some widening of the original bands and new slip bands were observed in the undeformed regions. During the initial portion of the linear strain hardening (stage II), finer primary slip lines appeared in the undeformed regions and at higher strains the primary $\{1\bar{1}2\}$ slip lines became somewhat wave and cross-slip to the $\{2\bar{1}3\}$ planes was evident. The coarse slip lines indicate that the strain hardening rate within the primary slip bands is relatively small (in agreement with the observed stress–strain behaviour).

The slip line characteristics for the simultaneous operation of the $(\bar{2}11)$ $[\bar{1}\bar{1}\bar{1}]$ and (211) $[1\bar{1}\bar{1}]$ slip systems during compressive deformation of crystals with orientation C are shown in Fig. 5a. Comparison of the slip lines observed during the operation of the two independent $(\bar{2}11)$ $[\bar{1}\bar{1}\bar{1}]$ and (211) $[1\bar{1}\bar{1}]$ slip systems with those observed during the operation of the single $(\bar{2}11)$ $[\bar{1}\bar{1}\bar{1}]$ slip system,

a  b

Fig. 5. Slip traces of crystals with orientations a) C and b) B strained in compression at 237.5 °K

a 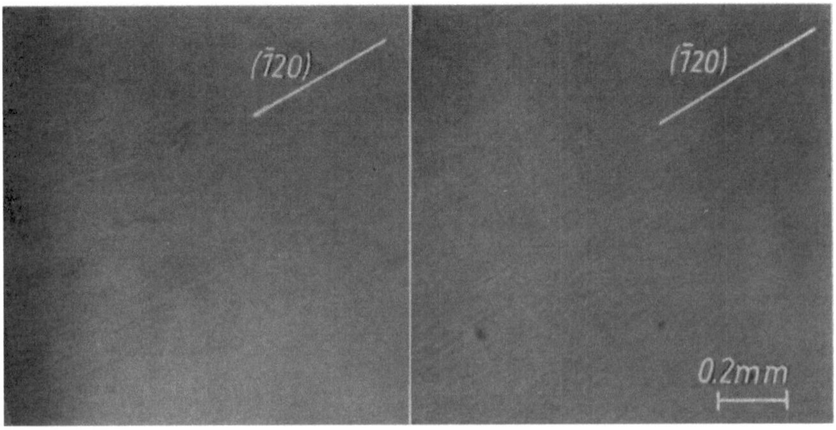 b

Fig. 6. Slip traces of crystals with orientations a) B and b) C strained in tension at 237.5 °K. $\gamma = 5\%$

Fig. 5b, reveals no differences in the slip line morphology. In addition, crystals deforming by the simultaneous operation of two independent slip systems (orientation C) exhibited a relatively low stage II hardening, identical to that for the operation of a single slip system (orientation L), Fig. 3. Furthermore, the deformation behaviour in tension exhibited by crystals with orientation R (deformed by the simultaneous operation of the $(1\bar{1}2)$ $[\bar{1}11]$ and $(\bar{1}\bar{1}2)$ $[111]$ slip systems) is identical to that exhibited by crystals with orientation N (deformed by the operation of a single $(1\bar{1}2)$ $[\bar{1}11]$ slip system), Fig. 2.

The slip lines corresponding to the $\{kh0\}$ $\langle 001\rangle$ system, Fig. 6, were initially extremely fine and homogeneous. With increased straining the lines became coarser and wavy. The fine and homogeneous slip lines are characteristics of deformation in which an effective strain hardening mechanism is operative, consistent with the observed stress–strain behaviour.

Crystals with orientations L and M exhibited a transition in the primary slip mode from $(1\bar{1}2)$ $[\bar{1}11]$ to $(\bar{1}20)$ $[001]$. The fine slip lines corresponding to the

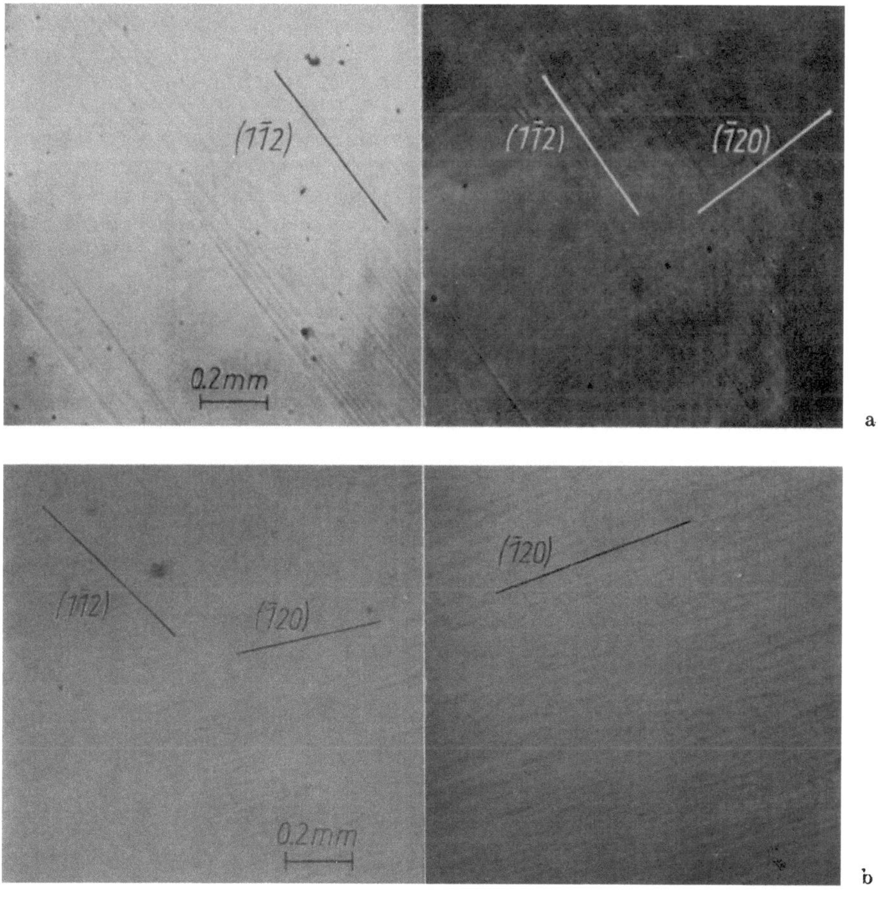

Fig. 7. Slip line characteristics during transition from $(1\bar{1}2)$ $[\bar{1}11]$ slip to $(\bar{1}20)$ $[001]$ slip at 237.5 °K. a) orientation L, b) orientation G

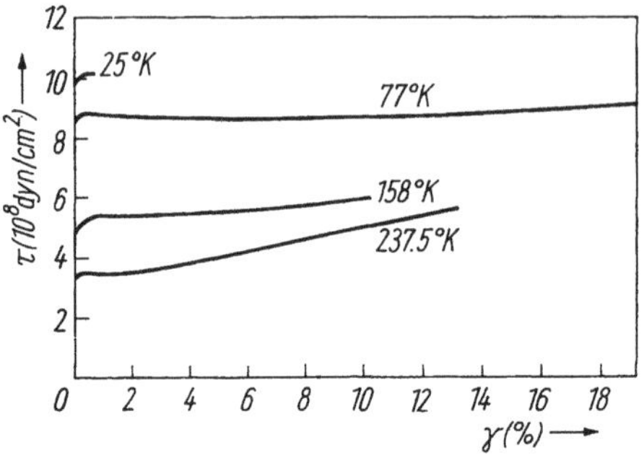

Fig. 8. Shear stress–shear strain of crystals with orientation Q deformed in tension as a function of temperature. Strain rate $\dot{\gamma} = 1.67 \times 10^{-5}\,s^{-1}$

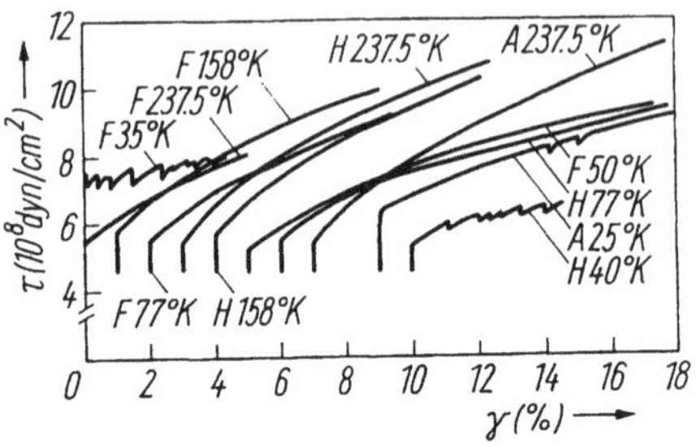

Fig. 9. Shear stress–shear strain of crystals with orientations A, F, and H deformed in tension as a function of temperature. $\dot{\gamma} = 1.67 \times 10^{-5} s^{-1}$

$(\bar{1}20)$ [001] slip appeared to extend outward from the $(1\bar{1}2)$ $[\bar{1}11]$ slip bands, Fig. 7. The appearance of these fine slip traces occurred at a resolved shear stress below that observed for yielding on the $(\bar{1}20)$ [001] system. This suggests that the [001] dislocation sources were activated in the vicinity of the primary $(1\bar{1}2)$ $[\bar{1}11]$ slip bands because of the local high internal stresses. A rapid increase in the strain hardening rate was observed after the onset of $(\bar{1}20)$ [001] glide.

The temperature dependence of the strain hardening curves is shown in Fig. 8 and 9 for crystals with orientations A, H, F, and Q. Crystals with orientation Q deformed initially by the $(1\bar{1}2)$ $[\bar{1}11]$ primary slip system and exhibited a decrease in the extent of both stage O and I, and an increase in the rate of linear hardening with increasing temperature above 77 °K, Fig. 8. The initial primary $(1\bar{1}2)$ plane slip traces became somewhat wavy and cross-slip to $(2\bar{1}3)$ planes was observed with increased deformation at and above 158 °K, Fig. 4. Crystals deformed at 298 °K exhibited wavy slip traces and the predominant slip planes appeared to be $\{2\bar{1}3\}$. At ≈ 25 °K, a sharp load-drop was observed after less than one percent prestrain, and was accompanied by the formation of deformation bands (not shown in Fig. 8). In addition to primary slip on the $(1\bar{1}2)$ $[\bar{1}11]$ system, minor amounts of slip on the $(\bar{1}20)$ [001] system were also observed in the vicinity of the deformation bands. The strong temperature dependence of the CRSS for slip on the $(1\bar{1}2)$ $[\bar{1}11]$ system below about 225 °K, Fig. 10, has been shown [20] to arise from the thermally activated motion of dislocations over Peierl's barriers. At higher temperatures, the CRSS is controlled by an athermal mechanism.

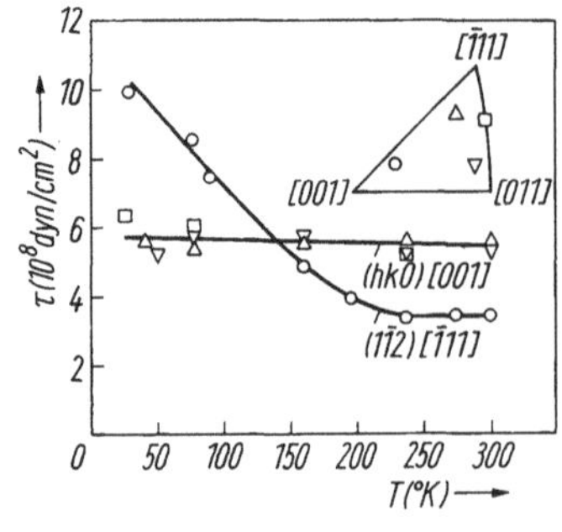

Fig. 10. Temperature dependence of critical resolved shear stress τ^* for $(1\bar{1}2)$ $[\bar{1}11]$ and (*hk*0) [001] slip systems. $\dot{\gamma} = 1.67 \times 10^{-5} s^{-1}$

a

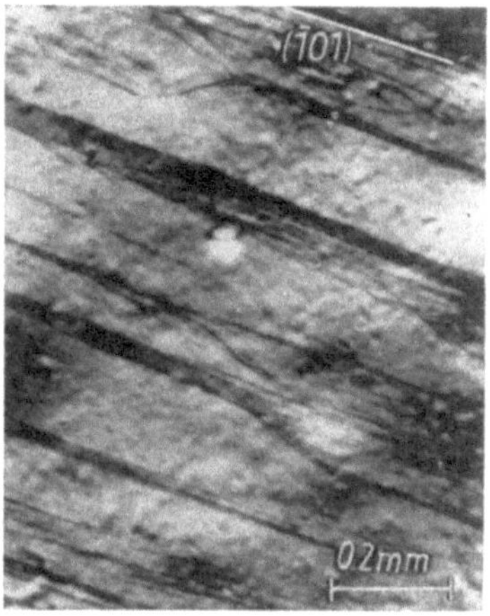

b

Fig. 11. Slip traces of crystals with orientation F strained in tension at a) 237.5 °K and b) 35 °K. $\gamma = 5\%$

The stress–strain curves of crystals with orientations A, H, and F was essentially independent of temperature from approximately 50 to 298 °K, Fig. 9. Below about 50 °K, crystals with orientations H and F exhibited a higher but erratic yield stress and highly serrated stress–strain curves, Fig. 9. In addition, slip traces corresponding to low stressed $\{h0l\}$ and $\{0kl\}$ planes were observed, Fig. 11, along with deformation bands. This observation was reconfirmed several times. When the stress at yielding was resolved on these planes, it was found that the CRSS was approximately the same as that on $\{hk0\}$ planes observed at temperatures above 50 °K. However, the CRSS on the highly stressed $\{hk0\}$ was now higher than that at temperatures above 50 °K. The non-crystallographic $\{hk0\}$ planes observed at and above 158 °K tended to $\{110\}$ planes at temperatures down to approximately 50 °K, particularly in orientations near the $[\bar{1}11]$ pole. Furthermore, while crystals with orientation A deformed by the operation of a single $\{hk0\}$ $\langle 001 \rangle$ slip system at higher temperatures, duplex slip of the types $\{hk0\}$ $\langle 001 \rangle$ and $\{h0l\}$ $\langle 010 \rangle$ was observed at about 25 °K, Fig. 12. Recently, Schulson and Teghtsoonian [21] also observed duplex slip of the types $\{hk0\}$ $\langle 001 \rangle$ and $\{h0l\}$ $\langle 010 \rangle$ at 77 °K, and single $\{hk0\}$ $\langle 001 \rangle$ slip at higher temperatures on tensile AuZn single crystals. Over the temperature range from approximately 25 to 298 °K, the CRSS for cube slip was independent of temperature, Fig. 10.

The present results clearly show that the slip and strain-hardening behaviour of β'-Ag-Mg crystals depends on the crystal orientation, temperature, and sense of the applied stress.

4. Discussion

The present results of the operative slip systems in β'-AgMg single crystals are in general agreement with those reported previously [1, 6, 22].

According to the results in Table 1, slip in β'-AgMg crystals oriented near the [001] pole should occur by the movement of dislocations of $\boldsymbol{b} = \langle 110 \rangle$ since these dislocations have a low value of S. This is not what is observed. The present results and those reported in an earlier paper show that crystals with orientations near the [001] pole deform by slip on the $\{112\}$ $\langle 111 \rangle$ system. It should

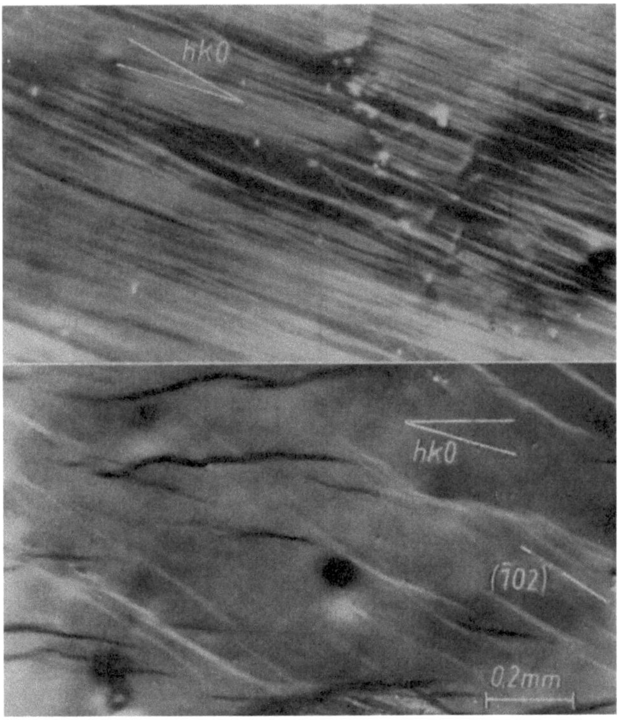

a

b

Fig. 12. Slip traces of crystals with orientation A strained in tension at a) 237.5 °K and b) 25 °K

be pointed out that both ⟨110⟩ and ⟨111⟩ dislocations will exist in β'-AgMg as pairs of $\frac{a}{2}$ ⟨111⟩ dislocations separated by a ribbon of APB. If the separation of the two $\frac{a}{2}$ ⟨111⟩ partials of the dissociated ⟨110⟩ dislocation is sufficiently large, its mobility would be expected to be comparable to the mobility of ⟨111⟩ dislocation. The separation calculated on the basis of an ordering energy of 98 erg/cm^2 was 4.6 Å for the two $\frac{a}{2}$ ⟨111⟩ partials making up a ⟨110⟩ screw dislocation and 28 Å for a screw dislocation of $\boldsymbol{b}$ = ⟨111⟩. Thus, the difficulty of movement of dislocations of $\boldsymbol{b}$ = ⟨110⟩ would imply that the separation between the partials is too small to consider the movement of such dislocations as pairs of $\frac{a}{2}$ ⟨111⟩ partials separated by a ribbon of APB.

The marked asymmetry of slip on the {112} planes suggests that each $\frac{a}{2}$ ⟨111⟩ partial in screw orientation undergo an asymmetric dissociation similar to that proposed to explain the asymmetry in slip behaviour observed in recent investigations on b.c.c. metals [23, 24]. Because of the very high stacking fault energies involved, this splitting is very minor and essentially leads to a decrease in the core energy of the dislocation with a corresponding asymmetric disposition of the core atoms. The asymmetric arrangement of the core atoms is assumed to be modified by the magnitude and direction of the applied shear stress in such a manner so as to cause the Peierls stress to increase as the core atoms are moved in the antitwinning direction [9]. This appears to produce slip on the {$hk0$} ⟨001⟩ system during tensile deformation of crystals oriented near the [011] and [1̄11] poles where the resolved shear stress on ⟨001⟩ dislocations is a large fraction of the applied stress. In addition, the results in Table 3 show that the CRSS for slip on the {112} ⟨111⟩ system in tension is lower than that in compression in crystals with orientations near the [001] pole. This is what one would expect since dislocation glide on {112} planes in the same sense as the twinning shear is easier than glide in the opposite sense. It is clear, then, that the calculations of dislocation mobility and energy presented in Table 1 do not enable one to predict all the slip systems in the compounds β'-AgMg. The ordering energy and the separation width of the $\frac{a}{2}$ ⟨111⟩ partials and the asymmetry in the Peierls stress on {112} planes appear to play an important role.

The wavy character of the slip lines corresponding to {*hk*0} planes and the fact that {*hk*0} planes tend to {110} at low temperatures suggest that {110} planes are the elementary planes on which dislocations of $\boldsymbol{b} = \langle 001 \rangle$ move. Thus {*hk*0} planes may only be apparent slip planes resulting from cross slip of dislocations of $\boldsymbol{b} = \langle 001 \rangle$ on {110} planes. This is in agreement with the results in Table 1 which indicate that {010} ⟨100⟩ slip occurs at a higher resolved shear stress than {$\bar{1}$01} ⟨010⟩. Recently, Mitchell, et al. [6] have suggested that dislocation sources activating slip on {*hk*0} ⟨001⟩ systems are produced by the stress-aided dissociation of superdislocations according to the reaction

$$a \langle 111 \rangle \rightarrow a \langle 110 \rangle + a \langle 001 \rangle \qquad [1, 8]\,. \tag{3}$$

It must be realized, however, that the dissociation of the a ⟨111⟩ dislocation according to equation (3) requires that the dislocation exists in a contracted form so that the two partials could be more easily forced together under stress. In β'-AgMg the two $\frac{a}{2}$ ⟨111⟩ partials are 28 Å apart and hence such a dissociation is unlikely to occur. It is then likely that the initially existing ⟨001⟩ dislocations act as sources of dislocations which activate slip on the {*hk*0} ⟨001⟩ systems during the tensile deformation of crystals oriented near the [011] and [$\bar{1}$11] poles.

It is evident from the results in Fig. 4 that slip on (2$\bar{1}$3) planes at and above 158 °K is activated by cross-slip from (1$\bar{1}$2) and (101) planes. The wavy character of the slip lines corresponding to (2$\bar{1}$3) planes at 298 °K also suggests the possibility that they might arise from intimate (1$\bar{1}$2) and (101) cross-slip. The results at 298 °K are in agreement with those of Rachinger and Cottrell in that {123} ⟨111⟩ system in an operative slip mode. However, the present observations indicate that at and below 298 °K the {123} ⟨111⟩ system is not an active primary slip system in terms of slip being activated by the operation of existing dislocation sources on these planes. The somewhat wavy appearance of the slip lines at 77 °K suggests that the screw segments of a ⟨111⟩ dislocations are capable of cross-slipping even at this low temperature. Loretto and Wasilewski [25] have recently reported similar observations in NiAl crystals stressed along ⟨001⟩ at 77 °K.

The results of the present investigation show that the high rate of strain hardening exhibited by β'-AgMg crystals corresponds to the motion of dislocations of $\boldsymbol{b} = \langle 001 \rangle$ which does not disorder the lattice. As a result, hardening mechanisms based on the formation of APB tubes which originate from the presence of nonaligned jogs on the moving superdislocations can not explain the observed high rate of strain hardening. On the other hand, the results show that crystals which slip by the movement of dislocations of $\boldsymbol{b} = \langle 111 \rangle$ exhibit a linear region (stage II) of relatively low rate of strain hardening. When the temperature is increased, the thermally activated component of the flow stress, τ^*, decreases and the maximum rate of strain hardening in stage II is observed when τ^* is zero. The observed dependency of the rate of strain hardening in stage II on the temperature, or on the thermally activated component of the flow stress τ^*, which in turn depends on the temperature, can be explained in the following manner: Schoeck [26] has shown that the formation of recombinations of moving superdislocations with attractive "trees" and the accumulation of stable APB tubes on edge superdislocations give rise to a large frictional drag on the movement of edge superdislocations through the lattice. This frictional dragging stress, τ_f, is independent of temperature.

There is also an additional athermal contribution to the flow stress from the internal stress, τ_G, produced, for example, by individual dislocations and dislocation groups in parallel glide planes. The applied shear stress that is required to cause plastic flow can then be represented by

$$\tau_A = \tau^* + \tau_f + \tau_G , \tag{4}$$

where τ^* is the effective stress required to aid thermal activation of dislocations over the Peierls barriers [20].

The results in Table 1 indicate that there is a difference in the mobilities of edge and screw segments of $a \langle 111 \rangle$ dislocations due to differences in the Peierls stress of edge and screw dislocations. The edge segments are more mobile than the screws.

At large τ^*, the athermal dragging stress is a small component of the applied stress and as a result it would not act as a barrier to the edge segments of the superdislocations. The highly mobile edge segmets, which are now acted on by the same τ^* as the screw segments, would then move relatively great distances, leaving behind the screw segments. Thus, the friction encountered by the edge segments at large τ^* is probably caused mainly by the Peierls barrier. On the other hand, at small τ^*, the athermal dragging stress is a large component of the applied stress and hence is a major barrier to dislocation motion. Therefore, at small τ^* there is a greater probability that the edge segments could be stopped by the accumulated APB tubes. Consequently, hardening in the primary system is expected to be higher at small τ^*. Furthermore, no secondary slip traces were generally observed during deformation in stage II. This is consistent with the APB tube forming model since only as small mount of secondary slip is necessary to account for the strain hardening.

The present results suggest that the low-temperature deformation by $\{hk0\} \langle 001 \rangle$ slip is controlled by an athermal mechanism. The athermal behaviour during $\{hk0\} \langle 001 \rangle$ slip extends to a much lower temperature than is observed in other CsCl lattices, e.g., NiAl [27] and AuZn [28], where a low-temperature thermally-activated mechanism determines the flow stress. The fine and homogeneous slip lines observed in the rapid-hardening range suggest that the primary dislocation sources are rapidly inactivated. The newly generated dislocation sources appear to be harder to operate. This appears necessary since easy generation of dislocations which are eventually stopped by, for example, dislocation interaction, would be expected to give substantial mean-free paths and hence would not produce the initial rapid hardening. The mechanism responsible for this hardening is not clear at the present. A detailed electron microscopy study is currently being made to observe the dislocation configuration and to reach a complete understanding of the athermal deformation behaviour during the operation of the $\{hk0\} \langle 001 \rangle$ system.

5. Summary

1. Single crystals of β'-AgMg deform in the temperature range from 25 to 298 °K by slip on the {112} ⟨111⟩, {110} ⟨111⟩, {123} ⟨111⟩, {120} ⟨001⟩, and {110} ⟨001⟩ systems.

2. The activation of dislocations of $\boldsymbol{b} = \langle 111 \rangle$ and $\langle 001 \rangle$ depends on the crystal orientation, sense of the applied stress and temperature.

3. Comparison of the observed slip systems with those predicted in β'-AgMg from anisotropic elasticity theory suggests that factors other than dislocation mobility and energy must be considered when predicting the slip systems. The ordering energy and the separation width of $\frac{a}{2}\langle 111 \rangle$ partials and the asymmetry in the Peierls stress on {112} planes appear to play an important role.

4. Crystals which exhibit a linear region (stage II) of a relatively low rate of strain hardening slip by the movement of dislocations of $\boldsymbol{b} = \langle 111 \rangle$, while those exhibiting a high rate of strain hardening slip by the movement of dislocations of $\boldsymbol{b} = \langle 001 \rangle$ which do not disorder the lattice.

Acknowledgements

The author would like to thank Dr. J. B. Mitchell for helpful discussions. The AgMg alloy used in the present study was supplied by Lawrence Radiation Laboratory, University of California, Livermore, California.

References

[1] W. A. Rachinger and A. H. Cottrell, Acta metall. **4**, 109 (1956).
[2] A. Ball and R. E. Smallman, Acta metall. **14**, 1517 (1966).
[3] P. A. Flinn, Trans. MS-AIME **218**, 145 (1960).
[4] S. G. Cupshalk and N. Brown, Acta metall. **16**, 657 (1968).
[5] W. L. Bragg and E. J. Williams, Proc. Roy. Soc. (London), Ser. A, **151**, 540 (1935).
[6] J. B. Mitchell, O. Aboelfotoh, and J. E. Dorn, Metal. Trans. **2**, 3265 (1971).
[7] D. J. Potter, Mater. Sci. Engng. **5**, 201 (1970).
[8] F. R. N. Nabarro, Adv. Phys. **1**, 385 (1952).
[9] J. E. Dorn and A. K. Mukherjee, Trans. MS-AIME **245**, 1493 (1969).
[10] D. L. Wood and J. H. Westbrook, Trans. MS-AIME **224**, 1024 (1962).
[11] P. M. Robinson and M. B. Bever, Acta metall. **13**, 647 (1965).
[12] W. L. Bragg, Proc. Phys. Soc. (London) **57**, 105 (1960).
[13] S. G. Cupschalk and N. Brown, Acta metall. **15**, 847 (1967).
[14] B. H. Kear, Acta metall. **12**, 555 (1964).
[15] R. G. Davis and N. S. Stoloff, Phil. Mag. **9**, 349 (1965).
[16] M. J. Marcinkowski and N. Brown, J. appl. Phys. **33**, 537 (1962).
[17] A. E. Vidoz and L. M. Brown, Phil. Mag. **7**, 1167 (1962).
[18] J. C. Terry and R. E. Smallman, Phil. Mag. **8**, 1827 (1963).
[19] M. J. Marcinkowski and M. Chessin, Phil. Mag. **10**, 837 (1964).
[20] A. K. Mukherjee and J. E. Dorn, Trans. MS-AIME, **230**, 1065 (1964).
[21] E. M. Schulson and E. Teghtsoonian, Phil. Mag. **19**, 155 (1969).
[22] V. B. Kurfman, Acta metall. **13**, 307 (1965).
[23] A. W. Sleeswyck, Phil. Mag. **8**, 1467 (1963).
[24] F. Kroupa, phys. stat. sol. **3**, K391 (1963).
[25] M. H. Loretto and R. J. Wasilewski, Phil. Mag. **23**, 13111 (1971).
[26] G. Schoeck, Acta metall. **17**, 147 (1969).
[27] R. J. Wasilewski, S. R. Butler, and J. E. Hanlon, Trans. MS-AIME, **239**, 1357 (1967).
[28] E. M. Schulson and E. Teghtsoonian, Trans. MS-AIME **245**, 1527 (1969).

(Received October 9, 1972)

phys. stat. sol. (a) **14**, 561 (1972)
Subject classification: 14.1; 10.1; 21; 21.1

Institute for Applied Physics, Kossuth Lajos University, Debrecen (a) and Research Institute for Non-Ferrous Metals, Budapest (b)

Temperature Dependence of Impurity Resistivity in Cold-Worked Dilute Al–Cr and Al–Mn Alloys between 78 and 300 K

By
F. J. KEDVES (a), E. KOVÁCS-CSETÉNYI (b), and L. GERGELY (a)

The temperature dependence of the impurity resistivity of dilute Al–Cr and Al–Mn alloys of different concentrations was measured in a cryostat between 78 and 300 K after cold-working. The main features of the T-dependence and the characteristic temperatures were found to be the same as for solid solutions. The difference of the extraresistivity in cold-worked and annealed states (solid solution) is positive and independent of temperature, and can be attributed to dislocations. In the presence of dislocations all the impurity atoms appear as dissolved in the solid.

Die Temperaturabhängigkeit des elektrischen Widerstandes von dünnen Al–Cr- und Al–Mn-Legierungen mit verschiedenen Konzentrationen wurde in einem Kryostat zwischen 78 und 300 K nach Kaltverformung gemessen. Der Verlauf der Temperaturabhängigkeit und die charakteristischen Temperaturwerte stimmen mit denen von dünnen festen Lösungen überein. Die Differenz zwischen den Extrawiderständen im kaltverformten und ausgeheilten Zustand (feste Lösung) ist positiv und unabhängig von der Temperatur und wird den Versetzungen zugeschrieben. Die Verunreinigungen befinden sich in fester Lösung, auch wenn Versetzungen vorhanden sind.

1. Introduction

The temperature dependence of impurity resistivity $(\Delta\varrho(T))$ in Al-based dilute 3d transition metal solid solutions has already been measured in a very wide temperature range [1 to 5]. The concept of localized spin fluctuation [6] has proved to be a good basis for the theoretical interpretation of the temperature dependence of $\Delta\varrho$ [7, 8]. Most recent measurements (between 78 and 930 K) in dilute Al-3d transition metal solid solutions [5], however, confirm the idea that there are localized moments on the impurities, i.e. the Al–Cr and Al–Mn systems are rather in the magnetic than in the non-magnetic limit at high temperatures (see also [9]).

There is only limited information on the effect of impurity clusters and impurities bound to dislocations on the temperature dependence of the extra resistivity [10, 11], and we have not found any publication for dilute Al alloys. The temperature dependence of the extra resistivity of dislocations was measured in Cu, Ag, and Au [12], and it was found that this resistivity increases rather steeply between 0 and 100 K and afterwards remains practically constant up to 300 K. Similar measurements have not been published up to now for pure cold-worked Al, but it is reasonable to suppose a similar behaviour in this metal, too.

The aim of this paper is to present some results on the impurity resistivity of cold-worked dilute Al–Cr and Al–Mn alloys between 78 and 300 K in order to investigate the effect of dislocations on the distribution of impurity atoms.

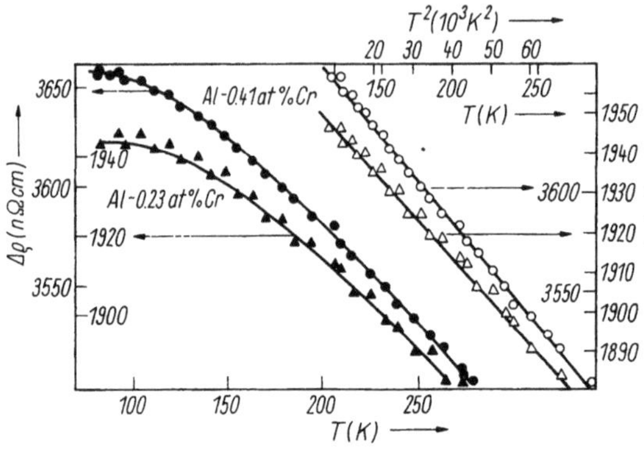

Fig. 1. Temperature dependence of the impurity resistivity of Al–Cr alloys. $\Delta\varrho$ vs. T^2 is shown in the insert by the corresponding open marks

2. Experimental

The alloys were prepared from 99.999% pure Al and 99.99% pure Cr and Mn by using master alloys. The cast rods (12 mm in diameter, cooled down rapidly, so that they solidified as dilute solutions) were cold-drawn (cross section reduction 99%) to wires of 1 mm diameter. The samples were then stored at room temperature for several weeks; so it is likely to suppose that the vacancies produced during cold-woking were annealed out [13] and therefore they contained only impurities and dislocations.

The measurements were carried out in a double chamber cryostat [14] by the conventional potentiometric method. The resistivity values have a mean error of about 1% [15].

3. Results

The temperature dependence of $\Delta\varrho$ in cold-worked dilute Al–Cr alloys is shown in Fig. 1. It can be seen from the insert that

$$\Delta\varrho(T) = \Delta\varrho(0)\left(1 - \left(\frac{T}{\theta_1}\right)^2\right) \tag{1}$$

holds in the whole investigated temperature range. The characteristic temperature $\theta_1 = (1300 \pm 100)$ K, which agrees with that of the solid solution [5] and with the results of Caplin and Rizzuto [1] measured between 1.5 and 4.2 K.

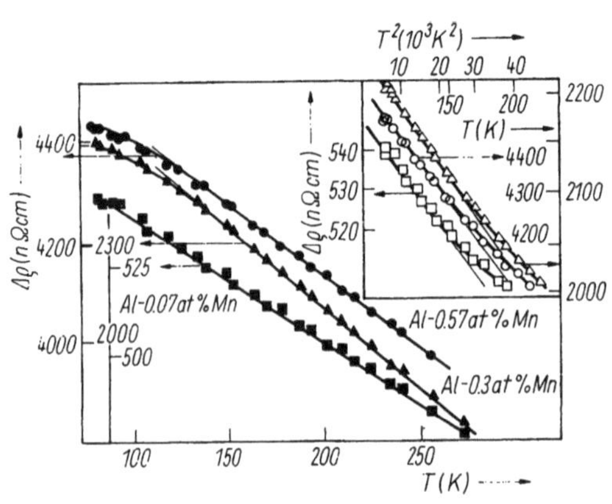

Fig. 2. Temperature dependence of the impurity resistivity of Al–Mn alloys. The resistivity in a T^2 scale is shown in the insert using the corresponding open marks

The temperature dependence of $\Delta\varrho$ in cold-worked dilute Al–Mn alloys is shown in Fig. 2. The behaviour can be described by equation (1) with $\theta_1 = (640 \pm 30)$ K, up to about 160 K. At higher temperatures $\Delta\varrho$ varies linearly with T as

$$\Delta\varrho(T) = A\left(1 - \frac{T}{\theta_2}\right) \tag{2}$$

with $\theta_2 = (1300 \pm 100)$ K, where A is a concentration dependent constant. It was found in previous investigations on solid solutions [3, 5] that $\theta_1 = (670 \pm 30)$ K and $\theta_2 = (1500 \pm 100)$ K. A consequent concentration dependence of θ_1 and θ_2 could not be established in these systems.

4. Discussion

Some previous results obtained for solid solutions [5] and the present data for a cold-worked (as-drawn) state are shown in Fig. 3. for the Al–0.41 at% Cr and Al–0.57 at% Mn. The full lines represent the temperature dependence of impurity resistivity in the solid solutions and the points indicated are taken from the resistivity measurements of the cold-worked alloys as

$$\Delta\varrho'_{\mathrm{cw}}(T) = \Delta\varrho_{\mathrm{cw}}(T) - B\,, \tag{3}$$

where B is found to be constant in the investigated temperature range, and its values are (240 ± 40) and (160 ± 40) nΩcm for Al–0.41 at% Cr and Al–0.57 at% Mn, respectively. These two contributions can be taken equal within the experimental error [15], however, the real value of the Al–0.41 at% Cr alloy is likely to lie lower because of precipitations since the concentration is nearly at the solubility limit. The above resistivity contributions can be attributed to dislocations, and it can be seen from Fig. 3 that they are independent of temperature, which supports the idea that the dislocation resistivity is independent of temperature between 78 and 300 K. This is in agreement with the results published by Basinski et al. [12] for pure metals. Using the value for the dislocation resistivity of aluminium [12],

$$\varrho_{\mathrm{d}} = 4.7 \times 10^{-10}\ \mathrm{n\Omega cm^3}\,, \tag{4}$$

the dislocation concentrations can be estimated to the order of magnitude of 10^{11} cm^{-2} for both alloys. These dislocation densities are reasonable for highly deformed metals.

The above results allow for further remarks. The extra resistivity of the cold-worked dilute alloys was always higher than the impurity resistivity of the an-

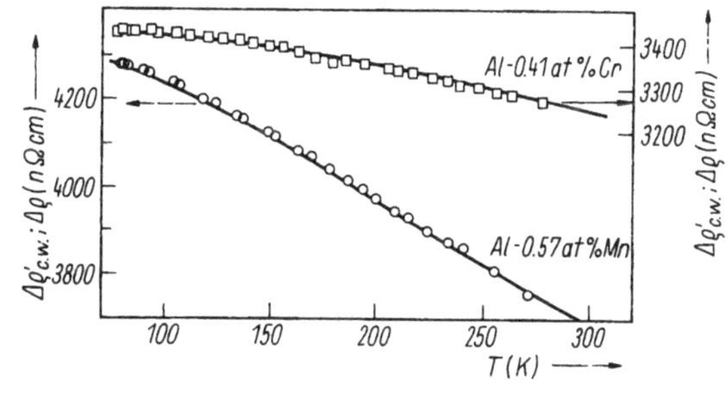

Fig. 3. Comparison of the temperature dependence of impurity resistivity in solid solution (full lines) and in coldworked state (squares and circles)

nealed alloys and the difference of these two extra resistivities could be attributed to dislocations. This leads to the conclusion that the impurity atoms in the solid solution are in cold-worked state, too, since the cold-worked sample has a *lower* extra resistivity if the dislocations act as heterogeneous nucleation centres (as in the case of e.g. Al–Si and Al–Fe systems [16]). The extra resistivity of dilute Al–Cr and Al–Mn containing precipitates is always lower than that of the solid solution for the same impurity concentrations, and the temperature dependence can disagree completely with the above case [17].

Some further measurements at higher temperatures are in progress which are likely to give more information on the annealing of dislocations in the presence of impurity ions.

Acknowledgements

The authors are indebted to the Research Institute for Non-Ferrous Metals, Budapest, for the support of this work and for the preparation and analysis of the samples. We want to express our thanks for the assistance in many measurements to Mrs. L. Vasváry, to our student Mr. I. Mudri, and for the technical help to Mr. I. Boross.

References

[1] A. D. CAPLIN and C. RIZZUTO, Phys. Rev. Letters **8**, 243 (1968).
[2] E. BABIĆ, R. KRSNIK, B. LEONTIĆ, Z. VUČIĆ, and I. ZORIĆ, Phys. Rev. Letters **27**, 805 (1971).
[3] E. KOVÁCS-CSETÉNYI, F. J. KEDVES, L. GERGELY, and G. GRÜNER, J. Phys. F (Metal Phys.) **2**, 499 (1972).
[4] E. BABIĆ, R. KRSNIK, B. LEONTIĆ, M. OČKO, Z. VUČIĆ, I. ZORIĆ, and E. GIRT, Solid State Commun. **10**, 691 (1972).
[5] F. J. KEDVES, M. HORDÓS, and L. GERGELY, Solid State Commun., **11**, 1067 (1972).
[6] N. RIVIER and M. J. ZUCKERMANN, Phys. Rev. Letters **21**, 204 (1968).
[7] M. J. ZUCKERMANN, J. Phys. F (Metal Phys.) **2**, L25 (1972).
[8] N. RIVIER and V. ZLATIĆ, J. Phys. F (Metal Phys.) **2**, L87 (1972).
[9] G. GRÜNER and A. ZAWADOWSKI, KFKI Rep. 72—24, March 1972.
[10] A. DREYFUSS-BOURQUARD, Solid State Commun. **8**, 1203 (1970).
[11] R. J. WHITE and G. A. SWALLOW, Phys. Letters (Netherlands) **35A**, 427 (1971).
[12] Z. S. BASINSKI, J. S. DUGDALE, and A. HOWIE, Phil. Mag. **8**, 1989 (1963).
[13] A. HEGAZI, I. KOVÁCS, and E. NAGY, phys. stat. sol. **33**, 131 (1969).
[14] L. GERGELY, D. NOVÁK, F. J. KEDVES, and S. CZELLÁR, Acta phys. et chem. Debr. XVII 99 (1971).
[15] F. J. KEDVES, L. GERGELY, M. HORDÓS, and E. KOVÁCS-CSETÉNYI, phys. stat. sol. (a) **13**, 685 (1972).
[16] K. R. VASSEL and E. KOVÁCS-CSENTÉNYI, Proc. Res. Inst. for Non-Ferrous Metals **9**, 323 (1971).
[17] E. KOVÁCS-CSETÉNYI, F. J. KEDVES, and L. GERGELY, to be published.

(Received October 6, 1972)

phys. stat. sol. (a) **14**, 565 (1972)

Subject classification: 20.1; 13.4; 22.5

Solid State Department, Physical Laboratory, University of Utrecht

Optical Absorption Spectra and Crystal-Field Calculations of Pr^{3+} and Nd^{3+} in $PbCl_2$

By

B. WILLEMSEN

The optical absorption spectra of Pr^{3+} and Nd^{3+} in single crystals of $PbCl_2$ have been studied in the spectral region 10000 to 29000 cm^{-1} at liquid-nitrogen and liquid-helium temperatures. The crystal-field splitting is interpreted in terms of crystal field potentials of symmetry C_s. For $PbCl_2$:Nd^{3+} the parameters B_2^0, B_2^2, B_4^0, B_4^2, B_4^4, B_6^0, B_6^2, B_6^4, and B_6^6 have been determined from the positions of 20 Stark components. The RMS deviation of the fit was 1 cm^{-1}. For $PbCl_2$:Pr^{3+} the parameters B_2^0, B_2^2, B_4^0, B_4^2, and B_4^4 were determined from 13 Stark components. The RMS deviation was 6 cm^{-1}. (The parameters B_l^m are interpreted in terms of crystal field theory). These crystal field parameters are compared with those obtained earlier from EPR studies.

Die Absorptionsspektren von Pr^{3+} und Nd^{3+} in Einkristallen von $PbCl_2$ werden im Bereich von etwa 10 000 bis 29 000 cm^{-1} bei den Temperaturen von flüssigem Stickstoff und flüssigem Helium untersucht. Die Aufspaltungen der Terme der Nd^{3+}- und Pr^{3+}-Ionen werden mit einem Kristallfeldpotential der Symmetrie C_s interpretiert. Für $PbCl_2$:Nd^{3+} werden die Parameter B_2^0, B_2^2, B_4^0, B_4^2, B_4^4, B_6^0, B_6^2, B_6^4 und B_6^6 von 20 beobachteten Kristallfeldkomponenten berechnet. Die mittlere Abweichung zwischen den gemessenen und mit diesen Parameterwerten berechneten Aufspaltungen beträgt 1 cm^{-1}. Für $PbCl_2$:Pr^{3+} werden die Parameter B_2^0, B_2^2, B_4^0, B_4^2 und B_4^4 von 13 beobachteten Komponenten berechnet. Die mittlere Abweichung beträgt 6 cm^{-1}. Die Parameter wurden mit der Kristallfeldtheorie interpretiert und mit früheren EPR-Untersuchungen verglichen.

1. Introduction

The splitting of the energy levels of rare-earth ions in crystals is caused by the interaction between the rare-earth ion and its surroundings. In the simplest model the ion is considered to be in a static electric field produced by the neighbouring ions, the so-called crystal field. In some cases the conditions for this model are well satisfied, in others not at all. Recently, the EPR spectra of Eu^{2+} and Gd^{3+} incorporated in $PbCl_2$ [1, 2], and Eu^{2+} in $PbBr_2$ [3], have been investigated. It is evident that the zero-field splitting of the ^8S ground state of Eu^{2+} and Gd^{3+} can be interpreted within the framework of crystal field theory, if simultaneously a small distortion of the crystalline environment around the rare-earth ion is taken into account [3]. A check on the reliability of this simple model should be an analysis of the absorption spectra of rare-earth ions incorporated in $PbCl_2$. If crystal field theory works well in this case, it must be possible to describe the Stark splitting of the energy levels by parameters which are in accordance with those employed for the interpretation of the EPR spectra. So, primarily, our aim is to pay attention to B_2^0 and B_2^2 which are related to b_2^0 and b_2^2, the leading parameters in the spin Hamiltonian which has been used for the analysis of the EPR spectra [1 to 3]. From the ionic conductivity data of the crystals it was possible to conclude [4] that the Pr^{3+} ions replace the lead ions substitutionally. The point-group symmetry at the lead ion site is C_s [5], so that the degeneracy of the free-ion levels is completely lifted except for the Kramers degeneracy. The analysis of the spectra in terms of crystal field potentials is complicated by the necessity to use a large number of crystals field parameters, because the point-group symmetry is very low.

2. Experimental

Single crystals of optical quality, doped with Pr^{3+} and Nd^{3+}, were grown in our Laboratory as described before [6]. In view of our experience with Eu and Sm [7], we estimate the concentration of Pr^{3+} or Nd^{3+} to be 0.01% by weight.

The absorption spectra were measured with a Leiss single monochromator, equipped with interchangeable Bausch & Lomb gratings. The energy range covered, was 10000 to 50000 cm^{-1}. The resolving power $\lambda/\Delta\lambda$ was $\approx$20000 at 500 nm. The light source was a Philips' tungsten lamp, type W2KGV211. An RCA IP28 photomultiplier served as a detector for wavelengths shorter than 600 nm; in the near-infrared a Philips' 150CVP was used.

The crystals were fixed in a holder against the bottom of the container in a cryostat, which could be filled with liquid helium or nitrogen.

The exprimental error in the observed splittings is about 1 cm^{-1}.

The crystals were cleaved perpendicular to the *a*-axis. In general, the spectrum was taken with the light directed along the *a*-axis. Polarization was obtained with polaroid sheets, taking the vector $\boldsymbol{E}$ parallel to one of the crystallographic axes.

The numerical computations were performed on an EL X-8 computer.

3. Results and Discussion

3.1 $PbCl^2 : Nd^{3+}$

The absorption spectrum of $PbCl_2$:Nd^{3+} consist of a number of closely spaced groups of sharp lines, which have a width of about 5 cm^{-1} at liquid-helium temperature (LHT). At liquid-nitrogen temperature (LNT) an extra number of absorption lines was observed, which could be ascribed to transitions from an excited level of the $^4I_{9/2}$ ground state. The levels found at LHT have an energy close to the energy of the levels of Nd^{3+} in $LaCl_3$ [8]. The correct number of $(J + \frac{1}{2})$ components was observed for a number of spin-allowed transitions in the spectral region below $\approx$ 20000 cm^{-1}. The components of the $^4G_{5/2}$ and $^2G_{7/2}$ levels constitute one group of lines. Seven lines are expected, but eight rather strong, and a number of weak lines were observed, even at LHT. The extra lines have been interpreted as vibronic lines. As the energy difference between the $^4G_{5/2}$ and $^2G_{7/2}$ levels is small, viz. 150 cm^{-1}; the assignment of the components was difficult. The assignment proposed, is based on a calculation of splittings.

Multiplets with energy higher than 29000 cm^{-1} cannot be observed as a result of the absorption of the host.

3.2 $PbCl^2 : Pr^{3+}$

In the absorption spectrum the following multiplets were observed: $^1D_2$, $^3P_0$, $^3P_1$, $^1I_6$, and $^3P_2$. Going from liquid-nitrogen temperature to liquid-helium temperature, a number of extra lines appears in each multiplet. These extra lines can be ascribed to vibronic transitions [9]. The strong vibronic lines at low temperature are often difficult to distinguish from pure electronic transitions. This complicates the analysis of the spectrum at liquid-helium temperature. A way out of these difficulties is to observe the spectrum at liquid-nitrogen or at even higher temperatures for then the vibronic lines are much weaker [10]. We will now consider the different multiplets. Four of the nine components of the $^3H_4$ ground level could be detected by observing the $^3H_4 \rightarrow {}^3P_0$ transitions at LNT. At LHT, however, the electronic $^3H_4 \rightarrow {}^3P_0$ transition is accompanied by about twelve other lines. A group of weak vibronic lines is observed at the high-energy side of the lowest $^3H_4 \rightarrow {}^3P_0$ transition. At the low-energy side strong lines appear which nearly coincide with excited levels of $^3H_4$. The $^3H_4 \rightarrow {}^3P_1$ transition is also complicated by many vibronic lines. The line corresponding to the transition of the lowest component of the $^3H_4$ level to the highest component of the $^3P_1$ level, is seriously broadened at LHT by overlapping vibronic transitions.

The transitions to the $^1D_2$ and $^3P_2$ multiplets are also accompanied by vibronic lines. The transition to the $^1D_2$ level is the least complicated of all. The spin-forbidden transitions are relatively weak. We did not observe all the components of the $^1I_6$ level. In Table 1 the differences between the wave numbers

Table 1

Difference between the wave numbers of some vibronic transitions and the electronic $^3H_4 \rightarrow {}^3P_0$ transition at 20530 cm^{-1} at liquid-helium temperature

cm^{-1}	intensity	cm^{-1}	intensity
9	strong	58	medium
23	strong	65	weak
33	strong	75	weak
41	weak	234	medium
53	medium		

of some vibronic lines, and the electric $^3H_4 \rightarrow {}^3P_0$ transition at 20530 cm^{-1} at LHT, are presented.

It is surprising that the absorption spectrum of $PbCl_2:Pr^{3+}$ at LHT shows many strong vibronic lines, while the absorption spectrum of $PbCl_2:Nd^{3+}$ does not do so. Satten [10, 11] has pointed out that the occurrence of strong vibronic lines may be due to the interaction of vibrations with electronic transitions within the rare-earth ion, if the frequency of these vibrations is comparable to the separations of the Stark levels of the multiplets of the rare-earth ion. One reason for the complications in the absorption spectrum of $PbCl_2:Pr^{3+}$ may be the existence of vibrational levels, which nearly coincide with the $^3H_4$ crystalline Stark levels.

The Raman spectra of our pure $PbCl_2$ crystals is known [12], the infrared spectrum, however, is not. The observed Raman-active vibrations with wave numbers of 32 and 62 cm^{-1}, possibly interact with the excited components of the $^3H_4$ multiplets at 34 and 57 cm^{-1} of Pr^{3+} [20]. The first excited level of $^4I_{9/2}$ of Nd^{3+} is situated at 80 cm^{-1}. This value does not correspond with a strong Raman line.

It should be noted that incorporation of rare-earth ions may influence the Raman-active vibrations of the lattice or give rise to local vibrational modes. For a more complete analysis the infrared-active vibrations should also be known.

4. Crystal Field Calculations

The crystalline potential field surrounding a particular rare-earth ion on a lead-ion site can be written as

$$V(r, \theta, \varphi) = \sum_{l=0}^{\infty} \sum_{m=-l}^{+l} F_l^m \, r^l \, Y_l^m(\theta, \varphi) \,. \tag{1}$$

This crystalline potential field contributes a term H_c to the total Hamiltonian. H_c can be written as

$$H_c = -e \sum_k V(r_k, \theta_k, \varphi_k) \,. \tag{2}$$

r_k, θ_k, and φ_k are the coordinates of the k-th 4f electron. The perturbing Hamiltonian for a monoclinic crystal field can be written as

$$\begin{aligned} H_c = {} & \alpha \sum_{\substack{m=0 \\ \text{even}}}^{2} (B_2^m O_2^m + B_2^{-m} O_2^{-m}) + \\ & + \beta \sum_{\substack{m=0 \\ \text{even}}}^{4} (B_4^m O_4^m + B_4^{-m} O_4^{-m}) + \\ & + \gamma \sum_{\substack{m=0 \\ \text{even}}}^{6} (B_6^m O_6^m + B_6^{-m} O_6^{-m}) \,, \end{aligned} \tag{3}$$

where α, β, and γ are operator equivalent constants [13] and the $O_l^{\pm m}$ are angular momentum operators. The B_l^m are related to the crystal-field parameters C_l^m [3].

$$B_l^{\pm m} = -e \, C_l^{\pm m} (1 - \sigma_l) \langle r^l \rangle = A_l^{\pm m} (1 - \sigma_l) \langle r^l \rangle \,, \tag{4}$$

where the factor $(1 - \sigma_l)$ takes into account the shielding of the 4f electrons by polarization of the other closed shells.

The coefficients $B_l^{\pm m}$ are treated as empirical parameters, whose values are adjusted to agree with the experimental data. As the point-group symmetry is C_s, fifteen parameters must be determined. Transitions between the four lowest multiplets of Nd^{3+} could not be measured with our apparatus. So we used the $^4F_{3/2}$, $^4F_{5/2}$, $^4F_{7/2}$, $^4G_{7/2}$, and $^4G_{9/2}$ multiplets and the two lowest levels of $^4I_{9/2}$ for the fitting procedure. Multiplets, which were not completely known or where strong J mixing was expected, were not used in the fitting procedure. The number of energy differences available is fourteen. The reference coordinate system can be chosen in such a way that $B_2^{-2} = 0$ [1 to 3]. The choice of $B_2^{-2} = 0$ does not imply that the other parameters B_l^{-m} become zero. From our earlier EPR experiments [2, 3] we got the impression that the parameters B_l^{-m} have small values. Therefore, we have restricted ourselves to the determination of the nine B_l^{+m} parameters, which could be readily obtained from the fourteen experimental energy differences. This procedure gives an excellent fit with a RMS deviation of 1 cm^{-1} [20]. It is unknown, however, how much the values of B_l^{+m} ($m \neq 0$) are affected by neglecting the B_l^{-m} parameters. In the actual calculations starting values of the B_l^m parameters are needed. These were adopted from crystal-field calculations used for the interpretation of the EPR spectra of Eu^{2+} and Gd^{3+} in $PbCl_2$ [3]. As the positions of the centres of the multiplets of Nd^{3+} in $PbCl_2$ coincide within 15 to 50 cm^{-1} with those of Nd^{3+} in $LaCl_3$, the intermediate coupling values for the operator equivalent constants computed by Dieke [8] were used. His calculations were based on the analysis of the spectrum of Nd^{3+} in $LaCl_3$ by Eisenstein [14].

For the $^4G_{5/2}$ and $^2G_{7/2}$ multiplets the assignment proposed, was deduced from the calculated splittings. The agreement is poor, probably, because J mixing has not been taken into account.

From the calculated splittings the positions of the centres of the multiplets were estimated in those cases were the components were incompletely known.

For the Pr^{3+} ion the situation is less favourable. Only three multiplets could be measured completely, viz. $^1D_2$, $^3P_1$, and $^3P_2$ [20]. Now we can only determine the B_l^{+m} with $l \leqq 4$, because $J \leqq 2$. The levels belong to two representations characterized by the crystal quantum number μ. Usually, the polarization of the transitions gives the representation to which a particular level belongs. Such an interpretation gives erroneous results in the fitting procedure. This leads to the conclusion that the selection rules are violated. Such an effect is well known for cases where vibrations participate [10, 11]. That vibrations participate in this case has been argued above. A priori calculations were performed using parameters obtained from the Nd^{3+} spectrum. This led to the assignment of the levels. A subsequent fitting procedure yielded a reasonable fit with a RMS deviation of 6 cm^{-1} [20]. For the operator equivalent constants we used the intermediate coupling values derived by Dieke [8] from the free-ion levels [15]. The B_l^m parameters are collected in Table 2.

Table 2

The B_l^m parameters in cm^{-1}

	$PbCl_2 : Nd^{3+}$	$PbCl_3 : Pr^{3+}$
B_2^0	-41 ± 1	-41.9 ± 0.3
B_2^2	62 ± 1.5	74 ± 1.5
B_4^0	-0.9 ± 0.4	-4.6 ± 0.1
B_4^2	162 ± 8	66 ± 2
B_4^4	-106 ± 15	85 ± 2
B_6^0	-1.51 ± 0.01	
B_6^2	-66 ± 10	
B_6^4	226 ± 3	
B_6^6	35 ± 30	

5. Discussion

From the empirical B_l^m parameters we have deduced the A_l^m parameters. For $\langle r^l \rangle$ the values given by Freeman and Watson [16] have been used. Values for σ_2 were estimated by linear interpolation of the values given by Gupta et al. [17], viz. 0.754 for Pr^{3+}, and 0.73 for Nd^{3+}. The screening factors σ_4 and σ_6 were adopted from Sternheimer et al. [18]. The values of A_l^m are collected in Table 3.

Table 3
Crystal field parameters A_l^m in cm^{-1}/a_0^l

	$PbCl_2 : Nd^{3+}$ (optical)	$PbCl_2 : Pr^{3+}$ (optical)	$PbCl_2 : Gd^{3+}$ (EPR)	Pb^{2+} site in $PbCl_2$ (calculated)
A_2^0	-151 ± 3	-152 ± 1	-86 ± 9	-82 ± 9
A_2^2	229 ± 2	267 ± 6	360 ± 40	255 ± 20
A_4^0	-0.4 ± 0.2	-1.80 ± 0.05		-3.1
A_4^2	74 ± 4	26 ± 1		48
A_4^4	-52 ± 8	33.0 ± 0.7		
A_6^0	-0.126 ± 0.01			-0.21
A_6^2	-6 ± 1			
A_6^4	19.0 ± 0.3			
A_6^6	3 ± 3			

From a comparison of the parameters of Nd^{3+} and Pr^{3+} it is evident that the agreement between the values of A_2^0 is very good. For the A_4^m parameters, however, there are discrepancies, that for A_4^4 being the most serious. The calculation of the A_l^m parameters for a lead-ion site in lead chloride in the point-charge approximation has been given in a previous paper [3]. The results are included in Table 3. Notwithstanding the simple electrostatic model, the agreement is satisfactory. Also the values of A_2^0 and A_2^2, deduced from the EPR spectra of Gd^{3+} in $PbCl_2$ [2], are given in Table 3. The differences between the values of the A_l^m parameters can easily be attributed to small distortions of the environment [3]. As the ionic radii (values according to Ahrens) of Pr^{3+} (1.06 Å) and of Nd^{3+} (1.04 Å) are smaller than those of Pb^{2+} (1.20 Å), it is probable that, for example, the three adjacent Cl^- ions in the *bc*-plane move a little into the direction of the rare-earth ion. A displacement of 0.07 Å is sufficient to bring the calculated value of A_2^0 in accordance with the experimental value ($A_2^{0\prime} = -154 \pm 10$), whereas the value of A_2^2 increases only slightly ($A_2^{2\prime} = 260 \pm 20$).

For the interpretation of the EPR spectrum of Gd^{3+} in $PbCl_2$ we had to adopt a more complicated model for the distortion of the environment [3]. This model included the fact that the Gd^{3+} ion is not exactly situated on the position of the substituted Pb^{2+} ion, a displacement of the three adjacent Cl^- ions in the *bc*-plane into the direction of the Gd^{3+} ion, and a small rotation of one of these ions around the Gd^{3+} ion. This seems not unreasonable in view of the even smaller radius (0.97 Å) of this ion. So we conclude that there is a reasonable agreement between the values of A_2^0 and A_2^2 found from the optical absorption measurements, and those employed in the interpretation of the EPR spectra within the frame-work of crystal field theory. This is in accordance with the opinion of Watson and Freeman [19], that the noncubic $l = 2$ terms in the potential field are almost completely determined by electrostatic effects. For $l = 4$ and $l = 6$ wave function overlap and covalency play an important role [19]. This may be the reason why the agreement for the values of A_4^0 and A_6^0 is less satisfactory.

Acknowledgements

The author would like to thank Prof. Dr. G. Blasse and Prof. Dr. J. H. van Santen for encouraging this work and many stimulating discussions. He is indebted to Dr. P. Glaudemans for supplying the least square fitting program used in the fitting procedure. He is grateful to Mr. J. N. H. van Hoof for his part in the experimental work.

References

[1] Q. H. F. Vrehen, Thesis, Utrecht 1963.
Q. H. F. Vrehen and J. Volger, Physica (Utrecht) **31**, 845 (1965).
[2] H. C. W. Beyerinck and B. Willemsen, Physica (Utrecht) **47**, 515 (1970).
[3] B. Willemsen and W. C. Hommels, phys. stat. sol. (a) **10**, 183 (1972).
[4] K. J. de Vries, Thesis, Utrecht 1965.
[5] W. Nieuwenkamp and J. M. Bijvoet, Z. Krist. **84**, 49 (1932).
[6] B. Willemsen, J. Solid State Chem. **3**, 567 (1971).
[7] B. Willemsen, phys. stat. sol. (a) **11**, K157 (1972).
[8] G. H. Dieke, Spectra and Energy Levels of Rare-Earth Ions in Crystals, Interscience Publishers, New York 1968.
[9] K. H. Hellwege and A. Roever, Z. Phys. **114**, 564 (1939).
[10] J. B. Gruber and R. A. Satten, J. chem. Phys. **39**, 1455 (1963).
[11] R. A. Satten, J. chem. Phys. **27**, 286 (1957).
[12] B. Willemsen, J. inorg. nuclear Chem. **33**, 3963 (1971).
[13] K. W. H. Stevens, Proc. Phys. Soc. **A65**, 209 (1952).
[14] J. C. Eisenstein, J. chem. Phys. **39**, 2134 (1963).
[15] H. M. Crosswhite, G. H. Dieke, and W. J. Carter, J. chem. Phys. **43**, 2047 (1965).
[16] A. J. Freeman and R. E. Watson, Phys. Rev. **127**, 2058 (1962).
[17] R. P. Gupta, B. K. Rao, and S. K. Sen, Phys. Rev. **A3**, 545 (1971).
[18] R. M. Sternheimer, M. Blume, and R. F. Peierls, Phys. Rev. **173**, 376 (1968).
[19] R. E. Watson and A. J. Freeman, Phys. Rev. **156**, 251 (1967).
[20] B. Willemsen, Thesis, Utrecht 1972.

(Received October 9, 1972)

phys. stat. sol. (a) **14**, 573 (1972)

Subject classification: 13.3 and 16; 22.4

2. Physikalisches Institut
der Rheinisch-Westfälischen Technischen Hochschule Aachen

Photoconductivity from Electronic Surface States on Prism Surfaces of Zinc Oxide Crystals

By

H. LÜTH and G. HEILAND

On (1$\bar{1}$00) surfaces of ZnO crystals photoconductivity has been observed at 77 K in the spectral range between 0.2 and 2.5 eV (band gap 3.4 eV). The photoconductivity was measured in ultrahigh vacuum. However, it is sensitive to the pretreatment of the surfaces with gases at room temperature. The spectra exhibit two characteristic bands which are interpreted in terms of an excitation of electrons from two sets of surface states at about 170 and 450 meV below the conduction band. Thus, photoconductivity can be used for spectroscopy of surface states.

An (1$\bar{1}$00)-Oberflächen von ZnO-Kristallen wurde bei 77 K Photoleitung im Spektralbereich von 0,2 bis 2,5 eV beobachtet (Bandabstand 3,4 eV). Die Messungen erfolgten im Ultrahochvakuum. Dabei wird die Photoleitfähigkeit durch eine vorhergehende Gasbehandlung der Oberflächen bei Zimmertemperatur beeinflußt. Die Spektren zeigen zwei charakteristische Banden, die durch Anregung von Elektronen aus zwei Gruppen von Oberflächenzuständen bei Energien von etwa 170 und 450 meV unter dem Leitungsband erklärt werden. Die Photoleitung kann also zur Spektroskopie von Oberflächenzuständen dienen.

1. Introduction

For the study of electronic surface states on clean and gas-covered surfaces of semiconductors spectroscopic methods are required which are sensitive to the small densities of such states. Widely used is the field effect, which in the case of ZnO has rendered some information on the electronic surface structure [1]. Spectroscopic methods using light can yield more direct information on energies of surface states without knowledge of surface potential and Fermi level. Up to now first results have been obtained from the spectral distributions of internal reflection [2], of surface photovoltage [3], and of photoconductivity [4]. The last method is used also in this work to study the (1$\bar{1}$00) prism surfaces of ZnO.

ZnO is an n-type semiconductor with a direct band gap of about 3.4 eV [5]. The surface of the crystals is sensitive to atomic hydrogen and to oxygen [5]. Hydrogen produces surface donors. The surface conductivity, thus, can be increased to about $10^{-4}\ \Omega^{-1}$. By adsorbed oxygen electrons are bound and the surface conductivity is decreased. In sufficiently high dose oxygen produces a depletion layer. The surface conductivity $\Delta\sigma$ often has been found to be a significant parameter for studies of ZnO surfaces. A definition of $\Delta\sigma$ has been given previously [6].

2. Experimental

The ZnO crystals were grown without additions from the vapour phase in this laboratory. They crystallize as hexagonal prisms with a length of about 30 mm and a thickness of about 2 mm. The electrical conductivity at room temperature varies between 10^{-1} and $10^{-3}\ \Omega^{-1}\ \text{cm}^{-1}$. Gold contacts were evaporated onto the prism surfaces. The current flows parallel to the c-axis, i.e. the axis of the prisms. Surfaces as grown were cleaned by annealing at 400 °C in ultrahigh vacuum ($p < 10^{-10}$ Torr). The study of cleaved surfaces required special precautions: Since the exciting light passes through the crystals all irradiated surfaces must be prepared equally. Three different cleavages parallel to the c-axis shaped the crystal to a trigonal prism which was irradiated normally to the c-axis. The dark surface conductivity $\Delta\sigma$ was varied by annealing in vacuum and by transient exposure to oxygen or atomic hydrogen (tungsten filament or discharge) at room temperature.

All measurements were performed at 77 K. The optical set-up consists of a tungsten-iodine lamp or a globar, and a double pass Perkin-Elmer grid monochromator E-1. The light was chopped with $13\ \text{s}^{-1}$ and the photoconductivity signal detected by a lock-in amplifier (PAR/HR-8). Only for lifetime measurements other frequencies were used. The total incident light intensity was measured by means of a gauged thermocouple.

3. Results

Spectral distributions of photoconductivity were measured in the range of photon energies between 0.2 and 2.5 eV, i.e. for energies below band gap (Fig. 1). The photoconductivity is given in terms of a surface conductivity $\delta\sigma$ because of its dependence on surface pretreatment. The reduction to constant photon flux can be done in a simple manner, since the photoconductivity signal increases proportional to the incident light intensity within at least two orders of magnitude. This was proved for three photon energies (1.1, 0.67, and 0.28 eV) after

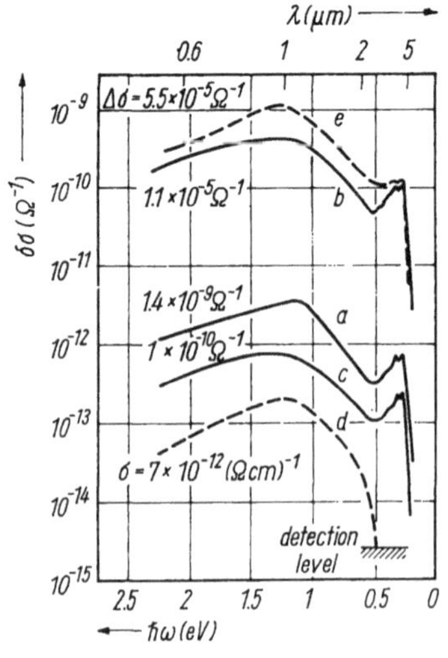

Fig. 1. Spectral distribution of the photoconductivity $\delta\sigma$ for various dark surface conductivities $\Delta\sigma$ measured in vacuum at 77 K. All spectra are related to the incident photon flux at 1.1 eV of $6 \times 10^{14}\ \text{cm}^{-2}\ \text{s}^{-1}$. Crystal 8: (1$\bar{1}$00) surfaces as-grown (full lines) (a) after annealing in vacuum at 400 °C; (b) after treatment with atomic hydrogen at room temperature; (c) after treatment with oxygen at room temperature, 0.1 Torr for about 10 min. Crystal 11: cleaved (1$\bar{1}$00) surfaces (dashed lines) (d) after cleavage in air; only the bulk conductivity σ is given since there is no observable surface conductivity $\Delta\sigma$ (depletion); (e) after treatment with atomic hydrogen at room temperature

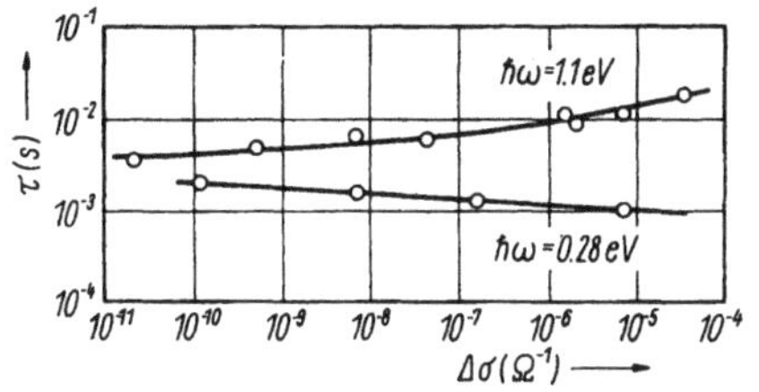

Fig. 2. Dependence of lifetime τ on dark surface conductivity $\Delta\sigma$ of crystal 8 at 77 K for photon energies at the two maxima of the curves in Fig. 1

different surface pretreatments. For the calculation of dark surface conductivity $\Delta\sigma$ the lowest conductance of the oxygen-treated crystal (depletion layer) was taken as reference. The absolute amount of the photoconductivity varies over a range of three orders of magnitude for dark surface conductivities between 10^{-10} and $10^{-4}\ \Omega^{-1}$. The shape of the spectra consisting of two structures with maxima at about 0.3 and 1.2 eV does not remarkably change with varying dark conductivity. Only curve (d) does not exhibit the peak at 0.3 eV. However, treatment with atomic hydrogen produces a high surface conductivity and the photoconductivity increases. Spectrum (e) now resembles the distributions obtained on prism surfaces as-grown, e.g. curve (b).

Since photoconductivity is governed by excitation and recombination, an additional knowledge about one of these mechanisms at least is useful for the interpretation. Therefore lifetime measurements were made for two photon energies (0.28 and 1.1 eV) and different dark surface conductivities. Because of the linear dependence of the photoconductive signal on the incident light intensity one single lifetime is defined for a fixed photon energy. The results of the measurements are shown in Fig. 2. The lifetime was determined from the frequency dependence of the photoconductivity (chopper frequencies: 13, 70, 100, 160 s^{-1}). For the two maxima of the spectrum two different lifetimes are obtained in the order of 10^{-2} to 10^{-3} s. With a variation of the dark surface conductivity over six orders of magnitude these lifetimes only change within half an order of magnitude.

4. Discussion

The photoconductivity distribution with two maxima indicates two excitation processes. The band near 0.3 eV exhibits a fine structure which must be attributed to oscillatory photoconductivity [7]. By means of this effect the threshold energy of photoconductivity is determined to (170 ± 30) meV. The margin of error in this value arises from the uncertainty about the recombination process in the oscillatory photoconductivity [8]. Extrapolating the broad structure with a maximum at 1.2 eV one arrives at a transition energy of (450 ± 100) meV. Since no theoretical model is available at present this extrapolation leads only to an approximate value. The interpretation in terms of at least two different excitation processes is supported by the difference in lifetimes (Fig. 2). The Fermi level in the bulk of the undoped ZnO crystals lies about 160 meV below the edge of the conduction band. The obtained transition energies of 170 and 450 meV lead from the upper edge of the valence band only to occupied states. The photoconductivity, therefore, must be explained by an excitation of electrons from states at about 170 and 450 meV below the conduction band.

The dependence of photoconductivity on dark surface conductivity (Fig. 1) shows that the effect must be interpreted as a surface phenomenon. Because of

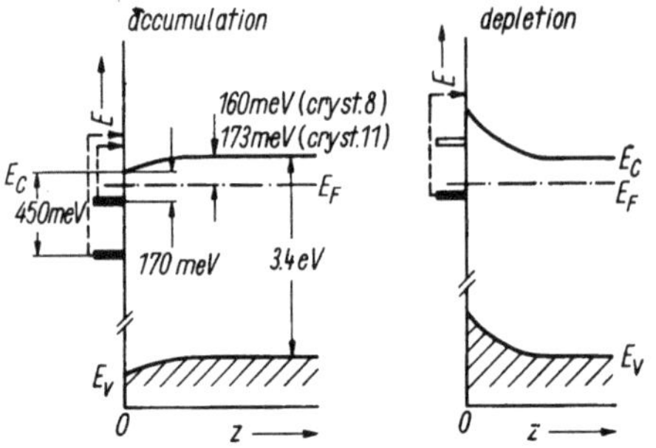

Fig. 3. Band scheme near the surface for 77 K. E_F Fermi energy, E_C and E_V edges of conduction band and valence band, respectively. Excitation processes are indicated by means of dashed arrows. z is the distance from the surface

the weak dependence of the two lifetimes (Fig. 2) on surface conductivity the variation of photoconductivity over four orders of magnitude with changing surface conductivity cannot be explained by a surface-sensitive recombination process. The excitation of electrons, therefore, must happen in a region close to the surface. If we assume an excitation of electrons from surface states below the conduction band into the space charge layer some further experimental details are understood. Curve (d) in Fig. 1 does not exhibit a maximum at 0.3 eV. Due to the preceding exposure to the oxygen of the air there is a depletion layer at the surface. As can be seen from Fig. 3 the upper set of surface states lies above the Fermi level and is not occupied by electrons. Therefore, on depleted surfaces photoconductivity from these states cannot be observed. If the photoconductive signal at 0.3 eV would be due to an excitation from bulk states inside the space charge layer the relative heights of the two maxima should depend on band bending, i.e. on surface conductivity. This is not observed.

Further evidence for the interpretation of the spectra in terms of excitation from surface states is given by preliminary photoconductivity measurements [9] on the polar surfaces of ZnO [6]. The spectra are somewhat different from the results of this paper on prism surfaces. This rules out an excitation from bulk states.

Concerning the surface states at (170 ± 30) meV there is good agreement with field effect measurements. On prism surfaces Kohl finds a maximum in the density of states at 180 meV below the conduction band [10]. It amounts to about 10^{13} cm^{-2}. From optical spectroscopy like photoconductivity the density of states cannot be derived as long as no theory for the transition probability from surface states is available.

The strong dependence of photoconductivity on dark surface conductivity is not completely understood at present. A similar effect has been found for the intrinsic photoconductivity of ZnO crystals near 3.4 eV with excitation of electrons out of the valence band [1]. In this spectral region the photoconductivity is induced within the penetration depth of band gap light (10^{-5} cm). The dependence on surface conductivity can be explained by the trapping action of surface states near the Fermi level which are detected on the same surfaces by field effect measurements [1]. A trap mechanism might also be involved in the photoconductivity with excitation out of surface states as reported in this work.

It should be emphasized that the two sets of surface states (Fig. 3) do not represent all electronic states on the surface. E.g. for explanation of band bending further sets of states ar needed.

5. Conclusion

Photoconductivity measurements allow a very sensitive spectroscopy of electronic surface states on semiconductors. Contrary to other optical methods, e.g. internal reflection, photoconductivity yields the reference point for the energy scale: The energetic position of the surface states is given with respect to the band which carries the photocurrent. The accuracy in the determination of transition energies is not always high since superposition of several excitation processes can lead to broad structures hiding sharp threshold energies.

Acknowledgement

The financial support of the Landesamt für Forschung (Nordrhein-Westfalen) is gratefully acknowledged.

References

[1] G. Heiland, J. Phys. Chem. Solids **6**, 155 (1958).
[2] G. Chiarotti, G. Del Signore, and S. Nannarone, Phys. Rev. Letters **21**, 1170 (1968).
[3] C. L. Balestra, J. Lagowski, and H. C. Gatos, Surface Sci. **26**, 317 (1971).
[4] W. Müller and W. Mönch, Phys. Rev. Letters **27**, 250 (1971).
[5] G. Heiland, E. Mollwo, and F. Stöckmann, Solid State Phys. **8**, 191 (1959).
[6] G. Heiland and P. Kunstmann, Surface Sci. **13**, 72 (1969).
[7] H. Lüth, Surface Sci., to be published.
[8] R. Ulbrich, Phys. Rev. Letters **27**, 1512 (1971).
[9] H. Lüth, Surface Sci., to be published.
[10] D. Kohl, Diploma work, II. Phys. Inst., RWTH Aachen, 1968.

(Received September 22, 1972)

phys. stat. sol. (a) 14, 579 (1972)

Subject classification: 20.3; 13.4; 22.4.3

Department of Physics, University of Hull

Evaluation of the Nature of the Radiative Emission Processes from Lattice Defect Centres in Cadmium Telluride Together with Their Associated Quenching Mechanisms

By

F. J. BRYANT, D. H. J. TOTTERDELL, and W. E. HAGSTON

The cathodoluminescence emission spectrum of cadmium telluride has been studied (i) at various temperatures and (ii) before and after electron radiation damage which produced cadmium displacements only or tellurium displacements only. The radiation damage results allow the nature of the composite emission band at ≈1.55 eV (≈ 800 nm) to be determined. Possible temperature-dependent mechanisms causing a reduction in the emission intensity have been considered. Such studies have been used to show that the 1.425 eV (870 nm) emission results from a free-to-bound transition involving a cadmium defect site and is quenched by thermal promotion of electrons from valence band to acceptor centre. Similarly, the higher energy component of the ≈ 1.55 eV emission is quenched by non-radiative internal transitions whereas the lower energy component is quenched as a result of a decrease of the radiative transition probability with increasing temperature.

Es wurde das Emissionsspektrum von Kadmiumtellurid 1. bei verschiedenen Temperaturen und 2. vor und nach Elektronenstrahlungsdamage, das entweder nur Kadmiumverlagerungen oder nur Tellurverlagerungen hervorruft, untersucht. Die Ergebnisse des Strahlungsdamage gestatten eine Bestimmung der physikalischen Natur der zusammengesetzten Emissionsbande bei ≈ 1,55 eV (≈ 800 nm). Mögliche temperaturabhängige Mechanismen die eine Verringerung der Emissionsintensität verursachen, werden berücksichtigt. Solche Untersuchungen werden benutzt, um zu zeigen, daß die 1,425 eV (870 nm)-Emission von einem Übergang von freien zu gebundenen Zuständen herrührt, der eine Kadmiumstörstelle einschließt und durch eine thermische Anregung von Elektronen aus dem Valenzband zu der Akzeptorstörstelle getilgt werden kann. In ähnlicher Weise wird die energetisch höhere Komponente der ≈ 1,55 eV-Emission durch nichtstrahlende, innere Übergänge getilgt, während die niederenergetische Komponente als Ergebnis einer Verringerung der Übergangswahrscheinlichkeit mit steigender Temperatur getilgt wird.

1. Introduction

By combining the results obtained by previous authors on the threshold energies for producing atomic displacements [1, 2, 3] with work on heat treatments of samples in cadmium or tellurium vapour [4] the present authors [5] were able to assign the three emission transitions observed near the band edge to the following configurations:

(i) The 1.587 eV (≈ 781 nm) band results from an exciton bound to a neutral acceptor centre (which involves a cadmium vacancy).

(ii) The 1.572 eV (≈ 788 nm) band results from the transition of a conduction electron into the ground state of the same acceptor centre.

(iii) The 1.554 eV (≈ 798 nm) band (and its phonon replica at 1.532 eV (≈ 809 nm)) results from the transfer of an electron from the conduction band into the first charge state of this same acceptor.

A partial experimental check on the correctness of these assignments was provided by the explanation they afford of the anomalous temperature variation of the intensity of emission of the ≈ 1.55 eV composite band. Here, non-radiative Auger dissociation of the 1.587 eV band results in a centre which is the ground state of the 1.554 eV emission. Thus, whether the intensity of the latter increases or decreases initially with increasing temperature is dependent on the presence or absence of the 1.587 eV band. These conclusions are amply borne out by the experiment [5].

It is therefore fairly conclusively demonstrated that at least one of the electronic processes responsible for the 1.554 eV emission involves the transition of an electron from the conduction band into a shallow acceptor centre. That this is not necessarily the only transition responsible for emission in this region was already apparent in the previous work [5]. By using an accurate theory [6] for the band edge shifts, calculations show [5] that the shift of the 1.554 eV band in a crystal in which the 1.587 eV emission was absent should be about 2 meV if the temperature was raised from 4 to 50 K. However the experimentally observed value of about 4.5 meV was much greater than this. It has been pointed out [5] that this difficulty could be overcome by the introduction of the possibility of a donor–valence band transition occurring at a similar energy and with a comparable intensity to the previously considered transition.

This tentative conclusion deduced previously [5] is given further support by the investigations of different authors [7 to 10] who suggest that a complex involving a tellurium vacancy and a halogen impurity give rise to a defect level at about (E_c — 0.06) eV. Such a centre would give rise to an emission at about 1.545 eV.

In the following, the emission at about 1.55 eV is investigated in greater detail and experiments are described which show that this band is indeed a composite band formed from at least two sub-bands. One of these is shown to involve the cadmium vacancy centre previously described, whilst the other is shown to involve a tellurium displacement. Furthermore, the relative ordering of these two bands in energy position is conclusively achieved. By measuring emission intensities as a function of temperature, we are able to discuss the nature of the mechanisms responsible for reducing the emission intensity of the 1.554 eV band. Similarly investigations of the intensity of the 1.545 eV band allow us to make an unambiguous identification of the nature of the electronic process responsible for this emission.

2. Experimental Procedures

The cadmium telluride samples were ground from single crystals grown in this department by the Bridgman method. Analysis of the single crystals revealed about 1 ppm of Ca, Cu, Fe, and Mg. The samples were embedded in a thermally conducting silver paste on a copper substrate which was firmly mounted on one of the faces of the cold finger of a multisample liquid helium cryostat [11].

The liquid helium cryostat was evacuated ($\leq 5 \times 10^{-6}$ Torr) by the same system used to evacuate the drift tube of the Van de Graaff accelerator which provided the electrons used for radiation damage or cathodoexcitation. The electron beam energy was stable to ± 2 keV and could be varied between 100 and 400 keV [12]. The exciting electrons were incident normally on the sample and the emission was observed at 45° to this direction. The emission was chopped at 800 Hz and, after passing through a Bausch and Lomb monochromator having a 500 mm grating blazed at 1.0 μm, was detected by a system consisting of a CVP 150 red-sensitive photomultiplier, low-noise amplifier, phase sensitive detector and potentiometric recorder. Sample temperatures were measured using a gold–iron versus chromel thermocouple in the range 4 to 77 K and a copper–constantan thermocouple above 77 K.

3. Experimental Results

The radiation sensitive properties of the broad cathodoluminescence emission shown in Fig. 1 has been the subject of a separate investigation [13]. The centre responsible for this broad emission is present in most as-grown material and was particularly strong in the original samples which were relatively impure. From electron irradiation studies it was concluded that this defect was complex and involved the cadmium vacancy and an impurity [13]. Fig. 1 shows this broad band emission with the sample held at different temperatures. By observing the peak emission intensity as a function of temperature for several fixed sample temperatures it has been possible to plot Fig. 2. Application of theory given in Section 4.1.1 to the results shown in Fig. 2 yields an activation energy of about 53 meV for this broad cathodoluminescence emission.

The cathodoluminescence edge emission spectrum of cadmium telluride has been shown to depend on the preparation conditions or post-growth treatment of the sample [4]. Furthermore, it is generally apparent that post-growth treatment of the sample (i.e. firing in excess of one or other of the host constituents) is often necessary before cathodoluminescence emission intensity can be used successfully to monitor the changes in the defect structure which can be caused by electron radiation damage. The effect of tellurium displacements on the edge emission can be revealed effectively by damaging, above the tellurium displacement threshold energy, a sample at 77 K which has been previously fired in excess tellurium. Although the threshold energy for tellurium displacement (340 keV) is above that for cadmium displacement (235 keV) it is possible to produce only tellurium displacements by irradiating with the sample at

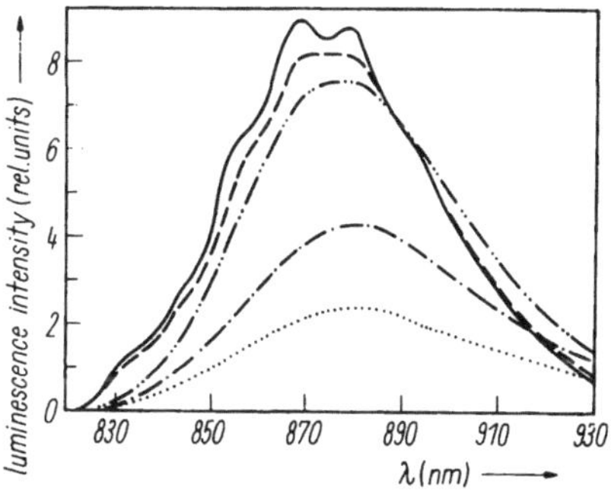

Fig. 1. Cathodoluminescence of the 1.425 eV (870 nm) emission of cadmium telluride excited by 13.3 μA cm^{-2} of 100 keV electrons: ——— at 8 K; – – – at 30 K; –··–··– at 83 K; –·–·–·– at 116 K; ········· at 134 K

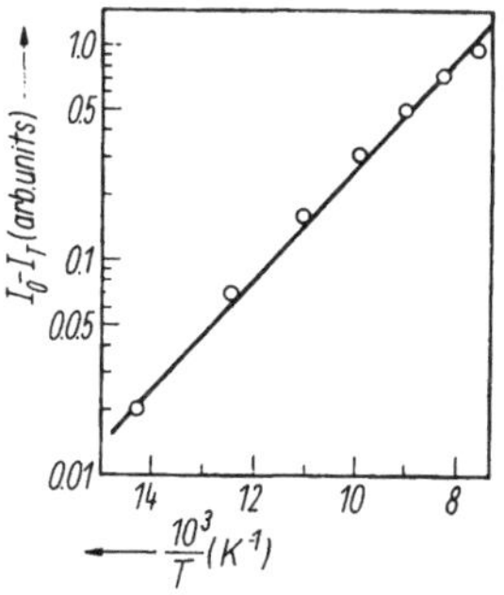

Fig. 2. Plot of log ($I_0 - I_T$) against $10^3/T$ for the 1.425 eV (870 nm) cathodoluminescence emission. $I_0(I_T)$ is the peak emission intensity at temperature 0 K (T K)

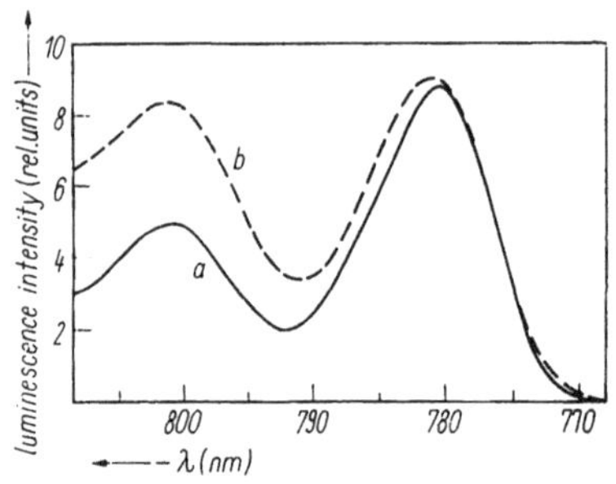

Fig. 3. The edge emission spectrum of tellurium fired cadmium telluride at 8 K excited by 13.3 μA cm⁻² of 100 keV electrons. (a) Before radiation damage; (b) after 1.0×10^{18} electrons cm^{-2} of 400 keV energy whilst the sample was at a temperature of 77 K

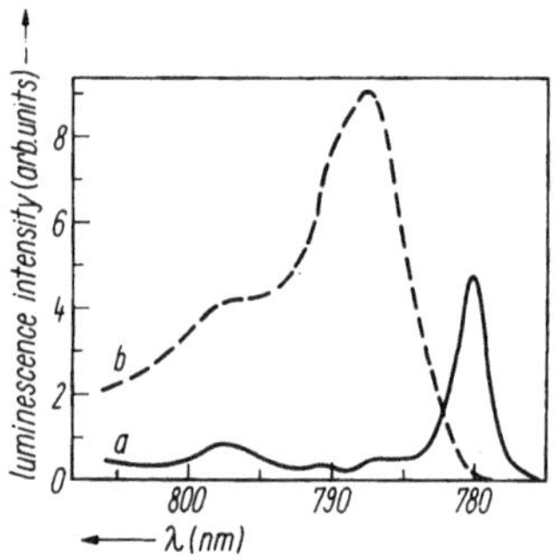

Fig. 4. The edge emission spectrum of cadmium fired cadmium telluride at 8 K excited by 13.3 μA cm^{-2} of 100 keV electrons. (a) Before radiation damage, (b) after 3.0×10^{17} electrons cm^{-2} of 330 keV energy whilst the sample was at a temperature of 8 K

a temperature of 77 K or higher, because the radiation-induced cadmium displacements are not stable at such temperatures. Cadmium displacements are produced at 8 K during any irradiation capable of producing tellurium displacements but the cadmium displacements anneal before a sample temperature of 77 K is reached. Fig. 3 shows the effect of tellurium displacement on the sample. Enhancement of the composite emission band at about 1.55 eV is observed and, although difficult to distinguish, appears to be due to the lower energy component of the composite band. As would be expected the emission at 1.587 eV is found to be virtually unaffected by tellurium displacement in the lattice because it results from the annihilation of an exciton bound to a cadmium vacancy [5].

By irradiating with electrons of energy in the range 235 to 340 keV (i.e. between the two displacement threshold energies) it is possible to displace cadmium atoms only. The effect on the edge emission of the displacement of cadmium atoms only is shown in Fig. 4. The exciton emission at 1.587 eV (781 nm), although resulting from excitons bound to cadmium vacancies, is completely destroyed. On the other hand, the emission at 1.572 eV (788 nm), which results from an electron transition to a cadmium vacancy [5], is enhanced very marked-

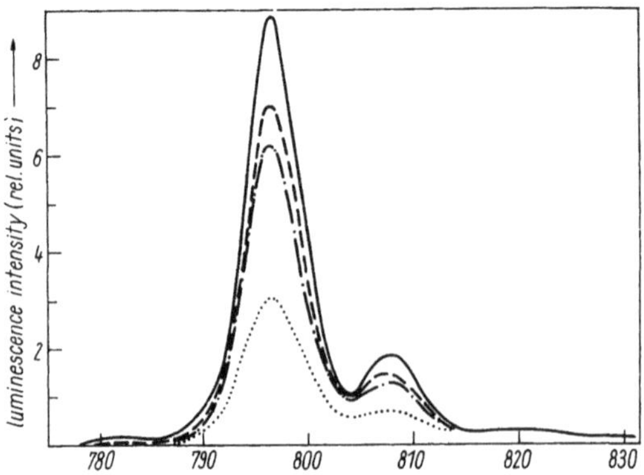

Fig. 5. The edge emission spectrum of cadmium telluride excited by 13.3 μA cm^{-2} of 100 keV electrons. ——— at 8 K; — — — at 20 K; —·—·— at 23 K; ····· at 31 K

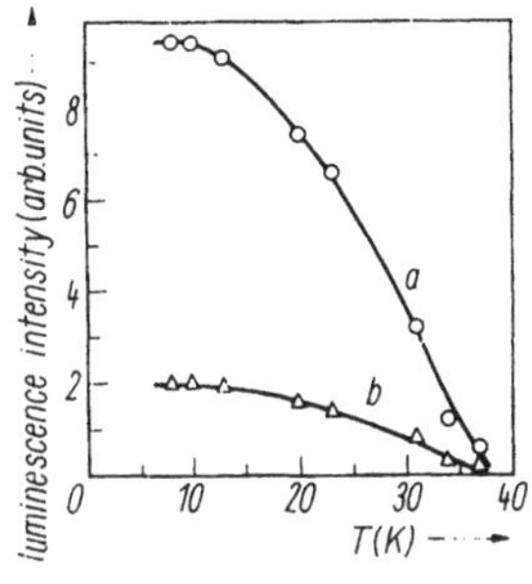

Fig. 6. Emission intensity dependence upon temperature (excited by 13.3 μA cm^{-2} of 100 keV electrons) (a) 1.554 eV (798 nm) emission; (b) 1.532 eV (809 nm) emission

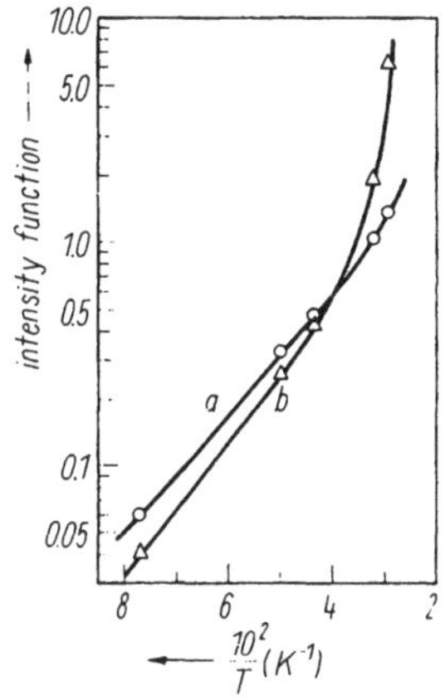

Fig. 7. Possible quenching mechanisms for the $\approx$ 1.55 eV ($\approx$ 800 nm) emission of cadmium telluride excited by 13.3 μA cm^{-2} of 100 keV electrons. (a) $\log(I_0 - I_T)$ against $10^2/T$ (K^{-1}), (b) $\log[(I_0/I_T)-1]$ against $10^2/T$ (K^{-1})

ly. The effect of the cadmium displacement on the composite band at about 1.55 eV (about 800 nm) is again difficult to elucidate but there is clearly a large enhancement which appears to occur towards the higher energy region of the composite band.

The cathodoluminescence edge emission spectrum at different temperatures is shown in Fig. 5 for a sample in which there was no 1.587 eV (781 nm) emission present initially. The temperature dependences of the peak intensities of the composite band ($\approx$1.55 eV) ($\approx$800 nm) and the 1.532 eV (809 nm) phonon replica are shown in Fig. 6. Using the values of Fig. 6 for the 1.55 eV ($\approx$ 800 nm) emission band it is possible to apply the theory given in Section 4.1.1 (which assumes thermal ejection of carriers from shallow centres) and the theory given in Section 4.1.3 (which assumes that the emission is quenched by internal non-radiative transitions). The result of this application of theory is shown in Fig. 7, neither plot yields a linear relationship over the whole of the range. The activation energy determined from the plot of $\log(I_0 - I_T)$ against $1/T$ is about 6 meV. The plot of $\log(I_0/I_T - 1)$ against $1/T$ yields values of activation energy also generally near 6 meV, but the value rises as high as $\approx$ 33 meV at the very highest temperatures where the accuracy of the intensity measurements is beginning to decrease noticeably.

4. Theory

4.1 Effects of temperature on the emission intensity

In this section various mechanisms which produce a decrease in the emission intensity of any given luminescence transition as the temperature is increased are investigated. Although it is possible that any or all of the processes to be described could be occurring at one and the same time, it is nonetheless convenient to treat each mechanism separately. This is not only relevant to later discussion but is also of great value in understanding the behaviour of the actual transitions involved, since the activation energies for the various processes are usually so widely different that one of them tends to dominate the others in any given temperature region. The values of optical activation energies compared in this paper with thermal activation energy were determined not from peak intensity positions but from the high energy edge of an emission band.

4.1.1 Change in the equilibrium occupancy of shallow centres

To illustrate the type of process considered here we will take a specific example. Suppose the luminescence transition occurs as a result of a transfer of an electron from the conduction band into a shallow acceptor centre. At $T = 0$ K all the acceptor centres will be unoccupied but with increasing temperature a certain fraction of them will be occupied at thermal equilibrium by electrons from the valence band. As a result the emission intensity will decrease. If $N_A(n_A)$ is the total concentration of acceptors (occupied acceptors), and the degeneracy of the holes in the valence band is weak, we have at equilibrium

$$n_A^2 = N_A\, N_C \exp\left[-\frac{E_A}{kT}\right], \tag{1}$$

where E_A is the acceptor ionization energy,

$$N_C = 4\left[\frac{2\,\pi\, m_h\, kT}{h^2}\right]^{3/2}.$$

m_h is the effective mass of holes in the valence band. At a temperature T the emission intensity I_T is given by

$$I_T \propto [N_A - n_A];$$

$$I_T = k\, N_A\left[1 - \frac{n_A}{N_A}\right],$$

where k is a constant.
When $T = 0$, $n_A = 0$, $I_T = I_0 = k\, N_A$, thus

$$I_T = I_0\left[1 - \frac{n_A}{N_A}\right]. \tag{2}$$

Coupling equations (1) and (2) yields the relation

$$y = c - \frac{E_A\, x}{2\, k} - \frac{3}{4}\ln x\,, \tag{3}$$

where $y = \ln\,[I_0 - I_T]$, $x = 1/T$, and c is a constant. (Note, by definition, the activation energy E is the coefficient of x/k in (3), i.e. $E = E_A/2$.) Differentiating equation (3) with respect to x gives

$$\frac{\partial y}{\partial x} = -\frac{E_A}{2\,k} - \frac{3}{4}\,T\,. \tag{4}$$

If E_A is small, e.g. $E_A \approx 5$ meV, then $E_A/k \approx 60$ K, and for a small range of T close to $T = 0$ K we see that $\partial y/\partial x$ is sensibly a constant, i.e. we have a straight line. However, as T becomes larger (i.e. x gets smaller) the magnitude of $\partial y/\partial x$ will increase significantly and the straight line becomes a curve. Similarly we see that if E_A is large, e.g. $E_A \approx 100$ meV so that $E_A/k \approx 1200$ K, then the plot of y against x will give a straight line over a large range of temperature.

4.1.2 *Variation of radiative transition probability*

To illustrate the type of process considered here we take the case of a transition of an electron from the conduction band into a shallow acceptor. At $T = 0$ K all the electrons are at the bottom of the band and the radiative transition probability is the same for them all. At higher temperatures a certain fraction of the electrons occupy states at energies above the bottom of the band and we have to form an expression for the "average transition probability" over the corresponding energy range. It is this average transition probability that is of interest here. As we will see below, our results show that for the various centres and temperature ranges of interest to us, the average radiative transition probability for electrons in the conduction band into shallow acceptor centres is sensibly a constant. However this is not the case for transitions of holes from the valence band into shallow donors. Here it is found that only for holes in the energy range $0 \leq E \leq E_D/5$ (where E_D is the ionization energy of the donor) is the transition probability sensibly a constant. However for holes with energies $E > E_D/5$ the transition probability effectively vanishes. Within the approximations of the theory as given below, this means that the difference between the emission intensities I_0 and I_T is given by a factor α where α is the fraction of holes in the energy range $0 \leq E \leq E_D/5$. Ignoring the complication of "light" and "heavy" hole bands, this then allows us to write the relation

$$I_T = I_0\left[1 - p\exp\left(-\frac{E_D}{5\,kT}\right)\right], \tag{5}$$

where p is some constant less than unity. This means we have placed $\alpha = \left(1 - p\exp\left(-\frac{E_D}{kT}\right)\right)$. From equation (5) we obtain

$$y = \ln\left[I_0 - I_T\right] = c' - \frac{E_D\,x}{5\,k}, \tag{6}$$

where $c' = $ const and $x = 1/T$. (Note again that the activation energy E which is the coefficient of x/k is given by $E = E_D/5$.)

In order to prove these statements we note that the theory of the radiative emission intensity developed for transitions of a carrier into a shallow centre [14, 15] shows that the emission intensity at energy E is governed in the main by a factor $(1 + \eta)^{-4}$. In the specific case of a transition of an electron in the conduction band into a shallow acceptor centre, we find [15]

$$\eta\,\beta = E - (E_G - E_A), \tag{7}$$

where $\beta = (m_v/m_c)\,E_A$, E_A is the ionization energy of acceptor, $m_v(m_c)$ is the valence (conduction) band effective mass. (The only difference for the transition of a hole from the valence band into a neutral donor is that E_A is replaced by E_D and $\beta = (m_c/m_v)\,E_D$).

Because of the $(1 + \eta)^{-4}$ factor it is apparent that the emission intensity will begin to show a noticeable decrease only when $\eta \gtrapprox 1$. Hence the energies of the electrons in the conduction band for which the transition probability into the acceptor centre remains approximately a constant will be the values up to and including those for which $\eta \approx 1$. Substituting this value into equation (7) shows that this corresponds to electrons at an energy β above the bottom of the conduction band. Later we concern ourselves with two transitions in which $E_A \approx 40$ meV and $E_A \approx 120$ meV respectively. Since it is generally the case in group II–VI compounds that $m_v/m_c \gtrapprox 5$ we see that $\beta \gtrapprox 5\,E_A$ and has the values 200 and 600 meV respectively. As we are only concerned with temperature changes up to 100 K, then it is apparent that for the case of electrons in the conduction band, the transition probability into shallow acceptor centres remains approximately a constant for all electron energies.

On the other hand, in the case of transitions of holes into donor states we see that using $m_c/m_v \lessapprox 1/5$ together with $E_D = 40$ and 120 meV the transition probability remains constant for holes possessing energies up to 8 and 24 meV, respectively. In particular we see from equation (6) that if the transition at 1.554 eV (i.e. equivalent to an E_D of 40 meV) is due to a transition of a hole from the valence band into a shallow donor, a plot of y against x should lead to an activation energy $E = E_D/5 \lessapprox 8$ meV. Likewise if the transition at 1.425 eV (i.e. equivalent to an E_D of ≈ 120 meV measured at ligth energy edge) is due to such a transition, then a similar plot should yield an activation energy $E \lessapprox 24$ meV.

4.1.3 Non-radiative effects

As is well known, any excited state in a solid can collapse to the ground state and dissipate its energy non-radiatively in the form of phonons. It is this non-radiative transition probability which is discussed in this section. Because of its effect we can write

$$I_T = \frac{P_r}{P_r + P_{nr}}\, I_0, \tag{8}$$

where P_r is the radiative transition probability, P_{nr} is the non-radiative transition probability. We assume here that P_r remains constant with temperature but that P_{nr} has the form

$$P_{nr} = C \exp\left(-\frac{E}{kT}\right), \tag{9}$$

where C is some frequency factor which may depend on temperature and E is the activation energy. From equations (8) and (9) we obtain

$$y = \ln\left[\frac{I_0}{I_T} - 1\right] = \ln C' - \frac{E\,x}{k}, \tag{10}$$

where $C' =$ const (possibly temperature dependent) and $x = 1/T$. The form of equation (10) is exactly the same as that of equation (3) and therefore the same remarks apply to it as were made in connection with equation (3).

4.2 Other aspects of optical and thermal effects

In this section we consider the relationship between the optical activation energy (E_{opt}) and the thermal activation energy (E_{th}). According to the theory of Kubo [16] the expression for this ratio is

$$\frac{E_{opt}}{E_{th}} = \frac{\left(\frac{1}{\varepsilon_0} + \frac{15}{16}\,C\right)}{\left(\frac{1}{\varepsilon_0} + \frac{5}{16}\,C\right)}, \tag{11}$$

where

$$C = \frac{1}{\varepsilon_\infty} - \frac{1}{\varepsilon_0}$$

and ε_0 and ε_∞ are respectively the static and high frequency dielectric constants of the medium. Inserting the relevant values for CdTe gives $E_{opt}/E_{th} = 1.0079$, i.e. almost unity. Further substantiation for the validity of this theoretical value is provided by experiment [17] which also yields a value close to unity. In what follows this fact is used as a criterion in that if a given mechanism is presumed to be responsible for the decrease in the emission intensity from a specific centre with increasing temperature, then the optical and thermal activation energies should be approximately equal. If they turn out to be significantly different then it is assumed that the proposed mechanism for producing the decrease in intensity is not the correct one.

5. Discussion

As pointed out in previous work [5] the two holes bound to the double acceptor centre follow orbits which are much more diffuse than those anticipated on the basis of a simple hydrogenic model. Hence with increasing concentration of such centres a point will soon be reached at which the overlapping of these orbits occurs. When this happens the excitons bound to such centres will decay rapidly by the Auger process leaving the acceptor centre in its first charge state. This accounts for the observations depicted in Fig. 4 where, following radiation damage in which cadmium atoms only are displaced, the bound exciton emission at 1.587 eV is completely destroyed.

Similarly from the earlier work [5] evidence exists which indicates that the emission at about 1.55 eV could be composed of two sub-bands. One of these is a transition into an acceptor centre involving a cadmium vacancy, whilst the other centre was thought possibly to involve a tellurium vacancy [5]. Due to the close proximity in energy position of these bands the problem is to decide which one occurs at the higher energy. With regard to the assignment of the lower energy component of the composite band at about 1.55 eV to a centre involving a tellurium vacancy it is apparent from Fig. 3 that in a sample excited at low temperatures (8 K) the intensity of the emission at 1.55 eV is approximately half that of the 1.587 eV emission. However following radiation damage at 77 K in which only tellurium displacement occurs (due to the annealing of the displaced cadmium) excitation at low temperatures (8K) gives again the same intensity for the emission at 1.587 eV but increases that at about 1.55 eV by a factor of two. The fact that the emission at 1.587 eV remains the same following damage at 77 K shows that no additional cadmium vacancies are produced (since the centre responsible for this emission involves such vacancies [5]) whereas the increase of the emission at about 1.55 eV shows that one of the bands involves a tellurium displacement. Furthermore, from Fig. 3 the tellurium displacement causes enhancement of the 1.55 eV composite band to occur at the lower energy end which supports the assignment of the 1.545 eV band to a tellurium displacement.

From Fig. 4, before radiation damage the composite band ≈ 1.55 eV is much less intense than the bound exciton band at 1.587 eV. However after radiation damage involving cadmium displacements only, the emission intensity of the composite band is enhanced by a factor of six. Furthermore the enhancement takes place at the higher energy end of the band supporting the assignment of the 1.554 eV band to cadmium displacements.

In the case of the emission at 1.425 eV it is found that the plot of ln $[I_0 - I_T]$ against $1/T$ fits the experimental points over the whole of the temperature range and yields an activation energy $E = E_A/2 = 53$ meV. Allowing for the approximate equality of the thermal and optical activation energies it is apparent that if the transition involved was of the donor–valence band type then from optical data an activation energy $E = E_D/5 \lessapprox 24$ meV would be expected. On the other hand, if the transition was of the conduction band–acceptor type, then equation (4) shows that an optical activation energy of $E = E_A/2 = 60$ meV would be expected. Since this is in reasonable agreement with the experimental value it is assumed that the emission at 1.425 eV is due to a transition of an electron from the conduction band into an acceptor centre. Furthermore the physical mechanism producing quenching of emission with increasing temperature is the thermal promotion of holes out of the acceptor centres into the valence band.

For the emission band at about 1.55 eV neither log plot yields a straight line over the whole of the temperature range. However this is hardly surprising for various reasons. Thus the fact that this band is thought to be a composite band formed from two sub-bands each of which could have two different activation energies would lead to complications not allowed for in the theory. Similarly as pointed out in the theory section, if the activation energy is small (a few meV) then because of other possible T^n dependences which could occur, a log plot would only yield a straight line close to $T = 0$ K. It is for these reasons that we believe that the plots of Fig. 7 could be curves and not straight lines. Nonetheless, irrespective of which point on curve (a) of Fig. 7 is taken we find from the slope at that point that the corresponding activation energy $E = E_A/2$ [or $E_D/2$] $\lessapprox 6$ meV. This shows that the process responsible for the decrease in intensity is not the thermal emptying (or filling) of donors (or acceptors) as in the case of the 1.425 eV band, since this would require an (optical) activation energy $E = 1/2\, E_A$ [or $1/2\, E_D$] $= 20$ meV. From the previous section we see that an activation energy $E \lessapprox 6$ meV could possibly be consistent with a decrease in intensity of a donor–valence band transition due to the kinetic energy associated with an increasing fraction of the holes (see Section 4.1.2). Such a possibility would have to be assigned to the centre which results when a tellurium ion is displaced from its lattice site, since we know from earlier results that the component of this band produced by cadmium displacement is of the conduction band–acceptor type [5]. For this conduction band–acceptor transition it is impossible for an activation energy of $E \lessapprox 6$ meV to occur by the process of thermal emptying described in Section 4.1.1 or by the process of 4.1.2, so that the decrease in emission intensity must occur as a result of internal non-radiative transitions. The maximum activation energy (E) associated with this transition can be determined from curve (b) of Fig. 7 to be $E \lessapprox 33$ meV.

This conclusion in itself raises an extremely interesting point in that the present results together with those described previously [5] show that a portion of the emission intensity at about 1.55 eV together with the emission at 1.425 eV is due to a transition of an electron from the conduction band into an acceptor centre. Although both transitions are of the same type, our analysis shows that the intensity of transitions into the shallower centre decay rapidly due to non-radiative internal decay, whereas the transitions into the deeper acceptor centre are not subject to an analogous decrease in intensity. In the latter case the decrease results from the thermal filling of the acceptor states with electrons from the valence band. This highlights the importance in general of evaluating exactly the nature of the process producing the decrease in emission intensity. For example in the case of the green and blue copper emissions of zinc sulphide a variety of different experimental results indicate that both of these centres have the same ground state which lies within a few tenths of an eV of the valence band [18]. As a result it was anticipated that both centres should quench out at the same temperature due to thermal promotion of electrons from the valence band into the acceptor. However it was found that the blue centre quenches at a temperature 100 to 150 K below that of the green centre, and this behaviour was thought to be anomalous [19].

In view of our earlier results we see that this could possibly be explained by assuming that the excited state level of the blue emission is much more strongly coupled to the lattice than that producing the green emission. As a result the former decays by non-radiative transitions with an activation energy E_a, say, whereas the latter either (i) decays non-radiatively with an activation energy $E_b > E_a$ or (ii) decays by a different mechanism with an activation energy $E_b > E_a$.

6. Conclusion

The results presented above give further support to the previous assignment of the emission peaks at 1.587, 1.572, and 1.554 eV [5]. In addition they provide confirmatory evidence that the emission in the region of ≈ 1.55 eV is a composite band formed from two sub-bands. The higher energy component of this pair is due to a transition of an electron from the conduction band into an acceptor state which involves a cadmium vacancy. The lower energy component exhibits thermal quenching properties which are consistent with a donor–valence band transition. Furthermore, radiation damage studies indicate that this centre involves either a tellurium vacancy or a tellurium interstitial or both.

A combination of radiation damage studies with observations of the temperature variation of the emission intensity shows that the emission at 1.425 eV results from an electron in the conduction band recombining with an acceptor centre which involves a cadmium defect site.

Finally the results focus attention on the necessity for evaluating carefully the various processes responsible for any decrease of the luminescence intensity with temperature, especially when the latter observations are thought to give results at variance with models for luminescence processes developed in order to account for the results of related optical, electrical, and magnetic phenomena.

References

[1] F. J. Bryant and E. Webster, phys. stat. sol. **21**, 315 (1967).
[2] K. Matsuura, N. Itoh, and T. Suita, J. Phys. Soc. Japan **22**, 1118 (1967).
[3] F. J. Bryant, A. F. J. Cox, and E. Webster, J. Phys. C (Proc. Phys. Soc.) **1**, 1737 (1968).
[4] F. J. Bryant and D. H. J. Totterdell, Radiat. Eff. **9**, 115 (1971).
[5] F. J. Bryant, W. E. Hagston, and D. H. J. Totterdell, J. Phys. C (Solid State Phys.) **4**, 641 (1971).

[6] G. D. MAHAN, J. Phys. Chem. Solids **26**, 751 (1965).
[7] M. R. LORENZ and H. H. WOODBURY, Phys. Rev. Letters **10**, 215 (1963).
[8] H. H. WOODBURY and M. AVEN, General Electric Res. Rept. No. 64-RL (36599) (1964).
[9] M. R. LORENZ, B. SEGALL, and H H. WOODBURY, Phys. Rev. **134**, 751 (1964).
[10] F. A. KROGER, J. Phys. Chem. Solids **26**, 1717 (1965).
[11] F. J. BRYANT and C. J. RADFORD, Cryogenics **10**, 329 (1970).
[12] F. J. BRYANT, A. F. J. COX, and R. KIRBITSON, Nuclear Instrum. and Methods **63**, 195 (1968).
[13] F. J. BRYANT and D. H. J. TOTTERDELL, phys. stat. sol. (a) **10**, K75 (1972).
[14] W. P. DUMKE, Phys. Rev. **132**, 1998 (1963).
[15] D. M. EAGLES, J. Phys. Chem. Solids **16**, 76 (1960).
[16] R. KUBO, J. Phys. Soc. Japan **3**, 254 (1948).
[17] R. E. HALSTED and B. SEGALL, Phys. Rev. Letters **10**, 392 (1963).
[18] G. F. ALFREY and M. A. S. SWEET, J. Lum. **4**, 291 (1971).
[19] D. CURIE, Luminescence In Crystals, Methuen, London 1963 (p. 118).

(Received October 18, 1972)

phys. stat. sol. (a) **14**, 591 (1972)

Subject classification: 18 and 20.1; 22.1.1

Microwave Laboratory, Department of Physics and Astrophysics, University of Delhi

On the Procedures of Measuring Microwave Faraday Rotation in Semiconductors

By

B. R. SETHI

A brief review of the procedures used by the various investigators for the measurement of the microwave Faraday rotation in semiconductors using the circular waveguide transmission method has been presented along with the distinguishable features of their set-up. Expressions have been developed for the accurate evaluation of the Faraday rotation from the data representing the changes in the transmitted detected signals corresponding to the two senses of the magnetic field, taking into account the changes due to transverse magneto-absorption and other interfering effects. The suitability of these expressions for the evaluation of the Faraday rotation from the measured transmission data at room temperature as well as at liquid nitrogen temperature has been examined for germanium samples.

Es wird kurz über Verfahren, die von verschiedenen Autoren zur Messung der Mikrowellen-Faradayrotation in Halbleitern unter Verwendung von Transmissionsmethoden mit zirkularen Wellenleitern benutzt wurden und über Unterschiede ihrer Apparaturen berichtet. Ausdrücke wurden abgeleitet für die genaue Bestimmung der Faradaydrehung aus den Änderungen des gemessenen Signals entsprechend den beiden Richtungen des Magnetfeldes, wobei die transversale Magneto-Absorption und andere Störeinflüsse berücksichtigt werden. Die Eignung dieser Ausdrücke für die Berechnung der Faradayrotation aus den gemessenen Transmissionswerten wurden sowohl bei Zimmertemperatur als auch bei der Temperatur des flüssigen Stickstoffs an Germaniumproben überprüft.

1. Introduction

The study of the microwave Faraday effect in semiconductors has been of great interest to scientists and engineers during the last two decades for the determination of the transport and band parameters of the semiconductor, and for the design of certain non-reciprocal microwave components and devices. For the measurement of the Faraday rotation or ellipticity, one has to investigate the microwave field pattern after transmission through the sample. There are various techniques for the study of microwave Faraday effect such as cavity methods [1, 2], crossed-waveguides coupler method [3], and the circular waveguide transmission method [4]. In the first two techniques complicated electromagnetic field patterns are involved; whereas the last technique is based on the most direct version of the definition of Faraday effect.

In this paper, a brief review of the procedures of measuring Faraday rotation, using the circular waveguide transmission method, by various investigators has been presented along with the distinguishing features of their set-up. Expressions have been developed for evaluating accurately the Faraday rotation from the data representing the changes in the transmitted detected signals for both senses of the field, i.e. $\boldsymbol{B}_+$ and $\boldsymbol{B}_-$, taking into account the changes in the signal due to transverse magneto-absorption and other interfering effects.

The suitability of these expressions have been examined for evaluating the Faraday rotation from the measured transmission data at room temperature as well as at liquid nitrogen temperature in n-type germanium samples.

2. Brief Review of Procedures Used for Faraday Rotation Measurements

The circular waveguide transmission method makes use of the semiconductor sample in the form of a circular slice placed in a circular waveguide. On the application of the longitudinal magnetic field, the changes produced in the microwave signal, transmitted through the sample, are analysed either in the circular TE_{11} mode or in the rectangular TE_{10} mode for the evaluation of Faraday rotation.

Rau and Caspari [5] were the first to use the circular waveguide method for Faraday rotation measurements. They analysed the transmitted power with a detecting crystal kept in a rectangular waveguide without using any rectangular-to-circular transition for the mode conversion. Their method of measurement involves the study of the transmitted power around the minimum and to measure the shift when applying the magnetic field. This procedure, however, has been found to be insensitive for small rotations of the order of 2 to 3° due to the flatness of the minimum. The plotting procedure is laborious and the shift in the minimum unequal for the two senses of the field.

Furdyna and Broersma [4] modified the set-up by using a rectangular-to-circular transition for the smooth conversion of the microwave power from the rectangular TE_{10} mode to the circular TE_{11} mode. They used a quite sensitive detecting system comprising two fine probes, at right angle to each other, inserted into the circular waveguide. They used two matched crystals for detection and the difference of the two signals was fed into a sensitive meter. The probes were set symmetrical with respect to $\boldsymbol{E}$ so that in the absence of the magnetic field, the combined output of the two signals is zero. The rotation was measured from the imbalance produced in the signals at application of the field, due to the rotation of the field pattern. The method requires perfectly matched crystals and calibration of rotation in terms of the imbalanced current, and may lead to serious errors due to imperfect fabrication of the analysing head.

Kuno and Hershberger [6, 7] also made use of this circular waveguide method for the measurement of Faraday rotation in solid state plasma of InSb. They, however, used two transitions — one preceding the sample and the other following it. Their procedure was similar to that adopted by Rau and Caspari [5]. A similar set-up and procedure was later on used by van Dalen [8] also for the measurement of rotation and attenuation in n-type polycrystalline GaAs samples.

It was suggested by Allen [9] that the Faraday rotation can be measured accurately by utilising the linear portion of the angular distribution of the power curve around $\varphi = 45°$ (Fig. 1), where φ represents the angle between the detector and the initial direction of $\boldsymbol{E}$. As the slope of the curve in this region is maximum, he suggested that maximum sensitivity in rotation measurements can be achieved. Chai and Vogelhut [10] also utilized the same principle for the measurement of very small angles of Faraday rotation. They modulated the microwaves with a gyrator for suppressing the disturbing signals,

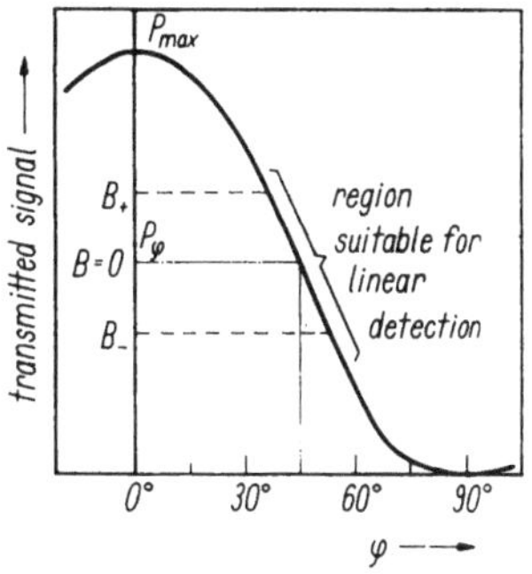

Fig. 1. The angular distribution of the detected signal in the circular waveguide after transmission through the semiconductor slice

and mixed the signal transmitted through the sample with the initial reference signal in a magic-tee balance mixer. The difference of signals was fed into a recorder through a phase lock-in amplifier. No explicit procedure was, however, given by them [9, 10] for the evaluation of Faraday rotation.

Making use of the linear detection technique, the author in this paper aims to describe the procedure and develop the expressions, from which the Faraday rotation can be evaluated accurately.

The procedure comprises firstly plotting the angular distribution of the transmitted microwave signal in the circular TE_{11} mode and then to note the changes at the application of the magnetic field for both senses, i.e. $\boldsymbol{B}_+$ and $\boldsymbol{B}_-$, keeping the analysing detector at the mid-point of the linear region and at the peak of the curve shown in Fig. 1. From these data, the Faraday rotation can be evaluated using the expressions developed in the next section.

3. Expressions for the Evaluation of Faraday Rotation

3.1 Simple treatment

In the absence of the magnetic field, let P represent the transmitted signal detected by the analyser set at the mid-point of the linear region; with the magnetic field applied, let ΔP represent the normalized change in the transmitted signal, i.e. the change in the signal divided by the initial transmitted signal.

For the sense $\boldsymbol{B}_+$ of the magnetic field, let A represent the normalized change in the transmitted signal. A part of this change, say $(\Delta P_1)_\theta$, is due to Faraday rotation, and the remaining change, say ΔP_2, is due to factors other than rotation, such as magneto-absorption etc., so that

$$A = (\Delta P_1)_\theta + \Delta P_2 \,. \tag{1}$$

For the reversed sense of the magnetic field, i.e. $\boldsymbol{B}_-$, B represents the normalized change in the transmitted signal. The part corresponding to Faraday rotation is $-(\Delta P_1)_\theta$, i.e. of the same order of magnitude as for $\boldsymbol{B}_+$ (because of the operation in the linear region of the curve) but of opposite sign (because of the rotation of the field pattern in the opposite sense). The remaining change, if considered solely due to magneto-absorption, will be the same in sign and magnitude, i.e. ΔP_2 is assumed to be a quadratic effect in B (which is justified at room temperature as revealed by the study [11, 12] regarding the magnetic field dependence of the ratio of transmitted amplitudes). Therefore,

$$B = -(\Delta P_1)_\theta + \Delta P_2 \,. \tag{2}$$

From equations (1) and (2), the normalized change in the transmitted signal due to Faraday rotation is

$$(\Delta P_1)_\theta = \frac{1}{2}(A - B)\,. \tag{3}$$

The Faraday rotation θ can be determined from this, making use of the slope of the linear region of the angular dependence of the detected power, i.e. by the relation

$$\theta = \frac{\delta\varphi}{(\delta P/P_\varphi)}(\Delta P_1)_\theta \tag{4a}$$

$$= \frac{1}{2}\frac{\delta\varphi}{\delta P}P_\varphi(A - B)\,, \tag{4}$$

where P_φ represents the signal level at the operation position of the detector (Fig. 1) in the linear region of the calibration curve.

3.2 More precise treatment

Equation (3) which represents the normalized change in the transmitted signal solely due to Faraday rotation, is based on the assumption that the other factors such as transverse magneto-absorption etc. on which the transmission depends cause an equal increase in the signal for both senses of $\boldsymbol{B}$. This has been assumed for simplification, and has been found to hold well for samples with moderate loss at room temperature only. In samples with higher conductivity and in measurements at lower temperatures, it has been found that the two senses of the magnetic field give rise to changes in the transmitted signal which differ by large amount. In such a general case, the equation (2) should be written as

$$B = -(\Delta P_1)_\theta + \Delta P_3\,, \tag{5}$$

where ΔP_3 represents the fractional change in the signal due to factors other than rotation. (In the simple treatment given in Section 3.1, ΔP_3 was assumed equal to ΔP_2).

From (1) and (5) one can write the change in the transmitted signal due to Faraday rotation as

$$(\Delta P_1)_\theta = \frac{1}{2}[(A - B) - (\Delta P_2 - \Delta P_3)]\,. \tag{6}$$

In order to find $(\Delta P_2 - \Delta P_3)$ one may make use of the changes in the transmitted power with the detector at $\varphi = 0°$ position, i.e. at the maximum of the calibration curve (Fig. 1). In this position, the rotation of the field pattern for both senses of the magnetic field causes an equal decrease in the transmitted signal.

For the sense $\boldsymbol{B}_+$, with the detector at $\varphi = 0°$, let the normalized change in the transmitted power be

$$D = -\Delta P_4 + \Delta P_5\,, \tag{7}$$

where $-\Delta P_4$ represents the fractional decrease in the signal due to rotation, and ΔP_5 the fractional increase in signal due to magneto-absorption etc.

For the reversed sense $\boldsymbol{B}_-$ of the magnetic field, assuming ΔP_6 to be the increase in the transmitted signal due to effects other than rotation, one can write the normalized change in the transmitted power as

$$F = -\Delta P_4 + \Delta P_6 . \tag{8}$$

From (7) and (8)

$$D - F = \Delta P_5 - \Delta P_6 . \tag{9}$$

Concentrating on the magneto-absorption part in the transmitted signals, one can correlate the changes detected by the analyser in the $\varphi = 45°$ and $\varphi = 0°$ positions, i.e. ΔP_2 can be related to ΔP_5 and ΔP_3 to ΔP_6 by the relations

$$\Delta P_2 = \left(\frac{P_\varphi}{P_{\max}}\right) \Delta P_5 , \qquad \Delta P_3 = \left(\frac{P_\varphi}{P_{\max}}\right) \Delta P_6 , \tag{10}$$

where $P_{\max}$ represents the maximum power level (Fig. 1), corresponding to the detector in the $\varphi = 0°$ position. From equations (10) and (9) follows

$$(\Delta P_2 - \Delta P_3) = \frac{P_\varphi}{P_{\max}} (D - F) \tag{11}$$

enabling one to rewrite equation (6) for the normalized change in the detected signal solely due to Faraday rotation as

$$(\Delta P_1)_\theta = \frac{1}{2} \left[(A - B) - \frac{P_\varphi}{P_{\max}} (D - F) \right] . \tag{12}$$

The Faraday rotation can be determined using (4a) and (12), i.e.

$$\theta = \frac{1}{2} \frac{\delta\varphi}{\delta P} P_\varphi \left[(A - B) - \frac{P_\varphi}{P_{\max}} (D - F) \right] . \tag{13}$$

Equation (13) represents the general form of the expression which can be used for the precise evaluation of the Faraday rotation in semiconductors.

As the present procedure makes use of the linear detection, which exists for a region of about 15 to 20° around $\varphi \approx 45°$ (Fig. 1), it is important to note that the rotation measurement beyond this value by this procedure is not justified. The larger values of rotation can be directly measured from the shift of the maximum of the curve.

4. Determination of Faraday Rotation and Discussions

The evaluation of the Faraday rotation has been done from the changes in the transmitted signals for the two senses of the magnetic field at room temperature in an n-type (110) oriented germanium sample of resistivity 9.13 Ωcm and of thickness 1.68 mm using 24.0 GHz radiation, making use of the simple formula (4) and the general formula (13). The results obtained are shown in Table 1. The rotation values obtained from the two expressions has been compared in order to see the difference, because equation (4) requires measurements of the transmitted signal with the detector only in $\varphi \approx 45°$ position, whereas equation (13) requires measurements in $\varphi \approx 45°$ as well as $\varphi = 0°$ positions. From Table 1, it can be noted that for measurements at room temper-

Table 1

Comparison of the Faraday rotation values evaluated from the measured changes in the transmitted signals for the two senses of the magnetic field at room temperature using equations (4) and (13), in an n-type (110) oriented germanium sample of resistivity 9.13 Ωcm and thickness 1.68 mm at 24.0 GHz

magnetic field B (Wb/m²)	Faraday rotation (deg)	
	calculated from the general expression (13)	calculated from the simplified expression (4)
0.20	3.54	3.56
0.30	5.47	5.54
0.39	7.68	7.74
0.50	10.22	10.25
0.60	13.53	13.47
0.70	16.55	16.60

ature the two approaches lead to results, which are very close to each other. This confirms the quadratic field dependence of the transverse magneto-absorption at room temperature. Further, as the use of the simplified expression (4) cuts down the labour involved in collecting the transmission data and the calculations to about half of that involved in using expression (13), the use of expression (4) is sufficiently accurate and justified in evaluating Faraday rotation at room temperature.

The evaluation of Faraday rotation has also been done from the measured changes in the transmitted signal at liquid nitrogen temperature (91 ± 6) K in an n-type (110) oriented germanium sample of resistivity 5.60 Ωcm at the temperature of investigation (36.4 Ωcm at room temperature) and thickness 1.41 mm at 24.0 GHz, using expressions (4) and (13). The values of the Faraday rotation obtained are shown in Table 2, which indicates a difference in the rotation values obtained from the two expressions. The rotation obtained

Table 2

Comparison of the Faraday rotation values evaluated from the measured changes in the transmitted signals for the two senses of the magnetic field at liquid N_2 temperature (91 ± 6) K using equations (4) and (13), in an n-type (110) oriented germanium sample of resistivity 5.60 Ωcm and thickness 1.41 mm at 24.0 GHz

magnetic field B (Wb/m²)	Faraday rotation (deg)	
	calculated from the general expression (13)	calculated from the simplified expression (4)
0.05	4.30	5.25
0.10	8.30	10.25
0.15	12.62	15.15
0.20	15.80	19.50
0.25	18.35	—
0.30	20.50	—

from the general formula (13) is smaller than the corresponding values from the simple formula (4). At fields above about 0.2 Wb/m^2, the changes in the transmitted signals at liquid N_2 temperature become so large, that the use of equation (4) becomes unjustified. It has been found by the author [13] that the Faraday rotation evaluated from equation (13) at different fields and the nature of variation of rotation with magnetic field show better agreement with the behaviour expected theoretically, which suggests that at such low temperatures the transverse magneto-absorption shows deviations from the quadratic field dependence, and hence the use of the simple expression (4) for the evaluation of rotation in such cases is not justified.

Acknowledgements

The author is thankful to Dr. G. P. Srivastava and Dr. P. C. Mathur for their interest in the problem and enlightening discussions; and grateful to Professor F. C. Auluck for providing research facilities. Thanks are also due to Mr. S. K. Agarwal for his helpful discussions.

References

[1] Y. Nishina and W. J. Spry, J. appl. Phys. **29**, 230 (1958).
[2] A. M. Portis and D. Teaney, J. appl. Phys. **29**, 1692 (1958); Phys. Rev. **116**, 838 (1959).
[3] A. Bouwknegt and J. Volger, Physica (Utrecht) **30**, 113 (1964).
[4] J. K. Furdyna and S. Broersma, Phys. Rev. **120**, 1995 (1960).
[5] R. R. Rau and M. E. Caspari, Phys. Rev. **100**, 632 (1955).
[6] H. J. Kuno and W. D. Hershberger, Proc. IEEE **54**, 978 (1966).
[7] H. J. Kuno, Ph. D. (Engg.) Thesis, University of California, Los Angeles 1966 (p. 93).
[8] P. A. van Dalen, J. appl. Phys. **41**, 3092 (1970).
[9] P. J. Allen, Rev. sci. Instrum. **25**, 394 (1954).
[10] S. Chai and P. Vogelhut, Rev. sci. Instrum. **37**, 1620 (1966).
[11] G. P. Srivastava, B. R. Sethi, and S. K. Agarwal, J. appl. Phys. **43**, 3748 (1972).
[12] J. K. Furdyna and M. E. Brodwin, Phys. Rev. **132**, 97 (1963).
[13] B. R. Sethi, Ph. D. Thesis, Department of Physics and Astrophysics, University of Delhi 1972 (p. 139 and 199).

(Received October 18, 1972)

phys. stat. sol. (a) **14**, 599 (1972)

Subject classification: 10.2; 22.5.2

AEG–Telefunken Forschungsinstitut and Physikalisches Institut der Johann-Wolfgang-Goethe-Universität, Frankfurt/Main, (a) and University of Oxford, Department of Metallurgy (b)

Dislocation Density in Highly Deformed NaCl Single Crystals [1])

By

J. HESSE (a) and L. W. HOBBS (b)

The controversy about the relation between the flow stress τ and the screw dislocation density N_s, at large values of N_s in NaCl is clarified. Using both etch pit and electron transmission techniques a relation of the type $\tau \sim \sqrt{N_s}$ is confirmed as has also been found for low N_s values. The origin of dislocation dipoles observed in transmission is discussed.

Es wird die Kontroverse über die Beziehung zwischen der Fließspannung τ und der Dichte der Schraubenversetzungen N_s bei großen Werten von N_s in NaCl geklärt. Sowohl mit der Ätzgrubentechnik als auch mit Elektronentransmissionsmethoden wurde die Beziehung von der Art $\tau \sim \sqrt{N_s}$ bestätigt, die auch bei niedrigen N_s-Werten gefunden wurde. Die Ursache der Versetzungsdipole wird diskutiert, die in Transmission beobachtet wurden.

1. Introduction

In this paper a comprehensive treatment of the relation between the flow stress τ and the screw dislocation density N_s in NaCl is presented. Previous work on this subject, using the etch pit technique, gave rise to various $\tau(N_s)$ relations in stage II of the stress–strain curve. A principal object of this work was, in fact, to assess the validity of the etch pit technique in NaCl.

In stage II of the stress–strain curve of $\langle 100 \rangle$-oriented crystals, deformed at room temperature in compression tests, $\tau(N_s)$ was first investigated by Davidge and Pratt [1] and Hesse [2]. A linear relation of the type

$$\tau = B\, N_s^p \tag{1}$$

was obtained, whereby the index p in N_s^p indicates the projection of the screw dislocation density on the $\{100\}$ surface. On $\langle 110 \rangle$ crystals, a similar relation to (1) was reported by Haasen and Hesse [3] and Hesse [2] for the initial phase of stage II.

Detailed etch pit counts on $\langle 110 \rangle$ samples were published by Matucha et al. [4]. Using high-magnification electron microscope studies of carbon replicas of the etched $\{100\}$ surface together with short etch times (seconds), they obtained a different result, namely a parabolic law of the type

$$\tau = A \sqrt{N_s^p} \tag{2}$$

[1]) Based on the Habilitationsschrift of J. Hesse accepted by the Naturwissenschaftliche Fakultät of the Johann-Wolfgang-Goethe-Universität, Frankfurt/Main, in 1971.

as found by others [1 to 3, 5 to 7] for both orientations in stage I. Further support for their result was obtained from parallel investigations of X-ray diffraction line broadening using the method of Wilkens [8, 9].

On the basis of these results, Matucha et al. [4] proposed that the discrepancy in the various published results with respect to $\tau(N_s^p)$ was due in [1, 2] to poor optical resolution of the small-diameter etch pits obtained at high dislocation densities. In order to test this hypothesis and at the same time to verify one or the other $\tau(N_s)$ relation, measurements on $\langle 100\rangle$ samples were repeated using an improved counting technique and the direct electron microscopic examination of dislocations by transmission of thin foils [10].

2. Etch Pit Investigations

The sample deformation procedure has been described elsewhere [2]. With the $\langle 100\rangle$ samples, however, the range of stresses for the corresponding etch pit counts was not very wide. This is due to the softness of this orientation [2] in which barrel-shaped samples are often obtained after compression thus complicating the definition of a central $\{100\}$ cross sectional area. On the other hand, the corresponding slip is smaller for $\langle 110\rangle$ samples thus permitting etch pit counts to be made in this case at larger stresses. The same pit etching procedure and the method of counting as recommended in [4] were used. The results are shown in Fig. 1 together with the results from [4], where the relation (2) was also obtained in stage II for $\langle 110\rangle$ samples, with the same A-value as in stage I. It was also found that $A_{\langle 100\rangle} > A_{\langle 110\rangle}$ (Fig. 1). We also verify the proposal of Frank [11] that $\tau(N_s^p)$ should be independent of impurity doping. Crystals with 33, 64, and 300 ppm Ca^{++} were deformed as described in [12] as well as crystals with 200 ppm Cd^{++}. The results are also shown in Fig. 1 confirming Frank's hypothesis. This result was also obtained in our previous work at low dislocation densities [12] and also by Reppich [13] on LiF. It can hence be concluded that the proposed reason [4] for the discrepancy in the hitherto published results, namely an inadequate etch pit resolution technique, is correct. In order to find out whether this error arose from the optical resolution alone or, in addition, from the different size of the etch pits, we varied the etch time t_e on $\langle 110\rangle$-oriented samples with $\tau = 500\ \text{p/mm}^2$ and determined N_s^p as a function of t_e. As can be seen from Fig. 2, N_s^p decreased up to 30% with increasing t_e over the t_e-range $10\ \text{s} \leq t_e \leq 7\ \text{min}$. The most marked variation in the measured value of N_s^p was observed at $t_e \approx 1.5$ min. It was also apparent that for etch times of this order, overlapping of neighbouring etch pits started. This could imply that pits smooth out and hence disappear. On the other hand it is conceivable that mechanically damaged surface layers have to be removed before N_s^p can be determined with any great accuracy. This possibility was however excluded by varying the thickness of the original surface layer prior to actual etching by a methanol–ethanol mixture [5] whereby no effect on the actual N_s^p-value was observed. Hence it is necessary to consider both the optical resolution as well as the etch pit diameter for an accurate N_s^p-determination. Both conditions were presumably not fulfilled in previous work [1, 2]. Furthermore it is possible that a similar effect was also responsible for the different surface and volume dislocation densities observed by Smirnov [14, 15] on LiF. Our results also imply that the linear relation $\tau = B\, N_s^p$ found for other ionic crystals, in particular Smirnov's [14, 15] on LiF are incorrect for the reason

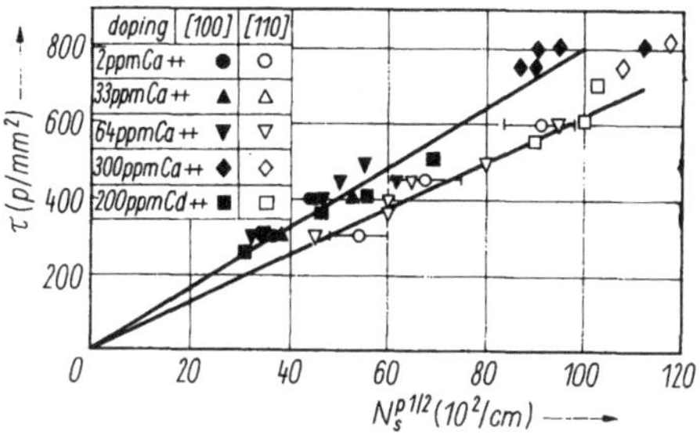

Fig. 1. Etch pit density N_s^p at screw dislocations in differently doped ⟨100⟩ and ⟨110⟩ crystals after room temperature deformation with $\dot{a} \approx 10^{-3}$/s in stage II (○ results from [4])

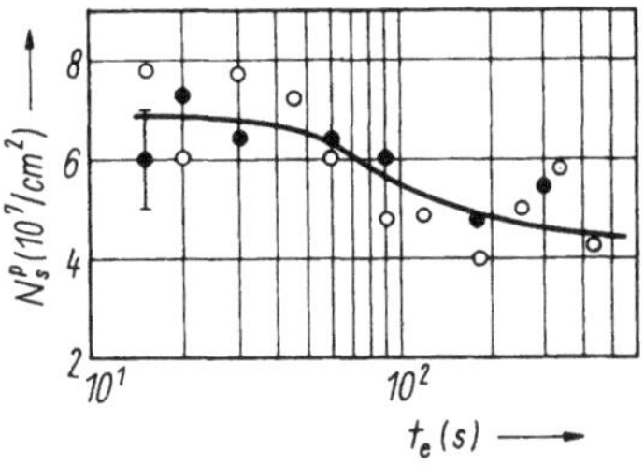

Fig. 2. Variation of etch pit density N_s^p at screw dislocations in ⟨100⟩ crystals at $\tau = 500$ p/mm² with etch time t_e (○ results from one specimen with repeated etching, ● results from various specimens for one etching only)

discussed above. The etch pit counts by Reppich [13] on LiF appear to confirm this. The results from Zimkin et al. [16] may require a similar re-interpretation. In LiF they observed slip on two orthogonal {110} ⟨110⟩ systems whereby, however, the deformation was carried mainly by one system. The investigated (forest) dislocation density N_f in the other system increased with a square law $\tau = \alpha\, G\, b \sqrt{N_f}$ with $\alpha \approx 0.75$. (Gilman and Johnston [17] measured the total dislocation density N_p in LiF as a function of strain, a, and similar to Deputat and Pawlowski [18] found a linear relationship between N_p and a. According to Reppich [13], however, the edge dislocation density N_e including an unknown number of dipoles, increases with τ more rapidly than N_s. Hence it is not easy to draw a comparison with these measurements.)

However, in the case of CaF_2, studied by Evans and Pratt [19], other effects than these must be involved. On samples deformed at room temperature to high dislocation densities ($N_s^p \approx 10^8/\text{cm}^2$) they again found $\tau = A \sqrt{N_s^p}$. At higher deformation temperatures, however, a region $\tau = B\, N_s^p$ was evident. The transition from the square to the linear form shifted to smaller N_s^p- or τ-values with increasing temperature. This cannot be related to the method of etch pit counting since this was the same throughout. It is hence possible that at higher temperatures other glide systems appeared with higher starting stresses, and thus that the resulting dislocations do not appear as etch pits on the chosen observation surface. It would be of interest therefore to determine whether a similar transition occurs in NaCl deformed above room temperature taking into account the well-known secondary slip systems and starting stresses [20].

3. Electron Transmission

One of us [10] was recently successful in solving the electron transmission problem in NaCl. An original application of this technique to dislocation counting will now be described in connecction with the $\tau(N_s)$ relation in stage II. Two ⟨110⟩ crystals, $4 \times 4 \times 18$ mm³, and with 2 ppm Ca^{++} were deformed at room temperature with $\dot{a} \approx 10^{-3}$/s to $\tau = 165$, 360 p/mm², respectively, as described in [2]. After unloading thin foils 0.1 mm thick were cut parallel to the {110} optically determined primary glide plane, thinned water-free to approximative 2000 Å and studied with 100 keV electrons at 10 °K.

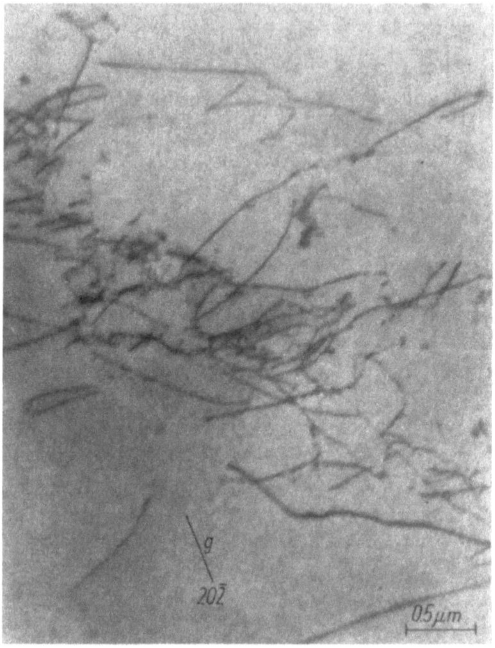

a

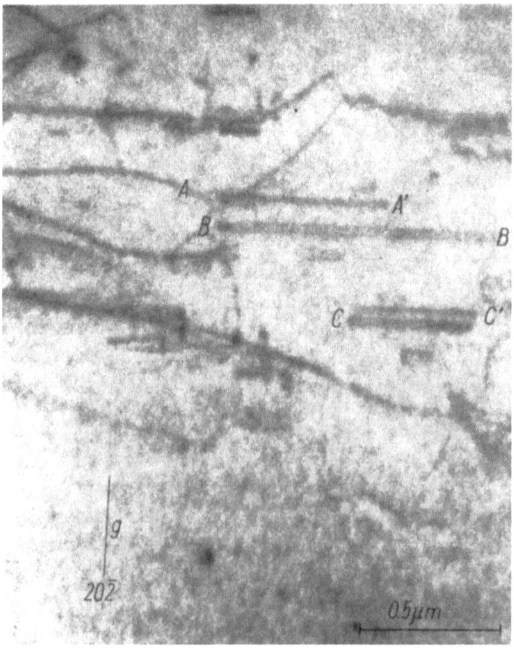

b

Fig. 3. Transmission electron micrographs of a $\langle 110 \rangle$ crystal after room temperature deformation with $\dot{a} \approx 10^{-3}$/s up to $\tau = 250$ p/mm² (foil plane parallel to the primary (101) slip plane)

The observed *dislocation array* will now be discussed. A non-uniform distribution was observed: It was common to find parallel, high dislocation density bands next to dislocation-free areas (see Fig. 3a). A preferred orientation of the dislocation lines parallel to the $\langle 100 \rangle$ direction was found to be independent of the distribution. A simple Burgers vector analysis was performed using the $\boldsymbol{G} \cdot \boldsymbol{b} = 0$ condition. From that it was concluded that most of the dislocations were near edge orientation. This confirms our earlier etch pit counts [2], since they lead to the same conclusion. With reference to our previous work [2] it is also interesting to note that dislocation pile-ups are not observable even in transmission. Of particular interest are the edge dislocation loops (dipoles) indicated in Fig. 3b by AA′, BB′, and CC′. Their existence was suggested from etch pit observations [17], crystal density measurements [1], and strain rate change experiments [2]. Such dipoles apparently exist in NaCl in large numbers: At $\tau = 360$ p/mm² we estimated $N_D \approx 10^7/\text{cm}^2$. (This value is smaller than that calculated by Davidge and Pratt [1] ($10^{10}/\text{cm}^2$) from crystal density measurements on $\langle 100 \rangle$ samples.) Despite an admitted lack of detailed experimental data, it is of interest to speculate about the origin of these dipoles. It seems to us that there are three different mechanisms for dipole formation which have to be considered: double cross slip [21], forest intersection [11], and stress field trapping [22]. The last named mechanism was favored by Ogawa [23] in MgO and by Hobbs [24] in KI crystals. The dipole BB′ in Fig. 3b shows component dislocations with a distance $d \approx 1000$ Å, in approximate agreement with Ogawa's [23] value $d \approx 400$ Å. However, the appearance of the dipoles in MgO and KI (in general, they are not pure edge dislocations and extend only over portions of the dislocation line) is somewhat different from those observed in NaCl (pure edge dislocation dipoles). We do not believe, therefore, that the stress field trapping mechanism is effective in NaCl. On the other hand, the observed pure edge character of the dipoles in NaCl is consistend with the forest intersection model, based on the intersection of dislocations of orthogonal $\{110\}\langle 110\rangle$ glide systems. Nevertheless we find it difficult to envisage that intersections can produce jogs of a sufficient size to get such widely spaced dipoles. However, we suppose that it is possible to produce such jogs for dislocations intersecting dislocation debris bundles on the orthogonal system. But, according to [2, 25], the required double slip was not observed in the crystals investigated here. The presence of a cross slip mechanism could be supported by the observation that the dipole formation seems to be easier in crystals deformed in $\langle 110\rangle$ orientation than in those deformed in $\langle 100\rangle$ orientation. In the latter case there is no resolved shear stress on the $\{100\}$ cross slip plane. A further support is the observation that the spacing of the dipole dislocations seems to be larger in $\langle 110\rangle$ than in $\langle 100\rangle$ crystals. To decide which of the last two discussed mechanisms is responsible for the creation of the dipoles, further work will be necessary. Especially, transmission experiments performed at higher electron optical magnification and using foils cut perpendicular to the primary glide plane are required.

The total *dislocation density* N, including N_D, was obtained from the total length of dislocation lines in the transmitted volume. The following mean values were obtained from consideration of slip bands and dislocation-free regions: $N = 5\times 10^7/\text{cm}^2$ at $\tau = 165$ p/mm² and $N = 2\times 10^8/\text{cm}^2$ at $\tau = 360$ p/mm². Consequently the total dislocation density increases proportional to τ^2, in agreement with our etch pit counts. Unfortunately, with respect to the short inspection time, it was not possible to isolate the screw dislocation component N_s in transmission using a Burgers vector determination in each case which would have allowed a direct comparison with the results of etch pit counts. However, one can estimate N_s from N measured in transmission, if one assumes that the relation $N_e/N_s = 2.7$ obtained from etch pit counts in stage I [2] applies also in transmission in stage II. (This is in fact supported by the dislocation distribution observed in electron transmission.) Then from $N = N_e + N_s$ we obtain $N_s = 1.3\times 10^7/\text{cm}^2$ at $\tau = 165$ p/mm² and $N_s = 5\times 10^7/\text{cm}^2$ at $\tau = 360$ p/mm². The etch pit counts (Fig. 1) yield for the same stresses $N_s = 8\times 10^6/\text{cm}^2$ and $N_s = 3.2\times 10^7/\text{cm}^2$, being by a factor 1.5 smaller in each case. Such a discrepancy was in fact predicted by Seeger [26]. On the other hand, our results in transmission were obtained from relatively few pictures and under the assumption that $N_e/N_s = 2.7$ obtained from etch pit observations in stage I is valid for transmission in stage II. Hence these differences still lie in the experimental error range. However, if N (electron transmission) is larger than N (pit), one can exclude that dislocations have left the layer during the chemical thinning (due to image forces). This observation should prove useful in future investigations.

4. Summary

Our results confirm a parabolic relationship $\tau = A \sqrt{N_s}$ between the flow stress τ and the screw dislocation density N_s for NaCl even at high degrees of deformation at various doping concentrations and for crystals with ⟨100⟩ and ⟨110⟩ orientations. This is in agreement with a work hardening theory by Frank [11] for ionic crystals. Also a linear relationship of the type $\tau = B N_s$ obtained elsewhere, was found to be due to an insufficient optical resolution at high etch pit densities, as was already pointed out by Matucha et al. [4]. In the same work hardening theory [11], a high density of edge dislocation dipoles was predicted to exist. Such dipoles were in fact found in electron microscopic transmission. However, it could not be determined, whether these dipoles arose by interaction of screw dislocations moving in orthogonal {110} slip planes as assumed in this theory [11] or by a double cross slip mechanism as proposed by Johnston and Gilman [21].

References

[1] R. W. DAVIDGE and P. L. PRATT, phys. stat. sol. **6**, 759 (1964).
[2] J. HESSE, phys. stat. sol. **9**, 209 (1965).
[3] P. HAASEN and J. HESSE, Disc. Faraday Soc. **38**, 238 (1964).
[4] K. H. MATUCHA, W. FRANZBECKER, and M. WILKENS, phys. stat. sol. **33**, 493 (1969).
[5] W. IN DER SCHMITTEN and P. HAASEN, J. appl. Phys. **32**, 1790 (1961).
[6] P. HAASEN and J. HESSE, The Relation between the Structure and Mechanical Properties of Metals, H. M. S. O., London 1963 (p. 137).
[7] N. N. NOVIKOV, Izv. Vyssh. Ucheb. Zaved. Fiz. **6**, 108 (1967).
[8] M. WILKENS, in: Proc. Conf. Fundamental Aspects of Dislocation Theory, Nat. Bureau of Standards, Washington 1969.
[9] M. WILKENS, phys. stat. sol. (a) **2**, 359 (1970).
[10] L. W. HOBBS, Proc. 25th Anniversary Meeting EMAG, Cambridge 1971 (p. 178).
[11] W. FRANK, Mater. Sci. Engng. **6**, 121 (1970); **6**, 132 (1970).
[12] J. HESSE, phys. stat. sol. **21**, 495 (1967).
[13] B. REPPICH, phys. stat. sol. **35**, 339 (1969).
[14] B. I. SMIRNOV, in: Reinstoffprobleme, Vol. 3, Akademie-Verlag, Berlin 1967 (p. 421).
[15] B. I. SMIRNOV and A. V. BORSENKO, Soviet Phys. — Solid State **9**, 1377 (1967).
[16] I. N. ZIMKIN, T. V. SAMOILOVA, and B. I. SMIRNOV, Soviet Phys. — Solid State **12**, 1386 (1970).
[17] J. J. GILMAN and W. G. JOHNSTON, J. appl. Phys. **30**, 129 (1959).
[18] J. DEPUTAT and Z. PAWLOWSKI, Bull. Acad. Polon. Sci., Ser. Sci. Tech. **14**, 419 (1966).
[19] A. G. EVANS and P. L. PRATT, Phil. Mag. **21**, 961 (1970).
[20] E. YU. GUTMANAS and E. M. NADGORNYI, phys. stat. sol. **38**, 777 (1970).
[21] W. G. JOHNSTON and J. J. GILMAN, J. appl. Phys. **31**, 632 (1960).
[22] J. WASHBURN, in: Electron Microscopy and Strength of Crystals, Ed. G. THOMAS and J. WASHBURN, Interscience Publishers, New York 1963 (p. 301).
[23] K. OGAWA, Phil. Mag. **14**, 619 (1966).
[24] L. W. HOBBS, to be published.
[25] J. HESSE and K. H. MATUCHA, Scripta metall. **6**, 865 (1972).
[26] A. SEEGER, Disc. Faraday Soc. **38**, 157 (1964).

(Received August 29, 1972)

phys. stat. sol. (a) **14**, 605 (1972)

Subject classification: 3 and 11; 1.5; 22.6

Institut für Angewandte Kernphysik, Kernforschungszentrum Karlsruhe

Impurity Incorporation during rf Sputtering of Silicon Oxide Layers

By

S. PETERSSON[1]), G. LINKER, and O. MEYER

Backscattering and channeling effect measurements have been used to analyse silica rf sputtered in argon atmosphere. Layers sputtered from fused quartz were found to be stoichiometric SiO_2 irrespective of variations of sputtering parameters during film growth. Variation of power density (deposition rate) does not influence the Ar content. The amount of Ar increases with increasing bias voltage. The lowest Ar content (0.4 mol%) was found at zero bias voltage. Traces of backing material used to establish good heat transfer between substrate and anode plate have been incorporated during film growth. The amount of impurities increased with increasing bias voltage. Contrary to Ar incorporation it has been observed that the In content decreased with increasing deposition rate.

Rückstreu- und Channelingmessungen wurden für die Analyse von in Argonatmosphäre HF-gesputtertem Siliziumoxid benutzt. Schichten, die von reinem Quarz gesputtert wurden, erwiesen sich als stöchiometrisches SiO_2, unabhängig von Änderungen der Sputteringparameter während des Schichtwachstums. Änderung der Leistungsdichte (Aufdampfrate) beeinflußt den Ar-Gehalt nicht. Der Anteil von Ar nimmt mit steigender Vorspannung zu. Der niedrigste Ar-Anteil (0,4 Mol%) wird ohne Vorspannung gefunden. Spuren des Trägermaterials, das für einen guten Wärmeübergang zwischen Substrat und Anode benutzt wurde, wurden während des Schichtwachstums eingebaut. Der Anteil der Verunreinigungen nahm mit steigender Vorspannung zu. Im Gegensatz zum Ar-Einbau wurde beobachtet, daß der In-Anteil mit steigender Aufdampfrate abnimmt.

1. Introduction

Physical properties of sputtered layers are affected by gaseous and other impurities incorporated during the sputtering process. Numerous work has been done to study the argon and nitrogen concentration in layers deposited by different sputtering techniques as a function of gas pressure, power density, bias voltage, and substrate temperature [1 to 4]. Besides the carrier gas necessary to maintain the plasma, other impurities may be accidentaly incorporated. These can come from bonding and backing materials, incomplete rf shielding or the incomplete coverage of the electrodes with target and substrate materials. It is obvious that impurities in the argon gas and in the target material should also be found in the sputtered layers. The incorporated impurity concentrations similar to the argon content will also depend on sputtering parameters. Various techniques have been used to study gas entrapment in sputtered layers. The X-ray fluorescence technique was used to determine the Ar content of rf sputtered SiO_2 films as a function of film thickness, rf power density, and substrate

[1]) Short-term visitor from Electronics Department, Institute of Technology, University of Uppsala.

temperature [3]. Further, laser degassing or flash heating techniques followed by a mass spectrometer analysis were used to determine Ar concentration in sputtered tantalum [1], tungsten [4], and nickel [2] films. Another suitable method of analysing films having thicknesses in the range of 500 to 5000 Å, is the backscattering and channeling technique using light ions of MeV energies. With this nondestructive technique, concentrations and depth distributions of all components of the thin film can be determined. The method has already been successfully applied to the analysis of dielectric layers, for example thermally grown SiO_2 layers [5] and Si_3N_4 layers deposited in a glow discharge [6]. In the present investigation, silicon oxide layers sputtered on single crystal Si substrates have been investigated in terms of stoichiometry, impurity content and influence of sputter etching on the substrate.

2. Experimental

The silicon oxide layers were sputtered on single-crystal silicon cut normally to the ⟨111⟩ direction. Prior to deposition the surfaces were mechanically polished and chemically cleaned using standard methods. In addition some samples were sputter etched for surface cleaning. Samples were mounted on a silicon oxide coated holder, the SiO_2 being used to prevent sputtering of substrate electrode material. For good thermal contact the samples were bonded to the substrate holder with a standard Ga–In–Sn eutectic. A 5 inch diameter fused quartz plate was used as a target material.

Substrate and target holders were water cooled with the target to substrate distance being kept constant at 7.2 cm during the sputtering process. A background pressure of 5×10^{-7} Torr was achieved. The purity of the argon was 99.9995% and a pressure of 5×10^{-3} Torr was usually maintained during sputtering. Prior to deposition the SiO_2 target was presputtered onto a removable shield, for 5 min. At zero bias and a rf power density of 2 Wcm^{-2}, the deposition rate was 100 Å/min. Thicknesses of most of the sputtered layers were in the range of 1200 to 1500 Å.

Backscattering and channeling effect measurements were performed using a beam of 2.8 MeV $^4He^+$ ions. The samples were mounted on a 2-axis goniometer for substrate alignment. Beam currents were measured with a current integrator and secondary electrons were suppressed. Particles backscattered through a laboratory angle of 168° were detected in a surface barrier detector and the system resolution was 16 keV. Detector signals were amplified and then stored in a multichannel analyzer. The energy to channel number conversion, which was calculated from the spectrum of a C–Al–Cu–Au calibration sample, was 4.5 keV per channel. Beam currents used were from 15 to 20 nA.

3. Analysis

In Fig. 1 a typical spectrum of a silicon oxide layer sputtered on a single crystal silicon substrate is shown. From well known collision kinetics and the experimental geometry, the energy scale can be converted to a mass scale. The energies of $^4He^+$ ions backscattered from atoms of different masses at the surface of the silicon oxide layer are well resolved in the energy spectrum, and the corresponding masses are indicated in the figure. By alignment of the incoming beam direction with the ⟨111⟩ axis of the silicon substrate, the number of back-

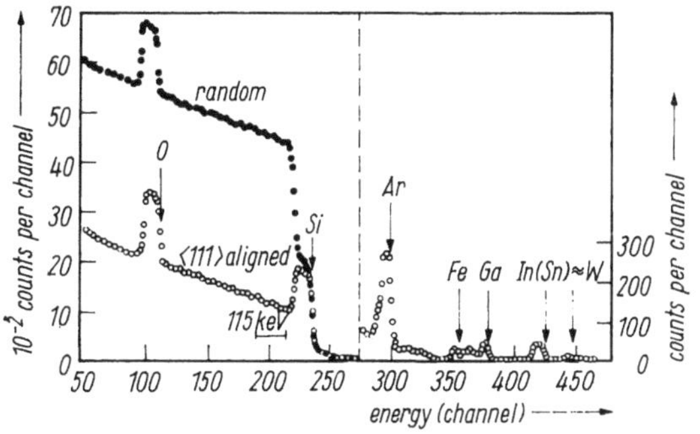

Fig. 1. Random and ⟨111⟩ aligned spectra for 2.8 MeV $^4He^+$ ions scattered from a silicon oxide layer deposited on ⟨111⟩-oriented silicon by rf sputtering (300 W, 80 V bias). The right-hand scale refers to the part of the spectrum right of the vertical dashed line

scattered particles from the substrate atoms is reduced. This alignment leads to an improved analysis of the silicon oxide layer. Normalization of the random spectrum to the aligned silicon substrate level, allows separation of $^4He^+$ particles backscattered from silicon atoms in the oxide layer and in the substrate close to the interface. With this procedure [7] the peak area for silicon in the oxide layer can be determined. This is important, as peak areas in the spectra are proportional to the total amount of atoms in a layer. For a Mass M:

$$A_M = f \sigma_M N_M \Phi \tag{1}$$

with A_M peak area, σ_M Rutherford scattering cross section, N_M total number atoms, Φ total incoming flux, f proportionality factor for the geometrical arrangement.

For a homogeneous distribution of atoms in a layer, the concentration ratio for two different masses M_1 and M_2 can be determined from equation (1):

$$\frac{N_{M_1}}{N_{M_2}} = \frac{A_{M_1}}{A_{M_2}} \frac{\sigma_{M_2}}{\sigma_{M_1}}. \tag{2}$$

If an absolute concentration is known (for example for the Si atoms in the substrate layer) other absolute concentrations can be calculated by equation (2). The energy scale in Fig. 1 can be converted to a depth scale if the specific energy loss is known. The energy resolution of the detecting system determines the ultimate depth resolution. Depending on the incident particle energy, the depth resolution is found to be in the range of 200 Å; within this accuracy depth profiles can be determined. A more detailed treatment of the analysis problems is given by Meyer et al. [5].

4. Results

The ratio of oxygen to silicon atoms, from [2], obtained by the analysis of 25 oxide layers sputtered from fused quartz, is found to be 2.05 ± 0.1. From this result it is concluded that stoichiometric amounts of each component has been sputtered and deposited during film growth. Over the ranges of power density and bias voltages used, no change in this stoichiometry was observed.

Sputter etching for 5 min at 400 W prior to deposition produces heavy radiation damage at the surface of a single crystal Si substrate. This is clearly seen in the ⟨111⟩-aligned spectrum in Fig. 2. A sharp peak is visible at the interface

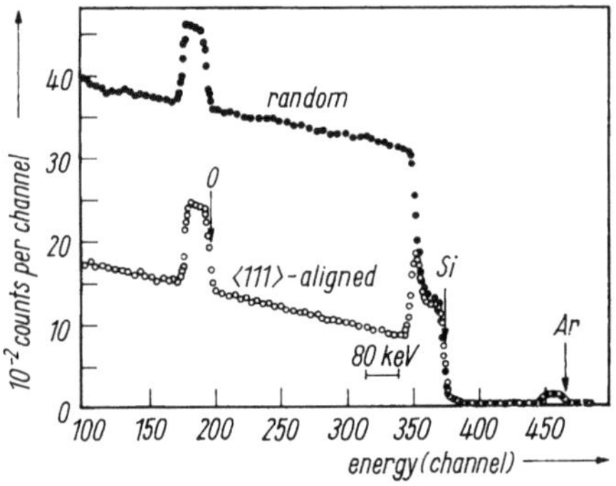

Fig. 2. Random and ⟨111⟩ aligned spectra for 2.8 MeV $^4He^+$ ions scattered from a silicon oxide layer deposited by rf sputtering (400 W, zero bias). The substrate was sputter etched prior to deposition

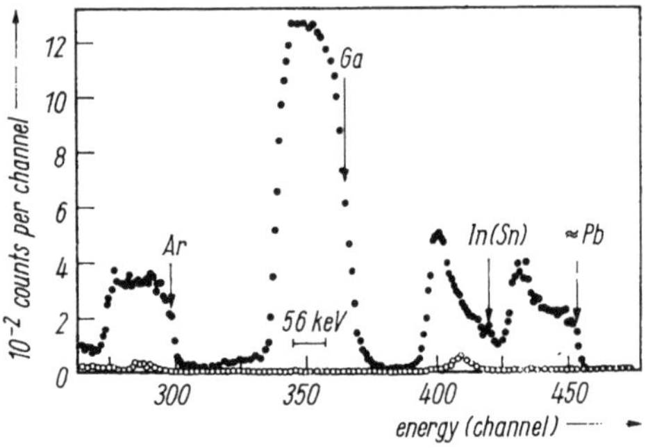

Fig. 3. Backscattered impurity spectra for 2.8 MeV $^4He^+$ of sputtered layers showing extremely high impurity content (closed circles) and reasonable low impurity content (open circles). ●: 170 V; 300 W, ○: 0 V; 100 W

between Si and SiO_2. This peak is due to scattering from disordered Si atoms at the substrate surface. The thickness of the damaged layer is estimated to be 300 to 400 Å, which is in agreement with results published by Davidson [8]. Far less disorder is detected when Si surfaces are chemically etched. A disordered layer can cause problems because the main assumption in the analysis is the existence of a sharp interface between the substrate and the sputtered layer. The plateau of the Ar peak indicates that Ar is equally distributed throughout the thickness of the deposited layer (see Fig. 2).

The impurity content in the oxide layer was found to be strongly dependent on the sputtering conditions. The large effect of bias sputtering on the impurity concentration is demonstrated in Fig. 3, where the high energy parts of two spectra are shown for comparison. At zero bias voltage, the oxide layer is rather clean and the impurity level is below 10 ppm in average. About 0.4% Ar are found in this layer. In contrast a 5000 Å thick oxide layer, produced with 15% bias (170 V), exhibits at least 4 big impurity peaks. Besides Ar, In and/or Sn (mass resolution in this mass range does not allow separation), a heavy mass with a mass number of about 200 ± 10 can be detected. These materials probably stem from the backing material that was used to obtain good heat conductivity between substrate and target electrode. It is assumed that during the sputtering process part of this material becomes gaseous, due to substrate heating and/or bias sputtering. The rectangular shape of the Ar peak indicates that the Ar is equally distributed throughout the SiO_2 layer, the same situation also occurs for the Ga peak. The In peak and the high mass peak however, show a marked decrease in concentration towards the surface of the oxide layer. This decrease of impurity concentration may by explained by either a drop in vapor pressure during film growth or by outdiffusion of impurities. This outdiffusion can occur because of enhanced layer heating during prolonged deposition times for bias sputtering.

In an initial experiment the depositions were performed using various power densities and the influence of deposition rate on impurity content was studied. By increasing the power density, to keep layer thicknesses to about 1200 Å, deposition times were reduced from 135 to 25 min. At 6% bias voltage the

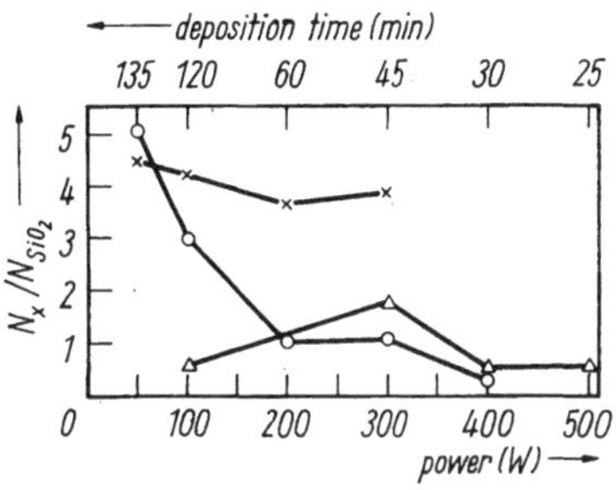

Fig. 4. Impurity content versus power density with bias voltage as a parameter. x: △ Ar (0% bias); × Ar (6% bias); ○ In (6% bias)

absolute bias voltage increases from 30 to 100 V, with increasing power density.

The results of these initial experiments are summarized in Fig. 4. The Ar content versus power density is shown for two sets of samples produced with zero and 6% bias voltage. In both cases only a little dependence of Ar concentration on power density is found but there is a marked difference in absolute values. This independence of Ar content on rf power density was also found by Hoffmeister and Zuegel [3].

For zero bias voltage, the Ar content was found to have values between 0.4 and 1.8 mol% (compared to the number of SiO_2 molecules in the oxide layer). The Ar concentration from 75% of the measured layers was found to be 0.4, 0.5, and 0.6 mol% so it is assumed that the value of 1.8% at 300 W power is exceptionally high. At 6% bias voltage, the Ar concentration slowly decreases with increasing power from 4.5 to 3.7 mol%. The only other pronounced impurity peak found in these samples was attributed to scattering from In(Sn) atoms. The In(Sn) content decreases with increasing power or decreasing deposition time. This result indicates that the process of entrapment is different for In(Sn) and Ar. The deposition rate of sputtered target atoms is found to be inversely proportional to the deposition rate of In and proportional to the deposition rate of Ar atoms.

A further experiment was to investigate the dependence of the impurity content on bias voltage, the results are shown in Fig. 5. A constant power density of 2 W cm^{-2} was used during sputtering. Equal layer thicknesses of about 1200 Å were obtained by increasing the deposition time with increasing bias voltage. The lowest Ar content was found at the zero bias voltage with steep increase of Ar content, occuring when the bias voltage is applied. To verify this the results discussed above are included (dashed curve). With in-

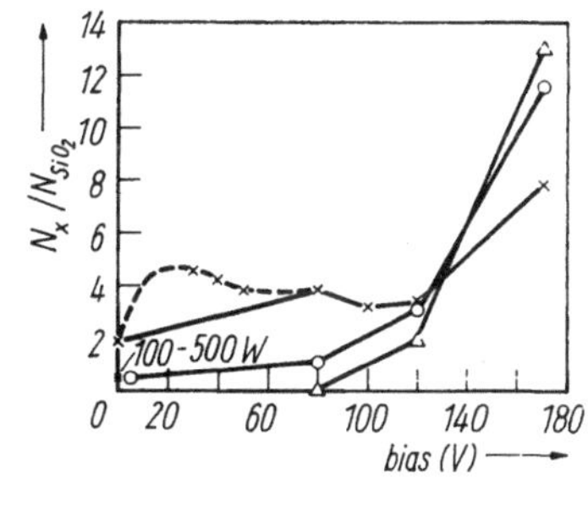

Fig. 5. Impurity content versus bias voltage at 300 W power. x: × Ar (data in %); ○ In (‰); △ Ga (‱)

creasing bias voltage up to 120 V, the Ar content decreases slightly from 4.5 to 3.1 mol%. For voltages greater than 120 V, a steep increase of Ar concentration is found. This steep increase may be due to an enhanced sticking probability for Ar ions with energies above 120 eV in the SiO_2 substrate.

Altering the bias voltage also appeared to change the concentration of Ga. A similar dependence to that of Ar with bias voltage was observed for the Ga and In(Sn) contents in the sputtered layer. The amount of Ga and In(Sn) increases strongly with increasing bias voltage (note the different scale factors); the same increase was also obtained for the high-mass impurity peak. An increase of the Fe content in sputtered SiO_2 layers on Si substrates with increasing bias voltage, was also found during bias sputtering when substrate holders were incompletely coated. It can be concluded from these results that bias sputtering, with bias voltages of about 100 V or above, will generally lead to an enhanced incorporation of impurities in the layer. For the layer with extremely high impurity content anneal experiments were performed. After 30 min anneal at 600 °C and 10^{-7} Torr, the Ar content dropped from 8 to 3 mol%. Further isochronal anneal treatments at 600 °C and 700 °C did not noticeably change the Ar concentration. This is in contrast to results presented by Hoffmeister and Zuegel [3] who found that almost all of the Ar was lost within a one hour anneal at 600 °C. The heavy mass impurity and the In(Sn) content was not affected by this anneal treatment. This indicates that the depth distributions of these impurities are probably due to a decrease of vapor pressure during deposition. A 30% decrease of Ga content was found however during heat treatment.

5. Discussions and Conclusions

The impurity content of silica layers, deposited by bombarding a fused quartz target with Ar ions generated in a rf glow discharge, was determined by means of backscattering and channeling effect measurements. The oxide layers were found to be stoichiometric SiO_2 irrespective of various sputtering parameters used during film growth. This result is in contrast to results obtained by the analysis of reactively sputtered silica layers [9]. Here it was found that the ratio of oxygen to silicon atoms strongly depends on the oxygen gas pressure. It was observed that sputter etching of the Si substrate prior to film deposition, caused severe damage at the substrate surface. This damage may cause electrical irregularities at the interface Si/SiO_2 and should be avoided.

At zero bias voltage, the Ar concentration was found to be in the range of 0.5 to 1.8 mol%. This is in agreement with results given by Hoffmeister and Zuegel [3] for rf sputtered SiO_2 layers. Applying small rf bias voltages to the substrate results in an increase of Ar content to about 4 mol%. This increase may be related to the absorption of low-energy Ar ions originated in the glow discharge, a mechanism which has been discussed by Lee and Oblas [4]. After increasing the bias voltage above 120 V a further increase of Ar content is observed. The absorption mechanism responsible for this enhanced increase may be due to entrapment of energetic neutral Ar atoms as discussed by Winter and Kay [2]. Rf power density variation showed only a little influence on Ar content. In all cases the entrapped Ar was found to be equally distributed throughout the layer thickness. Substrate holder material and backing material was found to be incorporated during the rf bias sputtering process. A minimum impurity concentration was found for layers produced at zero bias voltage.

It was found that the impurity concentration strongly increased for bias voltages above 120 V. In contrast to Ar, the number of In atoms trapped in the films decreased with increasing power. To summarise, we conclude that biased rf sputtering does not have any advantages in the production of pure SiO_2 films. If other properties, such as the enhanced mechanical strength are improved by bias sputtering, then moderate bias voltages (below 120 V) may be tolerated.

Acknowledgements

We gratefully acknowledge the assistance of Mr. A. Hyder who planned and performed the sputtering experiments. We also wish to thank Dr. M. Gettings and Prof. P. A. Tove for valuable discussions.

References

[1] D. Gerstenberg and C. J. Calbick, J. appl. Phys. **35**, 402 (1964).
[2] H. F. Winters and E. Kay, J. appl. Phys. **38**, 3928 (1967).
[3] W. Hoffmeister and M. Zuegel, Thin Solid Films **3**, 35 (1969).
[4] W. W. Lee and D. Oblas, J. Vacuum Sc. and Technol. **7**, 129 (1970).
[5] O. Meyer, J. Gyulai, and J. W. Mayer, Surface Sci. **22**, 263 (1970).
[6] O. Meyer and W. Scherber, J. Phys. Chem. Solids **32**, 1909 (1971).
[7] M. Kamoshida, I. U. Mitchell, and J. W. Mayer, Appl. Phys. Letters **18**, 292 (1971).
[8] S. M. Davidson, J. Mater. Sci. **7**, 473 (1972).
[9] G. Linker and O. Meyer, unpublished results.

(Received July 26, 1972)

phys. stat. sol. (a) **14**, 613 (1972)

Subject classification: 10.2; 22.5.2

Physikalisches Institut Weihenstephan der Technischen Universität München, Freising

Über die Verfestigung von Steinsalzeinkristallen durch Ultraschall [1]

Von

F. Asselmeyer und K. Hartmann

Getemperte Steinsalzeinkristalle wurden in [001] durch Wechseldrücke belastet, die mit Ultraschall von 20 kHz erzeugt und über Dehnungsmeßstreifen gemessen wurden. Die Wechseldruckeinwirkung führt zur Bildung von Versetzungen und Gleitbändern und zu starker Gleitbandverbreiterung. Der Nachweis der Versetzungen erfolgte mit Ätztechnik. Versetzungsmultiplikation und Gleitbandverbreiterung wurden in erster Linie auf doppeltes Quergleiten von Schraubenversetzungen zurückgeführt. Aus der Ätzgrübchenanordnung auf geätzten {100}-Spaltflächen folgte, daß schon bei relativ niedrigen Wechseldrücken alle primären Gleitsysteme im Mittel gleich stark betätigt werden. Durch Reaktionen von Versetzungen auf schräg zueinander liegenden Gleitsystemen werden unbewegliche Versetzungsstücke gebildet. Die Ultraschallbeanspruchung verfestigt die Steinsalzeinkristalle. Die Verfestigung wurde anhand von dynamischen Verfestigungskurven untersucht, die durch Stauchung der Proben in [001] gemessen wurden. Als Maß für die Verfestigung diente die kritische Schubspannung τ_k. Es wurde gefunden, daß τ_k mit wachsendem Wechseldruck zunächst leicht zunimmt, dann jedoch wieder abfällt, ein relatives Minimum durchläuft, um schließlich auf hohe Werte anzusteigen. Eine Abhängigkeit der kritischen Schubspannung von der Dauer der Wechseldruckeinwirkung in dem untersuchten Zeitraum zwischen 25 und 250 s war nicht zu erkennen.

Annealed sodium chloride single crystals were exposed along [001] to alternating pressures produced by ultrasonic waves of 20 kHz and measured by means of strain gauges. The alternating pressures lead to the formation of dislocations and glide bands and a strong broadening of these bands. The dislocations were revealed by etching. The multiplication of dislocations and the broadening of the glide bands were primarily ascribed to double cross gliding of screw dislocations. The configuration of etch pits on {100} cleavage faces shows that — averaged across the whole crystal — all primary slip systems are activated to the same extent. This result holds for relatively low alternating pressure already. Due to reactions of dislocations gliding on oblique systems immobile parts of edge dislocations are created. The sodium chloride single crystals are hardened by the exposure to ultrasonic stressing. The hardening was investigated by means of dynamical stress–strain curves measured by compression along [001]. The critical resolved shear stress τ_k served as a measure for hardening. It was found that τ_k at first increases slightly with increasing alternating pressure. Then, however, τ_k decreases down to a minimum and finally increases to high values. A dependence of τ_k on the exposure time of alternating pressure within the investigated intervall between 25 and 250 s could not be recognized.

[1]) Auszug aus der von der Fakultät für Brauwesen und Lebensmitteltechnologie der Technischen Universität München genehmigten Dissertation „Über die Verfestigung von Steinsalzeinkristallen durch Ultraschall" des Dipl.-Phys. Klaus Hartmann zur Erlangung des akademischen Grades „Dr. rer. nat.". Tag der Promotion: 20. 7. 1972.

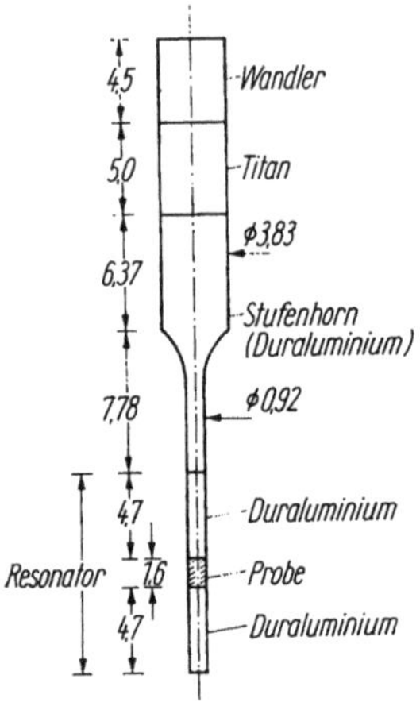

Fig. 1. Schemazeichnung der zur Beschallung verwendeten Apparatur. Maße in cm

1. Herstellung und Beschallung der Proben

Die Steinsalzeinkristalle wurden mit dem Kyropoulos-Verfahren in Luft hergestellt. Als Ausgangsmaterial diente NaCl krist. p. a. Aus den Kristallrohlingen wurden Spaltstücke herausgespalten, die auf einer Drehbank rund gedreht und danach 50 h lang bei 675 °C getempert wurden. Die Proben waren 1,6 cm lang und hatten Durchmesser von 0,92 cm. Die Probenachse lag in [001].

Die Beschallung der Proben, d. h. ihre Belastung mit Wechseldrücken, erfolgte mit einer Resonanzmethode. Die Beschallungsfrequenz betrug 20 kHz. Sie war in einem Bereich von $\pm$ 0,6 kHz variabel. Bei 20 kHz waren die Proben allein nicht resonanzfähig. Sie wurden deshalb mit Uhu-plus zwischen zwei Duraluminiumstäbe eingeklebt. Dieses resonanzfähige System wird im folgenden mit Resonator bezeichnet. Der Resonator war in ein ebenfalls aus Duraluminium bestehendes Stufenhorn eingeschraubt, das über ein metallisches Zwischenstück an den Ultraschallwandler gekoppelt war. Der Aufbau und die Abmessungen der zur Beschallung verwendeten Einrichtung gehen aus Fig. 1 hervor.

Beschallt wurde stets in Luft bei Zimmertemperatur und vertikaler Lage des Resonators. Beschallungsrichtung war die [001]-Richtung, also die Richtung der Probenachse. Eine statische Belastung der Proben während der Beschallung war vernachlässigbar klein. Sie betrug nur etwa 0,13 N/cm^2. Auf Grund des Verhältnisses von Resonatordurchmesser zu Schallwellenlänge, das wesentlich kleiner als eins war, bildete sich im Resonator keine reine Longitudinalwelle, sondern eine Stabdehnungswelle aus [1].

Nach der Beschallung wurden die Proben bei Zimmertemperatur mit Chloroform aus dem Resonator herausgelöst und untersucht.

2. Wechseldruckbestimmung in der Steinsalzprobe

Der bei der Beschallung in der Probe auftretende Wechseldruck wurde ausgehend von der Beziehung

$$p(x, t) = E \frac{\partial e(x, t)}{\partial x} = E\, \varepsilon(x, t) \tag{1}$$

bestimmt, in der $p(x, t)$ den Wechseldruck, $e(x, t)$ die Wechselverformung, $\varepsilon(x, t)$ die relative Längenänderung und E den Elastizitätsmodul bedeuten. x ist die Ortskoordinate, t die Zeit. Dämpfung konnte vernachlässigt werden. Der dadurch entstandene Fehler im Wechseldruck war kleiner als $3 \times 10^{-4}\%$. Dadurch erwies sich die Einführung eines komplexen E-Moduls als unnötig.

ε ist mit Dehnungsmeßstreifen (DMS) meßbar. Wird der DMS in eine symmetrische Wheatstonesche Viertelbrücke eingesetzt, dann ergibt sich unter der Voraussetzung eines sinusförmigen Wechseldrucks für die Wechseldruckamplitude an der Stelle x in Verbindung mit Gleichung (1) folgende Beziehung:

$$P(x) = \frac{4}{k} \frac{\sqrt{2}\, E}{U_S\, \varphi} U'_A(x) . \tag{2}$$

Darin sind U_S die Speisespannung der Wheatstoneschen Brücke, $U'_A(x)$ der Effektivwert der um den Faktor φ verstärkten Brückenausgangsspannung und k der charakteristische k-Faktor des DMS.

In Fällen, in denen DMS einsetzbar sind, kann die Wechseldruckamplitude in festen Körpern mit Hilfe von Gleichung (2) bestimmt werden, wenn der Elastizitätsmodul des betreffenden Körpers bekannt und Dämpfung zu vernachlässigen ist.

Es hat sich nun allerdings gezeigt, daß bei Steinsalzeinkristallen Messungen mit DMS nicht zu richtigen Ergebnissen führen. Die DMS lassen sich nicht zuverlässig aufkleben. Die Ursache ist nicht bekannt. Es war daher nicht möglich, den Wechseldruck in den Steinsalzproben im Resonator unmittelbar durch Aufkleben von DMS auf die Proben selbst zu ermitteln. Es mußte vielmehr ein Umweg eingeschlagen werden. Er führte über den Wechseldruck im metallischen Resonatorteil. Dort ist der Wechseldruck mit DMS meßbar. Daneben läßt sich die Wechseldruckverteilung im ganzen Resonator, also auch in der Steinsalzprobe, bis auf einen experimentell zu bestimmenden Faktor, der die Stärke der Schwingungsanregung beschreibt, berechnen. Vergleich von berechnetem und gemessenem Wechseldruck an einer beliebigen Stelle im metallischen Resonatorteil liefert diesen Faktor. Der Wechseldruck ist damit an jeder Stelle im Resonator, also auch an jeder Stelle in der Steinsalzprobe, eindeutig bestimmt.

Zur Berechnung der Wechseldruckverteilung im Resonator wurde dieser in drei Bereiche eingeteilt, wie es Fig. 2 zeigt. Dieser Figur sind auch die im folgenden verwendeten Bezeichnungen zu entnehmen.

Ausgehend von der eindimensionalen Wellengleichung wurde so vorgegangen, daß zunächst die Verteilung der Verformungsamplitude berechnet wurde, aus der sich die Verteilung der Druckamplitude über Gleichung (1) ergibt.

Unter der Voraussetzung einer harmonischen Schwingungserregung läßt sich die Wellengleichung durch einen Produktansatz in folgende gewöhnliche Differentialgleichung für die Amplitude $f(x)$ der Wechselverformung überführen:

$$\frac{d^2 f_k(x)}{dx^2} + n_k^2\, f_k(x) = 0 ; \tag{3}$$

n_k ist gegeben durch

$$n_k^2 = \omega^2 \frac{\varrho_k}{E_k} \tag{4}$$

mit ϱ_k Dichte und ω Kreisfrequenz. Der Index k kennzeichnet jeweils den betreffenden Resonatorbereich.

Zur Lösung von Gleichung (3) wurde ein Ansatz für eine stehende Welle herangezogen. Eine Randbedingung ist ein verschwindender Druck am freien Resonatorende ($x = x_3$).

An der Stelle der Ankopplung des Resonators an das Stufenhorn ($x = 0$) wurde die Bedingung $f_1\ (x = 0) = f_0$ eingeführt, wobei f_0 der oben erwähnte unbestimmte Faktor ist.

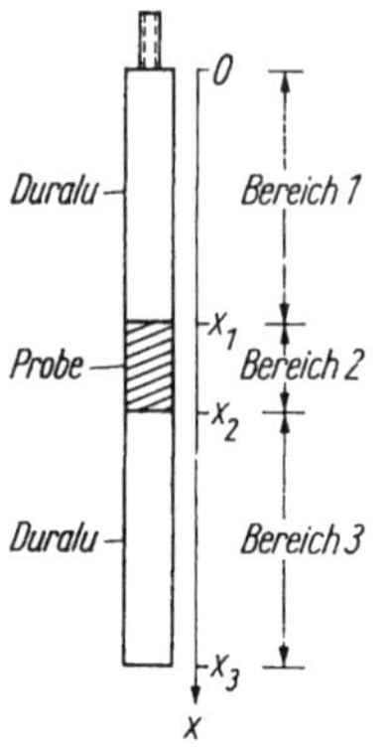

Fig. 2. Zur Berechnung der Wechseldruckverteilung im Resonator

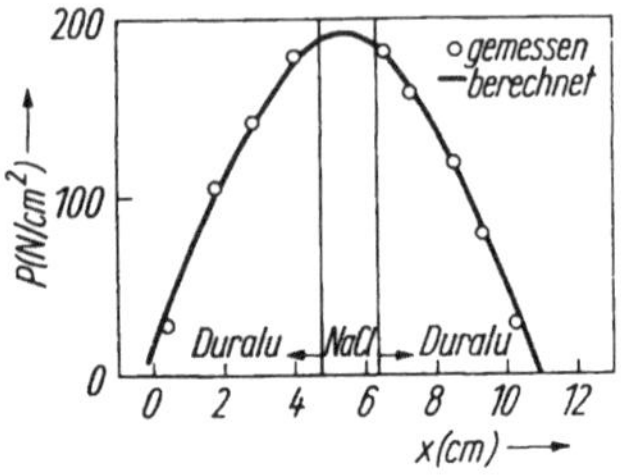

Fig. 3. Vergleich von berechneter und mit Dehnungsmeßstreifen gemessener Wechseldruckverteilung im Resonatur. ○ gemessen, ——— berechnet

An den Übergangsstellen zwischen Duraluminium und Steinsalz müssen außerdem die Kräfte und die Verformungsamplituden in den jeweils aneinander grenzenden Bereichen miteinander übereinstimmen.

Vergleicht man den aus der Rechnung resultierenden Wechseldruck an einer beliebigen Meßstelle $x = x_M$ im metallischen Resonatorteil (Bereich 1 oder 3) mit dem aus der Messung nach Gleichung (2) zu gewinnenden Wechseldruck an derselben Stelle, so ergibt sich ein Ausdruck für den Faktor f_0, mit dem sich die Verteilung der Wechseldruckamplitude in der Steinsalzprobe durch folgende Gleichung beschreiben läßt:

$$P_2(x) = \frac{4\sqrt{2}\, E_1\, U'_{AM}\, [\cos(n_1 x_1) \sin(n_2 (x_2 - x)) + q \sin(n_1 x_1) \cos(n_2 (x_2 - x))]}{k\, \varphi\, q\, U_S \sin(n_1 (x_3 - x_M))}. \tag{5}$$

U'_{AM} bezeichnet dabei den Effektivwert der an der Stelle $x = x_M$ gemessenen, um den Faktor φ verstärkten Brückenausgangsspannung. q stellt das Verhältnis von Wellenwiderstand in Duraluminium zu Wellenwiderstand in Steinsalz in [001] dar.

Auf Grund der geringen Probenlänge ist die Wechseldruckamplitude längs der Probenachse nahezu konstant. Als Maß für den Wechseldruck in der Probe wurde demgemäß der an einer nicht näher interessierenden Stelle auftretende Maximalwert verwendet. Er ist durch folgenden Ausdruck gegeben:

$$P_{2\,\max} = \frac{4\sqrt{2}\, E_1\, U'_{AM} \sin(n_1 x_1)}{k\, \varphi\, U_S \sin(n_1 (x_3 - x_M)) \cos\left[\operatorname{arctg}\left(\frac{1}{q} \operatorname{ctg}(n_1 x_1)\right)\right]}. \tag{6}$$

Alle im folgenden vorkommenden Wechseldruckangaben beruhen auf Gleichung (6). Sie sind mit einem Fehler von $\pm 6\%$ behaftet.

Zur Überprüfung des Verfahrens wurden die Wechseldruckverteilungen in den Bereichen 1 und 3 mit Gleichungen, die ähnlich aufgebaut sind wie Gleichung (5), berechnet und daneben auf der Grundlage von Gleichung (2) gemessen. Die für die Rechnung herangezogenen Stoffwerte wurden teils der Literatur entnommen, teils experimentell ermittelt.

Wie Fig. 3 zeigt, ist die Übereinstimmung zwischen gemessener und berechneter Druckverteilung gut. Es wurde davon ausgegangen, daß auch in der Steinsalzprobe die berechnete Verteilung und die tatsächlich vorliegende mit derselben Genauigkeit übereinstimmen.

3. Untersuchung beschallter Steinsalzeinkristalle mit Ätztechnik

Die beschallten Proben wurden nach dem Ablösen aus dem Resonator gespalten und sofort danach mit einer etwa halb gesättigten Lösung von $HgCl_2$ p. a. in Äthanol absolut p. a. geätzt.

In Fig. 4 ist eine Reihe von geätzten (100)-Spaltflächen beschallter Proben wiedergegeben. Die auf den primären Gleitsystemen wirkenden Wechselschubspannungen wurden mit Hilfe der oben beschriebenen Meßmethode unter Berücksichtigung des Orientierungsfaktors ermittelt. Die Beschallungsdauer betrug bei den gezeigten Proben einheitlich 12 s.

Fig. 4a zeigt eine weitgehend statistische Verteilung der Ätzgrübchen. Es sind nur einige vereinzelte in [010], [0$\bar{1}$1] und [011] verlaufende Gleitlinien zu erkennen. Ätzgrübchenanordnung und -dichte unterscheiden sich im wesentlichen nicht von denen in unbeschallten Proben.

Eine Wechselschubspannung von $\approx$ 90 N/cm² führt zu Ätzgrübchenverteilungen in der Art von Fig. 4b. Man erkennt eine große Zahl von scharf begrenzten, in [0$\bar{1}$1] und [011] verlaufende Gleitlinien und Gleitbänder hoher Ätzgrübchendichte und in [010] orientierte Bänder geringer Ätzgrübchendichte mit diffusen Rändern.

Bei einer weiteren Erhöhung der Wechselschubspannung auf $\approx$ 120 N/cm² wird eine starke Verbreiterung der $\langle 110 \rangle$-Bänder beobachtet (Fig. 4c). Die Bänder gehen stellenweise ineinander über und bilden größere Bereiche sehr hoher Ätzgrübchendichte, die von Bereichen geringerer Ätzgrübchendichte umgeben sind. Bänder in [010] sind kaum noch zu identifizieren.

Durch eine nochmalige Steigerung der Wechselschubspannung auf $\approx$ 150 N/cm² wird die Ätzgrübchendichte nicht mehr wesentlich vergrößert (Fig. 4d). Es ist nun fast nicht mehr erkennbar, daß die Bereiche hoher Ätzgrübchendichte ursprünglich aus $\langle 110 \rangle$-Bändern hervorgegangen sind.

Eine Wechselschubspannung über etwa 150 N/cm² bringt qualitativ keine Änderung mehr in der Verteilung und Dichte der Ätzgrübchen. Es scheint ein Sättigungszustand einzutreten.

Die Struktur von geätzten (010)-Spaltflächen unterscheidet sich qualitativ nicht von den beschriebenen (100)-Flächen, was auf Grund der Rotationssymmetrie der Proben auch nicht zu erwarten ist. Dagegen zeigen geätzte (001)-Flächen, also Flächen senkrecht zur Beschallungsrichtung, im wesentlichen nur Gleitbänder in [100] und [010]. Gleitbänder in $\langle 110 \rangle$-Richtungen fehlen.

Die beschriebenen Ätzgrübchenanordnungen ergeben sich auch auf $\{100\}$-Spaltflächen von Proben mit quadratischer oder nur wenig vom Quadrat abweichender Querschnittsfläche. Sie sind also nicht charakteristisch für runde Proben.

Ähnliche Bildfolgen wurden auch zur Untersuchung der Abhängigkeit der Ätzgrübchenanordnung von der Beschallungsdauer im Intervall zwischen 20 und 250 s und nahezu konstanter Wechselschubspannung aufgenommen. Die schon aus den Fig. 4a bis 4d ersichtliche stark heterogene Verteilung, die auf

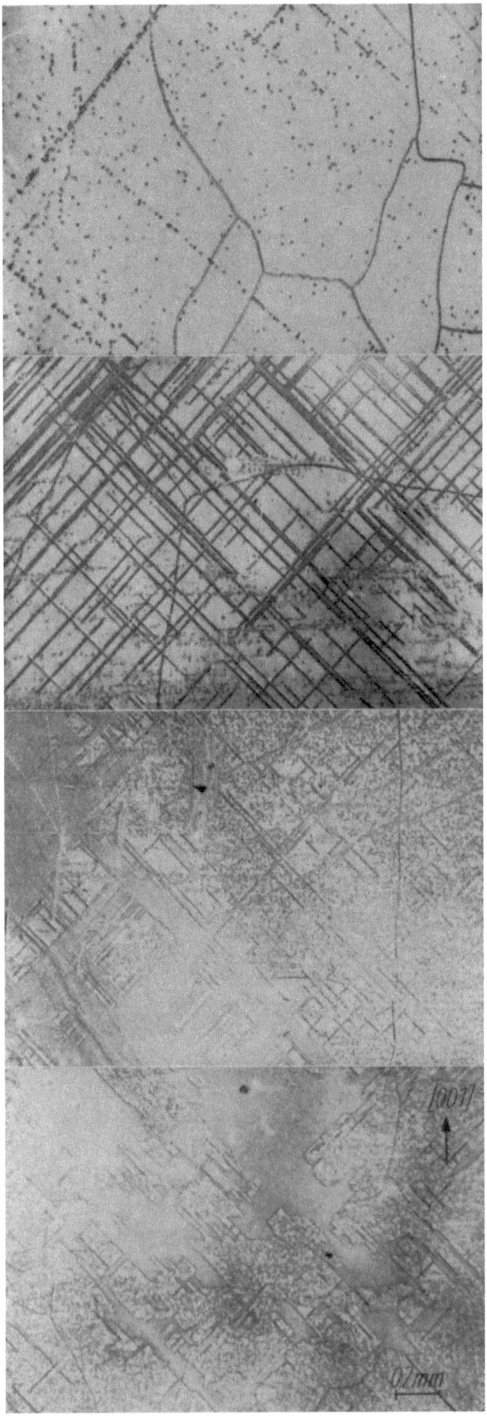

a
b
c
d

Fig. 4. Geätzte (100)-Spaltflächen beschallter Steinsalzeinkristalle; Wechselschubspannung: a) 35 N/cm², b)90 N/cm², c) 120 N/cm², d) 150 N/cm²

eine inhomogene Druckverteilung in den Proben hinweist, ließ jedoch eine eindeutige Aussage über eine Zeitabhängigkeit in dem betrachteten Zeitraum nicht zu.

Die Aussagen über die Druckabhängigkeit der Verteilung und Dichte der Ätzgrübchen können aus diesem Grunde ebenfalls kaum mehr als informatorischen Charakter besitzen.

Die Anordnung der Gleitbänder auf den betrachteten Spaltflächen läßt Rückschlüsse über die bei Beschallung betätigten Gleitsysteme zu. Das Fehlen von $\langle 110 \rangle$-Bändern auf (001)-Flächen kann mit einer Nichtaktivierung der latenten Systeme erklärt werden. Das gleichzeitige Auftreten von $\langle 110 \rangle$- und $\langle 100 \rangle$-Bändern auf (100)- und (010)-Flächen ist mit Mehrfachgleitung zu deuten. Dabei werden nach Aussage der Ätzgrübchenbilder alle primären Systeme gleichzeitig und bei höheren Wechselschubspannungen im Mittel gleich stark betätigt. Obwohl bei einer Druckeinwirkung in [001] die primären Systeme schubspannungsmäßig gleichberechtigt sind und man dementsprechend immer Mehrfachgleitung erwarten sollte, wurde von Matucha [2] mit einem elektronenmikroskopischen Abdruckverfahren festgestellt, daß bei gestauchten Steinsalzeinkristallen bei niedrigen Schubspannungen (Bereich I der Verfestigungskurve) stets nur zwei zueinander orthogonale Gleitsysteme aktiviert werden. Bei höheren Schubspannungen (Bereich 2) kommen dann die beiden anderen primären Systeme hinzu. Sie tragen jedoch nur etwa 10% zur gesamten Abgleitung bei. Begünstigt bleiben über die ganze Verfestigungskurve die beiden ursprünglich betätigten Systeme. Das Gleitverhalten von Steinsalzeinkristallen bei Beschallung unterscheidet sich somit wesentlich von dem Gleitverhalten von Proben, die nur in einer Richtung belastet werden.

Bei höheren Wechselschubspannungen wechseln die aktiven primären Gleitsysteme in beschallten Proben mehrmals innerhalb kleinster Kristallbereiche. Dies führt zu den charakteristischen, zerklüfteten und bizarren Begrenzungen der Bezirke hoher Ätzgrübchendichte, wie es beispielsweise aus Fig. 4d hervorgeht.

Nur in Ausnahmefällen werden Gleitbänder gefunden, die über die ganze Breite der Probe laufen und dies gewöhnlich auch nur bei relativ niedrigen Wechselschubspannungen. In der Regel sind die Gleitbänder kurz. Auf Grund der gleichzeitigen Aktivierung aller primären Systeme wird Gleitung auf einem System schnell von einem dazu orthogonalen oder schrägen System gestoppt. Fig. 5 ist hierfür ein typisches Beispiel.

Die Gesetze, nach denen die Auswahl der in bestimmten Kristallbereichen aktiven Gleitsysteme erfolgt, sind nicht bekannt. In erster Linie wäre an Schwankungen im Orientierungsfaktor zu denken, die die Gleichberechtigung der primären Systeme lokal aufheben. Die Ursache hierfür könnte in einer nicht exakt achsenparallelen Wechseldruckeinwirkung oder im Vorliegen unterschiedlich orientierter Kristallblöcke zu suchen sein.

Die geätzten Spaltflächen der Fig. 4a bis 4d zeigen, daß bei Beschallung eine starke, von der Höhe der Wechselschubspannung abhängige Versetzungsmulti-

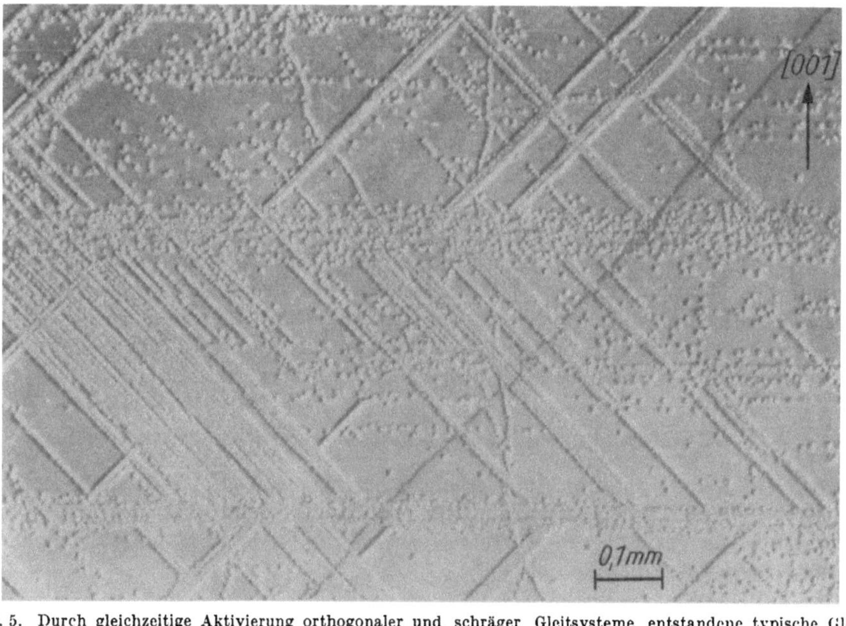

Fig. 5. Durch gleichzeitige Aktivierung orthogonaler und schräger Gleitsysteme entstandene typische Gleitbandanordnung auf einer (100)-Spaltfläche einer beschallten Probe

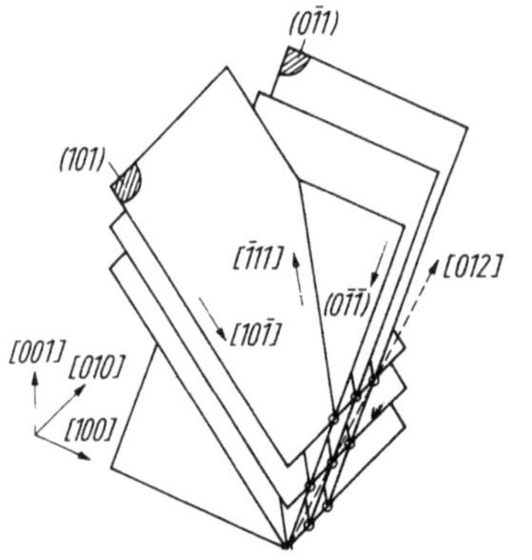

Fig. 6. Zur Bildung von unbeweglichen Versetzungsstücken auf Grund von Versetzungsreaktionen beim Schneiden schräger Gleitsysteme. ○ Durchstoßpunkte von ⟨111⟩-Versetzungen durch die (100)-Ebene

plikation stattfindet. Das auffallende Breitenwachstum der Gleitbänder (z. B. in Fig. 5) kann mit einem Frank–Read-Mechanismus allein nicht erklärt werden. Es deutet vielmehr auf doppeltes Quergleiten von Schraubenversetzungen als entscheidenden Prozeß hin. Ein Frank–Read-Prozeß wird zwar in Verbindung mit dem doppelten Quergleiten auftreten, wird jedoch nur eine sekundäre Rolle spielen.

Ein weiterer Hinweis auf doppelte Quergleitprozesse ist die stark unterschiedliche Ätzgrübchendichte in ⟨110⟩- und ⟨100⟩-Bändern, die schon oben erwähnt wurde. Die höhere Ätzgrübchendichte in ⟨110⟩-Bändern ist auf Stufenversetzungsdipole zurückzuführen. Eine Wechseldruckbelastung scheint offenbar die Bildung von Stufenversetzungsdipolen zu begünstigen.

Auf Grund der gleichzeitigen Aktivierung schräger Systeme werden Versetzungsreaktionen der folgenden Art ermöglicht [3]:

$$\frac{a}{2}[10\bar{1}] + \frac{a}{2}[011] = \frac{a}{2}[110]; \tag{7}$$

a ist die Gitterkonstante. Die gebildeten Versetzungen liegen in ⟨111⟩ [3]. Sie sind weitgehend unbeweglich und stellen deshalb starke Hindernisse für reguläre Gleitung auf {110} und für Quergleitprozesse dar. Wie aus Fig. 6 hervorgeht, liegen die Durchstoßpunkte paralleler ⟨111⟩-Versetzungsstücke durch {100}-Ebenen in ⟨100⟩, ⟨110⟩ und ⟨012⟩, wenn man von der Annahme ausgeht, daß der Abstand zwischen aktivierten Gleitebenen in sich schräg treffenden Gleitsystemen im Mittel derselbe ist. Auf geätzten {100}-Flächen werden demgemäß Begrenzungen von ⟨110⟩-Bändern in diesen Richtungen auftreten. Charakteristisch sind insbesondere Begrenzungen in ⟨012⟩. Sie wurden in der Tat auf geätzten (100)- und (010)-Spaltflächen beschallter Proben sehr häufig gefunden. Ein typisches Beispiel ist in Fig. 7 wiedergegeben.

4. Verfestigungskurven beschallter Steinsalzeinkristalle

Auf Grund der Zunahme der Versetzungsdichte und auch wegen der Bildung unbeweglicher ⟨111⟩-Versetzungsstücke war zu vermuten, daß beschallte Steinsalzeinkristalle verfestigt sind. Die Verfestigung wurde anhand von Verfestigungskurven untersucht.

Als Maß für die Verfestigung diente die kritische Schubspannung τ_k. Untersucht wurde die Abhängigkeit der kritischen Schubspannung von der Wechsel-

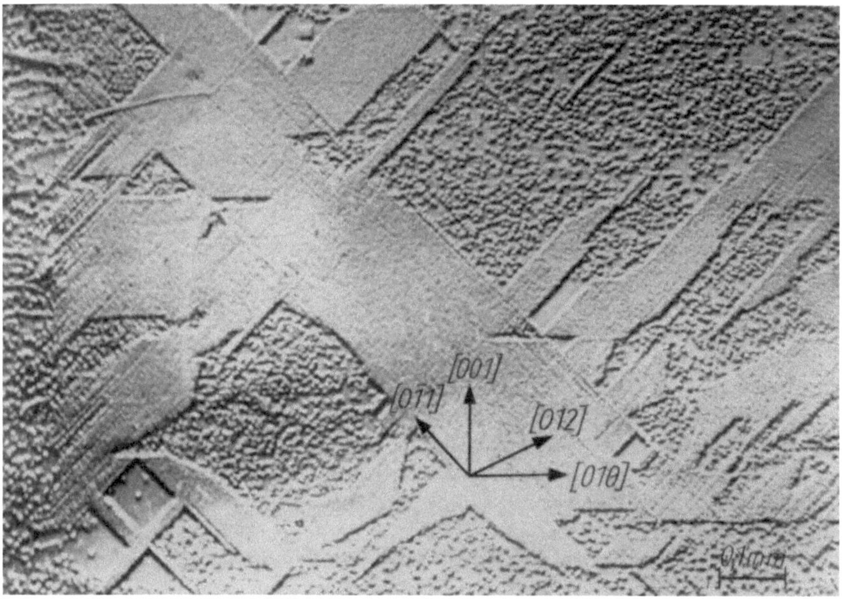

Fig. 7. Auf unbewegliche ⟨111⟩-Versetzungsstücke zurückführende Ätzgrübchenanordnung auf einer (100)-Spaltfläche einer beschallten Probe

schubspannung τ_B und der Beschallungsdauer t_B im Zeitintervall zwischen 25 und 250 s.

Die Verfestigungskurven wurden bei Zimmertemperature im dynamischen Stauchversuch in [001], also in Beschallungsrichtung, aufgenommen. Die Abgleitgeschwindigkeit betrug bei Stauchbeginn $1{,}9 \times 10^{-4}\ s^{-1}$.

In Fig. 8 ist τ_k in Abhängigkeit von der Wechselschubspannung τ_B aufgetragen. Die Beschallungsdauer betrug dabei 250 s. Wie man sieht, nimmt τ_k ausgehend von $\tau_{k0} = 60\ N/cm^2$ (kritische Schubspannung der unbeschallten Proben) mit wachsendem τ_B zunächst zu, durchläuft bei niedrigen Werten von τ_B ein relatives Maximum, scheint dann wieder abzunehmen, um schließlich nach Durchlaufen eines mehr oder weniger gut ausgeprägten relativen Minimums auf hohe Werte anzusteigen. Bei anderen Beschallungszeiten in dem untersuchten Zeitraum ist qualitativ derselbe Verlauf von $\tau_k = f(\tau_B)$ festzustellen.

Bemerkenswert an diesem Verlauf ist insbesondere das Minimum von τ_k. Es tritt in der Regel bei Wechselschubspannung von etwa $100\ N/cm^2$ auf. Im Minimum unterscheidet sich die kritische Schubspannung beschallter Proben kaum von der unbeschallter Proben. Dort liegt dementsprechend keine Ultraschallverfestigung vor.

Auf Grund der erheblichen Streuung der Meßwerte war dieses Minimum bei der Art der Darstellung von Fig. 8 häufig nicht deutlich erkennbar. Klarer geht es aus einer Darstellung hervor, die man durch Einführung eines sogenannten Verfestigungsquotienten η erhält. η wurde wie folgt definiert:

$$\eta = \frac{\tau_k - \tau_{k0}}{\tau_B}. \tag{8}$$

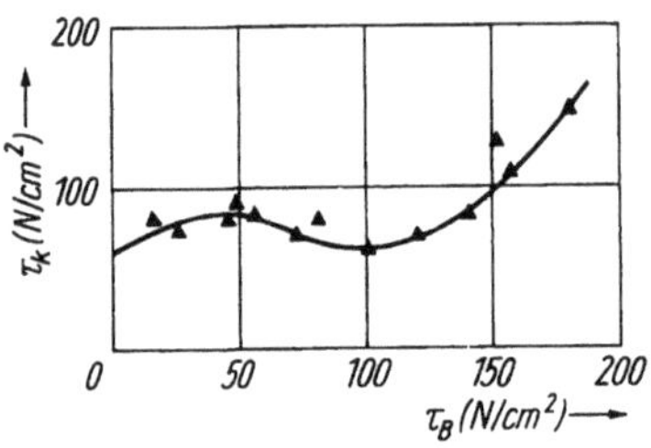

Fig. 8. Abhängigkeit der kritischen Schubspannung τ_k von der Wechselschubspannung τ_B; Beschallungsdauer $t_B = 250$ s

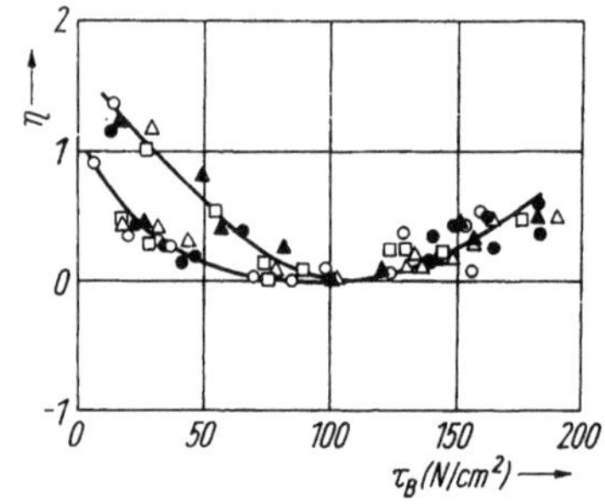

Fig. 9. Verfestigungsquotient η in Abhängigkeit von der Wechselschubspannung τ_B; Parameter: Beschallungszeit t_B. ○ $t_B = 25$ s, ● $t_B = 50$ s, △ $t_B = 100$ s, □ $t_B = 175$ s, ▲ $t_B = 250$ s

Fig. 9 zeigt die Abhängigkeit des Verfestigungsquotienten von der Wechselschubspannung. Parameter ist die Beschallungsdauer.

Auffallend in Fig. 9 ist zunächst, daß unterhalb des Minimums von τ_k zwei Kurvenäste vorliegen. In diesem Druckbereich werden offensichtlich zwei Verfestigungsmechanismen wirksam. Man ist versucht, dies auf die Aktivierung orthogonaler Systeme (unterer Zweig) und schräger Systeme (oberer Zweig) in diskreten Kristallbereichen zurückzuführen. Dagegen spricht allerdings, daß mit Ätztechnik in diesem Druckbereich Gleitlinien und Gleitbänder nur in geringem Umfange beobachtet werden. Eine befriedigende Erklärung für die Abhängigkeit der kritischen Schubspannung von der Wechselschubspannung in diesem Druckbereich konnte nicht gefunden werden.

Die Zunahme der kritischen Schubspannung bei Wechselschubspannungen über ≈ 100 N/cm² wurde in erster Linie auf unbewegliche $\langle 111 \rangle$-Versetzungsstücke der oben behandelten Art zurückgeführt. Diese Deutung wurde insbesondere durch die Anordnung der Gleitbänder auf geätzten $\{100\}$-Spaltflächen nahegelegt. Es ist in diesem Zusammenhang bemerkenswert, daß bei Steinsalzeinkristallen, die nur in einer Richtung verformt werden, der Bereich großen Verfestigungsanstiegs, nämlich Bereich 2 der Verfestigungskurve, ebenfalls bei einer Schubspannung von etwa 100 N/cm² beginnt (vgl. z. B. [4]).

Eine Abhängigkeit der kritischen Schubspannung von der Beschallungsdauer in dem untersuchten Zeitraum zwischen 25 und 250 s war nicht zu erkennen (vgl. Fig. 9).

Der Fraunhofer-Gesellschaft e. V. danken wir für finanzielle Unterstützung und für die Bereitstellung von Meßgeräten.

Literatur

[1] L. Bergmann, Der Ultraschall, 4. Aufl., S. Hirzel Verlag, Stuttgart 1954.
[2] K.-H. Matucha, phys. stat. sol. **26**, 291 (1968).
[3] B. H. Kear, A. Taylor und P. L. Pratt, Phil. Mag. **4**, 667 (1959).
[4] J. Hesse, phys. stat. sol. **9**, 209 (1965).

(Received October 23, 1972)

phys. stat. sol. (a) 14, 623 (1972)

Subject classification: 10.2; 22.1.3

Laboratoire de Physique des Défauts dans les Cristaux,
Université des Sciences et Techniques de Lille, Villeneuve d'Ascq

Low-Temperature Dislocation Substructure in the Early Stage of Deformation of Tellurium Single Crystals[1]

By

J. L. FARVACQUE, J. CRAMPON, J. C. DOUKHAN, and B. ESCAIG

Low temperature prismatic glide of tellurium is strongly thermaly activated. The evolution of dislocation substructure has been studied by means of Berg-Barrett topographs during the parabolic stage corresponding to the passage from microplasticity to macroplasticity. Topographs show that crude polygonisation walls develop at the very first stage of deformation throughout the sample. For *a* glide walls are tilt boundaries lying in glide planes. They are constituted by two families of *a* edge dislocations. For *c* glide, walls in basal planes are also tilt boundaries, but formed by one family of *c* edge dislocations. Electron microscopy confirms these results. Possible subgrain boundaries in tellurium are discussed from the point of view of their formation and their energy. It is found that these walls screen out long range stress fields which should be developed by dislocation multiplication.

Les glissements prismatiques du tellure, à basse température, sont fortement activés thermiquement. De ce fait, le passage de la microplasticité à la macroplasticité se fait au cours d'un stade parabolique étendu. On a observé, par la méthode de Berg-Barrett, l'évolution de la sous structure des dislocations le long de ce stade parabolique. Les topographies montrent que des murs de polygonisation se développent dans tout le cristal dès les premiers pourcents de la déformation. Pour le glissement *a*, les murs sont des sous-joints de flexion parallèles aux plans de glissement. Ils sont constituées de deux familles de dislocations coin. Pour le glissement *c*, les murs parallèles aux plans de base sont aussi des sous-joints de flexion, mais cette fois constitués d'une seule famille de dislocations *c* coin. La microscopie électronique confirme ces résultats. Les différents sous-joints possibles dans le tellure ont été discutés du point de vue de leur formation et de leur énergie. On trouve que ces murs écrantent les champs de contraintes à grande distance qui ont pu être développés par la multiplication des dislocations.

1. Introduction

This study deals with the evolution of the dislocation substructure at the beginning of plastic deformation in tellurium single crystals. Particular attention has been devoted to the passage from microplasticity to macroplasticity. Classical Berg-Barret reflexion topographs as described by Newkirk [1] have been extensively used. They clearly show that glide polygonisation appears during the early stage of plastic deformation at 77 °K.

The observed subgrain boundaries (which could better be called "walls") are responsible for tilts, the rotation axes of which have been determined by Wilkens technique [2].

---

[1]) This paper is part of the work done by Mr. J. L. Farvacque in partial fulfilment of his 3rd Cycle Doctor degree held in the Lille University on June 1972.

The principle of the Berg-Barret method is briefly recalled in Section 2. Section 3 is devoted to the description of the glide systems of tellurium. Experimental results are presented in Section 4 and the nature of polygonisation walls is discussed in Section 5.

2. The Berg-Barrett Method

There are two kinds of contrast on the X-ray reflexion topographs:

1. Contrast due to primary extinction phenomenon: a defect decreases primary extinction locally so that the reflective power is enhanced; on the topographs contrast lines appear as black lines on a grey background.

2. Contrast due to local rotation of the lattice (misorientation contrast). The local rotation is characterised by the rotation vector $\boldsymbol{\Omega}$ which can be decomposed into three parts:

$$\boldsymbol{\Omega} = \boldsymbol{\Omega}_1 + \boldsymbol{\Omega}_2 + \boldsymbol{\Omega}_3 \, .$$

$\boldsymbol{\Omega}_1$ normal to the reflecting plane, $\boldsymbol{\Omega}_2$ in the reflecting plane and the incident plane, $\boldsymbol{\Omega}_3$ normal to the incident plane.

Is is clear that no contrast is produced by a rotation α around $\boldsymbol{\Omega}_1$.

A rotation β around $\boldsymbol{\Omega}_2$ increases the incident angle by $\delta \approx \beta^2/2$ while the diffracted beam is turned by $2\,\beta \sin\theta$. For a deformed crystal the reflexion domain $\Delta\theta$ is of the order of 1′ while the Bragg angle θ is always chosen to be about 45° so that β must be as large as 1° to obtain $\delta\theta > \Delta\theta$. For β smaller than 1°, the diffracted beam is only displaced and gives rise to a "displacement contrast". In our study, this kind of contrast can be neglected because the lattice rotations β are of the order of some minutes and the corresponding displacements are smaller than the apparatus resolution (1.5 μm).

On the contrary, a rotation γ around $\boldsymbol{\Omega}_3$ produces an increase γ of the incident angle. As soon as γ is larger than $\Delta\theta$, the corresponding part of the crystal does not diffract any more and the topograph exhibits a black-white contrast. The incident beam divergency (1.6′ in our case) is a limitation to the occurrence of this kind of contrast on topographs. Tilts and twists are easily distinguished by rotation contrast. For a tilt subgrain boundary, the contrast line is parallel to $\boldsymbol{\Omega}_3$, for a twist subgrain boundary it is normal to $\boldsymbol{\Omega}_3$.

In our experiments we used the linear focus (7×0.07 mm^2) of a Cu anticathode. Owing to the small incident beam divergency and the strong absorption of tellurium ($\mu_a \approx 1300$ cm^{-1} for Cu K_α) exposure times had to be 5 to 20 h for $U = 40$ kV and $I = 15$ to 20 mA in the X-ray tube.

3. Plastic Deformation of Tellurium

The structure of Te is usually described as a hexagonal array of helicoidal chains with covalent bonds between atoms of the chain while bonds between atoms of adjacent chains are approximately four times weaker [3, 5].

At low temperature, dislocations only glide in planes parallel to the chain axis to avoid breaking covalent bonds. Tensile tests [6] confirm that glide occurs only in the first class of prismatic planes $\{10\bar{1}0\}$ with slip directions (Burgers vectors) $\boldsymbol{a} = 1/3 \langle 11\bar{2}0\rangle$ and $\boldsymbol{c} = \langle 0001\rangle$ which we call a and c glides. It is now well established that low temperature glides are strongly controlled by Peierls forces [7, 8].

For a glide which does obey Schmid's law, this dislocation–lattice interaction is a core effect. Helicoidal chains are strongly distorted in the core region of a

Fig. 1. Standard triangle indicating the different sample orientations

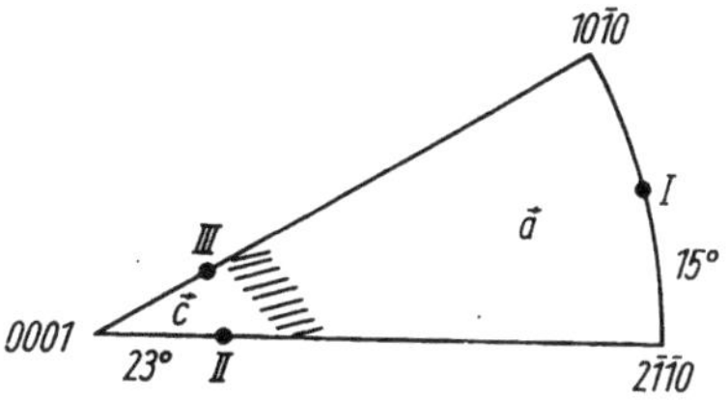

dislocations except for edge orientation. a edge dislocations are then much more numerous than a screws; they lie in Peierls valleys and their mobility is small as compared with screws. Edge glide occurs through double-kink nucleation so that a glide is strongly thermally activated [9].

For c glide which does not obey Schmid's law, dislocation–lattice interaction rather occurs through a sessile splitting of c screws on several $\{10\bar{1}0\}$ planes at a time. c edge dislocations which are not dissociated have a higher energy and a correlatively higher mobility than screws. This implies that c glide, as a glide, is strongly thermally activated.

Therefore, low temperature stress–strain curves exhibit for a and c glides a characteristic parabolic stage which does not allow us to measure any precise elastic limit. We study below the dislocation substructure evolution during the parabolic stage. Compression tests are performed at 77 °K on samples, the orientations of which are reported on the standard triangle, Fig. 1. Stress–strain curves for a and c glide are reported in Fig. 2. Spots labelled 1, 2, 3 represent the deformation states of the different samples observed.

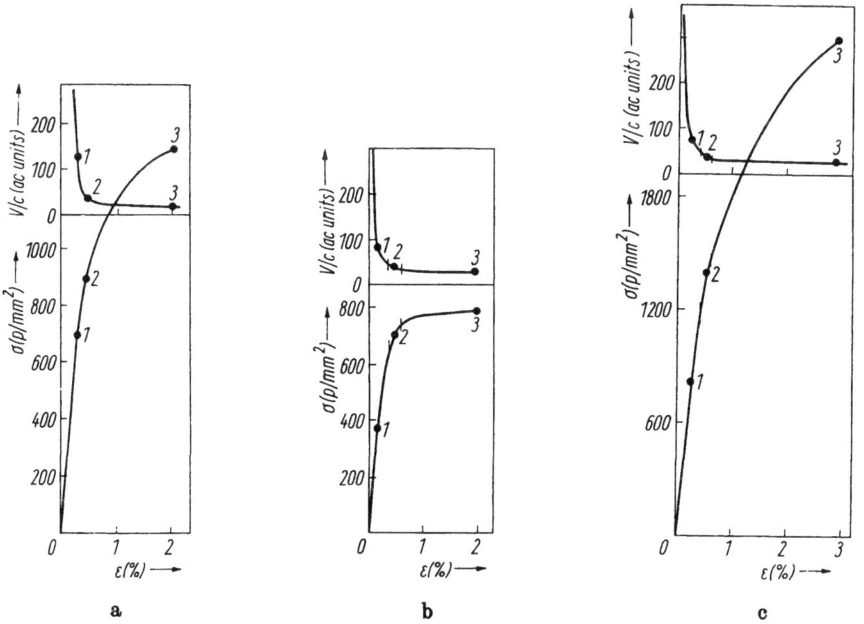

Fig. 2. Stress–strain curve at 77 °K a) for a glide; b) for c glide (orientation II); c) for c glide (orientation III). Spots labelled 1, 2, 3 represent the deformation state of the different observed samples

Fig. 3. *a* glide, sample 3, A face (31$\bar{4}$0), reflexion ⟨40$\bar{4}$0⟩

4. Experimental Results

4.1 a glide (orientation I in the standard triangle)

Topographs visualise glide planes and a rough polygonisation which becomes more and more effective as deformation proceeds from 1 to 3 on the deformation curve Fig. 2.

Face A of the most deformed sample shows a set of lines parallel to the *c* direction (Fig. 3). Some of these lines present locally a white edge which strongly suggests a rotation contrast around the [0001] direction. More precise studies with a small divergent beam reveal this rotation (Fig. 4a, b, c).

Face B of the same sample reveals only glide planes which appear as extinction contrast lines (Fig. 4b).

These observations are consistent with the presence of tilt subgrain boundaries parallel to the glide planes, their tilt axis being parallel to the *c* direction.

For smaller deformation ($\varepsilon = 0.3\%$, sample 2 on Fig. 2) the same tilt boundaries are observed in Fig. 5, but sample 1 (the less deformed one) exhibits the same contrast as for an unstrained sample (Fig. 6). Development of dislocation subgrain boundaries parallels the evolution of the activation volume [9] as shown in Fig. 2.

4.2 c glide (orientation II and III in the standard triangle)

Two cases have been studied for *c* glide. For orientation II, two of the three *c* glide systems have the same Schmid factor (the last one has a zero Schmid factor)

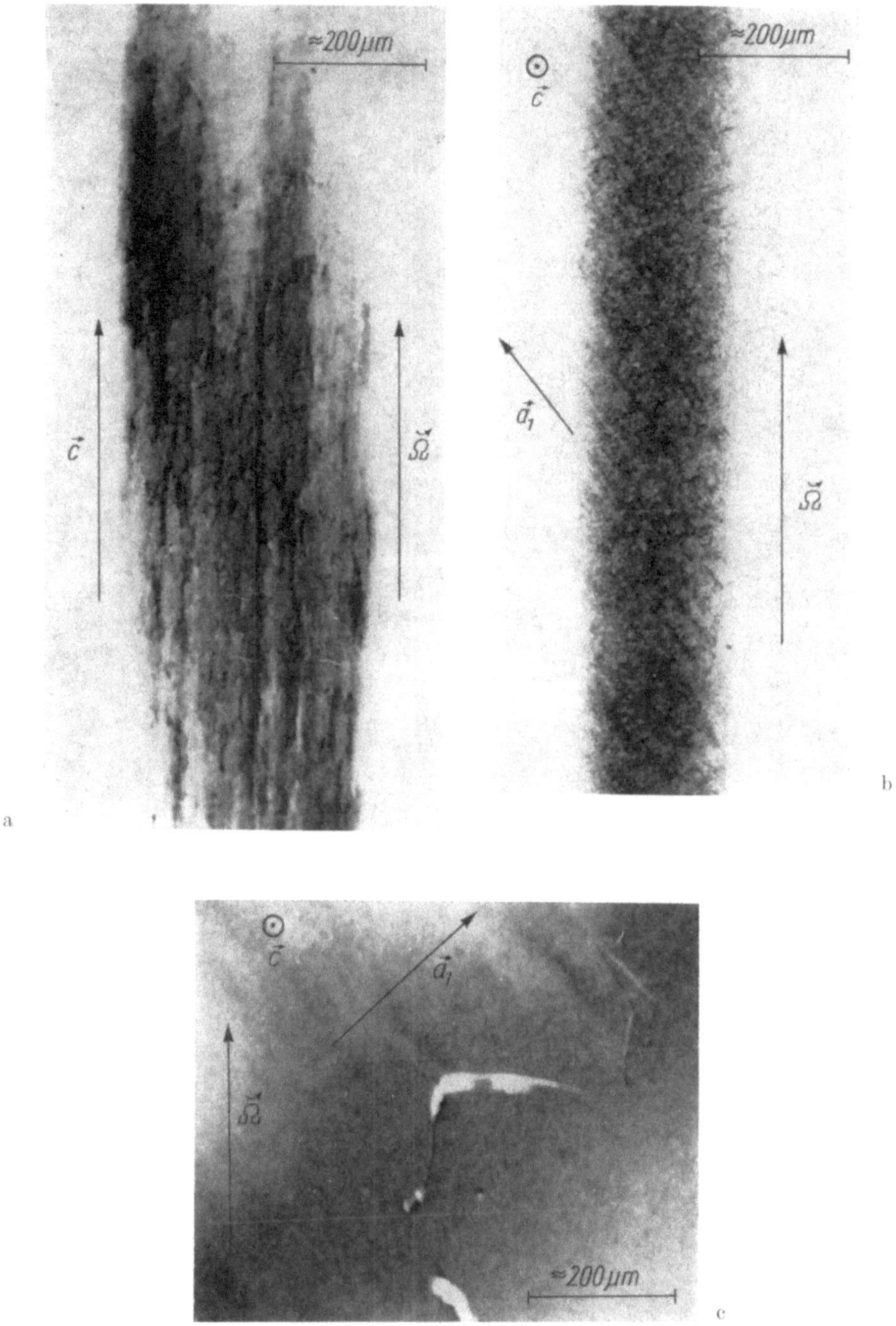

Fig. 4. Study of rotation contrast with a small divergency *a* glide; sample 3

Fig. 5. *a* glide; sample 2; A face (31$\bar{4}$0); reflexion ⟨40$\bar{4}$0⟩

Fig. 6. *a* glide; sample 1; A face (31$\bar{4}$0); reflexion ⟨40$\bar{4}$0⟩

Fig. 7. c glide, orientation II, sample 3; A face (01$\bar{1}$0); reflexion ⟨40$\bar{4}$0⟩

Fig. 8. c glide, orientation II, sample 3; B face ($\bar{8}$44$\bar{5}$); reflexion ⟨11$\bar{2}$3⟩

so that duplex glide is a priori possible in that case. For orientation III, one of the Schmid factors is larger than the two other ones.

We first comment on the results for orientation II. Topographs of the most deformed sample show (Fig. 7 and 8):

a) One set of thin and regularly spaced lines visualising glide planes. They indicate also that only one glide system has been activated. In our experiments

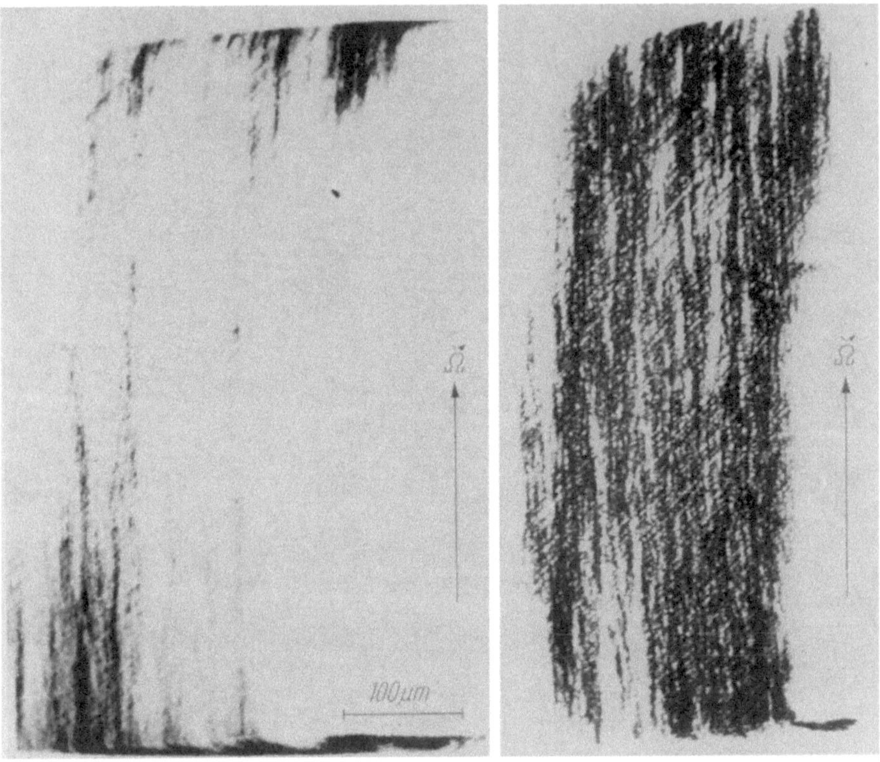

Fig. 9. Orientation II, sample 3; face B: Study of rotation contrast with a small divergency. 2′ tilt around the ⟨10$\bar{1}$0⟩ axis from a) to b)

Fig. 10. c glide, orientation II, sample 2; B face ($\bar{8}$44$\bar{5}$); reflexion ⟨11$\bar{2}$3⟩

Fig. 11. *c* glide, orientation II, sample 1; B face $(\bar{8}44\bar{5})$; reflexion $\langle 11\bar{2}3\rangle$

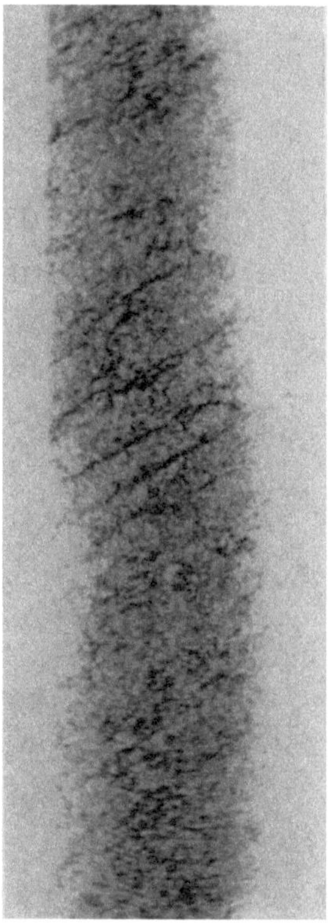

on orientation II, we always observed simple glide. This means that duplex glide is probably unstable.

b) Another set of thicker and more or less straight lines, along the traces of the basal plane, which indicate that polygonisation has begun to occur. Study of rotations confirms that these lines are tilt subgrain boundaries with tilt axis parallel to $[10\bar{1}0]$ (Fig. 9a and b) tilt amplitude is approximately 2′ large.

For smaller deformation in orientation II (sample 2) the polygonisation substructure is still observed, Fig. 10; sample 1 is like an unstrained crystal. As for *a* glide, substructure development parallels the activation volume [9] (Fig. 11).

Orientation III on the standard triangle corresponds to the maximum dissimilarity for the Schmid factors. Topographs show again the same kind of substructure as plastic deformation develops (Fig. 12). However, as compared with orientation II, the polygonisation seems to be somewhat smaller; for example it is not yet observed for the second sample on the stress–strain curve (Fig. 13).

5. Discussion

5.1 Nature of the observed polygonisation walls

Polygonisation generally develops in order to screen out long range stresses due to dislocation multiplication, so that the trend is toward finer and finer subgrain boundaries. Glide polygonisation is of course much rougher than annealing polygonisation, because dislocations are not allowed to climb, but the important point is that the same dislocation families are involved in both cases. The relief of long range stresses is reached only much more roughly and on a much more local scale than in the annealed substructure.

Therefore, we have looked for all the possible subgrain boundaries without long range stress fields, with the help of the Frank formula [10]. The following assumptions have been used:
(i) glide planes are prismatic planes of the first class $\{10\bar{1}0\}$; (ii) possible Burgers vectors are $\boldsymbol{c}$, $\boldsymbol{a}_1$, $\boldsymbol{a}_2$, and $\boldsymbol{a}_3 = -(\boldsymbol{a}_1 + \boldsymbol{a}_2)$; (iii) since a glide plane has only two possible Burgers vectors, which are perpendicular, only boundaries containing two dislocation families need to be considered; (iv) only plane boundaries are considered, which are defined by their normal $\boldsymbol{\nu}$ and their rotation axis $\boldsymbol{\Omega}$.

The results are summarised in Table 1 (see Appendix for details).

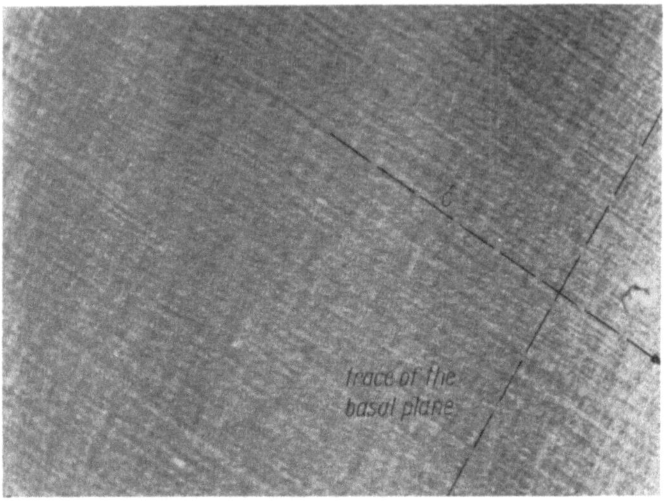

Fig. 12. c glide, orientation III, sample 3; Face (11$\bar{2}$0); reflexion ⟨12$\bar{3}$1⟩

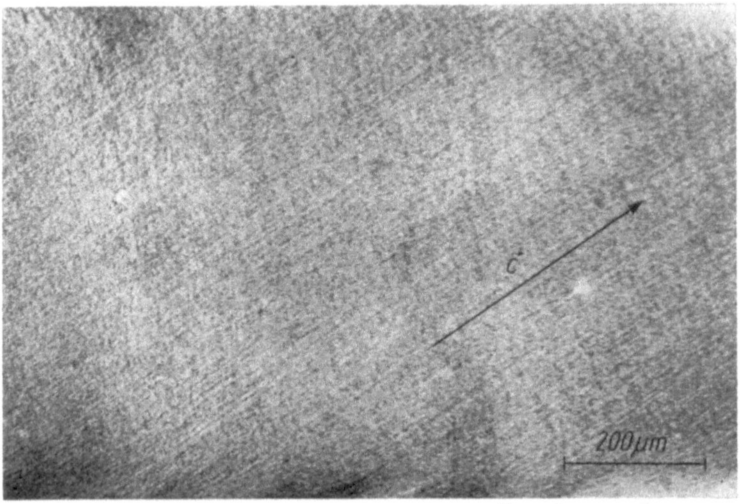

Fig. 13. c glide, orientation III, sample 2; Face (11$\bar{2}$0); reflexion ⟨12$\bar{3}$1⟩

5.1.1 Case of a glide

Polygonisation walls are found lying in the glide plane, with their tilt axis along the c direction:

$$\boldsymbol{\nu} = \frac{\boldsymbol{a}_1 \times \boldsymbol{c}}{|\boldsymbol{a}_1 \times \boldsymbol{c}|}, \qquad \boldsymbol{\Omega} = \frac{\boldsymbol{c}}{|\boldsymbol{c}|}.$$

Only one case in Table 1, case No. 5, fits with these values. This wall is built up with two parallel a dislocation families, of Burgers vectors $\boldsymbol{a}_1$ and $\boldsymbol{a}_2$:

a) One family is made of a_1 edge dislocations, glissile in the wall plane. These

Table 1

Possible subgrain boundaries without long range stress field in tellurium. Unit vectors $\boldsymbol{a}_1'$ and $\boldsymbol{a}_2'$ are vectors in the basal plane perpendicular to $\boldsymbol{a}_1$ and $\boldsymbol{a}_2$ lattice vectors, respectively.

	type	$\boldsymbol{v}$	$\boldsymbol{\Omega}$	dislocation families	density	nature
1	1 Burgers vector	$\boldsymbol{a}$	$\boldsymbol{c}$	one family of a edge dislocations	$\varrho = \theta\, a^{-1}$	tilt
2	1 Burgers vector	$\boldsymbol{c}$	$\boldsymbol{a}$	one family of c dislocation	$\varrho = \theta\, c^{-1}$	tilt
3	1 Burgers vector	$\boldsymbol{c}$	somewhere in (0001)	two families of c dislocations, lines a_1 and a_2	$\varrho_{c\,\mathrm{I}} = \frac{2\,\theta}{\sqrt{3}\,c} \sin(\boldsymbol{a}_2, \boldsymbol{\Omega})$ $\varrho_{c\,2} = \frac{2\,\theta}{c\sqrt{3}} \sin(\boldsymbol{a}_1, \boldsymbol{\Omega})$	tilt
4	2 Burgers vectors, dislocations having parallel lines	in $(10\bar{1}0)$ neither $\boldsymbol{a}$ nor $\boldsymbol{c}$	at the intersection of the boundary and the $(10\bar{1}0)$ plane	two families of Burgers vectors $\boldsymbol{a}$ and $\boldsymbol{c}$, having the same glide plane	$\varrho_c = \frac{\theta}{c^2}(\boldsymbol{v}, \boldsymbol{c})$ $\varrho_a = \frac{\theta}{a^2}(\boldsymbol{v}. \boldsymbol{a})$ $\varrho_{\mathrm{T}} = \varrho_a + \varrho_c$	tilt
5	2 Burgers vectors, line dislocations having parallel lines	in (0001) neither $\boldsymbol{a}_1$ nor $\boldsymbol{a}_2$	$\boldsymbol{c}$	two families of edge dislocations, Burgers vectors $\boldsymbol{a}_1$ and $\boldsymbol{a}_2$	$\varrho_{a_1} = \frac{2\,\theta}{a\sqrt{3}}(\boldsymbol{v}, \boldsymbol{a}_2')$ $\varrho_{a_2} = \frac{2\,\theta}{a\sqrt{3}}(\boldsymbol{v}, \boldsymbol{a}_1')$	tilt
6	2 Burgers vectors, dislocations having non-parallel lines	$\boldsymbol{a}_1'$	$\boldsymbol{a}_2'$	two families of screw dislocations, Burgers vectors $\boldsymbol{a}$ and $\boldsymbol{c}$	$\varrho_c = \theta\, c^{-1}$ $\varrho_a = \theta\, a^{-1}$	twist

are slip dislocations. Their density as expressed in dislocations per unit wall length is $\varrho_1 = \theta/a\sqrt{3}$, where θ is the tilt angle.

b) The other family is made of a_2 edge dislocations with density $\varrho_2 = 2\,\theta/a\sqrt{3} = 2\,\varrho_1$.

This kind of wall probably forms in the following way. Primary a slip develops numerous a_1 dislocations, mainly in edge orientation and hard to move due to strong Peierls forces (see Section 3). These dislocations fill up their slip plane, so building up long range stresses. It happens that such stresses are locally relieved when attracting neighbouring a_2 edge dislocations in those regions where they are available, gliding on their own prismatic plane. Note that this secondary a is favoured by applied stresses too. Thus Schmid factors are respectively 0.5 and 0.25 for primary slip and for the two other a slip systems in orientation I.

Finally it has to be emphasized how locally and somewhat erratically these walls are observed along slip lines. Indeed, this is consistent with our model in which an exact balance between a_1 and a_2 dislocations ($\varrho_2 = 2\,\varrho_1$) can only be achieved partially and locally, from place to place.

5.1.2 Case of c glide

Glide polygonisation occurs much more efficiently and much more homogeneously during c glide than during a glide. Here, walls are found lying in basal plane, with their tilt axis along a_1, the direction which is common to slip plane and to basal plane. This kind of wall is the case No. 2 Table 1; it is simply a wall of parallel c edge dislocations (the usual tilt boundary) with density $\varrho = \theta/c$.

It is easy to understand how readily such a wall forms from mobile edge parts of c loops left in the crystal. It is another problem to wonder why long sessile c screws which should be found very frequently after a few percent strain (Section 3), do not seem to play any role in the deformation polygonisation.

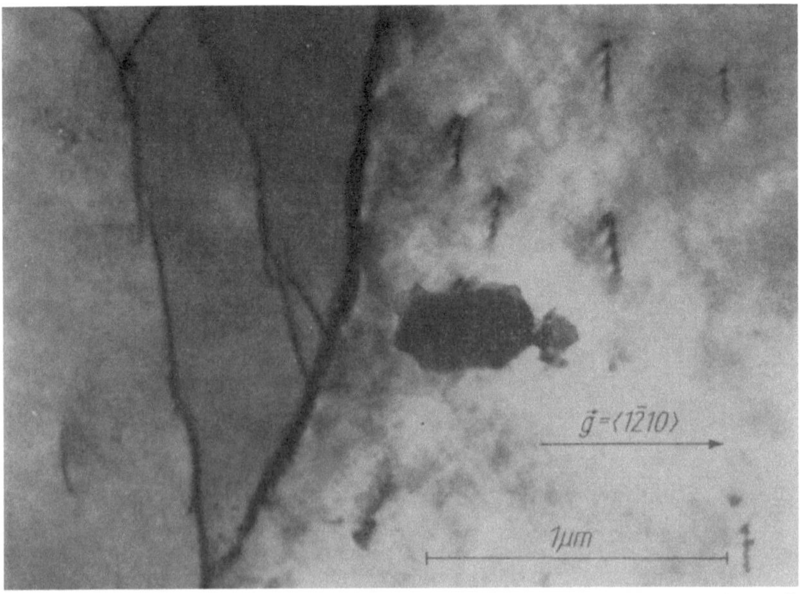

Fig. 14. a glide, 293 °K, γ = 15%, electron micrograph showing two kinds of a dislocations (plane (10$\bar{1}$0))

It may be because c screws are only screened by those a screw dislocations (case No. 6 Table 1) which are not easily produced due to their high core energy (Section 3), but for the time being this point is not very clear.

5.2 Connections with electron microscopy observations

Dislocation configurations have been observed after a few percent strain by transmission electron microscopy for a and c glide.

For a glide, the two-family walls have not been observed after 1 or 2% strain at 77 °K. However, these subgrain boundaries are frequently seen on samples 15% strained at room temperature (Fig. 14). Low temperature and small deformation are less favourable conditions for wall formation. On Berg-Barrett topographs also, they only appear in localised regions along some glide bands, consistently with the fact that they could be missed in electron microscopy.

For c glide, polygonisation is more effective and there is a good agreement between X-ray topographs and electron micrographs. A wall of c edge dislocations is shown in Fig. 15. Crude evaluation of dislocations in the walls gives a density of $10^8\,cm^{-2}$. Half of these dislocations are dipoles the symmetrical contrast of which is easily recognized. For the other half, densities of $+c$ and $-c$ are approximately equal. It is to be noticed that a 10% excess of one type ($+c$ for example) is enough to produce a tilt rotation of about 1′ which is the mean misorientation observed on X-ray topographs.

Electron micrographs show the roughness of the "subgrain boundaries" which should be better described as thin layers of high dislocation densities the thickness of which may be 1 to 5 μm. Inside the layer localised excess of one type of dislocations gives rise to the misorientations observed on Berg-Barret topographs.

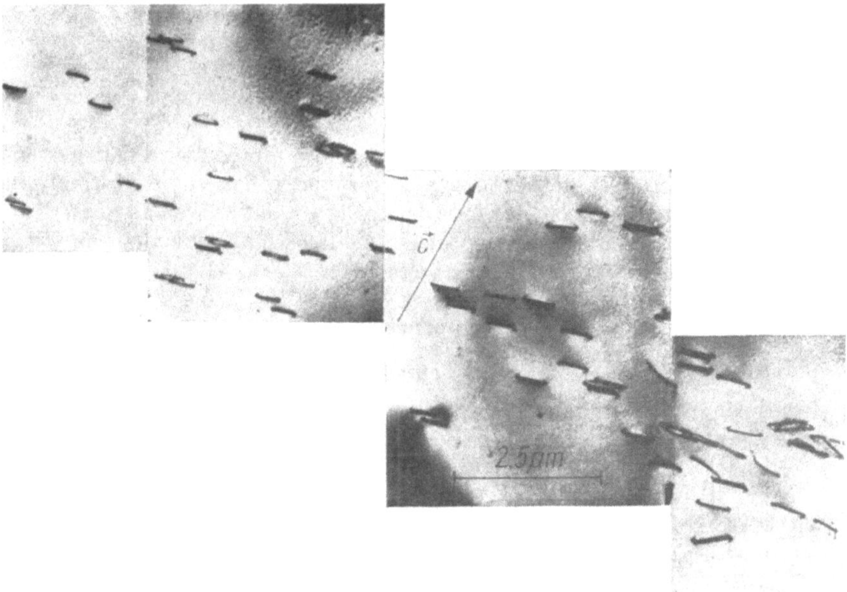

Fig. 15. c glide, 77 °K, $\gamma = 2\%$ (orientation II), electron micrograph showing a c edge dislocation wall (plane $(11\bar{2}0)$)

On the other hand, etch pits reveal the slip lines but fail to show polygonisation walls probably too diffuse to be clearly evidenced by this technique.

5.3 Remarks

Tilt boundaries which appear for low deformations do not necessitate a high mean dislocation density. Topographs show that spacing between walls is 15 to 25 μm while their rotation angle is $\theta \approx 10^{-3}$ rad corresponding to a linear dislocation density $\theta/b \approx 10^4$ cm^{-1} (all these dislocations in the wall have the same sign). A short computation leads to 10^7 cm^{-2} as a minimum mean dislocation density in the deformed sample.

Polygonisation walls appear just before the initiation of a new stage with linear workhardening. Wilkens [2] also observed similar polygonisation just before stage II in deformed copper.

6. Conclusion

The Berg-Barrett technique is a non-destructive test which allows a very simple identification of slip systems. Moreover it has been possible with this technique to follow throughout the sample the rough polygonisation which appears in tellurium in the first stage of plastic deformation.

This polygonisation does not develop in the same way for the three orientations examined. For example walls are already numerous at 2% strain for orientation II (**c** glide) while for orientation III (**c** glide too), walls only begin to appear at the same 2% strain. This point emphasizes that knowledge of deformation alone is not sufficient information for defining uniquely the plastic state of the crystal and a conventional elastic limit (flow stress at a given ε, 0.2% for example) should have no physical meaning.

On the contrary we observed that polygonisation never appears before the activation volume has reached its minimum value (plateau) so that we can think of the elastic limit as that flow stress which corresponds to the volume minimum [9]; i.e. to the beginning of polygonisation.

This physical definition is fundamental for other studies such as the influence of temperature, purity, etc. on the elastic limit.

The origin of polygonisation walls in Te at low temperature can be partly understood. It is clear that these walls correspond to layers of relatively high dislocation densities which cluster in order to screen out as well as possible long range stress fields developed by slip dislocations. On the level of the sample, these layers can be assimilated to tilt boundaries.

Appendix

Possible subgrain boundaries in tellurium

Possible subgrain boundaries without long range stresses are investigated, using the simple geometrical Frank theory [10] which is based on the equation

$$\boldsymbol{d}(\boldsymbol{r}) = (\boldsymbol{r} \times \boldsymbol{\Omega})\, \theta\, .$$

In the equation, θ is the relative misorientation between two subgrains, $\boldsymbol{\Omega}$ the rotation axis, $\boldsymbol{r}$ any vector chosen in the boundary plane, and $\boldsymbol{d}$ is the $\boldsymbol{r}$-dependent vector defined as

$$\boldsymbol{d} = \sum_i n_i\, \boldsymbol{b}_i\, .$$

Table 2

	one family boundaries		two family boundaries								
Burgers vectors	c	a_1	c, a_1				a_1, a_2		a_1, a_1	c, c	
unit line vectors	t	t	$t_c = t_a$		$t_c \neq t_a$		$t_1 = t_2$	$t_1 \neq t_2$	$t_1 \neq t_2$	$t_1 \neq t_2$	
general properties	$\Omega = t$, $\nu \parallel c$	$\Omega = t$, $\nu \parallel a_1$	$\Omega = t$, $\nu a_1' \neq 0$		$\Omega = a_1'$, $\nu a_1' \neq 0$	$t_c a_1 = 0$, $t_a c = 0$	$\Omega = t$, $\nu c = 0$	$\Omega \parallel c$, $\nu c \neq 0$; $t_1 a_2 = 0$, $t_2 a_1 = 0$	$\Omega a_1 = 0$, $\nu = a_1$	$\Omega c = 0$, $\nu = c$	
unit normal of the prismatic glide planes	a_1'	a_1'	a_1', a_1'	a_1', a_2'	a_1', a_1'	a_1', a_2'	a_1', a_1'	a_1', a_2'	a_1', a_1'	a_1', a_1'	a_1', a_2'
particular properties	$t \parallel a_1$	$t \parallel c$	—	$\Omega = t \parallel c$	$\nu = \Omega = a_1'$		$\Omega = t \parallel c$	$t_1 = t_2 \parallel c$ impossible —	$t_1 = t_2 \parallel c$ impossible —	$t_1 = t_2 \parallel a_1$ impossible —	$t_1 \parallel a_1$, $t_2 \parallel a_2$
boundary plane	basal plane	normal to prismatic glide plane	normal to prismatic glide plane	$\nu \parallel a_1$	prismatic glide plane	$\nu = a_1'$*)	in zone with c				basal plane
dislocation character	edge	edge	mixed dislocation	$f_c = 0$	two perpendicular screw families		two edge families				two edge families
boundary character	pure flexion	pure flexion	pure flexion		pure twist		pure flexion				pure flexion
number	2	1	4	1	6		5				3

*) In this case it can be shown [7] that $\nu \parallel (a_1' \times c) \times (a_2' \times a_1)$, i.e. $\nu = a_1'$.

Here, summation is extended over all the dislocations which are crossed over by $\boldsymbol{r}$ in the boundary plane, among which a number n_i has Burgers vector $\boldsymbol{b}_i$.

Systematical investigation is summarized in Table 2. The subgrain boundary is defined by its unit normal $\boldsymbol{\nu}$ and its rotation axis $\boldsymbol{\Omega}$. A dislocation family is defined as a set of parallel identical dislocations of unit line vector $\boldsymbol{t}$, and of Burgers vector $\boldsymbol{c}$ or $\boldsymbol{a}$; these vectors are represented in the two first lines of Table 2 for all the various possibilities, together with corresponding general properties for the grain boundary they form, in the third line, as given in [10]. In the fourth line are given the possible combinations of slip planes of these families; slip planes are represented by their unit normal for each of the families; because slip planes are assumed to be only prismatic slip planes this unit normal is defined as

$$\boldsymbol{a}_i' = \frac{\boldsymbol{c} \times \boldsymbol{a}_i}{|\boldsymbol{c} \times \boldsymbol{a}_i|}, \qquad i = 1, 2$$

$\boldsymbol{a}_1$ and $\boldsymbol{a}_2$ being the 120° basal lattice vectors, $\boldsymbol{a} = \frac{1}{3}\langle 11\bar{2}0\rangle$. Under the assumption that subgrain boundaries form only by gliding, giving the slip planes allows generally the determination of $\boldsymbol{t}$ vectors since $\boldsymbol{t}$ must lie in the boundary ($\boldsymbol{\nu}\,\boldsymbol{t} = 0$ if $\boldsymbol{\nu}$ is the boundary normal) and in the slip plane ($\boldsymbol{a}_i'\,\boldsymbol{t} = 0$) at a time. Finally, dislocation densities can be computed for each family in the boundary. In the two family case of Burgers vectors $\boldsymbol{b}_1$ and $\boldsymbol{b}_2$ one has

$$\varrho_1 = \frac{|\boldsymbol{\nu} \times \boldsymbol{b}_2|}{|\boldsymbol{b}_1 \times \boldsymbol{b}_2|} \quad \text{and} \quad \varrho_2 = \frac{|\boldsymbol{\nu} \times \boldsymbol{b}_1|}{|\boldsymbol{b}_1 \times \boldsymbol{b}_2|}.$$

References

[1] J. B. NEWKIRK, Trans. AIME **215**, 483 (1959).
[2] M. WILKENS, Canad. J. Phys. **45**, 567 (1967).
[3] A. VON IPPEL, J. chem. Phys. **16**, 372 (1948).
[4] J. S. BLAKEMORE, D. LONG, K. C. NOMURA, and A. NUSSBAUM, Progr. Semicond. **6**, 37 (1962).
[5] P. GROSSE, Springer Tracts in Modern Physics, Vol. 48, Chap. I, Springer-Verlag, 1969.
[6] J. DI PERSIO, J. C. DOUKHAN, and G. SAADA, phys. stat. sol. **42**, 281 (1970).
[7] J. DI PERSIO, J. C. DOUKHAN, and G. SAADA, phys. stat. sol. **42**, 297 (1970).
[8] J. C. DOUKHAN and B. ESCAIG, phys. stat. sol. (a) **7**, 441 (1971).
[9] J. CRAMPON, to be published (1972).
[10] F. C. FRANK, Rep. Symp. Plastic Deformation of Crystalline Solids, Carnegie Institute of Technology, Pittsburg 1950 (p. 150).

(Received June 23, 1972)

phys. stat. sol. (a) 14, 639 (1972)

Subject classification: 1.1 and 8; 22.8

Solid State Physics Laboratory, Delhi

Variation of Lattice Parameter with Temperature and Thermal Expansion of the Compound Cu_2SnSe_3

By

B. B. SHARMA and F. R. CHAVADA

The temperature variation of the lattice parameter of the compound Cu_2SnSe_3 has been studied in the temperature range 30 to 600 °C by X-ray diffraction methods. The linear coefficient of thermal expansion at room temperature (30 °C) is 20.4×10^{-6} deg^{-1} and it is found to decrease with increasing temperature.

Die Temperaturänderung des Gitterparameters der Verbindung Cu_2SnSe_3 wurde im Temperaturbereich 30 bis 600 °C durch Röntgenstrahlenbeugungsmethoden untersucht. Der lineare Koeffizient der thermischen Ausdehnung bei Zimmertemperatur (30 °C) beträgt $20{,}4 \times 10^{-6}$ grd^{-1} und nimmt mit steigender Temperatur ab.

1. Introduction

Cu_2SnSe_3 belongs to a group of compounds represented by the general formula $A_2^{I}B^{IV}C_3^{VI}$ (superscripts denote group in the periodic table and subscripts the number of atoms). These compounds are ternary analogues of IV group elements having tetrahedral bonding [1]. In an earlier paper [2] we reported the temperature variation of lattice parameters for another compound, Cu_2GeSe_3, belonging to the same group. In this communication we report some results of lattice parameter measurements for Cu_2SnSe_3 in the temperature range 30 to 600 °C.

This compound is known to have a sphalerite type lattice but there is some discrepancy in the reported values of the lattice constant [3, 4]. The value of the thermal expansion coefficient α obtained in the present experiment by X-ray diffraction methods is much larger as compared to the reported [5, 6] value obtained by dilatometric methods.

2. Experimental

The compound was synthesized from spectroscopically pure elements obtained from Johnson-Matthey UK. The constituent elements taken in stoichiometric ratio were sealed in an evacuated (pressure $\approx 10^{-4}$ Torr) quartz capsule which was kept at 1000 °C for 6 h with a continuous vibration mixing arrangement. The ingot obtained after cooling the melt was cut and polished. A microscopic examination of the polsihed surface and microhardness test at different places confirmed that the material is single phase. The microhardness number obtained with a PMT-3 (Russian) instrument is (250 ± 10) kp/mm^2. This is in good agreement with [7]. A cooling curve was also drawn and it gave a sharp freezing point at 694 °C, in agreement with [4].

For X-ray powder diffraction work a part of the ingot was powdered and passed through a 400 mesh sieve. This powder was annealed at 350 °C for 100 h. The lattice constant at room temperature was obtained using a 114.46 mm diameter Debye-Scherrer camera employing asymmetric mounting of the film. High temperature work was done on a 19 cm diameter Unicam camera. The procedure followed for temperature measurement was the same as described earlier [2]. All experiments were performed with CuK_α radiation.

3. Results

The line quality in the high-angle region of the powder pattern was not very good. Even at room temperature only one $\alpha_1\,\alpha_2$-doublet occurring at 75° Bragg anlge could be resolved. With increasing temperature the line quality rapidly deteriorated and no line beyond 60° Bragg angle could be measured above 200°C.

All the lines could be indexed on the base of a cubic unit cell with $a = (5.6877 \pm \pm 2)$ Å. The lattice parameters were calculated by least squares analysis using all lines beyond 30° Bragg angle. The extrapolation function used was

$$\frac{1}{2}\left(\frac{\cos^2\theta}{\sin\theta} + \frac{\cos^2\theta}{\theta}\right).$$

The probable error in the lattice parameter was estimated in the following way. All the Bragg angles were corrected using the value of the drift constant obtained from least squares analysis. The lattice parameter was then obtained from each individual powder line. The standard deviation of these lattice parameter values from that obtained from least squares analysis was taken as a measure of the probable error.

Fig. 1 shows a plot of lattice parameter values against temperature. The estimated error has been shown by arrow marks. The observed points in Fig. 1 were fitted to a parabolic curve, equation (1), by least squares analysis:

$$a_T = 5.6851\,(1 + 2.083 \times 10^{-5}\,T - 6.922 \times 10^{-9}\,T^2)\,, \tag{1}$$

where a_T is the lattice parameter at T °C. The curve in Fig. 1 is actually a plot of equation (1). The linear coefficient of thermal expansion at different temperatures was obtained from the relation $\alpha_T = (\mathrm{d}a_T/\mathrm{d}T)/a_T$ by differentiating equation (1). Fig. 2 shows a plot of α versus temperature. The values of lattice

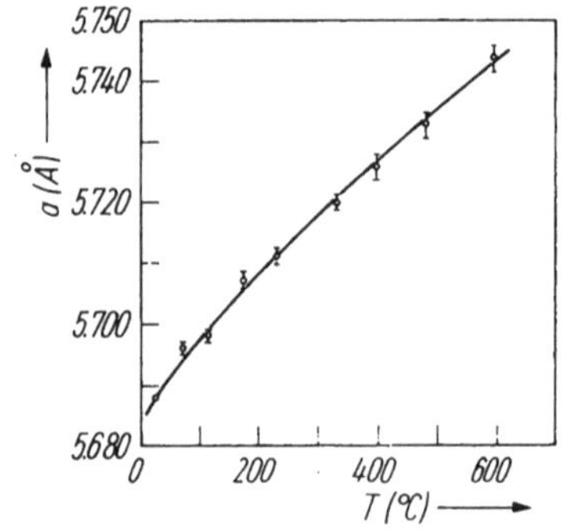

Fig. 1. Variation of lattice constant a with temperature

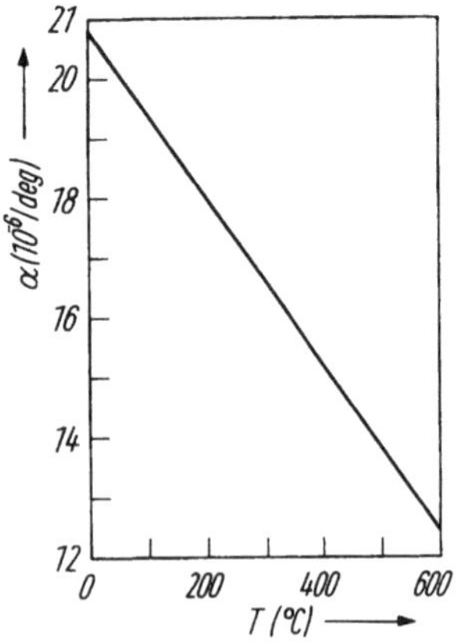

Fig. 2. Variation of α with temperature

Fig. 3. Variation of unit cell volume with temperature

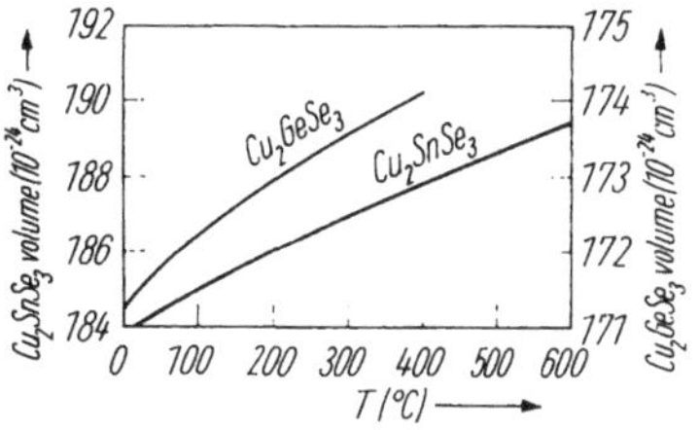

parameter a and thermal expansion coefficient at room temperature (30 °C) are given below:

$$a = (5.6877 \pm 2)\ \text{Å}\,,$$

$$\alpha = 20.4 \times 10^{-6}\,\text{deg}^{-1}\,.$$

The reported value of α determined from dilatometric measurements [5, 6] is $8.9 \times 10^{-6}\,\text{deg}^{-1}$.

4. Discussion

The results of the present study on Cu_2SnSe_3 are similar to those obtained earlier for Cu_2GeSe_3 [2]. In both cases α falls with increasing temperature. For the sake of comparison, the unit-cell volumes of the two compounds (it may be recalled that Cu_2GeSe_3 has a tetragonal symmetry while Cu_2SnSe_3 is cubic) have been plotted as functions of temperature in Fig. 3. The data for Cu_2GeSe_3 have been taken from our earlier work [2]. In case of Cu_2GeSe_3, the decrease of α with increasing temperature was attributed to the possible existence of a superstructure in the lattice, a fact which has since been confirmed [8]. In the case of Cu_2SnSe_3 also a superstructure has been reported [3] but we did not observe any superstructure lines in the powder photographs even after prolonged exposures. The density measurements show that the cubic unit cell with $a = 5.6877$ Å contains four units with the formula 1/3 (Cu_2SnSe_3), indicating random distribution of Cu and Sn atoms. Normally α is expected to increase with temperature because of the increased lattice anharmonicity. However, in many simple tetrahedral compounds and elements α is known to behave anomalously [9] especially at low temperatures. The ternary compounds investigated by us, also have tetrahedral bonding but present a very complicated system for theoretical analysis, especially in the absence of sufficient experimental data on their thermal and elastic properties.

No convincing explanation could be found for the very large diference between the values of α obtained from X-ray diffraction methods and the values obtained from dilatometric methods reported earlier. A similar discrepancy was reported earlier for Cu_2GeSe_3 [2].

Acknowledgements

We are very grateful to Dr. V. V. Agashe and Dr. A. K. Sreedhar for many valuable suggestions and to Mr. Hari Singh for his help in measurements. We also express our sincere gratitude to the Director of the Solid State Physics Laboratory for his encouragement during the work and permission to publish it.

References

[1] N. A. GORYUNOVA, Chemistry of Diamond Like Semiconductors, Chapman and Hall, Ltd., 1965.
[2] B. B. SHARMA, phys. stat. sol. (a) **2**, K13 (1970).
[3] W. B. PEARSON, Hand Book of Lattice Spacings and Structure of Metals and Alloys, Vol. 2, Pergamon Press, London 1967 (p. 886).
[4] G. K. AVERKIEVA et. al., Soviet Research in New Semiconducting Materials, Ed. D. N. NASLEDOV and N. A. GORYUNOVA, Consultants Bureau, 1965 (p. 26).
[5] L. I. BERGER and A. E. BALENSKAYA, Soviet Phys. — Solid State **6**, 1023 (1964).
[6] R. A. ANNAMAMEDOV et. al., Chemical Bonds in Semiconductors and Thermodynamics, Ed. N. N. SIROTA, Consultants Bureau, 1968 (p. 240).
[7] K. HIRONO and M. KONO, Japan. J. appl. Phys. **7**, 54 (1968).
[8] E. PARTHE and J. GARIN, Monatshefte für Chemie **102**, 1197 (1971).
[9] S. I. NOVIKOVA, Semiconductors and Semimetals, Vol. 2, Ed. R. K. WILLARDSON and A. C. BEER, Academic Press, London 1966 (p. 33).

(Received August 18, 1972)

phys. stat. sol. (a) 14, 643 (1972)

Subject classification: 20.1; 22.9

Department of Chemistry, University of Queensland, St. Lucia

Optical Parameters from Reflectance Data

A New and Versatile Approach

By

M. G. SCEATS and G. C. MORRIS

A method is described which can replace both a Kramers-Kronig and a dispersion analysis as a means of obtaining optical parameters from reflectance data of condensed systems. It has advantages whenever a large number of data points must be sampled and when the absorption band profile is used to interpret quasiparticle dynamics. The efficiency of the method arises because evaluation of the phase shift upon reflection is better performed by multiplication in the time domain than by convolution in the frequency domain as is done by a Kramers-Kronig method. The efficiency of the method is illustrated and using reflectivity data of an anthracene crystal, the errors in its use, while still comparable to the less efficient traditional methods are shown to be insignificant compared with the experimental errors in such data.

Es wird über eine Methode berichtet, die eine Kramers-Kronig-Inversion sowie eine Dispersionsanalyse ersetzen kann, um optische Parameter aus Reflexionsdaten von kondensierten Systemen zu bestimmen. Es ergeben sich Vorteile, wenn eine große Anzahl von Datenpunkten behandelt werden muß, und wenn man die Form der Absorptionsbanden benutzt, um die Dynamik von Quasipartikeln zu interpretieren. Die Leistungsfähigkeit der Methode besteht darin, daß die Phasenverschiebung bei der Reflexion besser ausgeführt wird durch Multiplikation im Zeitgebiet als durch Konvolution im Frequenzgebiet wie bei der Kramers-Kronig-Methode. Beim Benutzen der Reflexionsdaten eines Anthrazenkristalls wird die Leistungsfähigkeit der Methode gezeigt und obgleich Fehler, die mit den älteren und weniger wirksamen Methoden vergleichbar sind, noch vorhanden sind, sind diese im Vergleich mit den Versuchsfehlern der Originaldaten von geringer Bedeutung.

1. Introduction

The frequency dependence of the complex dielectric permittivity $\varepsilon(\omega)$ provides an effective probe of both the coupling processes in the solid of the photon–crystal eigenmodes and of the scattering processes undertaken by the exciton or polariton [1, 2]. To obtain $\varepsilon(\omega)$ when the photon–exciton are strongly coupled, the primary experimental determination is the frequency dependence of the normal-incidence reflection coefficient $R(\omega)$ over a limited frequency range because

$$\varepsilon(\omega) = \left(\frac{1 + \sqrt{R(\omega)}\, \mathrm{e}^{-i\theta(\omega)}}{1 - \sqrt{R(\omega)}\, \mathrm{e}^{-i\theta(\omega)}}\right)^2, \tag{1}$$

where $\theta(\omega)$ is the phase shift upon reflection. The reduction of the $R(\omega)$ data to the real and imaginary parts of $\varepsilon(\omega)$ has previously been achieved by either a Kramers-Kronig transform [3, 4] or a dispersion analysis [5 to 7]. Such methods have disadvantages in many cases traceable to the fact that $\varepsilon(\omega)$ is a very strong function of $R(\omega)$. In particular if narrow structure in $R(\omega)$ accompanies broad structure, the narrow structure in $\varepsilon(\omega)$ does not appear as the result of either method unless a large number of data points are sampled assuming Nyquist sampling. Such structure has been observed in the ultraviolet reflection spectra of molecular crystals [8, 9] and of semiconductors [10, 11] and in the infrared reflection spectra of a number of materials [7, 12, 13]. Ideally in transforming $R(\omega)$ to $\varepsilon(\omega)$, the data points should be sampled with a resolution consistent with that of the spectrometer used. But this could often mean a sample of some 10^3 to 10^4 data points from the spectrum. Experimentally this is not difficult as the $R(\omega)$ spectrum is usually produced on a chart recorder, but tedious manual procedures are often necessary to convert this analogue data into digital form for computer processing. A preferable method is to perform such a conversion electronically and store the digital data on tape [14]. The greater wealth of data points (about 10^4 for the crystal systems we have been investigating) requires an efficient algorithm to use the data without escalating computer costs and to ensure that even a large computer can handle the processing effectively.

As part of our studies on exciton dynamics in molecular crystals we have developed an approach to analysis of reflectance data which is of general applicability in cases where a Kramers-Kronig transform or a dispersion analysis have been used previously. In many circumstances where a large number of samples is required, e.g. when spectral fine structure is present, our method is advantageous in maximizing efficiency and accuracy. Further, the method is versatile in that it is of use in the common situation where the spectral data is used to investigate scattering processes within the solid. Information on such processes is contained in the band profile of $\varepsilon(\omega)$. Previously analysis [15] of the temperature dependence of parameters describing this profile has been used to interpret the processes occurring by comparison of the results with those expected from simple models, e.g. $\varepsilon(\omega)$ would have either a Lorentzian or a Gaussian profile dependent on whether the exciton–phonon coupling is either weak or strong. However information on exciton dynamics may be better deduced by use of correlation functions [16, 17]. If $\varepsilon(\omega)$ is known, then the correlation function $C(t)$ describing the dynamics of the absorption process is

$$C(t) = \int_{-\infty}^{\infty} (\omega' + \omega_0) \operatorname{Im} [\varepsilon (\omega' + \omega_0)] \, e^{i\omega' t} d\omega' , \tag{2}$$

where $\omega' = \omega - \omega_0$, and ω_0 is the average frequency $\int_{-\infty}^{\infty} \omega \operatorname{Im} [\varepsilon(\omega)] \, d\omega$. The correlation function may be analysed statistically or compared to results generated from theory since $C(t)$ is closely related to the double time retarded Green's function of the system [18]. Comparison with theory in the time domain relaxes requirements that the theoretical results have simple Fourier transforms. The same Fourier transform algorithm used to generate $\varepsilon(\omega)$ from $R(\omega)$ can be used to generate $C(t)$ from $\varepsilon(\omega)$.

The basis of the method we propose is the Fourier transformation of the $R(\omega)$ data to the time domain, wherein the data is desymmetrized. This allows, on return to the frequency domain, the generation of a phase shift. Whilst in the time domain spectral noise may be eliminated by filtering. The same algorithm allows deconvolution with the instrument function to be simply applied. Along with the phase shift, the corrected and smoothed $R(\omega)$ may be generated.

In this paper, we briefly describe the current methods of analysis, consider our proposed method, and illustrate its use and advantages over previous methods by an application.

2. Current Methods and Their Efficiency

2.1 Kramers-Kronig analysis

The Kramers-Kronig transform

$$\theta(\omega) = -\frac{\omega}{\pi} \int_0^\infty \frac{\ln R(\omega')}{\omega'^2 - \omega^2} \, d\omega' \tag{3}$$

allows the calculation of $\theta(\omega)$ from $R(\omega)$ and hence $\varepsilon(\omega)$ using (1) provided $R(\omega)^{1/2} e^{-i\theta(\omega)}$ describes both a linear and a causal response. The effects of spatial dispersion [19] or bulk inhomogeneities [20] may invalidate the transform which may be used to probe such effects. The integral in (3) is usually calculated from the sampled points using Simpson's rule in the region of interest (say $\omega = \omega_a$ to $\omega = \omega_b$). If N points are sampled, a minimum of about N^2 operations are required to evaluate the integral. In cases where fine structure is present N is of the order 10^3 or 10^4, making excessively large computing costs.

2.2 Dispersion analysis

The second method used previously is dispersion analysis. An artificial reflectance $R_{calc}(\omega)$ is calculated from models of $\varepsilon(\omega)$ which contain adjustable parameters. The error function $\sum_{n=1}^{N} [R(\omega_n) - R_{calc}(\omega_n)]^2$ is minimized using optimization theory [21]. A number of models may be used, e.g. Lorentzian [5, 6], Gaussian [22], and general dispersion relationships [7]. Problems associated with this analysis are many [21], e.g. minimization of the number of parameters, multimodality of the error function, constraint to causal solutions, symmetrization of gradients of the error function for all parameters, mutual interaction of parameters, initial estimate of parametres, efficiency of the optimization program, nature of the error hypersurface and applicability of the model used. It is difficult to estimate the efficiency of a dispersion analysis but others [7] have found it preferable to a Kramers-Kronig analysis of certain reststrahlen data.

3. Proposed Method of Analysis

3.1 Theory

Consider a weak electromagnetic field $\boldsymbol{E}(t)$ which induces an electric displacement $\boldsymbol{D}(t)$ in the crystal. The value of $\boldsymbol{D}(t)$ depends [23] on all values of $\boldsymbol{E}(t')$ for $t' \leq t$. If $F(t)$ is the response function of the system, then

$$\boldsymbol{D}(t) = \boldsymbol{E}(t) + \int_0^\infty F(t') \, \boldsymbol{E}(t - t') \, dt' . \tag{4}$$

For an isotropic system $F(t)$ is a causal function obeying

$$F(t) = 0 \qquad \text{for } t < 0 ,$$

$$F(t) \quad \text{real, finite} \qquad \text{for } t > 0 .$$

Taking Fourier components

$$D(\omega) = \int_{-\infty}^{\infty} D(t)\, e^{-i\omega t}\, dt\,,$$

$$E(\omega) = \int_{-\infty}^{\infty} E(t)\, e^{-i\omega t}\, dt$$

and defining the complex dielectric permittivity $\varepsilon(\omega)$ as

$$D(\omega) = \varepsilon(\omega)\, E(\omega)$$

then, using the convolution theorem,

$$\varepsilon(\omega) = 1 + \int_{0}^{\infty} F(t)\, e^{-i\omega t}\, dt\,. \tag{5}$$

Considering the analytic properties of the causal function $\varepsilon(\omega)$ in the complex frequency plane, it can be shown that $\varepsilon(\omega)$ must have poles below the real axis in order to be analytic on and above the real axis. Because $\varepsilon^{-1}(\omega)$ is also a causal function, the zeros of $\varepsilon(\omega)$ must also lie below the real axis. In this case $\varepsilon(\omega)$ is a minimum phase shift function [24] as is

$$\frac{\varepsilon^{1/2}(\omega) - 1}{\varepsilon^{1/2}(\omega) + 1}\,.$$

Defining

$$\alpha(\omega) = -\frac{1}{2} \ln R(\omega) \tag{6}$$

then

$$\alpha(\omega) + i\,\theta(\omega) = -\ln\,[R^{1/2}(\omega)\, e^{-i\,\theta(\omega)}] = -\ln\left(\frac{\varepsilon^{1/2}(\omega) - 1}{\varepsilon^{1/2}(\omega) + 1}\right).$$

Because $[\varepsilon^{1/2}(\omega) - 1]/[\varepsilon^{1/2}(\omega) + 1]$ is a minimum phase shift function, then $\alpha(\omega) + i\,\theta(\omega)$ is a causal function, being analytic on and above the real axis. Therefore, along the real axis

$$\alpha(\omega) + i\,\theta(\omega) = \int_{-\infty}^{\infty} G(t)\, e^{-i\omega t}\, dt\,,$$

where

$$G(t) = 0 \qquad t < 0\,,$$

$$G(t) \quad \text{real, finite} \qquad t > 0\,.$$

Because $G(t)$ is real, $\alpha(\omega)$ is a real even function of ω while $\theta(\omega)$ is a real odd function of ω. Breaking $G(t)$ up into even and odd parts,

$$G(t) = E(t) + O(t)\,.$$

But

$$O(t) = \text{sgn}(t)\, E(t)\,,$$

where

$$\text{sgn}(t) = 1 \qquad t > 0\,, \qquad = -1 \qquad t < 0\,. \tag{7}$$

Hence

$$G(t) = E(t) + \operatorname{sgn}(t)\, E(t)$$

and

$$\alpha(\omega) = \int_{-\infty}^{\infty} E(t)\, e^{-i\omega t} dt\,,$$

$$\theta(\omega) = -i \int_{-\infty}^{\infty} E(t) \operatorname{sgn}(t)\, e^{-i\omega t}\, dt\,, \tag{8}$$

$$E(t) = \int_{-\infty}^{\infty} \alpha(\omega)\, e^{i\omega t}\, d\omega\,. \tag{9}$$

Hence, from (6), $\theta(\omega)$ is the Fourier transform of the product of $E(t)$ and sgn(t). $E(t)$ can be calculated from $R(\omega)$ using (6) and (9), while sgn(t) is defined by (7). Using these relationships the spectrum of the phase change may be calculated from the reflectance spectrum. Because $R(\omega)$ is known and sampled over a limited range of ω, the actual relationships used in place of (8) and (9) are discrete Fourier transforms. Errors will be introduced because of the folding of the reflectance sample points implied by the discrete Fourier transform. However, these errors are simply eliminated.

3.2 Comparison with the Kramers-Kronig transform

The relationships developed above are simply related to the Kramers-Kronig transform by the convolution theorem. From (8) and (9),

$$\theta(\omega) = -i \int_{-\infty}^{\infty} \left[\int_{-\infty}^{\infty} E(t)\, e^{-i\omega' t}\, dt\right] \left[\int_{-\infty}^{\infty} \operatorname{sgn}(t)\, e^{-i(\omega-\omega')t}\, dt\right] d\omega'\,.$$

Now

$$\int_{-\infty}^{\infty} \operatorname{sgn}(t)\, e^{-i(\omega-\omega')t}\, dt = \frac{i}{\pi\,(\omega' - \omega)}$$

and

$$\int_{-\infty}^{\infty} E(t)\, e^{-i\omega' t}\, dt = -\frac{1}{2} \ln R(\omega')\,.$$

Therefore,

$$\theta(\omega) = -\frac{1}{2\pi} \int_{-\infty}^{\infty} \frac{\ln R(\omega')}{\omega' - \omega}\, d\omega'$$

which is equivalent to (3) because $R(\omega)$ is an even function about $\omega = 0$. Hence the Kramers-Kronig transform is a convolution integral. We suggest that evaluation of $\theta(\omega)$ by multiplication in the time domain can be performed more efficiently than by convolution in the frequency domain as is done by the Kramers-Kronig method.

3.3 Comparison of efficiency

3.3.1 General comparisons

The Fourier transformations of sampled data required by (8) and (9) may be performed extremely efficiently if N, the number of samples, is a highly composite number. The discrete Fourier transform

$$E(j) = \sum_{k=0}^{N-1} \alpha(k)\, e^{+(2\pi i/N) jk} \quad j = 0, 1, \ldots, N-1 \tag{10}$$

may be calculated using conventional methods (e.g. Simpson's rule) which would require a minimum of N^2 operations. However only $N \log_2 N$ operations are required if N is a power of 2 when the Cooley-Tukey algorithm [25] is used. Basically this important algorithm takes advantage of any factors of N, being most efficient when N is a power of 2. The lowering of processor computation time is very significant if N is large. For example, the gain in efficiency over the conventional methods is approximately 100 for $N = 1024$ and 630 for $N = 8192$. The use of the algorithm has the following benefits:

(i) Feasibility of performing transformations on large numbers of data points.

(ii) Efficient hardware [26] and software [27] based on the algorithm are available for performing the transformations.

(iii) Round-off noise is minimized [28].

(iv) If computer memory is limited, data may be stored in auxiliary memory devices. The algorithm has been adapted to minimize the number of passes between files [29].

Transformation to the time domain has the following benefits:

(i) Data may be smoothed by simple multiplication in the time domain with a low-pass filter function, i.e. tedious convolution in the frequency domain may be avoided.

(ii) The shape of narrow structure in the reflectance spectrum may be corrected for distortion which results from spectral slit broadening. If $R(\omega)$ is the measured reflectance, then

$$R(\omega) = \int_{-\infty}^{\infty} R^T(\omega')\, S^0(\omega - \omega')\, d\omega' , \tag{11}$$

where $R^T(\omega)$ is the true reflectance and $S^0(\omega)$ is the instrument function of the spectrometer which may be approximated by a triangular function in most cases. If $X(t)$ and $X^T(t)$ are the transforms of $R(\omega)$ and $R^T(\omega)$ then

$$X^T(t) = \frac{X(t)}{S'(t)} ,$$

where $S'(t)$ is the Fourier transform of $S^0(\omega)$. Thus the distortion correction is applied by transforming $R(\omega)$ to $X(t)$, dividing by $S'(t)$, and transforming back to the frequency domain. The same computer programs are used to perform these operations except that $R(\omega)$ instead of $-(1/2) \ln R(\omega)$ is analysed and the filter function to replace $\mathrm{sgn}(t)$ is $\left(\frac{\sin \pi b t}{\pi b t}\right)^2$ where b is the slit width. Only $3 N \log N$ operations are required using the Cooley-Tukey algorithm compared with N^2 operations using conventional deconvolution methods.

3.3.2 *Specific application*

Let $\alpha(k)$ be values of $-(1/2) \ln R(\omega_k)$ sampled at frequencies $\omega_k = \omega_a + k\,\Delta\omega$; $k = 0, \ldots, N-1$. The discrete Fourier transform $E(j)$, $j = 0, \ldots, N-1$ of $\alpha(k)$ is given by (10). Since $\alpha(k)$ is real, $E(j)$ has an even real part and odd imaginary part, i.e.

$$\left.\begin{aligned} \mathrm{Re}\, E(i) &= \mathrm{Re}\, E(N - i - 1) \\ \mathrm{Im}\, E(i) &= -\mathrm{Im}\, E(N - i - 1) \end{aligned}\right\} \quad i = 0, \ldots, \frac{N}{2} - 1 .$$

The imaginary part arises because samples were taken only in the positive frequency domain. Now sgn (j) is defined as

$$\operatorname{sgn}(j) = 1 \qquad j = 0, \ldots, \frac{N}{2} - 1\,,$$

$$\operatorname{sgn}(j) = -1 \qquad j = \frac{N}{2}, \ldots, N - 1\,.$$

If $E(j)$ were multiplied by $\operatorname{sgn}(j)$ and the $(-i)$-Fourier transform taken, then N values of $\theta(\omega_k)$ would be evaluated with effective resolution $\Delta\omega$.

$$A(\omega_k) = i\,\theta(\omega_k) = \sum_{j=0}^{N-1} E(j)\operatorname{sgn}(j)\,\mathrm{e}^{-(2\pi i/N)jk}\,, \qquad k = 0, \ldots, N-1\,. \tag{12}$$

However, if smoothed values of $\alpha(\omega_k)$ and $\theta(\omega_k)$ were required, then multiplying $E(j)$ by $H(j)$, where

$$H(j) = 1 \qquad j = 0, \ldots, \frac{N}{2} - 1\,,$$

$$H(j) = 0 \qquad j = \frac{N}{2}, \ldots, N - 1\,,$$

followed by taking the $(-i)$-Fourier transform on the remaining $N/2$ non-zero points, yields

$$A(\omega_k) = \alpha(\omega_k) + i\,\theta(\omega_k) = 4.0 \sum_{j=0}^{N/2-1} E(j)\,\mathrm{e}^{-(4\pi i/N)jk}\,, \qquad k = 0, \ldots, \frac{N}{2} - 1\,. \tag{13}$$

The effective resolution is now $2\,\Delta\omega$. The first method requires $2\,N\,\mathrm{lb}\,N + N/2$ operations while the second method requires $N\,\mathrm{lb}\,N + N/2\,\mathrm{lb}\,N/2$ operations, compared with N^2 for a Kramers-Kronig inversion.

4. Discussion. Possible Errors

The errors which could arise in transforming reflectivity data occur from incomplete spectral information, from computer round-off and from aliasing. The last is peculiar to our method. We show here the magnitude of such errors and indicate that the transformation errors are relatively insignificant compared with the experimental errors in measuring $R(\omega)$.

Consider first the aliasing errors. The discrete Fourier transform assumes that the spectral data points $R(\omega_k)$ are periodic, repeating at frequencies

$$\omega_k = \omega_\mathrm{a} + k\,\Delta\omega + N\,(\omega_\mathrm{b} - \omega_\mathrm{a})\,,$$

where

$$N = 0, \pm 1, \pm 2, \ldots\,.$$

This causes distortion in $E(j)$, which is maximum around $j = N/2 - 1$ because $E(j)$ is also periodic. This distortion is carried across to $\theta(\omega_k)$ being most severe around $\omega = \omega_\mathrm{a}$ and $\omega = \omega_\mathrm{b}$. These aliasing or fold-over errors are minimized by taking $(\omega_\mathrm{b} - \omega_\mathrm{a})$ and N as large as possible. Their effect on $\theta(\omega_k)$ is similar to the effect of structure in $R(\omega)$ on the low-energy side of ω_a and the high-energy side of ω_b. A simple correction function for each case is considered below. Following Bode [30], if $R'(\omega)$ simulates high-energy structure then its contribution to $\theta(\omega)$ is given by

$$\theta^{H}_{corr}(\omega) = -\frac{\omega}{\pi}\int\limits_{\omega_b}^{\infty}\frac{\ln R'(\omega')}{\omega'^2-\omega^2}\,d\omega' .$$

If $\omega' \gg \omega$, then

$$\theta^{H}_{corr}(\omega) = \omega\left(-\frac{1}{\pi}\int\limits_{\omega_b}^{\infty}\frac{\ln R'(\omega')}{\omega'^2}\,d\omega'\right).$$

High-energy structure gives rise to a linear phase distortion. If $R''(\omega)$ simulates low-energy structure, the contribution to $\theta(\omega)$ is given by

$$\theta^{L}_{corr}(\omega) = -\frac{\omega}{\pi}\int\limits_{0}^{\omega_a}\frac{\ln R''(\omega')}{\omega'^2-\omega^2}\,d\omega' .$$

If $\omega' \ll \omega$, then

$$\theta^{L}_{corr}(\omega) = \frac{1}{\omega}\left(+\frac{1}{\pi}\int\limits_{0}^{\omega_a}\ln R''(\omega')\,d\omega'\right).$$

Low-energy structure gives distortion varying inversely as ω. Hence the total distortion correction may be represented as $X\,\omega + Y/\omega$. However, if $(\omega_b - \omega_a) \ll \omega$ then this is approximately a constant C.

If the effects of incomplete data are analysed by Roessler's method [31], which is equivalent to extrapolation by constant reflectance of e^{+2A} and e^{-2B} to low and high energies then the total corrected phase angle becomes

$$\theta(\omega_j) = \theta(\omega_j) + \frac{A}{\pi}\ln\left|\frac{\omega_a+\omega_j}{\omega_a-\omega_j}\right| + \frac{B}{\pi}\ln\left|\frac{\omega_b+\omega_j}{\omega_b-\omega_j}\right| + X\,\omega_j + \frac{Y}{\omega_j}, \quad (14)$$

$$\theta(\omega_j) \approx \theta(\omega_j) + \frac{A}{\pi}\ln\left|\frac{\omega_a+\omega_j}{\omega_a-\omega_j}\right| + \frac{B}{\pi}\ln\left|\frac{\omega_b+\omega_j}{\omega_b-\omega_j}\right| + C . \quad (15)$$

How does this error compare with experimental errors? The errors may be evaluated by considering the ultraviolet reflection of an anthracene crystal. Using a Lorentzian line dielectric function model based on Helmholtz-Kettler oscillators, $R(\omega_k)$ and $\theta(\omega_k)$ may be calculated because

$$\omega_k = \omega_a + k\,\Delta\omega \qquad k = 0, \ldots, N-1 ,$$

$$\varepsilon(\omega_k) = \varepsilon_0 + \frac{4\,\pi\,f\,\omega_0^2}{\omega_0^2 - \omega_k^2 - i\,\gamma\,\omega_k},$$

$$R(\omega_k)^{1/2}\,e^{-i\,\theta(\omega_k)} = \frac{\varepsilon^{1/2}(\omega_k) - 1}{\varepsilon^{1/2}(\omega_k) + 1}.$$

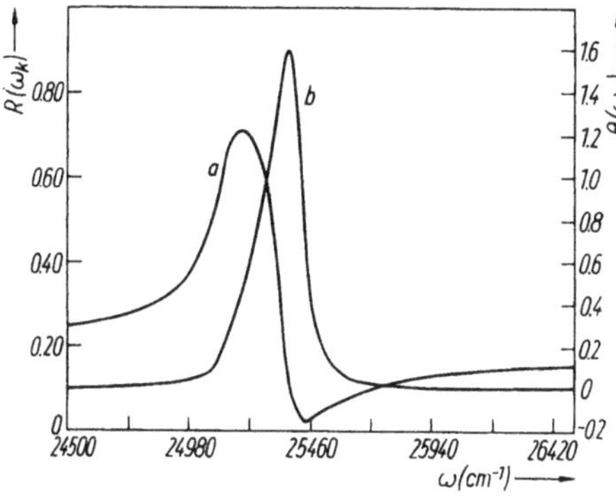

Fig. 1. (a) Normal incidence reflectance $R(\omega_k)$; (b) phase change on reflection $\theta(\omega_k)$.
Calculated for a system of uncoupled Helmholtz-Kettler oscillators with $\varepsilon_0 = 6.4$, $\omega_0 = 25127\ cm^{-1}$, $f = 0.01$, $\gamma = 75\ cm^{-1}$, $\Delta\omega = 8\ cm^{-1}$. The graphs are for clarity of presentation full-line interpolations between the 256 data points sampled at 8 cm^{-1} intervals

Appropriate values of the parameters [32] are $\varepsilon_0 = 6.4$, $f = 0.01$, $\omega_0 = 25127\ cm^{-1}$, $\gamma = 75\ cm^{-1}$, $\omega_a = 24500\ cm^{-1}$, $\Delta\omega = 8\ cm^{-1}$. Let the number of data points sampled (N) be 256.

Graphs of $R(\omega_k)$ and $\theta(\omega_k)$ are shown in Fig. 1.

The $(+i)$ discrete Fourier transform of $-(1/2) \ln R(\omega_k)$ for $k = 0, \ldots, N-1$ was evaluated to give $E(j)$ and using (13) the $(-i)$ discrete Fourier transform of $4\,E(j)$ for $j = 0, \ldots, N/2 - 1$ was evaluated to give $A(k)$, $k = 0, \ldots, N/2 - 1$. Now, $\theta(\omega_k) = \operatorname{Im} A(k)$ and $R^0(\omega_k) = e^{-2 \operatorname{Re} A(k)}$ where $\omega_k = \omega_a + 2\,\Delta\omega\,k$.

$\theta(\omega_k)$ is plotted in Fig. 2, curve (a). $R^0(\omega_{2k})$ was indistinguishable from $R(\omega_k)$ to at least six figures indicating that round-off noise was not important. The corrections for distortion from (15) were then applied. Fig. 2, curve (b) shows the correction for $C = 0$ and Fig. 2, curve (c) for $C \neq 0$. The constants, A, B, C were evaluated from simultaneous equations using equation (15) at three points within (ω_a, ω_b) where $\theta^0(\omega_j)$ is known. If $C = 0$ only two points are required; or if equation (14) was used four points would be required. Compare $\theta(\omega_k)$ in Fig. 1 and in Fig. 2(c). No noticeable variation occurs. It can be seen that the aliasing error is significant but is easily eliminated with the use of a non-zero value of C.

Experimental errors were simulated by adding a constant reflectance, viz. a 2% variation in $R(\omega)$ and a 0.2% variation in $R(\omega)$, i.e.

(i) $$R = R^0 + 0.002\,,$$

(ii) $$R = R^0 + 0.02\,.$$

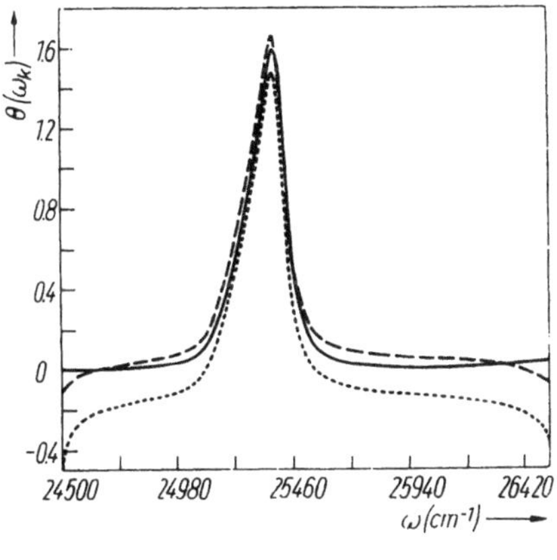

Fig. 2. Phase change $\theta(\omega_k)$ calculated from 256 samples of $R(\omega_k)$ using the Fourier transform method. As in Fig. 1, the graphs are full-line interpolations.
Uncorrected for both limited data and aliasing, i.e., $A = 0$, $B = 0$, $C = 0$ (- - - - - - - -); corrected for limited data but uncorrected for aliasing, A, B non-zero, $C = 0$ (— — — — —); corrected for both limited data and aliasing, A, B, C non-zero (————————)

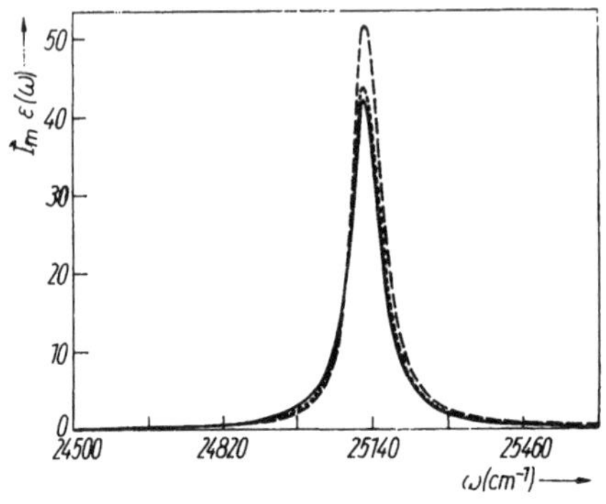

Fig. 3. Simulation of the effects upon the imaginary part of the dielectric permittivity $\varepsilon(\omega)$ caused by experimental errors in measurement of $R(\omega)$. Im $\varepsilon(\omega)$ calculated from $R(\omega)$, corresponding to no experimental error in $R(\omega)$ (————); Im $\varepsilon(\omega)$ calculated fom $R(\omega) + 0.002$, corresponding to 0.2% error in $R(\omega)$ (– – – – – –); Im $\varepsilon(\omega)$ calculated from $R(\omega) + 0.02$ corresponding to 2.0% error in $R(\omega)$ (– – – – – –)

Graphs of Im $\varepsilon(\omega_k)$ are shown in Fig. 3 for each of these cases. The effect of experimental inaccuracies are obvious. The errors are shown in Table 1. It is seen that in most cases where the reflection coefficient cannot be measured to better than 0.01, the contribution to the distortion from the method of analysis is swamped by the contributions arising from experimental inaccuracies. A similar conclusion has also been reached for the Kramers-Kronig analysis [33].

Table 1

Percentage errors induced in the peak values of the real and imaginary parts of both the dielectric permittivity and the refractive index by incorrect experimental measures of reflectivity

	perfect data $R^0(\omega)$	incorrect experimental data	
		$R^0(\omega) + 0.002$	$R^0(\omega) + 0.02$
R_{max}	—	0.28	2.8
θ_{max}	−0.43	−2.22	−14.7
Re ε_{max}	1.90	3.98	22.2
Im ε_{max}	4.20	6.84	29.8
Re n_{max}	1.60	2.91	13.9
Im n_{max}	0.40	0.53	10.1

5. Conclusions

It has been shown that a new method of analysis of reflectance spectra gives results with errors comparable to a conventional Kramers-Kronig analysis. By using an efficient algorithm for computing the discrete Fourier transform a gain of $2\,N \log_2 N + N/2$ or $N \log_2 N + N/2 \log_2 N/2$ in processor time may be achieved if N, the number of data points, is a power of 2. Gains still greater than 1 may be achieved if N has any factors whatsoever. This new method has other advantages in that round-off noise is reduced, the input data easily filtered for noise elimination or deconvoluted with the spectrometer instrument function. The efficient algorithm used may allow real-time generation of optical profiles using fast Fourier transform hardware or software. The increase in efficiency allows more data points to be sampled without great increase in cost allowing the spectra of sharp structure on a varying background to be analysed efficiently over a large spectral region.

Further, the method is versatile and will be of use when the optical absorption profile is being used as a probe of exciton dynamics in a condensed system. The same Fourier transform algorithm can be used.

Acknowledgements

This work was supported by the Australian Research Grants Committee and by the Commonwealth Scientific and Industrial Research Organization (a scholarship to M.G.S.).

References

[1] J. J. Hopfield, Phys. Rev. **112**, 1555 (1958).
[2] A. S. Dayvdov, Theory of Molecular Excitons, Plenum Press, New York 1971.
[3] F. C. Jahoda, Phys. Rev. **107**, 1261 (1957).
[4] F. Stern, Solid State Phys. **15**, 288 (1963).
[5] W. G. Spitzer, D. Kleinman, and D. Walsh, Phys. Rev. **113**, 127 (1959).
[6] H. W. Verleur, J. Opt. Soc. Amer. **58**, 1356 (1968).
[7] D. W. Berreman and F. C. Unterwald, Phys. Rev. **174**, 791 (1968).
[8] G. C. Morris, S. A. Rice, and A. E. Martin, J. chem. Phys. **52**, 5148 (1970).
[9] M. S. Brodin, S. V. Marisova, and S. A. Shturkhetskaya, Ukr. fiz. Zh. **13**, 249 (1968).
[10] H. R. Philipp and H. Ehrenreich, Phys. Rev. **129**, 1550 (1963).
[11] M. Cardona and D. L. Greenaway, Phys. Rev. **133**, A1685 (1964).
[12] J. R. Jasperse, A. Kahan, J. N. Plendl, and S. S. Mitra, Phys. Rev. **146**, 526 (1966).
[13] G. Anderman and D. A. Dows, J. Phys. Chem. Solids **28**, 1307 (1967).
[14] G. C. Morris, M. G. Sceats, and D. R. Sweatman, to be pulished.
[15] Y. Toyozawa, Progr. theor. Phys. (Kyoto), Suppl. **12**, 111 (1959).
[16] R. L. Fulton, J. chem. Phys. **55**, 1386 (1971).
[17] W. L. Greer and S. A. Rice, Adv. chem. Phys. **17**, 229 (1970).
[18] D. N. Zubarev, Soviet Phys. — Uspekhi **3**, 320 (1960); Uspekhi fiz. Nauk **71**, 71 (1960).
[19] T. Skettrup and I. Balsev, Phys. Rev. **B3**, 1457 (1971).
[20] I. Filiński, phys. stat. sol. (b) **49**, 577 (1972).
[21] D. J. Wilde, Optimum Seeking Methods, Prentice Hall, Inc., Englewood Cliffs (N.J.), 1964.
[22] P. M. Grant and J. C. Suits, Appl. Phys. Letters **14**, 172 (1969).
[23] A. S. Davydov, Quantum Mechanics, Ed. D. ter Haar, Addison-Wesley Publ. Co., Inc., Reading (Mass.) 1965 (Section 112).
[24] A. Papoulis, The Fourier Integral and Its Applications, McGraw-Hill Publ. Co., 1962 (p 204).
[25] J. W. Cooley and J. W. Tukey, Math. Comput. **19**, 297 (1965).
[26] G. D. Bergland, IEEE Trans. Comput. **21**, 366 (1972).
[27] J. W. Cooley, P. A. W. Lewis, and P. D. Welch, IEEE Trans. Educ. **12**, 27 (1969).
[28] C. J. Weinstein, IEEE Trans. Audio and Electroacoustics, **17**, 209 (1969).
[29] M. Drubin, IEEE Trans. Comput. **20**, 1552 (1971).
[30] H. W. Bode, Network Analysis and Feedback Amplifier Design, Van Nostrand Co., New York 1945 (p. 307).
[31] D. M. Roessler, Brit. J. appl. Phys. **16**. 1119 (1965); **16**, 1359 (1965); **17**, 1313 (1966).
[32] G. C. Morris, S. A. Rice, M. G. Sceats, and A. E. Martin, J. chem. Phys. **55**, 5610 (1971).
[33] P. O. Nilsson and L. Munkby, Phys. kondens. Materie **10**, 290 (1969).

(Received June 20, 1972)

phys. stat. sol. (a) 14, 655 (1972)

Subject classification: 13.4 and 14.3.3; 22.1.2

Department of Electrical Engineering and Materials Research Laboratory, University of Illinois, Urbana

Thermal Emission Rates and Activation Energies of Electrons and Holes at Cobalt Centers in Silicon[1])

By

L. D. YAU, W. W. CHAN, and C. T. SAH

The thermal ionization rates of electrons and holes at cobalt centers in silicon have been measured using the dark-capacitance and dark-current transient techniques. Using only n^+-p diodes, the cobalt charge states, the thermal emission rates including their field dependences are determined. The thermal emission rate data fitted to $e^t = A \exp(-\Delta E/kT)$ gives: (i) $\Delta E = E_T - E_V = (377 \pm 1)$ meV and $A = (4.43 \pm 0.33) \times 10^{13}\ s^{-1}$ for the donor level, (ii) $\Delta E = E_C - E_T = (569 \pm 3)$ meV and $A = (2.59 \pm 0.38) \times 10^{12}\ s^{-1}$ for the acceptor level.

Es wurden die thermischen Ionisierungsraten von Elektronen und Löchern an Kobaltzentren in Silizium mit der Methode des Abklingens der Dunkelkapazität und des Dunkelstromes gemessen. Bei ausschließlicher Benutzung von n^+-p-Dioden wurde der Ladungszustand von Kobalt und die thermischen Emissionsraten einschließlich ihrer Feldabhängigkeiten bestimmt. Die thermischen Emissionsraten werden an $e^t = A \exp(-\Delta E/kT)$ angepaßt und ergeben: 1. $\Delta E = E_T - E_V = (377 \pm 1)$ meV und $A = (4,43 \pm 0,33) \times 10^{13}\ s^{-1}$ für das Donatorniveau, 2. $\Delta E = E_C - E_T = (569 \pm 3)$ meV und $A = (2,59 \pm 0,38) \times 10^{12}\ s^{-1}$ für das Akzeptorniveau.

1. Introduction

Cobalt levels in silicon have been previously studied [1 to 6]. A summary of the reported energy levels is given in Table 1. Irmler [1] attempted to overcompensate n-type and p-type silicon substrates with cobalt. He observed overcompensation in the n-type substrates, but no evidence of overcompensation in the p-type substrates. This observation led Irmler to conclude that the two cobalt levels are acceptors. From concentration–temperature-dependence measurements Ghandhi et al. [3] found an energy level 0.44 eV above the valence band. Penchina et al. [4], also employed the compensation technique to determine the charge states of the cobalt levels in silicon [3]. Their conclusion on the charge states agreed with Irmler's. Later reports of Moore et al. [5], showed that the initially 4000 Ωcm p-type and 14000 Ωcm n-type substrates were both compensated to about 50000 Ωcm p-type. They concluded that the lower cobalt level is actually a donor. The energy levels were obtained from thermal (resistivity, Hall coefficient) and optical measurements.

In this report, the charge states of the cobalt center are determined by using the thermally stimulated capacitance (TSCAP) [7] to compare the net space charge of the Co-doped diode with that of the control diode. The analysis of the data in Section 4 shows that the lower level is a donor level, and the upper

[1]) Work supported by the Advanced Research Projects Agency and the Air Force Office of Scientific Research.

Table 1
Reported cobalt levels in silicon

type of level	$E_C - E_T$ (eV)	$E_T - E_V$ (eV)	measurement techniques	authors
acceptor	0.39		thermal	Irmler [1]
acceptor		0.37		
acceptor		0.5	optical	Holonyak [2]
—		0.44	thermal	Ghandhi et al. [3]
acceptor	0.53		optical	Penchina et al. [4]
acceptor		0.35		
acceptor	0.53		thermal and optical	Moore et al. [5]
donor		0.40		
acceptor	0.52		optical	Chang et al. [6]
donor		0.38		

level an acceptor level. Thermal ionization energies are obtained from a least square fit to the equation

$$e^t = A\left(\frac{T}{300}\right)^n \exp\left(-\frac{\Delta E}{kT}\right). \tag{1}$$

The result is given in Section 5.

2. Device Fabrication Procedure

In this report, only the p-type substrate or the n^+–p diodes are needed using optical excitation to trap minority carriers initially [8]. For the three devices presented in this paper, the starting material was a 15 Ωcm boron-doped silicon, grown by Czochralski technique with a (111) surface. The silicon wafer was originally 1 inch in diameter and 14 mils thick. After standard cleaning and n^+ junction diffusion, the whole wafer was divided into several smaller pieces. For cobalt doping, a 99.999% cobalt wire was vacuum evaporated on the back of the silicon slices and cobalt diffusion immediately followed in an open tube furnace with argon ambient. The first Co-doped diode labelled IA2A-3 was diffused at 1100 °C for 20 h and the ratio of the active cobalt centers to the shallow boron impurity is $N_{TT}/N_A \approx 0.1$. The second Co-doped diode labelled IA2D-5 was diffused at 1150 °C for 15 h and this gives $N_{TT}/N_A \approx 0.3$. The spatial distribution of $N_{TT}(x)$ and $N_A(x)$ is given in Section 6. The control diode (no cobalt deposited on the back of the substrate) labelled IA2C-2 was heat treated at 1150 °C for 15 h in a different clean furnace tube to simulate the temperature cycle of IA2D-5. Diodes IA2D-5 and IA2C-2 will be used in Section 4 for charge state determination of the cobalt levels in silicon. 3 mil diameter contact holes were cut on both the heat treated control and Co-diffused slices. 0.2 μm of 99.999% aluminum was vacuum evaporated on the n^+ side of the slices, and after etching the aluminum contacts, the slices were lapped to a final thickness of 4 mils. A second aluminum film was evaporated on the lapped p-type side

Fig. 1. Geometry of the Co-doped and the control diodes used (vertical scale expanded)

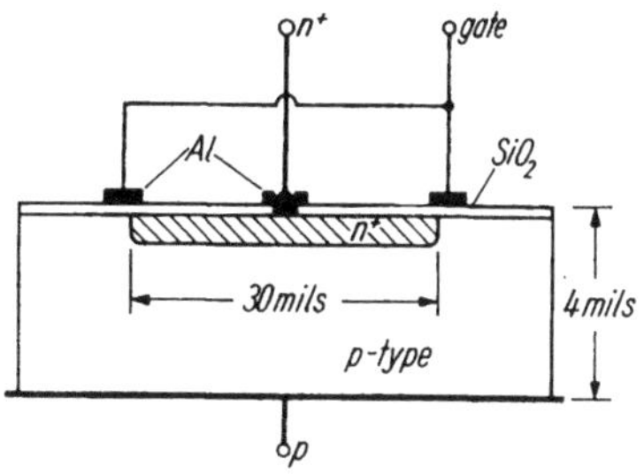

of the slices and sintering was carried out at 525 °C for 30 min to improve the ohmic contacts. The diced devices were die bonded on gold plated TO-5 headers at 450 °C and gold leads were ultrasonically bonded to the terminals.

Fig. 1 shows the finished n^+–p diode structure. The gate is used for suppressing surface effects at the edge of the junction. Only a small area of the n^+–p junction is metallized for ohmic contact. This allows optical excitation of the traps in the depletion region.

3. Scanning the Bandgap for Deep Levels

The use of thermally stimulated capacitance (TSCAP) technique to scan for deep levels in the bandgap was previously reported [7] and applied in references [9, 10]. This technique is used to scan for deep levels in the Co-diode IA2D-5 and the control diode IA2C-2. The results of the TSCAP scans are shown in Fig. 2 to 4.

In Fig. 2, the diode was first cooled to 77 °K in the dark with zero bias to fill the deep traps below the midgap with holes. Then a reverse bias V_R is applied to sweep out the mobile carriers while the trapped holes remain frozen at 77 °K. When heat is applied, the trapped holes are thermally released, causing the junction capacitance to change. The junction capacitance was measured by a Boonton B71A meter. The analogue output of the capacitance meter was electronically differentiated to give $dC(t)/dt$. $C(t)$ is the time dependent junction capacitance given by $C(t) = K_s \varepsilon_0 A/W(t)$, where $K_s \varepsilon_0$ is the silicon dielectric constant, A the junction area, and $W(t)$ the depletion width which changes as the trapped holes are thermally released. For comparison, the result of the scan for the Co-doped diode IA2D-5 and control diode IA2C-2 are shown in Fig. 2. The positive dC/dt hump around 140 °K in the Co-doped diode is due

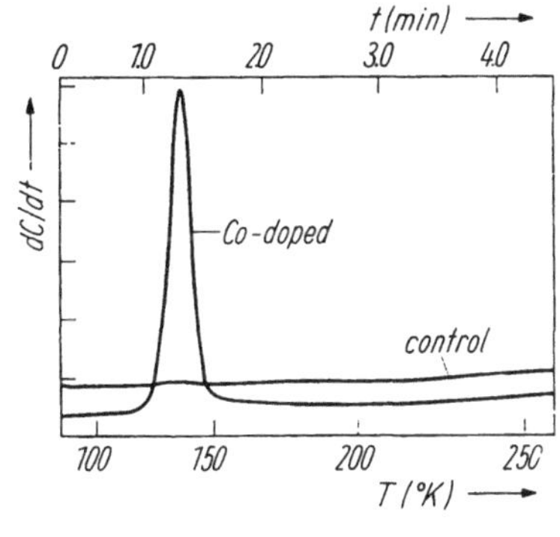

Fig. 2. Differentiated TSCAP of Co-doped diode IA2D-5 and control diode IA2C-2. The diodes were initially cooled to 77 °K at zero bias, and then at $t = 0$ switched to $V_R = 10$ V. A nearly constant heating rate was subsequently applied

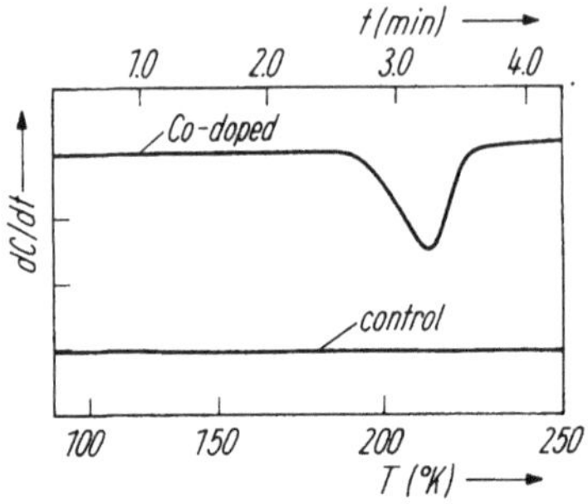

Fig. 3. Differentiated TSCAP of Co-doped and control diodes. The diodes were initially illuminated with $\lambda = 1.15\ \mu$m light ($t < 0$) at 77 °K and $V_R = 10$ V. The heat was applied after the light was turned off

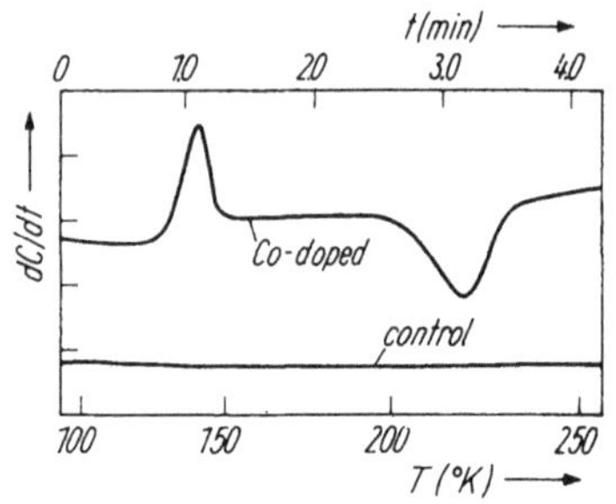

Fig. 4. Differentiated TSCAP of Co-doped and control diodes initially illuminated with $\lambda = 1.00\ \mu$m ($t < 0$) at 77 °K and with $V_R = 10$ V. Heat was applied after the light was turned off

to holes released from the lower level. This level will be shown in the next section to be a donor level. The control diode show no significant deep level activity.

To scan for deep levels above the midgap or minority carrier traps in p-type silicon where $e_p^t \ll e_n^t$, the same n$^+$–p diode is cooled down to 77 °K with a constant reverse bias V_R applied at all times. The deep levels are then optically filled with minority carriers (electrons) using $\lambda = 1.15\ \mu$m initially. When the trap charge states reach a steady state value, the light is removed and the trapped carriers will stay frozen. Then the sample is heated at a nearly constant dT/dt. As previously reported, the differentiated TSCAP gives positive dC/dt for majority carrier emission and negative dC/dt for minority carrier emission [9]. In Fig. 3, $\lambda = 1.15\ \mu$m was used initially and the Co-doped diode IA2D-5 shows only one deep level due to minority carrier (electron) emission at $T \approx$ ≈ 220 °K. It will be shown in the next section that this is an acceptor level. The initial optical excitation may also be set at a different wave length so that majority carrier (hole) emission from the lower Co level and minority carrier (electron) emission from the upper Co level can both be observed. This is demonstrated in Fig. 4 using $\lambda = 1.00\ \mu$m initially. In this particular scan, the cobalt centers were initially distributed in three charge states: those initially in the positive charge state (lower Co level) contribute to the positive dC/dt at $T \approx$ ≈ 140 °K: those initially in the neutral charge state do not contribute to any dC/dt; and those initially in the negative charge state (upper Co level) contribute to the negative dC/dt at $T \approx 220$ °K.

4. Charge State Determination

The absolute values of the capcitances in the TSCAP scans of Fig. 2 and 3 are plotted in Fig. 5. All curves were measured at $V_R = 10$ V. Since the junction capacitance is directly related to the net space charge, the charge states of the cobalt centers can be determined.

In Fig. 5, the two lower curves show that when the traps are initially filled with holes ($T < 120$ °K), the junction capacitance of the Co-doped diode is lower than the junction capacitance of the control diode. As the temperature increases, the trapped holes are released and the two capacitances merge. This implies that the dC/dt peak near 140 °K observed in Fig. 2 is due to holes released from a *donor* level. The upper Co-diode capacitance curve was obtained by

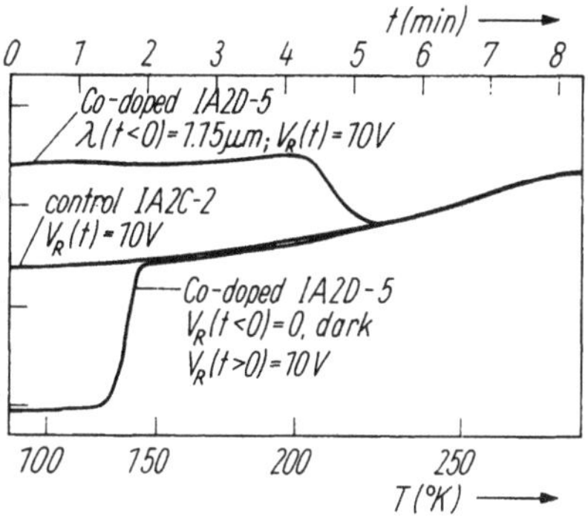

Fig. 5. TSCAP of Co-doped and control diodes under different initial conditions shown in the figure. The control diode is not affected by the initial condition

optically filling the deep traps with electrons initially. The trap just identified as donors is also filled with electrons, but $e_p^t \gg e_n^t$ for this level, therefore no carrier emission is observed at 140 °K. Since the upper Co-doped diode has a higher capacitance than the control diode until the electrons from the upper Co-level are all emitted near 225 °K, the upper Co-level is an *acceptor* level.

5. Thermal Emission Rate Measurements at Constant Temperatures

The hole and electron thermal emission rates are measured from n^+–p diodes. The hole emission rate e_{p-1}^t is measured in the dark by switching momentarily the bias V_R, to zero or some intermediate value V_I [11]. This initial step is to fill up the deep traps (cobalt donor in this case) with majority carriers (holes). When the bias is switched back to V_R the trapped holes will be thermally emitted to the valence band. This results in a simultaneous current flow and a change in the depletion width which give rise to the current and the capacitance transients, respectively. The time constants above 100 ms was obtained from the capacitance transients and the time constants below 100 ms by the current transients.

The electron thermal emission rate, e_{n1}^t, from the acceptor level is measured in a similar manner except that the acceptor level was initially filled with electron by optical excitation [8]. The photon energy used was $\lambda = 1.15$ μm. After setting the initial condition optically, the electron thermal emission rate e_{n1}^t is then measured in the dark similar to the dark current and capacitance transients in determining e_{p-1}^t previously.

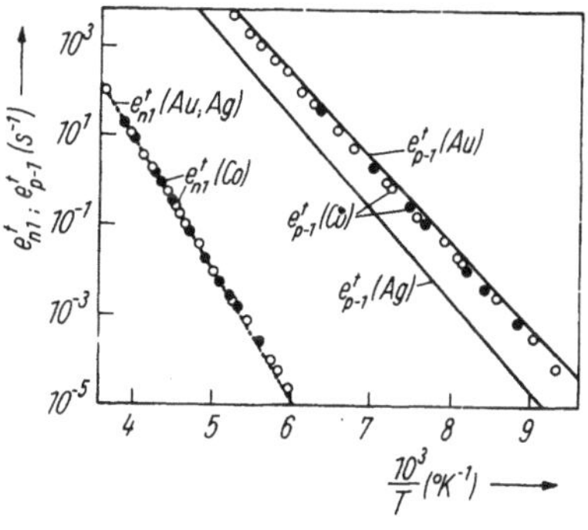

Fig. 6. Hole and electron thermal emission rates of cobalt centers in silicon at $\langle \varepsilon \rangle = 10^4$ V/cm. The thermal emission rates of gold and silver are also given in straight lines for comparison. ○ IA2D-5; ● IA2A-3

The result of the thermal emission rate measurements is shown in Fig. 6. The measured values of e^t_{p-1} and e^t_{n1} are shown as open and closed circles for two Co-doped diodes. The average electric field used for all measurements in Fig. 6 was 10^4 V/cm. Previously reported thermal emission rates for silver [12, 13] and gold [14, 15] in silicon are shown in straight lines for comparison. The values of e^t_{n1} for the three impurities are similar. On the other hand, the hole emission rates e^t_{p-1} for the three impurities are distinctly different. The activation energies and the pre-exponential factors resulting from a least square fit of the cobalt data points to equation (1) is given in Table 2 for different values of n.

Table 2
Pre-exponential factors and activation energies of cobalt centers in silicon

n	acceptor level		donor level	
	A (10^{12} s^{-1})	$E_C - E_T$ (eV)	A (10^{13} s^{-1})	$E_T - E_V$ (eV)
−2	8.76 ± 1.41	0.604 ± 0.003	7.97 ± 0.70	0.402 ± 0.001
−1	4.77 ± 0.73	0.587 ± 0.003	5.95 ± 0.48	0.390 ± 0.001
0	2.59 ± 0.38	0.569 ± 0.003	4.43 ± 0.33	0.377 ± 0.001
1	1.41 ± 0.20	0.552 ± 0.003	3.30 ± 0.24	0.366 ± 0.001
2	0.767 ± 0.10	0.534 ± 0.003	2.46 ± 0.19	0.352 ± 0.001

6. Electric Field Dependence

The electric field dependence of the hole (majority carrier) thermal emission rate from the donor level was measured by graphically subtracting two capacitance transient decay curves in the same way applied to Au centers [16]. These curves are obtained by switching the reverse bias V_R to an intermediate value $V_I < V_R$, then back to V_R for one curve, and to $V_I + \Delta V$ then back to V_R for the second curve. The resulting difference curve, given by equation (12) in [16], is truly exponential if $\Delta V \ll V_I$ and $N_{TT} \ll N_A$. The electric field value which corresponds to the time constant of the difference curve was evaluated graphically. For this purpose, the impurity profiles of the diodes were measured by using the transient and steady state junction capacitance methods [16]. The result of the profile measurement is shown in Fig. 7a. The corresponding electric field for $V_R = 70$ V is shown in Fig. 7b where the cobalt centers are assumed to be in their neutral charge state. To narrow down the spread of the electric field which is represented by the difference curve, ΔV must be kept small. In practice, ΔV should be large enough to provide sufficient accuracy in obtaining the difference curve.

The preceding procedure cannot be applied to the trapped minority carriers (electrons), and a new technique, which is a variation of the general technique discussed previously, is presented here. With the aid of Fig. 7b, the electron thermal emission rate from the acceptor traps between $0 < x < W_I$ is measured as follows: At $t < 0$, the bias V_R is switched down to $V_I < V_R$. A monochromatic extrinsic light with $\lambda = 1.15$ μm is used to excite valence electrons into the traps in $0 < x < W_I$. In the region $x > W_I$, the majority carriers (holes) fill up the traps. At $t = 0$, the light is removed and simultaneously the voltage is switched back to V_R. At $t = 0^+$, there is an initial fast transient resulting

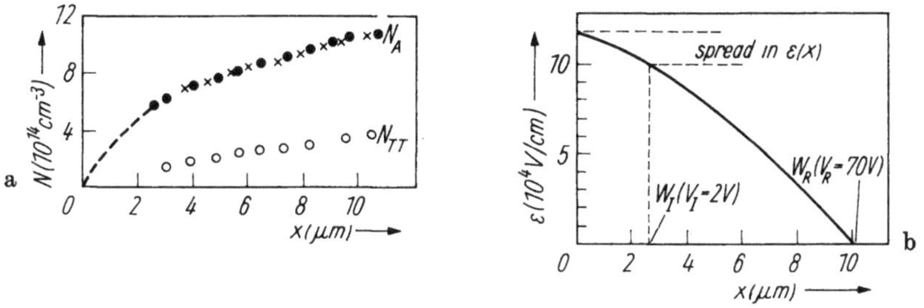

Fig. 7. Impurity profiles and electric field distribution in n^+–p diodes used. a) Impurity profiles of Co-doped (●, ○ IA2D-5) and control (× IA2C-2) diodes, showing the shallow and deep impurity concentrations. b) Spatial dependence of the electric field in the control and Co-doped diodes (with neutral trap charge) for $V_R = 70$ V

from the emission of trapped holes at the donor level in the region $W_I < x < W_R$. But as one observes in Fig. 6, the initial fast transient has a time constant 6 to 7 orders of magnitude faster than the electron emission time constant. Therefore, the electron thermal emission in the region $0 < x < W_I$ can be determined from the slower capacitance decay. At $t = 0^+$, the trapped electron distribution in $0 < x < W_I$ is denoted by $n_T(0^+, x)$. The charge induced at W_R due to $n_T(0^+, x)$ is

$$q\,N_A(W_R)\,\Delta W = \int\limits_0^{W_I} \frac{q\,n_T(t, x)\,x}{W_R}\,dx \tag{2}$$

which can be derived from the general result, equation (1) in [16], since $\Delta W \ll W_R$, the corresponding change in the capacitance is

$$\Delta C(t) = -\frac{K_s\,\varepsilon_0\,A}{W_R^2}\,\Delta W = \Delta C(0^+)\,e^{-t/\tau}, \tag{3}$$

where from equation (2)

$$\Delta C(0^+) = -\frac{K_s\,\varepsilon_0\,A}{W_R^3\,N_A(W_R)} \int\limits_0^{W_I} n_T(0^+, x)\,x\,dx \tag{4}$$

and $\tau = 1/e_{n1}^t$. The spatial variation of N_{TT} does not affect the time constant as long as $n_T(0^+, x) \ll N_A(x)$, which is a less stringent requirement than $N_{TT} \ll N_A$.

The measured electric field dependences of e_{p-1}^t and e_{n1}^t are shown in Fig. 8. The emission rates are normalized to the low field values. Due to the low value of N_A ($\approx 10^{15}$ cm^{-3}) in the diodes used, the maximum field is limited only to

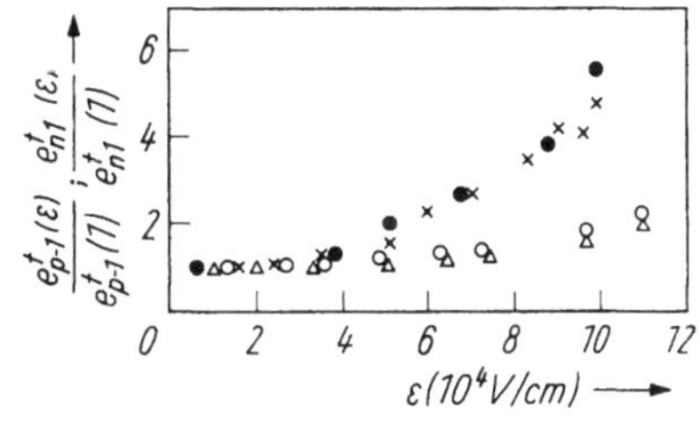

Fig. 8. Electric field dependences of the thermal emission rates of cobalt centers in silicon. e_{p-1}^t: ● $T = 190.0$ °K, × $T = 127.6$ °K; e_{n1}^t: ○ $T = 214.0$ °K, △ $T = 204.0$ °K

$\approx 10^5$ V/cm at which point premature breakdown occurs. For $\varepsilon \leqq 10^5$ V/cm, the electric field dependences of the normalized emission rates has very little temperature dependence in the temperature ranges shown in Fig. 8.

References

[1] H. IRMLER, Z. Naturf. **13a**, 557 (1958).
[2] N. HOLONYAK, JR., Proc. IRE **50**, 2421 (1962).
[3] S. K. GHANDHI, K. E. MORTENSON, and J. N. PARK, Proc. IRE **53**, 635 (1965).
[4] C. M. PENCHINA, J. S. MOORE, and N. HOLONYAK, JR., Phys. Rev. **143**, 634 (1966).
[5] J. S. MOORE, M. C. P. CHANG, and C. M. PENCHINA, J. appl. Phys. **41**, 5282 (1970).
[6] M. C. P. CHANG, C. M. PENCHINA, and J. S. MOORE, Phys. Rev. **B4**, 1229 (1971).
[7] C. T. SAH, W. W. CHAN, H. S. FU, and J. W. WALKER, Appl. Phys. Letters **20**, 193 (1972).
[8] L. D. YAU and C. T. SAH, Appl. Phys. Letters, **21**, 157 (1972).
[9] C. T. SAH, W. W. CHAN, and H. S. FU, presented at the American Physical Society Meeting in Atlantic City, March 1972.
[10] J. W. WALKER and C. T. SAH, phys. stat. sol. (a) **11**, 513 (1972).
[11] C. T. SAH, L. FORBES, L. L. ROSIER, and A. F. TASCH, JR., Solid State Electronics **13**, 759 (1970).
[12] L. D. YAU and C. T. SAH, phys. stat. sol. (a) **6**, 561 (1971).
[13] L. D. YAU, C. F. SMILEY, and C. T. SAH, phys. stat. sol. (a) **13**, 457 (1972).
[14] C. T. SAH, L. FORBES, L. L. ROSIER, A. F. TASCH, JR., and A. B. TOLE, Appl. Phys. Letters **15**, 145 (1969).
[15] A. F. TASCH, JR. and C. T. SAH, Phys. Rev. **81**, 800 (1970).
[16] C. T. SAH and H. S. FU, phys. stat. sol. (a) **14**, 59 (1972).

(Received September 25, 1972)

phys. stat. sol. (a) 14, 663 (1972)

Subject classification: 1.1; 22.1.2

Department of Physics, University of Alberta, Edmonton

Effects of a Non-Systematic Reflection on Anomalous Absorption in Electron Diffraction

By

C. D. CANN and S. S. SHEININ

The effects of the ($\bar{1}33$) non-systematic reflection on anomalous absorption in (220) images in Si have been studied as a function of the deviation of the ($\bar{1}33$) reflection from its Bragg condition. It was found that the presence of the ($\bar{1}33$) reflection resulted in marked changes in the absorption coefficients of the Bloch waves contributing to the (220) image. These changes were such that the effects of anomalous absorption were significantly less for small positive deviations of the ($\bar{1}33$) reflection from its Bragg condition than for small negative deviations. These predictions were confirmed experimentally in the case of thickness extinction contours. An analysis of the variation of the absorption coefficients with deviation from the Bragg condition of the non-systematic reflection in terms of Bloch wave channelling was also carried out. These results showed a good correlation between the magnitude of the absorption coefficient and the extent to which the Bloch wave intensity is concentrated at the atomic positions.

Der Einfluß der nichtsystematischen ($\bar{1}33$)-Reflexion auf die anomale Absorption von (220)-Diagrammen in Si wurde als Funktion der Abweichungen der ($\bar{1}33$)-Reflexionen von ihren Braggbedingungen untersucht. Es wurde gefunden, daß die Anwesenheit der ($\bar{1}33$)-Reflexion beträchtliche Änderungen der Absorptionskoeffizienten der Blochwellen ergibt, die zum (220)-Diagramm beitragen. Diese Änderungen sind von der Art, daß die Effekte der anomalen Absorption für kleine positive Abweichungen der ($\bar{1}33$)-Reflexion von ihrer Braggbedingung beträchtlich kleiner sind als für kleine negative Abweichungen. Diese Vorhersagen wurden experimentell für den Fall der Dicken-Extinktionskurven bestätigt. Eine Analyse der Änderungen der Absorptionskoeffizienten mit der Abweichung von der Braggbedingung der nichtsystematischen Reflexion mit dem Blochwellenchannelling wurde ebenfalls durchgeführt. Diese Ergebnisse zeigten eine gute Korrelation zwischen der Größe des Absorptionskoeffizienten und dem Ausmaß, wie die Blochwellenintensität an den Atomlagen konzentriert ist.

1. Introduction

The dynamical theory of electron diffraction describes the interaction of high-energy electrons with a crystal in terms of Bloch waves. Experimental observations such as the disappearance of thickness fringes in thick crystals [1] indicate that the amplitudes of these Bloch waves are attenuated at different rates as they propagate through the crystal. These rates of attenuation are described in the theory in terms of parameters called Bloch wave absorption coefficients. The contrast effects associated with the differences in these coefficients are referred to as being due to anomalous absorption. Taking anomalous absorption into account has played an important role in obtaining good agreement between contrast calculations based on the dynamical theory and experiment.

Recent work by one of the authors [2] has shown that, in the case where only systematic reflections are assumed to be present, the effects of anomalous absorption can be drastically altered by tilting the crystal. This behaviour results primarily from the fact that changes in Bloch wave excitation occur under these circumstances. Thus, for example, when the lowest-order reflection, g, of a systematic set is in the exact Bragg condition, pronounced effects of anomalous absorption can normally be observed. The reason for this is that the two Bloch waves which make a significant contribution to the diffracted beam intensity at this orientation have widely differing absorption coefficients. If the crystal is tilted outside the reflection $3\,g$ in the Bragg condition, on the other hand, the effects of anomalous absorption are found to be small. The explanation for this behaviour lies in the fact that in thick crystals, the two Bloch waves which make a significant contribution to the diffracted beam intensity have nearly equal absorption coefficients.

In practice, the assumption that only systematic reflections are present is open to question since non-systematic reflections must to some extent also be excited. Both Ayroles [3] and Lehmpfuhl [4] have shown that such reflections can have a significant effect on the combination of Bloch waves contributing to image intensity. Their results, therefore, indicate the possibility that anomalous absorption effects in the presence of non-systematic reflections might differ appreciably from those obtained by assuming that only systematic reflections are excited. The present investigation was undertaken with a view to exploring this possibility.

2. Theoretical Background

In the dynamical theory of electron diffraction [5, 6] solutions to the Schrödinger equation for a high-energy electron in a crystal potential are found in terms of Bloch waves of the form

$$B^j(\boldsymbol{r}) = \sum_g C_g^j \exp\{2\,\pi\,i\,(\boldsymbol{k}^j + \boldsymbol{g})\cdot\boldsymbol{r}\}\,, \tag{1}$$

where the sum is over all reciprocal lattice vectors considered, C_g^j is a Fourier coefficient, and $\boldsymbol{k}^j$ is the Bloch wave vector. Inelastic scattering is introduced phenomenologically in the theory by adding a small imaginary component $i\,U'(\boldsymbol{r})$ to the lattice potential $U(\boldsymbol{r})$. This component, when taken into account by using first-order perturbation theory, results in the addition of an imaginary part $i\,q^j$ to k^j where, assuming the $B^j(\boldsymbol{r})$'s to be normalized,

$$q^j = \frac{1}{2\,k^j}\int B^j(\boldsymbol{r})^*\;U'(\boldsymbol{r})\;B^j(\boldsymbol{r})\;\mathrm{d}\tau\,. \tag{2}$$

The differences that are commonly found between these Bloch wave absorption coefficients, q^j, have been explained by a number of authors in terms of the channelling of Bloch waves [1, 7, 8]. In general, the intensity $B^j(\boldsymbol{r})^*\,B^j(\boldsymbol{r})$ of a Bloch wave as it propagates through a crystal is not evenly distributed but instead is localized or channelled into certain regions either at or between the atomic positions. Since the potential $U'(\boldsymbol{r})$ is positive and localized at the atoms, it can be seen by examining equation (2) that a Bloch wave whose intensity is also concentrated at these positions will have a larger positive value of q^j than a wave whose intensity is a maximum between the atoms.

The importance of these differences in the absorption coefficients of the Bloch waves on thickness fringe contrast can be seen by examining the following expressions for the amplitude and intensity of a diffracted beam. In the case of the g-th reflection the amplitude at a crystal depth z is given by the expression

$$\psi_g(z) = \sum_{j=1}^{N} \varphi_g^j(z) \exp(2\pi i \gamma^j z), \tag{3}$$

where the summation is over all Bloch waves excited, γ^j is the difference between the z-components of the Bloch wave vector $\boldsymbol{k}^j$ and the incident wave vector $\boldsymbol{K}$ and

$$\varphi_g^j(z) = C_0^j C_g^j \exp(-2\pi q^j z). \tag{4}$$

From these two expressions it can be seen that the contribution of a particular Bloch wave to the diffracted beam amplitude at a depth z is determined by $\varphi_g^j(z)$ and, therefore, depends on both the product $C_0^j C_g^j$ and the exponential term $\exp(-2\pi q^j z)$. The effects of the differing absorption coefficients can now be understood by following the changes in the relative values of the φ_g^j's of the important Bloch waves which occur as crystal thickness increases. Consider, for example, the familiar case of a low-order reflection, g, in the exact Bragg condition. Under these circumstances, only two of the Bloch waves, numbered 1 and 2 [9], are strongly excited and the intensity of the diffracted beam g at a depth z is given by the expression

$$\psi_g(z)^* \psi_g(z) = \{\varphi_g^1(z)\}^2 + \{\varphi_g^2(z)\}^2 + 2\varphi_g^1(z)\varphi_g^2(z) \cos\{2\pi(\gamma^1 - \gamma^2) z\}. \tag{5}$$

Since, near the top surface of the crystal, $\varphi_g^1 \approx \varphi_g^2$ it can be seen from this expression that excellent thickness fringe contrast will be obtained. However, because the absorption coefficient of Bloch wave 1 is usually five to ten times greater than that of Bloch wave 2, $\varphi_g^1 \ll \varphi_g^2$ in thick crystals and equation (5) shows that poor fringe contrast will result. This change in fringe contrast with crystal thickness due to differing absorption coefficients is well known and is referred to as being due to anomalous absorption [1]. (It should be noted that the φ_g^j's can, in general, be positive or negative. However, only magnitudes will be considered in this paper since the effect of the sign is to introduce a phase factor which does not affect the fringe contrast at a given depth in the crystal.)

For a more general crystal orientation than that considered above, the magnitudes of the φ_g^j's of the important Bloch waves need not be equal at the top surface of the crystal. However, the effect of the differences in the absorption coefficients is to still change the relative magnitudes of the φ_g^j's with increasing thickness. These relative magnitudes will be defined in this paper by the ratio φ_g^i/φ_g^j where i and j are the important Bloch waves. In the following sections these ratios will be used to predict the effects of anomalous absorption on thickness fringe contrast.

3. Calculations

The first problem in undertaking an investigation involving non-systematic reflections is that of the choice of orientations to be used. For most orientations, a number of strong non-systematic reflections will be excited, making the analysis of the results extremely complicated. In order to avoid this problem, the authors attempted to find an orientation for which the effects of a single non-systematic refection could be isolated. This situation was obtained by choosing an orientation in Si at which the (220) and ($\bar{1}33$) reflections were close to their Bragg conditions. In order to verify that no other reflections would have a significant effect on the final results, both 3 and 23 beam calculations of the Bloch wave parameters were carried out. Good agreement was obtained between the results of both sets of calculations indicating that, in this case, the simpler 3 beam approximation is adequate for predicting the effects of a non-systematic reflection on anomalous absorption. In presenting the results which follow, the ($\bar{1}33$) and (220) reflections will be referred to as the non-systematic and systematic reflections respectively. This designation is, of course, completely arbitrary and is used, as is the convention in the literature, to indicate that the reciprocal lattice vectors concerned are not co-linear.

In order to determine possible effects of the ($\bar{1}33$) reflection on anomalous absorption in the (220) dark-field image, the relative values of both the absorption coefficients and the φ_g^j's of the excited Bloch waves must be considered. Since it would be expected that the magnitude of these effects would depend on $\Delta\theta_{\bar{1}33}$, the deviation of the ($\bar{1}33$) reflection from its Bragg condition, the q^j's and φ_g^j's were computed for a range of values of this parameter from $-0.4\ \theta_{\bar{1}33}$ to $+0.4\ \theta_{\bar{1}33}$ where $\theta_{\bar{1}33}$ is the ($\bar{1}33$) Bragg angle. These calculations were carried out for an accelerating voltage of 150 kV assuming the (220) reflection to be in its Bragg condition. Because of the particular configuration of reflections used, it was found that at $\Delta\theta_{\bar{1}33} \approx 0.05\ \theta_{\bar{1}33}$ two of the Bloch waves are degenerate. Sprague and Wilkens [10] have shown that when absorption is taken into account in this case the two branches of the dispersion surface concerned become smeared out in a small region around the degeneracy point. As a result, defining Bloch wave parameters in this region is of questionable significance. This has been indicated by the dashed lines in the plots of the q^j's and φ_g^j's appearing in Fig. 1 and 3. It should be noted, however, that this degeneracy is of importance over a quite small range of values of $\Delta\theta_{\bar{1}33}$ and thus, does not affect the general results to be presented here.

3.1 Bloch wave absorption coefficients

As was discussed earlier in Section 2, anomalous absorption effects arise from differences in the Bloch wave absorption coefficients. The values of these coefficients for the three Bloch waves excited are shown in Fig. 1 as a function of $\Delta\theta_{\bar{1}33}$. As can be seen the coefficients show a strong dependence on deviation from the Bragg condition of the ($\bar{1}33$) reflection indicating that the presence of the non-systematic reflection does, in fact, have a pronounced effect on the values of the absorption coefficients obtained. In order to interpret the variation of the q^j's with $\Delta\theta_{\bar{1}33}$ a study has been carried out of the channelling of the Bloch waves in the crystal. Fig. 2 shows the calculated Bloch wave intensity

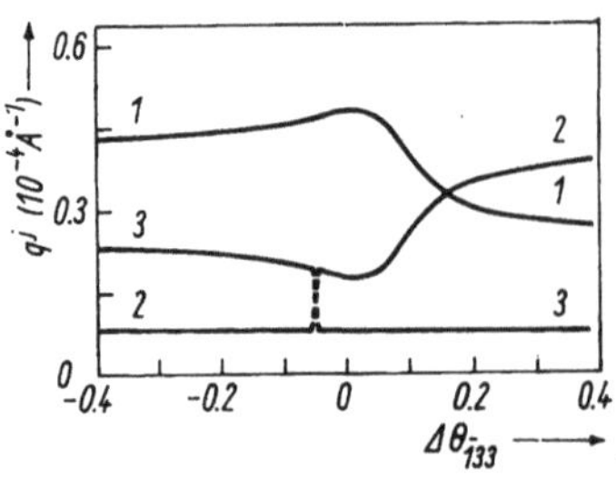

Fig. 1. The variation of the Bloch wave absorption coefficients, q^j, with $\Delta\theta_{\bar{1}33}$, the deviation of the ($\bar{1}33$) reflection from its Bragg condition. $\Delta\theta_{\bar{1}33}$ is given in fractions of $\theta_{\bar{1}33}$, the ($\bar{1}33$) Bragg angle

distributions across a section of crystal perpendicular to the direction of current flow for the three Bloch waves at orientations of $-0.3\ \theta_{\bar{1}33}$, $0.02\ \theta_{\bar{1}33}$, and $0.3\ \theta_{\bar{1}33}$ respectively [11]. The light regions in these displays correspond to zones of high current flow. The lines extending vertically from the top and bottom of each display and those extending diagonally from the sides show the locations of (220) and ($\bar{1}33$) planes, respectively. The black disks show the positions of the atomic rows which lie perpendicular to the plane of the displays.

As can be seen in Fig. 2a, at $\Delta\theta_{\bar{1}33} = -0.3\ \theta_{\bar{1}33}$ the intensity of Bloch wave 1 is concentrated along the (220) planes and, moreover, is localized near the atoms, thus resulting in it being highly absorbed. Bloch wave 2 is concentrated between the (220) planes and, therefore, is weakly absorbed while the intensity distribution of Bloch wave 3 is nearly uniform resulting in moderate absorption. At $\Delta\theta_{\bar{1}33} = 0.02\ \theta_{\bar{1}33}$, as seen in Fig. 2b, the intensity of Bloch wave 1 is now very strongly concentrated at the atomic positions resulting in even higher absorption than at $\Delta\theta_{\bar{1}33} = -0.3\ \theta_{\bar{1}33}$. In the case of Bloch wave 2 the intensity, although mainly lying along the (220) planes, is strongly concentrated between the atoms and, thus, a relatively low absorption coefficient is expected. For Bloch wave 3 weak absorption should be obtained as its intensity is principally located between the (220) planes. At $\Delta\theta_{\bar{1}33} = 0.3\ \theta_{\bar{1}33}$ the intensity of Bloch wave 1, as seen in Fig. 2c, is nearly uniformly distributed with the result that this wave is now only moderately absorbed. Bloch wave 2 is no longer as strongly concentrated between the atoms on the (220) planes as at $\Delta\theta_{\bar{1}33} = 0.02\ \theta_{\bar{1}33}$ with the result that it has become more strongly absorbed. Bloch wave 3 is concentrated between the (220) set of planes and, therefore, is weakly absorbed. These results indicate, therefore, that the non-systematic reflection has a pronounced effect on the channelling characteristics of the Bloch waves and that, as in the two-beam and systematic cases, these characteristics offer a physical explanation for the differences in the Bloch wave absorption coefficients.

3.2 *Effects of non-systematic absorption coefficients on thickness fringe contrast*

In Section 2 it was noted that the effects of anomalous absorption on thickness fringe contrast are determined by the variation of the relative magnitudes of the φ_g^j's of the important Bloch waves with crystal thickness. Before examining these variations, it is of interest, however, to look at the φ_g^j's at the top surface of the crystal as shown in Fig. 3a. It can be seen from these results that for large negative values of $\Delta\theta_{\bar{1}33}$, $\varphi_g^1 \approx \varphi_g^2 \gg \varphi_g^3$ indicating that the important Bloch waves are 1 and 2. For large positive values of $\Delta\theta_{\bar{1}33}$, on the other hand, $\varphi_g^2 \approx \varphi_g^3 \gg \varphi_g^1$ indicating that the important Bloch waves are 2 and 3. In the limit as $|\Delta\theta_{\bar{1}33}|$ becomes large, the two-beam situation, for which the important Bloch waves are 1 and 2, should be approached. Thus, for large negative values of $\Delta\theta_{\bar{1}33}$, the Bloch wave numbering in the non-systematic case is consistent with the two-beam case while for large positive values of $\Delta\theta_{\bar{1}33}$ it is not. This effect of non-systematic reflections on the Bloch wave numbering system will be discussed in a later paper by the authors.

It is next of interest to look at the variation of the relative values of the φ_g^j's with crystal thickness for small positive and negative values of $\Delta\theta_{\bar{1}33}$ (values of $\Delta\theta_{\bar{1}33}$ will be expressed in fractions of a Bragg angle of the ($\bar{1}33$) reflection). Fig. 3a, b show that for $\Delta\theta_{\bar{1}33} = -0.1$, the values of φ_g^1/φ_g^2 near the top surface of the

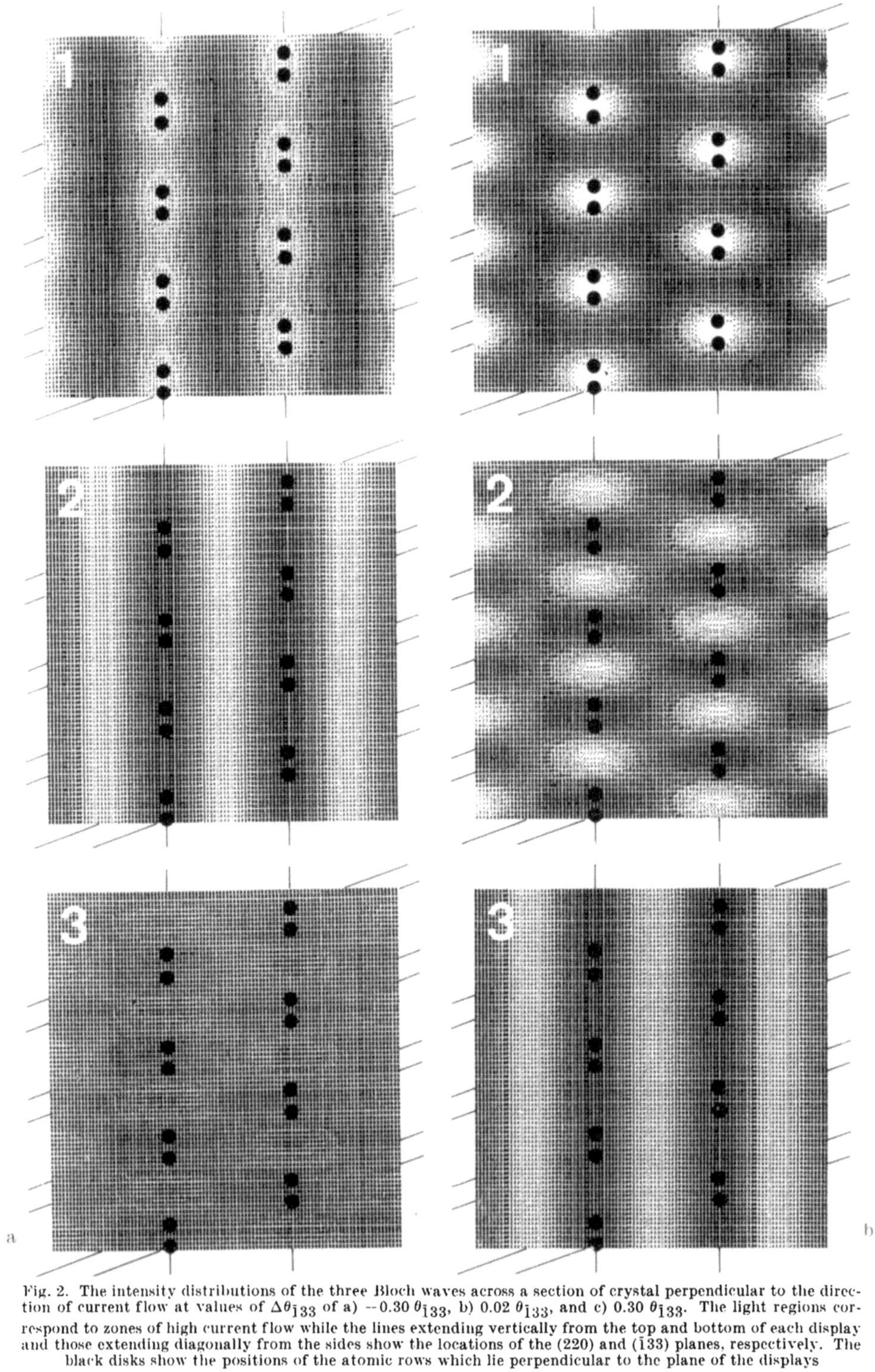

Fig. 2. The intensity distributions of the three Bloch waves across a section of crystal perpendicular to the direction of current flow at values of $\Delta\theta_{\bar{1}33}$ of a) $-0.30\ \theta_{\bar{1}33}$, b) $0.02\ \theta_{\bar{1}33}$, and c) $0.30\ \theta_{\bar{1}33}$. The light regions correspond to zones of high current flow while the lines extending vertically from the top and bottom of each display and those extending diagonally from the sides show the locations of the (220) and ($\bar{1}$33) planes, respectively. The black disks show the positions of the atomic rows which lie perpendicular to the plane of the displays

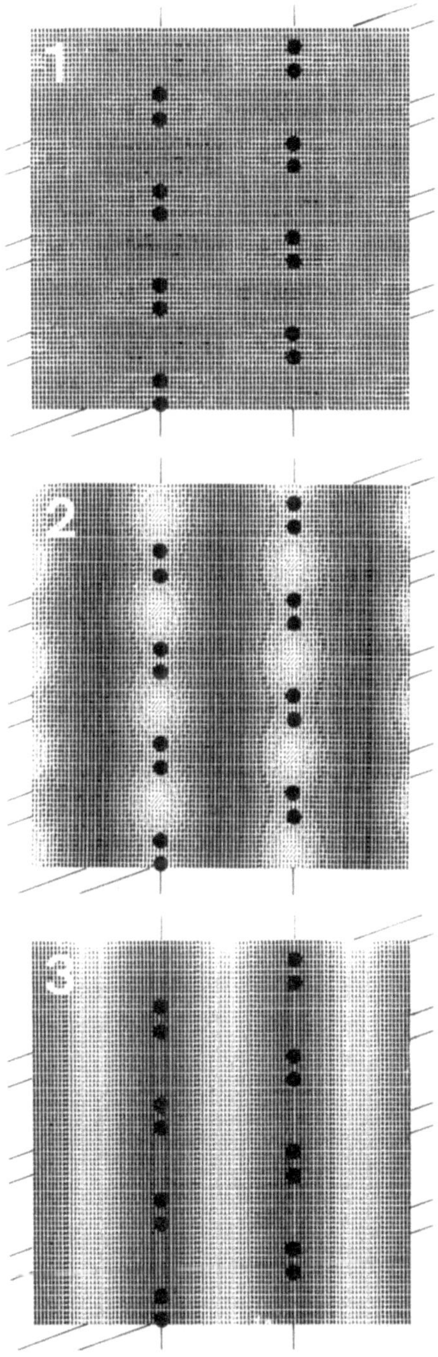

crystal and at a depth of 10000 Å are about 1 and 0.1, respectively. At $\Delta\theta_{\bar{1}33} = +0.1$ the corresponding values of φ_g^2/φ_g^3 are about 0.7 and 0.25, respectively. These ratios indicate that a marked decrease in thickness fringe contrast occurs with increasing thickness for $\Delta\theta_{\bar{1}33} = -0.1$ (φ_g^1/φ_g^2 decreases a factor of 10) while for $\Delta\theta_{\bar{1}33} = = +0.1$ the decrease in contrast is relatively small (φ_g^2/φ_g^3 decreases by a factor of 3). As a consequence, thickness fringes in thick crystals exhibit considerably better contrast at small positive values of $\Delta\theta_{\bar{1}33}$ than for small negative values even though in thin crystals somewhat worse contrast is obtained. Since a decrease in fringe contrast with crystal thickness is associated with anomalous abs orption (see Section 2), these results indicate that for small negative values of $\Delta\theta_{\bar{1}33}$, anomalous absorption effects are much more pronounced than for small positive values.

The enhanced thickness fringe contrast at small positive values of $\Delta\theta_{\bar{1}33}$ noted in the previous paragraph should result in thickness fringes being observed in thicker crystals than for small negative values. That this is, in fact, so can be seen from Fig. 4 and 5 which show theoretical and experimental thickness fringes for a range of values of $\Delta\theta_{\bar{1}33}$ around the Bragg condition of the ($\bar{1}$33) reflection. The theoretical results were obtained by calculation of diffracted beam intensity as a function of depth in the crystal for different values of $\Delta\theta_{\bar{1}33}$ and using a two-dimensional display to plot the results [12]. The experimental results show thickness fringes in a crystal bent so that approximately the same range of values of $\Delta\theta_{\bar{1}33}$ was obtained. It can be seen that there is good agreement between theory and experiment verifying that effects of anomalous ab-

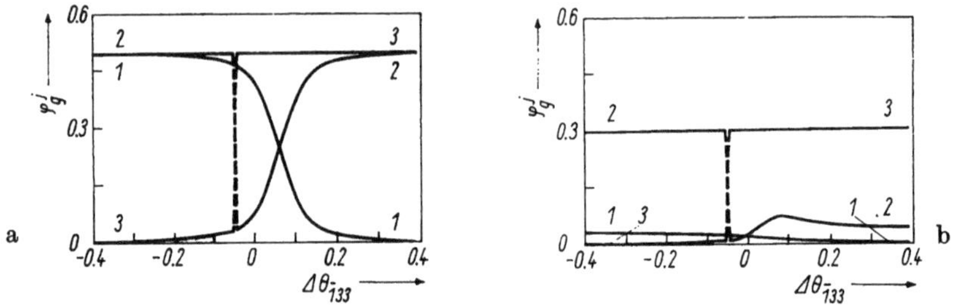

Fig. 3. The variation of the φ_g^j's, the contributions of the different Bloch waves to the (220) diffracted beam amplitude, with $\Delta\theta_{\bar{1}33}$ at a) the top surface of the crystal and b) a depth of 10000 Å

sorption are more pronounced for small negative values of $\Delta\theta_{\bar{1}33}$ than for small positive values.

Finally the effects of anomalous absorption obtained close to the Bragg condition of the non-systematic reflection were compared with those obtained in the two-beam limit (i.e., for large values of $|\Delta\theta_{\bar{1}33}|$). It was found that, under these circumstances, the values of φ_g^1/φ_g^2 near the top surface of the crystal and at a depth of 10000 Å are 1 and 0.12 respectively. A comparison of these values with those for the small values of $\Delta\theta_{\bar{1}33}$ quoted above indicate that anomalous absorption effects for large values of $|\Delta\theta_{\bar{1}33}|$ are very similar to those obtained for small negative values. For small positive values of $\Delta\theta_{\bar{1}33}$, on the

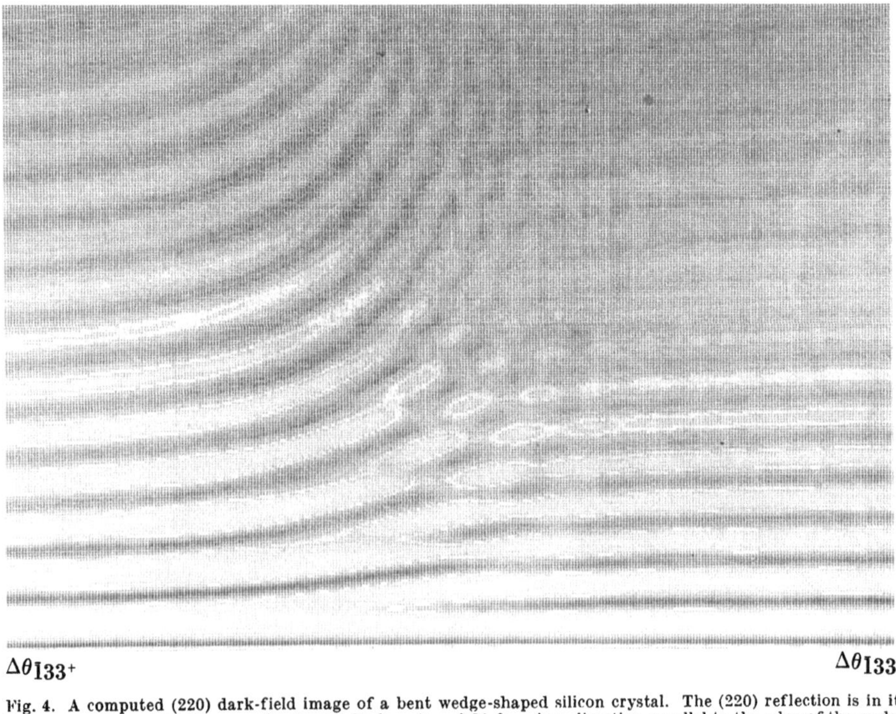

$\Delta\theta_{\bar{1}33}{}^{+}$ $\Delta\theta_{\bar{1}33}{}^{-}$

Fig. 4. A computed (220) dark-field image of a bent wedge-shaped silicon crystal. The (220) reflection is in its Bragg condition while $\Delta\theta_{\bar{1}33}$ varies from 0.21 $\theta_{\bar{1}33}$ to -0.21 $\theta_{\bar{1}33}$ in a direction parallel to the edge of the wedge

other hand, the marked changes in effect of anomalous absorption which have already been noted, occur. This behaviour is consistent with the results in Fig. 1, which show that small changes in absorption coefficient occur for negative values of $\Delta\theta_{\bar{1}33}$ whereas marked changes occur for positive values.

A final point of explanation with regard to the complex thickness fringes is seen at $\Delta\theta_{\bar{1}33} \approx 0$ in Fig. 4 and 5. Complex fringes have been observed before in both systematic and non-systematic cases [3, 12 to 15] and have been explained in terms of the presence of at least three important Bloch waves. That there are also three important Bloch waves at the orientation considered in this paper can be seen from Fig. 3a, b.

4. Conclusions

(i) The results obtained in Section 3.1 show that the presence of the ($\bar{1}$33) non-systematic reflection has a pronounced effect on the absorption coefficients of the Bloch waves excited when the (220) reflection is in its Bragg condition. This effect can be explained by the strong dependence of the channelling characteristics of the Bloch waves on $\Delta\theta_{\bar{1}33}$.

(ii) The variations of the absorption coefficients with $\Delta\theta_{\bar{1}33}$ is such that for small positive values of $\Delta\theta_{\bar{1}33}$ anomalous absorption effects in the case of thickness contours are significantly less than those for small negative values. This results in thickness contours being visible in thicker regions of the crystal for $\Delta\theta_{\bar{1}33}$ positive than for $\Delta\theta_{\bar{1}33}$ negative.

$\Delta\theta_{\bar{1}33}+$ $\Delta\theta_{\bar{1}33}-$

Fig. 5. An experimental (220) dark-field image of a bent wedge-shaped silicon crystal. The (220) reflection is approximately in its Bragg condition while $\Delta\theta_{\bar{1}33}$ varies from positive to negative in moving from left to right along the edge of the wedge

(iii) A comparison of the effects of anomalous absorption obtained when the $(\bar{1}33)$ non-systematic reflection is close to its Bragg condition with those obtained for large values of $|\Delta\theta_{\bar{1}33}|$ (i.e., in the two-beam limit) has been carried out. The results indicated that thickness fringes for small negative values of $\Delta\theta_{\bar{1}33}$ should be very similar to those obtained in the two-beam limit. Thus, the effects of a non-systematic reflection on thickness fringe contrast can, to a good approximation, be avoided even though the reflection may be close to its Bragg condition. For small positive values of $\Delta\theta_{\bar{1}33}$, the effects of anomalous absorption are considerably smaller than in the two-beam limit and the enhanced contrast in thick crystals referred to in (ii) results.

Acknowledgements

The authors wish to thank A. E. B. Monk for assistance in the computer programming. They are also grateful to P. S. Turner and K. Z. Botros for helpful discussions and to the National Research Council of Canada for financial assistance.

References

[1] H. HASHIMOTO, A. HOWIE, and M. J. WHELAN, Proc. Roy. Soc. **A269**, 80 (1962).
[2] S. S. SHEININ, phys. stat. sol. **38**, 675 (1970).
[3] R. AYROLES, J. Microscopie **11**, 193 (1971).
[4] G. LEHMPFUHL, Z. Naturf. **27a**, 425 (1972).
[5] H. BETHE, Ann. Phys. (Germany) **87**, 55 (1928).
[6] P. B. HIRSCH, A. HOWIE, R. B. NICHOLSON, D. W. PASHLEY, and M. J. WHELAN. Electron Microscopy of Thin Crystals, Butterworths, London 1965.
[7] A. HOWIE, Phil. Mag. **14**, 223 (1966).
[8] F. REYNAUD, J. Microscopie **11**, 179 (1971).
[9] C. J. HUMPHREYS and R. M. FISHER, Acta cryst. **A27**, 42 (1971).
[10] J. A. SPRAGUE and M. WILKENS, Proc. 7th Internat. Conf. Electron Microscopy, Grenoble 1970, Vol. 1 (p. 95).
[11] A. J. F. METHERELL and M. S. SPRING, Proc. 7th Internat. Conf. Electron Microscopy, Grenoble 1970, Vol. 1 (p. 105).
[12] M. S. SPRING and J. W. STEEDS, phys. stat. sol. **37**, 303 (1970).
[13] S. S. SHEININ, phys. stat. sol. **21**, 247 (1967).
[14] J. W. STEEDS, phys. stat. sol. **38**, 203 (1970).
[15] R. UYEDA and M. NONOYAMA, Japan J. appl. Phys. **4**, 498 (1965).

(Received September 25, 1972)

phys. stat. sol. (a) **14**, 673 (1972)

Subject classification: 9; 22.6

Department of Nuclear Engineering, North Carolina State University, Raleigh

Xenon-133 Diffusion and Trapping in Single-Crystal Uranium Dioxide[1]

By

J. C. CARTER[2], E. J. DRISCOLL[3], and T. S. ELLEMAN

The diffusion of ^{133}Xe in single-crystal UO_2 was studied following the recoil injection of the ^{133}Xe from an external fissionable source. This technique was selected to minimize the diffusion path to a free surface and decrease the possibility of gas-atom trapping during the diffusion anneals. Gas release results consistent with classical diffusion solutions were obtained over the entire heating cycle for each release curve and no "burst effect" was observed. The results were consistent with $D = 2.88 \times 10^3 \exp[(-4.78 \pm 0.15)\,\text{eV}/kT]\,\text{cm}^2\,\text{s}^{-1}$ over the temperature range 1065 to 1300 °C. Increasing the recoil concentration to 3×10^{12} fission fragments cm^{-2} (3×10^{-8} fission atom fraction) produced anomalies in the gas release curves which probably resulted from gas-atom trapping. Using parameters measured elsewhere for rare gas trapping in UO_2 and CaF_2, it was shown that many of the rare gas diffusion coefficients reported in the literature for UO_2 appear to be too low as a consequence of gas-atom trapping.

Die Diffusion von ^{133}Xe in UO_2-Einkristallen wurde im Anschluß an die Rückstoßinjektion von ^{133}Xe aus einer äußeren Quelle untersucht. Diese Technik wurde gewählt, um den Diffusionsweg zu einer freien Oberfläche zu minimalisieren und verringert die Möglichkeit des Anhaftens der Gasatome während der Diffusionstemperung. Ergebnisse für Gasabgabe in Übereinstimmung mit klassischen Diffusionslösungen wurden über den gesamten Heizzyklus für jede Abgabekurve erhalten und kein "burst effect" beobachtet. Die Ergebnisse waren konsistent mit $D = 2{,}88 \times 10^3 \exp[(-4{,}78 \pm 0{,}15)\,\text{eV}/kT]\,\text{cm}^2\,\text{s}^{-1}$ im Temperaturbereich 1065 bis 1300 °C. Ein Anstieg der Rückstoßkonzentration auf 3×10^{12} Spaltteilchen cm^{-2} (3×10^{-8} Spaltungsatomanteil) ergab Anomalien in der Gasabgabekurve, die vermutlich vom Anhaften der Gasatome herrührten. Mit an anderer Stelle gemessenen Parametern für Anhaften von Edelgasen in UO_2 und CaF_2 wird gezeigt, daß viele der Edelgasdiffusionskoeffizienten, die in der Literatur für UO_2 angegeben werden, infolge des Anhaftens der Gasatome, zu klein zu sein scheinen.

1. Introduction

Numerous measurements [1 to 12] of rare gas release from UO_2 have been carried out for use in fuel swelling and fission product release models. Unfortunately, the gas release curves have frequently been nonclassical and compilations of calculated diffusion coefficients have shown variations of four orders of magnitude or larger at selected temperatures [13, 14]. Other studies have shown that internal porosity [7], impurities [15], excess oxygen [12, 16], and radiation damage [6, 9, 17] can affect gas release behavior, and it appears that combinations of these effects have produced the wide range of diffusion coefficients measured for UO_2.

---

[1]) Work supported under USAEC Contract AT (40-1)-3508.

[2]) Current address: Tennessee Valley Authority, Knoxville, Tennessee 37902.

[3]) Current address: U.S. Army Engineer School, Ft. Belvoir, Virginia.

Trapping of gas atoms at radiation-induced defect clusters is a particularly important mechanism for inhibiting rare gas release from solids. MacEwan and Morel [9] observed a decrease of three orders of magnitude for Xe diffusion coefficients in UO_2 when the fission density was increased from 10^{14} to 10^{18} fissions cm^{-3}. Other investigators [18 to 22] have demonstrated radiation-induced trapping in a variety of solids, and there appears to be little doubt that this mechanism can lead to anomalous rare gas diffusion at relatively low fission densities or fast neutron fluences. From measurements of ^{133}Xe release from irradiated CsI, Elleman et al. [21] showed that high trap concentrations can result in apparently classical gas release curves but diffusion coefficients which are significantly lower than those obtained in the absence of trapping. The gas atoms move from trap to trap and the release rate from traps becomes the controlling factor in gas transport rather than random walk between lattice sites. Thus, the observation of classical release kinetics and temperature dependence is not sufficient evidence to eliminate the possibility that the diffusion coefficients are affected by trapping.

Trapping effects can be minimized by using ion injection or fission recoil to introduce the diffusing gas atoms into a solid. Since the gas atoms initially lie near the surface with these methods and move only relatively short distances before release at a free surface, there is less likelihood of a trap being encountered than with a solid where the gas is initially uniformly distributed. Ong and Elleman [23] have applied the simple trapping model proposed by Hurst [24] to the case of a recoil-injected gas and have concluded that trapping effects do not have a major effect on the gas release curves at release fractions below 0.2 when the recoil range is no greater than ten times the trapping mean free path, but trapping can be a major influence when this ratio reaches 100.

The present study concerns the measurement of ^{133}Xe diffusion in stoichiometric UO_2 where the diffusing xenon has been introduced through fission recoil from an external fissionable source to minimize trapping effects. By using fully enriched uranium foils and depleted, single crystals of UO_2, the recoil injected ^{133}Xe becomes the only significant contributor to rare gas release. The purpose of the investigation was to determine if gas release measurements under conditions where trapping was minimized would give diffusion coefficients consistent with the literature and, further, to investigate whether the wide spread in literature diffusion coefficients can be explained in terms of consistent trapping parameters.

2. Experimental

Single crystals of UO_2 depleted in ^{235}U were obtained from Battelle Northwest Laboratories. The crystals were hand-picked from a fused melt and heated in hydrogen at 1750 °C to assure stoichiometry. Density was measured by mercury displacement as 10.94 g cm^{-3} (theoretical density 10.98 g cm^{-3}) and the oxygen-to-metal ratio was measured as 2.0006 to 2.001. Total impurities as determined by spectrographic analysis were less than 200 ppm. Slab specimens were cut with a diamond wire saw and one surface was polished with fine-grit SiC suspended to ethylene glycol. The samples were vacuum-annealed at 1300 °C for 20 h to remove surface strains.

90% ^{235}U enriched uranium metal foils were sandwiched between UO_2 crystals and KI pellets and irradiated under vacuum in the 10 kW NCSU reactor.

Fission densities were kept low to minimize trap production and most irradiations were terminated at a thermal fluence of 1×10^{14} n cm^{-2} (equivalent to 3.6×10^{11} fission recoils cm^{-2} or 2.3×10^{-9} fission atom fraction). All samples were stored under vacuum for 10 to 12 h prior to a diffusion anneal to remove the surface-absorbed rare gas which was found to occur and which contributed to misleading gas release results when this step was omitted. The total ^{133}Xe in a specimen was monitored directly for both UO_2 and KI with a Ge(Li) solid-state detector set to record only the 0.081 MeV photopeak. A KI crystal was then sealed at 1×10^{-2} Torr in a previously degassed gas-release quartz tube, the ^{133}Xe driven from the samples by melting the crystal, and the equilibrium counting rate determined. Since it was not feasible to release all of the ^{133}Xe from the UO_2 specimens, this technique was employed to determine the counting efficiency for ^{133}Xe under geometry conditions identical to those used in the UO_2 experiments. These results were used in combination with the solid-state detector results for KI and UO_2 to predict the ^{133}Xe counting rate for complete release from UO_2 which was needed for the calculation of gas release fractions.

Xenon-133 release from UO_2 was measured at constant temperature by continuously recording the counting rate of ^{133}Xe external to the specimen as a function of time over a period of 40 h. Alumina or recrystallized alumina tubes replaced quartz for the higher annealing temperatures, and the system was thoroughly purged with helium prior to evacuation to remove oxygen. Each crystal was inspected at the end of a run for evidence of film formation or evaporation. Xenon-133 release was measured only over the temperature range 1065 to 1300 °C since below 1065 °C ^{133}Xe release rates were undesirably low while above 1300 °C the surface evaporation losses became sufficiently large to interfere with diffusion release. Surface evaporation was monitored by weight loss, and approximately 0.5 μm was lost from the surface after 40 h at 1300 °C. This loss rate was consistent with the values predicted from extrapolation of higher-temperature evaporation data in which a loss of 0.22 μm would be predicted at 1300 °C.

3. Results

Gas release results at 3.6×10^{11} fission fragments cm^{-2} were reproducible and consistent with classical diffusion solutions over the entire release curve. Fig. 1 shows the agreement between four experimental curves and the best-fit theoretical curves obtained for constant values of D. The slight divergence of theoretical and experimental curves at 1300 °C after 6 h of heating is believed to

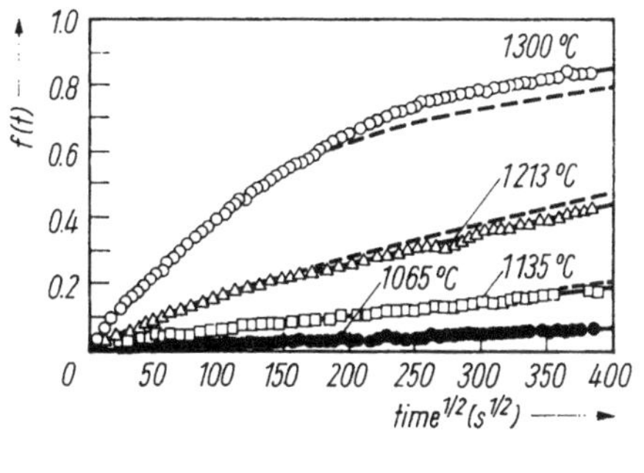

Fig. 1. Typical release characteristics of ^{133}Xe, $f(t)$, from UO_2 single crystals (3×10^{11} fission fragments cm^{-2}). Solid line: experimental; dashed line: theoretical best fit

reflect the gas release resulting from evaporation. In contrast to many earlier investigations, no "burst release" or other deviations from classical release were observed. The calculated curves in Fig. 1 were obtained from the classical diffusion solutions for gas release from a fission-recoil-injected slab which have been summarized by DiCola and Matzke [25].

Values of D obtained from the classical diffusion solution are summarized in Fig. 2 as a plot of ln D versus T^{-1}. The results are consistent with

$$D = D_0 \exp\left(-\frac{\Delta H}{kT}\right),$$

$$D_0 = \left(2.88 \begin{array}{c} +\ 13.4 \\ -\ \ 2.4 \end{array}\right) \times 10^3\ \mathrm{cm^2\ s^{-1}}\,,$$

$$\Delta H = (4.78 \pm 0.15)\ \mathrm{eV}\,.$$

Gas release experiments at an order of magnitude higher fission recoil concentration (3×10^{12} fission fragments $\mathrm{cm^{-2}}$, equivalent to 3×10^{-8} fission atom fraction) exhibited distinct deviations from classical gas release kinetics. The initial slopes of the gas release curves were often consistent with the lower-concentration release curves for the first two to three hours, but the experimental curves then dropped significantly below the predicted values at longer heating times. Fig. 3 illustrates this effect in experiments run at 1065 and 1213 °C. The dashed line shows the predicted release while the points represent observed release. This behavior is consistent with gas trapping in the solid, and is also more representative of the bulk of gas release experiments found in the literature. The specimens in Fig. 1 and 3 were cut from the same boule and differed only in fission recoil concentrations.

Specimens cut from one crystal gave apparent diffusion coefficients which were significantly lower than other measured values at 3×10^{12} fission fragments $\mathrm{cm^{-2}}$. Visual examination of these specimens showed that they contained large numbers of small voids, apparently a result of gas precipitation during cooling

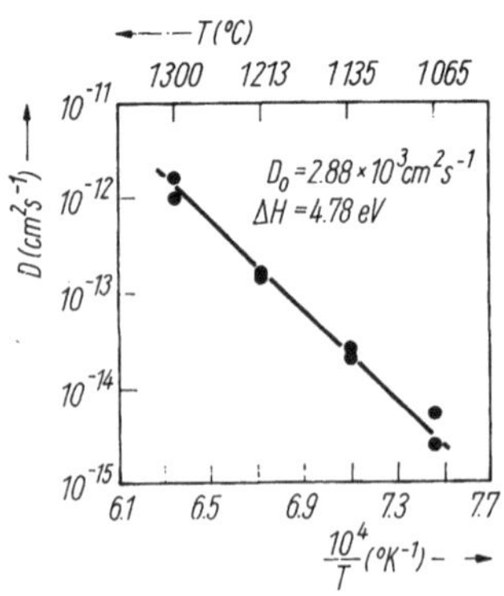

Fig. 2. Diffusion coefficients D for ^{133}Xe in UO_2 single crystals

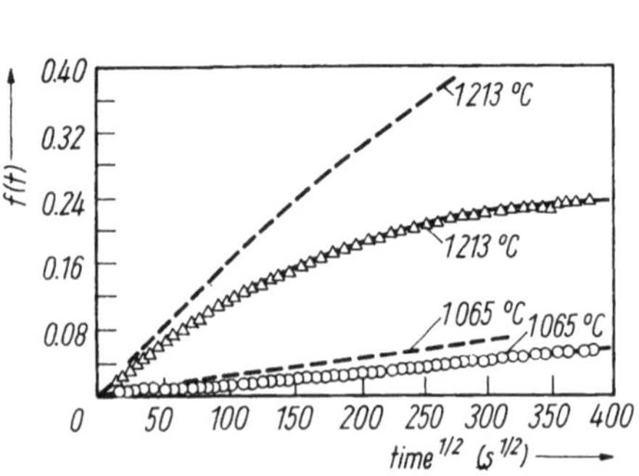

Fig. 3. Typical release characteristics of ^{133}Xe, $f(t)$, from UO_2 single crystals (3×10^{12} fission fragments $\mathrm{cm^{-2}}$). ○, △ experimental; dashed line: predicted curve from Fig. 2 results

from a melt. The concentration of pores in the size range from several tenths to one micron was 3×10^{11} pores cm^{-3} which was approximately one per cent of the minimum trap concentration required with Hurst's model [24] to explain the observed lowering of apparent diffusion coefficients. It is likely that many of the pores were below visible limits and the true concentration was perhaps closer to 10^{13} pores cm^{-3}. MacEwan and Stevens [7] have observed pore concentrations of 2×10^{13} cm^{-3} and typical pore sizes of 30 to 300 Å in unirradiated, sintered UO_2 specimens. They further observed that these trap concentrations were sufficient to account for observed trapping at low-radiation exposures. The observation of high void concentrations in the present work illustrates that even single-crystal materials can exhibit sufficient pore densities to measurably reduce rare gas release rates and give misleading diffusion coefficients.

4. Discussion

4.1 Comparison with other work

Fig. 4 summarizes the ^{133}Xe diffusion results in UO_2 which have been previously reported in the literature. Only experiments in which the experimental techniques have been well defined have been included. The dashed line gives the results of the present study which is observed to fall in the upper range of measured diffusion coefficients. Trapping of gas atoms at defects lowers the gas release rates and results in low apparent diffusion coefficients when classical diffusion solutions are used. In most of the experiments tabulated in Fig. 4, the authors reported deviations from classical release kinetics which could be attributed to trapping effects in the solid. The observation of relatively high values of D in the present study is consistent with the hypothesis that many earlier measurements were influenced by gas trapping and exhibited anomalous values of D. The absence of a "burst effect" and the observation of classical diffusion kinetics in the present study which is virtually unique among the reported UO_2 literature further supports this interpretation.

Exposure of heated UO_2 to traces of oxygen results in hyperstoichiometric UO_2 and accelerated rare gas diffusion. Oxygen analyses were not performed on the single-crystal specimens following diffusion anneals but considerable effort was devoted to excluding oxygen, and it is not believed that the higher-than-average diffusion coefficients can be attributed to hyperstoichiometric UO_2. On several occasions, a small amount of air was allowed to enter the sample chamber during a diffusion anneal and a dramatic increase in gas release was immediately observed. Constant values of D were observed at all stages of heating in the present work, and it seems unlikely that the sample was oxidizing during heating.

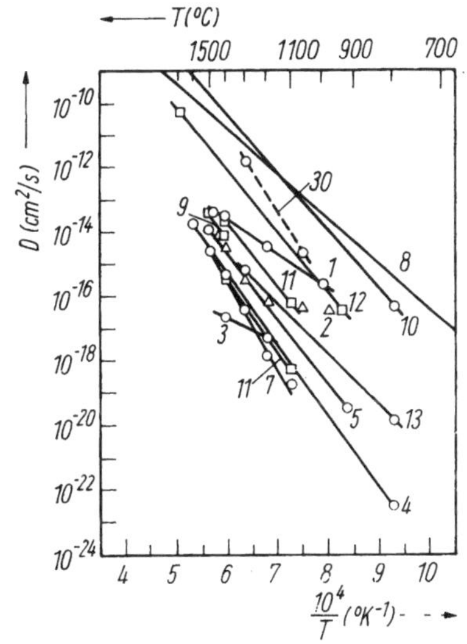

Fig. 4. Summary of literature values for ^{133}Xe diffusion D in UO_2

The predicted diffusion coefficient for ^{133}Xe at the UO_2 melting point appears to be consistent with earlier predictions which are based upon an interstitial mechanism for rare gas migration. Norgett and Lidiard [26] have presented arguments supporting interstitial migration of rare gases in solids and Felix and Lagerwall [27, 28] have stated that the diffusion coefficient at the melting point should be about 10^{-5} cm^2 s^{-1} if this mechanism is correct. Other solids in which diffusion is believed to occur through an interstitial process have exhibited general, consistent agreement with this diffusion coefficient at their melting points [27, 28]. The current results give a predicted D at the UO_2 melting point of 0.6×10^{-5} cm^2 s^{-1} which is consistent with the prediction by Felix.

4.2 Trapping effects

The spread in reported diffusion coefficients for rare gases in UO_2 appears to result principally from the combined effects of trapping in internal pores, trapping at radiation-induced defects, and uncertainty in the interpretation of non-classical gas release kinetics. Hyperstoichiometry, impurity effects, and faulty surface area measurements may also be contributing factors in selected cases. It is of interest to investigate the extent to which the spread in results can be explained solely in terms of trapping at radiation-induced defects, since this effect has been clearly identified in UO_2 and other solids as a major source of error in diffusion studies.

The generalized trapping model presented by Hurst [24] permits the calculation of gas release rates from a fissionable solid when the gas diffusion constant, trap concentration, and average retention time in the traps are known. MacEwan and Morel [9] and Spindler and Lindner [11] have measured values of these trapping parameters for UO_2. Unfortunately, their results cover only isolated conditions and it is not possible to determine a clear pattern for either the trap concentrations or gas retention times. Ong and Elleman [29] have carried out more extensive measurements of trapping for ^{133}Xe diffusion in single-crystal CaF_2 at several fission fragment concentrations and were able to represent trap concentrations and average retention times for gas atoms in traps as exponential functions of temperature for a given recoil concentration. Calcium fluoride has often been used as a model substance for uranium dioxide in past investigations as the crystal structures, lattice parameters, ionic radii, and rare gas diffusion activation energies are similar for the two solids. The trapping parameters determined for CaF_2 have, therefore, been adjusted to the few values measured directly for UO_2 and used to estimate trapping parameters in UO_2 at selected temperatures and fission densities. These calculated parameters have been applied to Hurst's model along with the measured diffusion coefficients in the present paper to calculate the expected gas release curves for the experimental conditions employed by past investigators.

Apparent diffusion coefficients were then calculated from these curves, assuming that classical diffusion solutions without trapping were applicable, and these apparent diffusion coefficients were compared with the values of D reported in literature.

The trapping parameters reported by Ong and Elleman [29] for CaF_2 differed by several orders of magnitude from the values reported by MacEwan and Morel [9] and Spindler and Lindner [11] for UO_2 at equivalent reduced temperatures. An arbitrary displacement of the CaF_2 trapping results parallel to the temperature axis was made to provide closer agreement between the two sets of data. There was no fundamental justification for this adjustment other than the recognition that it is unrealistic to expect exact correspondence between UO_2 and CaF_2 and it is probably best to ascribe the greatest reliability to the few direct measurements which have been carried out for UO_2. The other essential features of the CaF_2 trapping values such as temperature and fission density dependence were retained as measured.

The diffusion coefficients for Xe diffusion in UO_2 calculated with these adjusted trapping parameters were in reasonable agreement with a significant number of the measured diffusion coefficients reported in the literature. For the results in Fig. 4 which appear to be low as a result of trapping, the agreement between calculated and observed values of the diffusion coefficient was within one order of magnitude in two-thirds of the fifty-two values checked for different temperatures and fission densities. The difference exceeded two orders of magnitude for only four points of the fifty-two.

In several cases where there was disagreement between calculated and experimental diffusion coefficients, it was apparent that different values of the diffusion coefficient could be inferred from the curves generated with Hurst's model, depending upon which portion of the curve was selected for fitting to the classical diffusion solutions. The range of possible diffusion coefficients often included the literature-reported value for the selected temperature and fission density. In other cases, it was found that the literature value of D could be obtained by varying either the trap concentration or gas retention time by no more than a factor of ten, a variation which is clearly possible in view of the uncertainties involved in estimating the trapping parameters. In most of the literature papers, there was insufficient information included on the experimental techniques to permit a comparison between reported and calculate diffusion coefficients over identical regions of the gas release curve.

The principal conclusion which can be drawn from this analysis is that trapping at radiation-induced defects is a probable cause of most of the discrepancies observed in rare gas diffusion measurements for UO_2. The fact that trapping parameters measured for CaF_2 and adjusted to correspond with the limited trapping values available for UO_2 will give observed diffusion coefficients in approximate agreement with a number of literature values lends support to this conclusion. A more accurate analysis of trapping is dependent upon the measurement of rare gas trapping parameters in UO_2 as a function of temperature and fission density.

5. Conclusions

1. The release of fission-recoil-injected ^{133}Xe from single-crystal UO_2 gives classical release kinetics without any indication of gas trapping. Measured diffusion coefficients are consistent with

$$D = \left(2.88 \begin{matrix} +13.4 \\ -\ \ 2.4 \end{matrix}\right) \times 10^3 \exp\left[-(4.78 \pm 0.15)\ \mathrm{eV}/kT\right]\ \mathrm{cm^2\ s^{-1}}$$

over the tempeature range 1065 to 1300 °C.

2. An increase of one order of magnitude in the fission recoil concentration to 3×10^{12} fission fragments cm^{-2} (equivalent to 3×10^{-8} fission atom fraction) results in nonclassical release curves which are apparently distorted by gas trapping.

3. A significant fraction of the literature diffusion coefficients for Xe in UO_2 which lie below the present results are approximately consistent with the values calculated from the present work and the gas trapping parameters obtained from studies on UO_2 and CaF_2. This implies that much of the early work was influenced by gas-atom trapping.

Acknowledgement

The assistance of Dr. H. J. Anderson of Battelle Northwest Laboratory in providing the single-crystal specimens used in this study is gratefully acknowledged.

References

[1] A. H. Booth and G. T. Rymer, Atomic Energy of Canada, Ltd. Rep. CRDC-720 or AECL-692 (1958).
[2] A. B. Auskern, USAEC Rep. UAPD-TM-185 and WAPD-TM-255 (1960).
[3] R. H. Barnes, M. Kangilaski, J. B. Melehan, and F. A. Rough, USAEC Rep. BMI-1533 (1961).
[4] D. Davies and G. Long, United Kingdom Rep. AERE-M-969 (1963).
[5] J. B. Melehan, R. H. Barnes, J. E. Gates, and F. A. Rough, USAEC Rep. BMI-1623 (1963).
[6] G. Frigerio and T. Gerevini, J. nuclear Mater. **16**, 76 (1965).
[7] J. R. MacEwan and W. H. Stevens, J. nuclear Mater. **11**, 77 (1964).
[8] N. Oi, Z. Naturf. **20a**, 1569 (1965).
[9] J. R. MacEwan and P. A. Morel, Nuclear Appl. **2**, 158 (1966).
[10] R. Soulhier, Nuclear Appl. **2**, 138 (1966).
[11] P. Spindler and R. Lindner, Z. Naturf. **21a**, 1723 (1966).
[12] W. Miekeley and F. W. Felix, J. nuclear Mater. **42**, 297 (1972).
[13] J. Belle, Uranium Dioxide: Properties and Nuclear Applications, USAEC, 1961 (p. 496).
[14] D. L. Morrison, Reactor Mater. **6**, 1 (1964).
[15] P. Schmeling, J. Phys. Chem. Solids **28**, 1185 (1967).
[16] R. Lindner and Hj. Matzke, Z. Naturf. **14a**, 582 (1959).
[17] G. Frigerio and T. Gerevini, J. nuclear Mater. **10**, 251 (1963).
[18] F. W. Felix, Proc. Brit. Ceram. Soc. **9**, 273 (1967).
[19] R. Kelly, Č. Jech, and Hj. Matzke, phys. stat. sol. **25**, 641 (1968).
[20] F. W. Felix, phys. stat. sol. **27**, 529 (1968).
[21] T. S. Elleman, C. H. Fox, and L. D. Mears, J. nuclear Mater. **20**, 89 (1969).
[22] Hj. Matzke and F. Springer, Z. Naturf. **25a**, 252 (1970).
[23] A. S. Ong and T. S. Elleman, Nuclear Instrum. and Methods **86**, 117 (1970).
[24] D. G. Hurst, Atomic Energy of Canada, Ltd., Rep. CRRP-1124 (1962).
[25] G. DiCola and Hj. Matzke, Nuclear Instrum. and Methods **57**, 341 (1967).
[26] M. J. Norgett and A. B. Lidiard, Phil. Mag. **18**, 1193 (1968).
[27] F. Felix, Hahn-Meitner Inst. (Berlin) HMI-B-93 (1970).
[28] F. W. Felix and S. Y. T. Lagerwall, phys. stat. sol. (a) **4**, 73 (1971).
[29] A. S. Ong and T. S. Elleman, J. nuclear Mater. **42**, 191 (1972).

(Received October 23, 1972)

phys. stat. sol. (a) 14, 681 (1972)

Subject classification: 14.3.4 and 16; 22.9

Institute of Physics and Energetics, Academy of Sciences of Latvian SSR, Riga

Photoionization of Quasicontinuous Traps in Tetracene Crystals

By

A. I. BELKIND and R. I. KALENDAREV

The spectral distribution of electron and hole photo-enhanced SCLC in tetracene crystals has been studied in the spectral range $h\nu \lessgtr 1.2$ eV. The photosensitivity observed has been related to photoionization of filled quasi-continuous traps. Agreement has been found for the measured spectral distribution of photocurrent and values obtained by a model suggested. Photoionization processes in molecular crystals are discussed.

Die spektrale Verteilung des photo-induzierten, raumladungsbegrenzten Elektronen- und Löcherstroms (SCLC) wurde an Tetraceneinkristallen im Spektralbereich $h\nu \lessgtr 1{,}2$ eV untersucht. Die beobachtete Photoempfindlichkeit wird mit der Photoionisation besetzter, quasikontinuierlicher Haftstellen verknüpft. Es wird Übereinstimmung der gemessenen Spektralverteilung des Photostromes mit Werten gefunden, die aus einem vorgeschlagenen Modell erhalten wurden. Die Photoionisierungsprozesse in Molekülkristallen werden diskutiert.

1. Introduction

Studies of photo-enhanced space-charge-limited currents (SCL) in anthracene and tetracene crystals have shown that long-wavelength photoconductivity beyond S–T absorption is determined by photoionization of traps [1 to 3]. It has been suggested by us [3] that in tetracene crystals within the spectral range $h\nu < 1.2$ eV traps with quasicontinuous energy distribution are emptied. The present paper is aimed at checking this assumption.

Quasicontinuous traps (QCT) filled with electrons are negatively ionized states of the basic molecules split off from the bottom of the lower conduction band as result of local deformation of the crystal lattice [4, 5]. Photoionization of filled QCT occurs through optical transition into the lower conduction band. The autoionization mechanism cannot be responsible for emptying of QCT since the lower excited state of the negative tetracene ion exceeds 1.2 eV [7]. Studying the electron delocalization processes under the condition of optical excitation provides additional information on photoionization processes in molecular crystals.

2. Calculation of Spectral Distribution of Photocurrent under the Condition of Electron QCT Photoionization

The probability P of a transition in a time unit from the initial state i into the final state f equals

$$P(E_\mathrm{i}, h\nu) = A\,|M|^2\,N_\mathrm{i}(E_\mathrm{i})\,N_\mathrm{f}\,(E_\mathrm{i} + h\nu)\,, \tag{1}$$

where N_i and N_f represent respectively initial and final state density within the unit energy range for energies E_i and $E_\mathrm{i} + h\nu$, $|M|^2$ is the square of transition matrix element. Energy is measured upwards from the bottom of the lower conduction band E_c'.

According to Spicer's model [8], verified for phthalocyanines and porphyrine [9, 10], the transition matrix element can be represented as a function of the initial and final states only, i.e.

$$|M|^2 = M_i(E_i)\, M_f\,(E_i + h\,\nu)\,. \tag{2}$$

This makes it possible to express the transition probability P through "optical densities" of initial states N_i^{opt} and final states N_f^{opt}:

$$P(E_i, h\,\nu) = A\; N_i^{opt}(E_i)\; N_f^{opt}\,(E_i + h\,\nu)\,, \tag{3}$$

where

$$N_i^{opt}(E_i) = M_i(E_i)\; N_i(E_i) \tag{4}$$

and

$$N_f^{opt}\,(E_i + h\,\nu) = M_f\,(E_i + h\,\nu)\; N_f\,(E_i + h\,\nu)\,. \tag{5}$$

2.1 "Optical density" of initial states

In case of strongly localized excitation as in molecular crystals the "effective optical matrix element" of initial states M_i is not dependent on photon energy [8]. Broadening of $N_i^{opt}(E_i)$ due to photoionization of host molecules in the crystal, as compared to $N_i(E_i)$, can be connected with vibrational excitation of the newly formed ion and the surrounding lattice [11 to 13]. Owing to lack of data we neglect possible vibrational excitation of the remaining empty trap. This can be justified to a certain extent by the following considerations.

If photoionization of QCT is accompanied by vibrational excitation of the trap, annealing may occur. However, extended illumination of the crystal in which the number of photons is comparable with the number of filled traps does not noticeably change the concentration and energy distribution of traps.

We assume that M_i = const and the "optical density" distribution of initial states is determined by the distribution of filled traps only

$$N_i^{opt}(E_i) = B\,N_i = B\,h(E_i)\,f(E_i)\,, \tag{6}$$

where B = const,

$$f(E_i) = \frac{1}{1 + \dfrac{1}{2}\exp\left(\dfrac{E_i - E_F}{kT}\right)} \tag{7}$$

is the Fermi distribution, and $h(E_i)$ the electron trap distribution. According to SCLC measurements in tetracene crystals [14, 15], the trap distribution can be approximately expressed by an exponential

$$h(E_i) = \frac{H_0}{kT_c}\exp\left(\frac{E_i}{kT_c}\right). \tag{8}$$

Here E_F is the quasi-Fermi level energy; H_0 the total number of QCT per unit volume; and kT_c a characteristic parameter.

2.2 "Optical density" of final states

We have no data on the distribution of the density of states in the lower conductivity band E_c'. For a parabolic band the density of states must increase proportionally to $(E_i + h\,\nu)^{1/2}$. Considering the possible change of the "effective optical matrix element" M_f we have attempted to describe the real distribution of the "optical density" of the final state by the expression

$$N_f^{opt}\,(E_i + h\,\nu) \sim (E_i + h\,\nu)^n\,, \tag{9}$$

where n is a parameter determined from comparison of calculation with experimental data.

Using (6) to (9), expression (3) for the transition probability can be rewritten as follows:

$$P(E_i, h\,\nu) \sim \frac{\exp\left(\dfrac{E_i}{kT_c}\right)}{1 + \dfrac{1}{2}\exp\left(\dfrac{E_i - E_F}{kT}\right)}\,(E_i + h\,\nu)^n\,. \tag{10}$$

Assuming that any trap of depth $-E_i < h\,\nu$ can be ionized with a probability $P(E_i, h\,\nu)$, the spectral distribution of photocurrent on illumination by constant photon flux can be expressed as

$$j^{cal}(h\,\nu, U) \sim U\int\limits_{-h\nu}^{0} P(E_i, h\,\nu)\,dE_i = U\int\limits_{-h\nu}^{0}\frac{\exp\left(\dfrac{E_i}{kT_c}\right)}{1 + \dfrac{1}{2}\exp\left(\dfrac{E_i - E_F}{kT}\right)}\,(E_i + h\nu)^n\,dE_i\,. \tag{11}$$

The energy of the quasi-Fermi level E_F has been calculated from SCLC theory [16]:

$$E_F = -\,kT_c\ln\left[\frac{(l+1)^2}{l\,(2\,l+1)}\,\frac{e\,H_0\,d^2}{\varepsilon_0\,\varepsilon\,U}\right]. \tag{12}$$

In these expressions $l = kT_c/kT$, d is the thickness of the crystal, e the electron charge; ε the HF dielectric permeability, and U the voltage.

Integral (11) cannot be expressed by means of elementary functions, hence the spectral distribution of photocurrent in the range $h\,\nu \lessapprox 1.2$ eV was calculated by means of a computer. Calculations were performed for $n = 0$; 0.5; 1 and for $kT_c = 0.095$ eV, $H_0 = 10^{17}$ cm^{-3} [15], $\varepsilon = 3$.

3. Experimental Results and Discussion

The photocurrents were measured by means of methods described in paper [3]. The spectral distribution of photocurrent within the range $h\,\nu \lessapprox 1.2$ eV is well described by expression (11) with $n = 0.5$ (Fig. 1).

Filled QCT are concentrated in the vicinity of the quasi-Fermi level. Hence ionization of discrete electron traps situated in the vicinity of this level can also lead to a dependence of the same type (11). In order to circumvent this ambiguity we use a specific property of QCT, the dependence of the distribution of filled traps on the applied voltage. Measured and calculated spectral photocurrent curves are shown on Fig. 2. The comparison shows that expression (11) correctly represents change of the spectral distribution of the photocurrent with increase of electron injection, but does not show agreement with experimental current–voltage curves. For medium field intensities (0.3 to 1.2) $\times\,10^4$ V/cm the relation between measured and calculated photocurrent depends linearly on

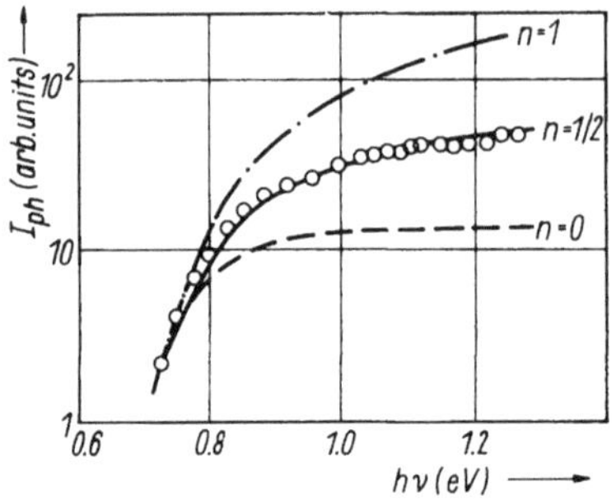

Fig. 1. Comparison of measured (circles) and calculated (lines) spectral distributions of photocurrent in a tetracene crystal at dark electron injection from an NaHg electrode for various values of parameter n. Voltage $U = 5$ V, thickness of crystal $d = 6$ μm

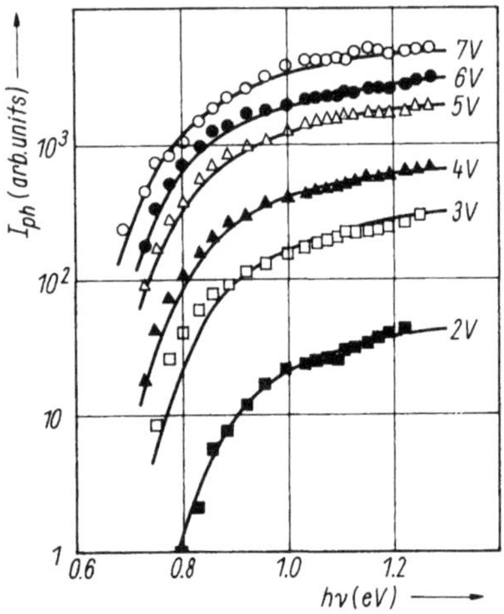

Fig. 2. Measured (various symbols) and calculated (solid lines) spectral distributions of photocurrent in a tetracene crystal for parameter $n = 0.5$ and for various levels of dark electron injection from a Na–Hg electrode. Crystal thickness $d = 6$ μm

voltage (Fig. 3) and does not depend on photon energy. This leads to the conclusion that the additional dependence of the photocurrent on voltage, not taken into account in deducing expression (11), is not connected with the transition probability. The dependence $\varphi(u) = j^{e}(h\nu, U)/j^{cal}(h\nu, U)$ can be interpreted as the effect of the electric field on the escape probability of an electron outside the capture range of the trap. It is necessary to admit in such a case that thermalization of an electron occurs within the capture range. The linear dependence of photogeneration yield on electric field intensity found by us has already been observed for intrinsic photogeneration [17, 18]. In this case it is determined by the effect of the electric field on the escape probability of the electron from the Coulomb field of the hole.

It follows from these results that photocurrent within the spectral range $h\nu \lessapprox 1.2$ eV can actually be explained by QCT photoionization and can be described by expression (11) with a parameter value $n = 0.5$, allowing for the dependence of optical trap ionization yield on electric field intensity which leads to

$$j(h\nu, U) \sim \varphi(U)\, U \int_{-h\nu}^{0} \frac{\exp\left(\frac{E_i}{kT_c}\right)}{1 + \frac{1}{2}\exp\left(\frac{E_i - E_F}{kT}\right)} (E_i + h\nu)^{1/2}\, dE_i\,. \qquad (13)$$

In this expression $\varphi(u)$ depends linearly on voltage (Fig. 3).

The integral in (13) is strongly field-dependent in the spectral region $h\nu \approx E_F$ and increases linearly with voltage at photon energies sufficient for ionization of any trap ($h\nu \gg E_F$). Hence, expression (13) gives a stronger increase of photocurrent with voltage than expression ($j \sim U^2$), obtained from the theory of photoenhanced SCLC [19].[1])

[1]) The strong field dependence of photocurrent, observed in [20] on illumination of anthracene with weakly absorbed light, is also rather connected with trap ionization than with photoinjection from the cathode.

Fif. 3. Electric field dependence of the ratio between measured and calculated photocurrents

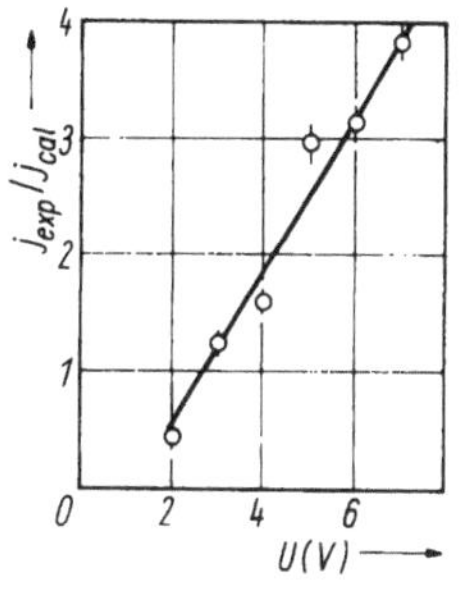

In our measurements an approximately square law current–voltage curve was observed only for tetracene singlet band light. Within the framework of our model of trap ionization the transition to a square law with increasing excitation energy might be linked with the fact that the electron is not able to dissipate all excess energy within the capture range of the trap, and the field dependence of photogeneration efficiency is weakened.

4. Some Remarks on Photoionization Processes in Molecular Crystals

The satisfactory agreement between calculated and measured spectral distributions of electronic photocurrent leads to two conclusions:

1. Photoionization of electronic QCT does not produce any marked optical broadening of the initial density of states.

2. The "optical density" of states at the bottom of the lower conduction band is proportional to the square root of electron energy E_e in the band

$$N_c^{opt} \sim E_e^{1/2} . \tag{14}$$

These conclusions provide a certain basis for suggesting certain peculiarities of photoionization processes in molecular crystals.

Paper [21] gives data on the spectral distribution of photoemission (PE) in the beyond-threshold region which is basically determined by "optical density" distribution of states at the top of the valence band

$$N_v^{opt} \sim E_h \tag{15}$$

and by the probability of electron escape through the potential barrier at the crystal surface

$$T \sim [h\nu - (E_{ph} + E_h)]^{0.5 \text{ to } 1} . \tag{16}$$

Here E_h is the hole energy in the valcence band, and E_{ph} the photoelectric threshold energy. Integration of the product of expressions (15) and (16) with respect to all initial-state energies leads to a power-law spectral distribution of PE in the beyond-threshold-region.

$$Y \sim (h\nu - E_{ph})^{2.5 \text{ to } 3} . \tag{17}$$

Since photogeneration of electron–hole pairs within the volume of the crystal is linked with ionization of the same initial state as PE, it has been considered necessary in paper [21] to introduce a "threshold" function T for the internal photoeffect in order to explain the observed coincidence of PE and photocurrent near the intrinsic photogeneration threshold. This function is connected with the electron escape probability from the Coulomb field of the hole. Such an approach presumes that the electron does not get thermalized within the capture range of the hole. This, however, is in contradiction with the independence of initial Coulomb recombination probability of photon energy in the beyond-threshold region of intrinsic photogeneration of anthracene [7, 18]. It seems feasible therefore, considering our analysis of QCT photoionization, to treat the "threshold" function, introduced in paper [21], as identical with the "optical density" of states at the bottom of the lower conduction band.

The "optical density" of final states at photoionization of host molecules, proceeding according to the autoionization mechanism [6], and at photoionization of QCT, proceeding through direct electron transition into the band, coincide. This is possible if the "effective optical matrix elements" for both processes are equally dependent on energy. Such a situation can arise if autoionization occurs from a vibrationally excited level of the molecule attained by optical transition, but not after partial loss of energy as a result of vibrational relaxation, as assumed in [17, 18, 22]. The latter is confirmed by the fact of conservation of the fastest peak in the kinetic energy distribution curves of photoelectrons in PE within a wide spectral range [9 to 12, 23].

It must be added that the distribution of the "optical density" of states N_c^{opt} as obtained by us assumes the absence of a wide forbidden energy gap between the lower and the wide conduction bands. The absence of such a gap may be due to the vibrational splitting of the lower conduction band [2, 24].

From expressions (14) and (15) for "optical densities" of initial and final states respectively, the intrinsic photocurrent in the beyond-threshold region can be obtained:

$$j \sim \int_0^{h\nu - E_g} E_h \, dE_h \int_0^{h\nu - E_g - E_h} E_e^{1/2} \, \delta\left[E_e - (h\nu - E_g - E_h)\right] dE_e \sim (h\nu - E_g)^{5/2} \,, \quad (18)$$

where E_g is the forbidden band energy, the delta function ensuring energy conservation of the transition.

As can be seen, this expression for the spectral distribution of intrinsic photocurrent is similar to that for the spectral distribution of PE (17).

This approximation agrees well with experimental data [22] for anthracene and tetracene, giving values for forbidden band energies 3.75 eV for anthracene and 2.8 eV for tetracene. The value for anthracene is in agreement with photoinjection measurements [25].

In the same way expressions can be found for the spectral distribution of photocurrent, obtained by photoionization of hole QCT. The latter might be suggested as optical transfer of an electron from the upper occupied basic molecular level to the trap level. In this case the "optical density" of the initial states coincides with the "optical density" of states in the valence band (expression (15)). The final state is determined by the distribution of filled hole traps. The QCT distribution is equal for holes and electrons in tetracene crystals [14, 15]. Hence, the spectral distribution of photocurrent at photoionization of hole QCT must also follow expression (11), only with the parameter value $n = 1$, according to the distribution of the "optical density" of states in the

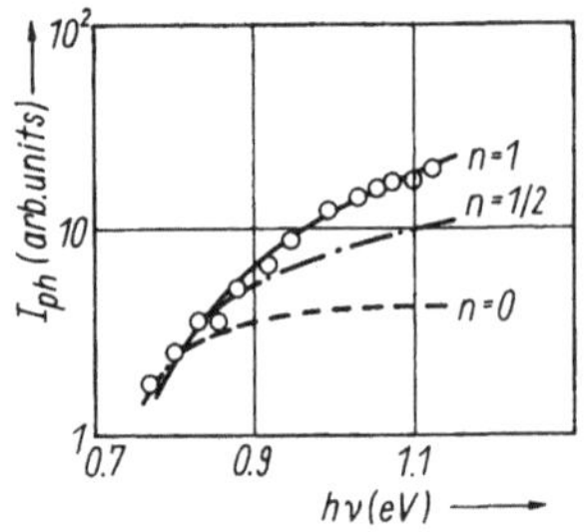

Fig. 4. Comparison of measured (circles) and calculated (lines) spectral distributions of photocurrent in a tetracene crystal at dark hole injection from a Hg electrode for various values of parameter n. Voltage $U = 1$ V, thickness of crystal $d = 2$ μm

valence band. Fig. 4 shows experimental and calculated spectral distribution curves for photoenhanced hole SCLC in tetracene at photon energies $h\nu < 1.2$ eV. The coincidence of the shapes of the curves confirms our assumption as regards the mechanism of charge carrier photogeneration in this spectral range.

According to this analysis photoionization of the host molecules in the crystal and photoionization of hole QCT are accompanied by vibrational excitation of the newly formed ion and the surrounding lattice. This excitation leads to a broadening of the narrow valence band. As result the distribution of the "optical density" of states at the top of the valence band is proportional to the hole energy.

5. Conclusion

The results shown above indicate that the Spicer model of a localized transition can be used for describing photoionization of QCT and the host molecules in the crystal.

In all cases the photoionization in tetracene crystals can be described as a process dependent on the product of the "optical densities" of initial and final states. The same "optical density" of states of the lower conduction band appears both in the processes of autoionization of host molecules in the crystal and in the processes of direct photoionization of electron QCT. Similarly, the same "optical density" of states of the valence band appears both in the processes of ionization of host molecules in the crystal and in the processes of ionization of hole QCT.

The "optical density" of states at the bottom of the lower conduction band is proportional to the square root of electron energy in the band. The "optical density" of states at the top of the valence band is proportional to the hole energy.

Acknowledgements

The authors are greatly indebted to G. Zavt, I. Kaulačs, and E. Silinsh for detailed discussion of highly relevant questions involved. It is a pleasure to acknowledge many fruitful discussions with Prof. N. Geacintov.

References

[1] M. Schadt, I. Zschokke-Gränacher, and E. Baldinger, Helv. phys. Acta **41**, 1349 (1968).
[2] A. Many, J. Levinson, and I. Teucher, Mol. Cryst. **5**, 273 (1969).
[3] R. I. Kalendarev, A. I. Belkind, and V. V. Aleksandrov, Izv. Akad. Nauk Latv. SSR, Ser. fiz. tekh. Nauk **5**, 49 (1970).
[4] J. Sworakowski, Mol. Cryst. Liquid Cryst. **11**, 1 (1970).
[5] E. A. Silinsh, phys. stat. sol. (a) **3**, 817 (1970).
[6] M. Pope, Trans. Faraday Soc. **51**, 7 (1971).
[7] R. Zahrodnik and P. Čarsky, J. phys. Chem. **74**, 1240 (1970).
[8] W. E. Spicer and R. C. Eden, Proc. 9th Internat. Conf. Phys. Semicond. (1) Moscow, Izd. Nauka, Leningrad 1969 (p. 68).
[9] B. H. Schechtman and W. E. Spicer, Chem. Phys. Letters **2**, 207 (1968).
[10] B. H. Schechtman, Ph.D. Dissertation, Stanford University, 1968.
[11] F. I. Vilesov, A. A. Zagrubskii, and D. Z. Garbuzov, Fiz. tverd. Tela **5**, 2000 (1963).
[12] M. Kochi, Y. Harada, T. Hirooka, and N. Inokuchi, Bull. Chem. Soc. Japan **43**, 2690 (1970).
[13] A. A. Zagrubskii and F. I. Vilesov, Fiz. tverd. Tela **13**, 2300 (1971).

[14] H. Baessler, G. Hermann, N. Riehl, and G. Vaubel, J. Phys. Chem. Solids **30**, 1579 (1969).
[15] R. I. Kalendarev and A. I. Belkind, Izv. Akad. Nauk Latv. SSR, Ser. fiz. tekh. Nauk **4**, 73 (1970).
[16] P. Mark and W. Helfrich, J. appl. Phys. **33**, 205 (1962).
[17] R. H. Batt, C. I. Braun, and J. F. Horning, J. chem. Phys. **49**, 1967 (1968).
[18] R. H. Batt, C. I. Braun, and J. F. Horning, Appl. Optics **3**, 20 (1969).
[19] W. Helfrich, phys. stat. sol. **7**, 863 (1964).
[20] J. Dresner, Phys. Rev. Letters **21**, 356 (1968).
[21] A. I. Belkind and S. B. Aleksandrov, phys. stat. sol. (a) **9**, 105 (1972).
[22] N. Geacintov and M. Pope, J. chem. Phys. **50**, 814 (1969)
[23] A. Terenin and F. Vilesov, Advances in Photochemistry, Vol. 2, Interscience Publishers, New York 1964 (p. 407).
[24] H. Baessler, N. Riehl, and G. Vaubel, phys. stat. sol. **26**, 607 (1968).
[25] H. Baessler, N. Riehl, and G. Vaubel, Mol. Crvst. Liquid Cryst. **9**, 249 (1969).

(Received September 13, 1972)

phys. stat. sol. (a) 14, 689 (1972)

Subject classification: 5 and 10.1; 21.1.1

Institute of Metal Physics, Academy of Sciences of the Ukrainian SSR, Kiev

Investigation of Carbide Phase in Strained Steel by the Method of Nuclear Gamma Resonance

By

V. N. GRIDNEV, V. G. GAVRILYUK, I. YA. DEKHTYAR, YU. YA. MESHKOV, P. S. NIZIN, and V. G. PROKOPENKO

The Mössbauer effect was used to investigate the cementite state in cold-worked steel. An intensive dissociation of cementite is found at plastic deformation of the steel with the structure of plate perlite. For deformation of 93% this dissociation amounts to 50%. The effect of dissociation was not observed in the steel with granular perlite. The fact that cementite dissociates at deformation may be explained by the increase of bond energy between C atoms and dislocations as compared to that for carbon in the cementite lattice. Alloying with the elements which decrease the bond of dislocations and substitutional impurities has the result that the cementite becomes much more stable at deformation. This is shown on the example of steel containing cobalt. The latter fact influences strongly the mechanical properties of a cold worked steel.

Эффект Мессбауэра применен для изучения состояния цементита в холоднодеформированной стали. Установлен интенсивный распад цементита при пластической деформации стали со структурой пластинчатого перлита. Степень распада составляет примерно 50% для деформации 93%. Эффект распада не наблюдается в стали с зернистым перлитом. Распад цементита при деформации объясняется тем, что энергия связи атомов углерода с дислокациями превышает энергию связи углерода в решетке цементита. Легирование элементами, уменьшающими связь дислокаций с примесями внедрения, существенно повышает устойчивость цементита при деформации, что показано на примере стали, содержащей кобальт. Последнее обстоятельство оказывает значительное влияние на механические свойства холоднодеформированной стали.

1. Introduction

Plastic deformation is one of the most effective means of steel hardening. The highest obtainable degree of deformation and hence the maximum strength depend on the plasticity resources which are determined by the state of two phases, namely ferrite and cementite. Whereas the ferrite structure in cold-worked steel was studied in detail by X-ray and electron microscopy methods [1 to 5], the information about the cementite state is very scarce. It is found that in the contact with the ferrite matrix the cementite is not a brittle phase component and can suffer high plastic deformations [6]. Maurer and Warrington [7] observed dislocation lines in the (010) plane of Fe_3C. This plane seems to be the glide plane in cementite. The thickness of the cementite plate and the interplate distance at drawing are shown to decrease proportional to the change of the specimen diameter [3, 8]. The dissociation of cementite at plastic deformation was investigated by some authors who used the methods of inner friction and magnetic measurements. The enlargement of Köster's peak (200 °C) as a result of plastic deformation [9, 10] indicates the increase of dislocation density and the concentration of C and N atoms in the solid solution. (These atoms were obtained from the dissociation of carbide and nitride phases.)

Quantitative measurements were made by Belous and Cherepin from the analysis of the thermo-magnetic curves of heating of deformed steels [11].

According to their results about 50% of the carbide phase may be dissociated at high plastic deformation. This untrivial conclusion cannot practically be supported by the methods of direct observation (metallography and electron microscopy). The comparatively small quantity of experimental data about changes in the carbide phase, the limited number of applicable investigation methods as well as their indirect character show the necessity in more direct means of evaluation of the cementite state in deformed steel.

The present work deals with the possibility to solve this problem using the Mössbauer effect. The high sensitivity of Mössbauer nuclei to their local neighbourhood allowed to use successfully the Mössbauer effect for the investigation of phase transformations during changes in composition and structure of the phase. The quantitative analysis was carried out in some cases. The Mössbauer effect in cementite extracted from steel was studied in [12 to 14]. The effective field H_e on Fe^{57} nuclei in stable cementite was evaluated. Its value amounts to 205 to 207 kOe. Owing to the difference between the fields at the Mössbauer nuclei in cementite and ferrite the multiplet spectra with $n > 6$ can be observed in steels. In the course of investigation of such spectra the separation of a carbide phase was observed during martensitic annealing [5]. These results allowed to expect that the changes in the cementite state during deformation of steel could be discovered in the Mössbauer spectra.

2. Experimental Procedure

A Fe + 1% C alloy containing 15 wt% cementite was annealed at 900 °C for 30 min with care taken against oxidation and decarbonization. Various degrees of deformation were obtained by rolling.

The final thickness of the film was 35 to 40 μm. The Mössbauer absorption spectra were obtained in a spectrometer with a system of electrodynamical motion type in the regime of constant acceleration. Source and specimens were at room temperature.

As a source of γ-quanta Co^{57} in Cr was used. The experimental spectra were treated with the computer M-220 to separate the multiplet lines.

Simultaneously magnetic measurements were carried out by the method outlined in [16].

The experimental data were used for the calculation of the cementite dissociation.[1])

3. Results and Discussion

Fig. 1 shows the spectrum of the initial specimen annealed as was mentioned above. Here the lines are clearly resolved for Fe^{57} in cementite. They correspond to the transitions $\pm 1/2 \rightarrow \pm 1/2$.

The computer treatment of the experimental spectrum allows to separate the positions of six lines which correspond to the cementite spectrum with the field $H_e = (205 \pm 5)$ kOe.

[1]) The authors are deeply grateful to M. V. Belous for proposing the method of calculation.

Fig. 1. Mössbauer absorption spectra in a Fe + 1% C alloy; a) annealed state, b) strained by 93%

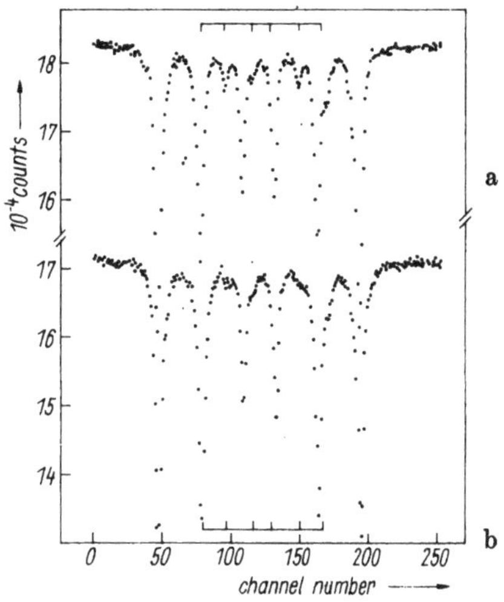

The data obtained agree with the results of [14]. Deformation by 36 and 93% leads to a marked decrease of the absorption for the cementite lines. We did not observe a significant change in other Mössbauer parameters but they are possible (e.g. due to the transition of Fe atoms from cementite to ferrite and the solution of C atoms in the ferrite lattice). This problem needs a more thorough consideration. To evaluate quantitatively the dissociation degree of cementite we used in each case the ratio between the integral intensities of cementite and ferrite spectra (see Table 1). The calculation showed that the cementite dissociation amounts to 40 and 50% for the deformation degrees investigated.

Fig. 2 gives the results of the magnetometric investigation of the dissociation of cementite at plastic deformation by drawing. The maximum effect is equal to 55% and coincides well with the data of NGR.

It is worth to consider how the form of the carbide inclusions influences their tendency to dissociation. From the data of electron microscopy and magnetic measurements [2, 4, 11], the coarse-granular cementite does not take part in plastic deformation and does not dissociate. In absorption spectra obtained by us for the steel with granular cementite we did not observe changes in the quantity of the carbide phase. This may serve as support for the above ideas.

The steady-state concentration of carbon in ferrite is less than 0.01%. The supersaturation of the solid solution at plastic deformation is hardly possible on a very large scale. Hence it may be expected that all the carbon obtained from cementite is practically connected with defects of crystal structure. The plastic deformation of ferrite is accompanied by a significant increase in the density of dislocations arranged mainly along the boundaries between ferrite and cementite plates [2 to 4].

Table 1

deformation degree (%)	$\frac{I^{Fe_3C}_{(1\,to\,6)}}{I^{\alpha\text{-}Fe}_{(1\,to\,6)}}$ *)	quantity of cementite in alloy (%)	carbon quantity in ferrite (%)
	Fe + 1% C		
0	0.280 ± 0.01	15.0	—
36	0.170 ± 0.01	9.0 ± 0.1	0.40 ± 0.07
93	0.140 ± 0.01	8.0 ± 0.1	0.50 ± 0.07
	Fe + 1% C + 6% Co		
0	0.180 ± 0.01	15.0	—
93	0.170 ± 0.01	14.0 ± 0.1	0.10 ± 0.07

*) $I_{(1\,to\,6)}$ is the sum of the integral intensities of six spectral lines.

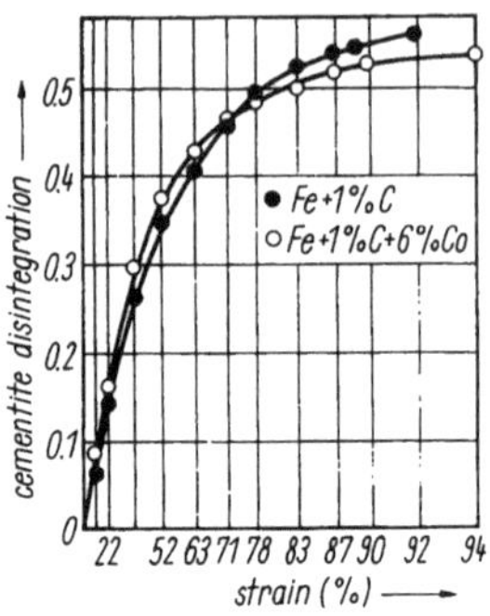

Fig. 2. Dissociation of cementite at plastic deformation of steel

The interphase boundary serves simultaneously as dislocation source [17] and as the main barrier to their motion. Lyubov and Shmakov [18] have proposed a mechanism for the dissociation of the second phase in the presence of dislocation pile-ups near particles.

The consideration of the results of microcalorimetric investigations may give a thermodynamic argument for the dissociation of cementite at deformation. The binding energy of carbon in cementite, according to the data of [19 to 21], amounts to 0.40 to 0.42 eV.

The binding energy between carbon atoms and a vacancy has been determined to 0.41 eV [22] while the interaction energy between carbon atoms and edge dislocations is 0.5 to 1.0 eV [23, 24]. From experiments on the amplitude dependence of the inner friction in iron the binding energy was determined for impurity atoms and dislocations to 0.76 to 0.78 eV [25]. Thus the concentration of dislocations at the boundary ferrite–cementite should lead to the isolation of C atoms from cementite and to the formation of atmospheres around dislocations. If this mechanism is valid then the dissociation of cementite at deformation should be influenced by elements which change the bond between

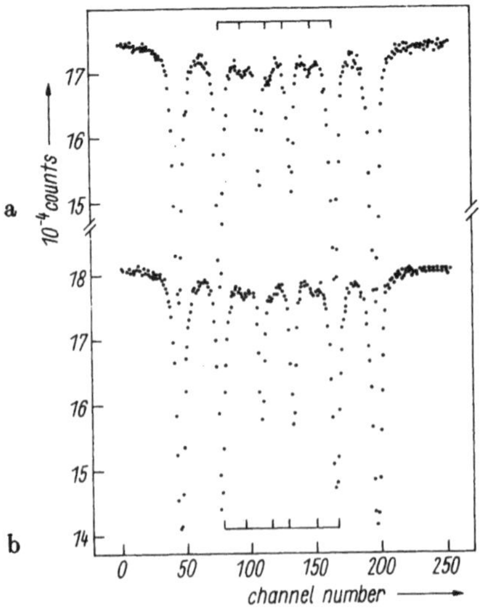

Fig. 3. Mössbauer absorption spectra in a Fe + 1% C + 6% Co alloy; a) annealed state, b) strained by 93%

Fig. 4. Mechanical properties of cold-worked steel

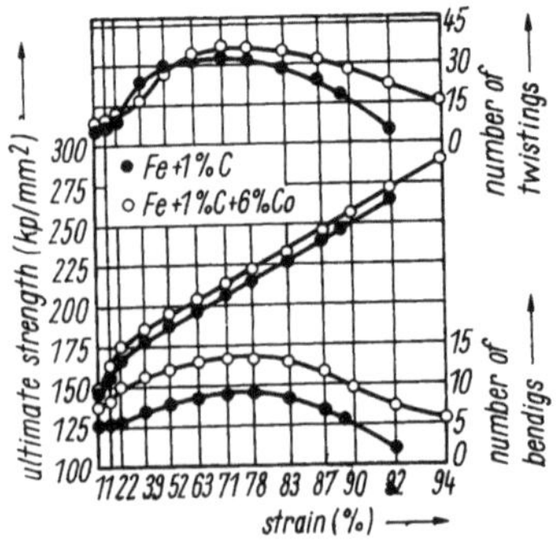

dislocations and carbon atoms. To check this possibility we used Co which decreases the binding energy between dislocations and interstitial impurities [25], but being arranged in the ferrite matrix, does not affect the crystal structure of cementite. We have investigated a Fe + 1% C + 6% Co alloy in cold-strained state. For differential magnetic measurements a Fe + 6% Co alloy was used for the compensation of the effect of Co on the ferrite magnetization. The results of magnetometric investigations show that Co does not influence the dissociation of cementite (Fig. 2). But these data are not confirmed by the NGR spectra. Comparison of the absorption spectra obtained for initial and deformed specimens, (Fig. 3, Table 1) makes it possible to conclude that cementite is much less inclined to dissociation at deformation in steel doped with Co.

This discrepancy is, to our mind, connected with the fact that the absence of Co in steel affects in a definite way the magnetic properties of the phases and this had not been taken into account in the corresponding magnetometric calculations. In fact, the ratio of integral intensities of the spectrum for cementite and ferrite in steel doped with Co is markedly lower than that in the unalloyed carbon steel (see Table 1). This may result from the decrease in the quantity of Fe^{57} nuclei in the cementite phase of alloyed steel due to the fact that a part of Fe atoms in the cementite lattice is replaced by Co atoms. If one could avoid the dissociation of the carbide phase accompanied by an undesired increase of the carbon concentration in ferrite and a decrease in the efficiency of the barrier strengthening mechanism, this would be rather favourable for the mechanical properties of cold-strained steel (Fig. 4). Plasticity, strength, and the deformation limit are much higher for steel specimens doped with Co than those for a common carbon steel.

4. Conclusions

1. The use of the Mössbauer effect allows to get new information about the carbide phase in cold-strained steel. The NGR spectra permit to evaluate directly the dissociation of cementite at deformation.

2. The dissociation degree of cementite can be changed by alloying the elements which affect the binding between dislocations and carbon atoms. This is illustrated by the example of steel doped with Co. This fact is of great importance for obtaining the desired mechanical properties of cold-strained steel.

References

[1] V. M. Kardonskii, G. V. Kurdyumov, and M. D. Perkas, Fiz. Metallov i Metallovedenie **2**, 244 (1963).
[2] V. M. Kardonskii, G. V. Kurdyumov, and M. D. Perkas, Metal Science and Heat Treatment **2**, 2 (1964).
[3] J. D. Embury and R. M. Fisher, Acta metall. **14**, 147 (1966).
[4] V. N. Gridnev and Yu. N. Petrov, in: Structure of Metal Alloys, Izd. Naukova dumka, Kiev 1966 (p. 4).
[5] K. L. Maurer and R. Rosegger, Berg- und Hüttenmänn. Mh. **111**, 412 (1966).
[6] K. E. Puttick, Iron and Steel Inst. **185**, 161 (1957).
[7] K. L. Maurer and D. H. Warrington, Phil. Mag. **134**, 321 (1967).
[8] V. N. Gridnev, V. G. Gavrilyuk, Yu. Ya. Meshkov, and V. V. Yaremchuk, in: Steel Cables, Vol. 4, Izd. Tekhnika, Kiev 1967 (p. 348).

[9] W. KÖSTER, L. BANGERT, and R. HAHN, Arch. Eisenhüttenwesen **25**, 569 (1964).
[10] W. KÖSTER and W. HORN, Arch. Eisenhüttenwesen **37**, 155 (1966).
[11] M. V. BELOUS and V. T. CHEREPIN, Fiz. Metallov i Metallovedenie **12**, 685 (1961).
[12] T. SHINZO, F. ITOH, M. TAKAKI, Y. NACAMURA, and N. SHIKAZONO, J. Phys. Soc. Japan **19**, 1252 (1964).
[13] M. RON, H. SHECHTER, and S. NIEDZWIEDZ, J. appl. Phys. **39**, 265 (1968).
[14] M. RON and Z. MATHALONE, Phys. Rev. **4**, 774 (1971).
[15] H. INO, T. MORIYA, F. FUJITA, Y. MACDA, Y. ONO, and Y. INOKUTI, J. Phys. Soc. Japan **25**, 88 (1968).
[16] V. G. PERMYAKOV, YU. V. NAIDICH, and M. A. RYBAK, Works laboratory **21**, 695 (1955).
[17] F. R. PICKERING, Iron and Steel Inst. **38**, 110 (1965).
[18] B. YA. LYUBOV and C. A. SHMAKOV, Fiz. Metallov i Metallovedenie **29**, 968 (1970).
[19] L. DARKEN and R. CURRY, Physical Chemistry of Metals, Chap. 16, McGraw-Hill Publ. Co., 1953 (p. 4).
[20] R. A. JOHNSON, G. I. DIENS, and A. C. DAMASK, Acta metall. **12**, 125 (1964).
[21] R. A. JOHNSON, Acta metall. **15**, 513 (1967).
[22] A. A. ARNDT and A. C. DAMASK, Acta metall. **12**, 341 (1964).
[23] A. H. COTTRELL, Rep. Conf. Plastic Deformation of Metals, Pittsburgh 1967 (p. 60).
[24] G. FRIEDEL, Dislocations, Izd. Mir, Moscow 1967 (p. 421).
[25] G. A. BERESNEV, V. I. SARRAK, and N. A. SHILOV, in: Problems of Metal Science and Physics of Metals, Vol. 8, Moscow 1964 (p. 157).

(Received October 5, 1972)

phys. stat. sol. (a) **14**, 695 (1792)

Subject classification: 14.3.3; 22.1.2

S. V. Starodubtsev Physico-Technical Institute,
Academy of Sciences of the Uzbekian SSR (a) and V. I. Lenin State University, Tashkent (b)

On the Theory of Transient Process after Reversal of a p–i–n Diode Current from Forward to Reverse Direction (I)

By

D. A. Aronov (a) and R. Mamatkulov (b)

Calculations are given related to the transient process of the reverse-resistance recovery after reversal of a p–i–n diode current from forward to reverse direction. The calculations have been performed for high-level forward injection with junction injection coefficients not equal to unity and considering the expansion of space-charge regions in pulling excess carriers out of the i-layer and heavily doped regions under the action of large reverse currents. By using Fourier series methods for the first (quasi-neutral) stage of the transient process, a general expression has been found for the carrier concentration distribution in the i-region.

В работе рассчитывается переходной процесс восстановления обратного сопротивления при переключении p–i–n-диода с прямого направления на обратное. Расчет проводится при большом уровне инжекции в прямом направлении с учетом отличия коэффициентов инжекции переходов от единицы и расширения областей объемного заряда в процессе вытягивания неравновесных носителей тока из i-слоя и сильнолегированных областей под действием сильного обратного тока. Используя методику рядов Фурье, для первая этапа переходного процесса — этапа квазинейтральности — найдено общее выражение для распределения концентрации носителей тока в i-области.

1. Present Situation and Statement of the Problem

The electrical properties of p–n–m and p–n–n$^+$ structures under steady-state conditions and with a forward current or voltage pulse at high injection level applied are well studied both theoretically and experimentally (references and a review of principal works are given in [1] and [2]). Most authors consider mainly the transient processes which occur in applying current or voltage pulses. Meanwhile, there are a few works describing investigations on the process of reverse-resistance recovery after the reversal of the direction of the current in a p–i–n diode from forward to reverse. Such investigations are of great importance for the development of highly effective semiconductor rectifiers and high-power switches since exactly the processes of accumulation and extraction of mobile carriers determine the switching properties of diodes.

The transient process of the blocking Ge p–n–n$^+$ diodes ДГЦ-23 and ДГЦ-25 with high injection level was studied for the first time by Barsukov [3] who found that at forward-current densities $j_F \approx 10$ to 10^3 A/cm^2 before the transient process of blocking, the response time of the diodes drops rapidly with increasing forward current: an increase in the current by two orders of magnitude causes the transient time to decrease by more than one order. The transient accumulation and extraction of charges (Si p–i–n diodes) were studied theoretically and experimentally by Gribnikov and Tkhorik [4] and by Spenke and co-workers [5]. In [4] the problem is discussed with due account for such factors as finite base length and recombination rate at the contact, decrease in the injection coefficients of the junction with increased forward current, and changes in the injection coefficient during the transient process as well as variations in the process of charge extraction during the reverse-resistance recovery depending upon the character of the transient accumulation. In contrast to references [3] and [4], paper [5] considers the reverse-resistance recovery in high-power and sufficiently long Si p–i–n diodes to be featured by a noticeable expansion of the junction space-charge regions by extracting carriers out of the high-resistivity i-layer. However, the authors of [5] considered only the case of forward current through a diode determined by the recombination of electrons and holes in the base. The obtained results are thus not applicable to the regions where injection leakage currents through the p–i and i–n junctions are of considerable value. It is known [6] that the current density range where the leakage currents are essential is by many orders of magnitude wider than the Hall region of currents discussed in [5]. The processes occurring after the reversal of a p–i–n diode current from forward to reverse direction with very high forward-current density (when it is necessary to account not only for the recombination of carriers in the base but also for the recombination of carriers in low-resistivity parts of the structure) remained outside the discussion in [3] to [5]. Such a discussion with due account for the expansion of space-charge regions by extraction and for injection coefficients of the junction differing from unity is the subject of this paper.

2. Fundamental Equations, Assumptions, Initial and Boundary Conditions

We consider a junction -type p–i–n diode with abrupt p–i and i–n junctions whose one-dimensional model is shown in Fig. 1. Let us assume, as in [5], that excess carrier lifetimes in the i-layer and the heavily doped regions remain constant during the whole transient process and are much shorter in the low-resistivity p- and n-regions than in the high-resistivity i-layer. Furthermore, we assume that the steady-state forward current corresponding to high-level injection flows through the diode during a sufficiently long time before the instant of switching. While the forward current exists, excess charge carriers accumulate in the base and heavily doped regions causing a time delay in the switching operation of the diodes.

It is known [5] that the accumulation of charges in the i-layer with the forward current flowing is caused by the predominance of recombination processes over generation when the current of holes arriving from the p-region meets the current of electrons arriving from the n-region. As it was said in [5], "in principle, this occurs not only in the middle region but also in the heavily doped regions". However, in the Hall current range (in power Si rectifiers with current densities $j_F \lessapprox 1$ to $10\ A/cm^2$), the processes in the heavily doped regions are negligible. Here we shall consider not only the current range where the processes in the heavily doped regions are negligible but also that where they

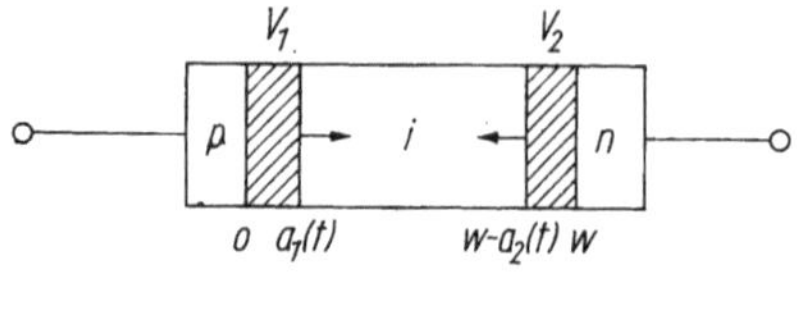

Fig. 1. Schematic model of a p–i–n diode

prevail. Just as the forward current in the range $j_F \sim \exp[qV/(2kT)]$ is caused by the annihilation of the charge accumulated in the i-layer, in the range $j_F \sim V^2$ it is caused by the annihilation of the charge accumulated in the heavily doped layers [6]. This means that the accumulation of charge in the heavily doped regions is possible even with a long i-layer ($w > L$) in the case of sufficiently large currents (in Si with $j_F \approx 10$ to 10^4 A/cm²). It is necessary to take this charge into account in current reversal from the forward to reverse direction because its value can turn out to be not less than that of the charge accumulated in the i-layer, and consequently it can play an important part in the process of carrier extraction.

The reverse current in a p–n junction diode is determined by the concentration of minority carriers in a certain region of the crystal. Therefore, after abrupt reversal of the diode current from forward to reverse direction at high injection level, the reverse current at the first instant can exceed considerably the saturation reverse current [7]. Moreover, after switching to reverse bias in a circuit with high-voltage power source and infinitely large (relative to the p–i–n diode forward resistance) series resistance (Fig. 2), the current may exceed considerably the forward current under steady-state conditions [5].[1]) Under the action of the large reverse current, most of the excess carriers accumulated in the base and the low-resistivity regions are extracted and this causes a shortening of the switching process.

Let us divide the whole blocking process of the diode into three stages as in [5]. At the stage I, $0 < t < t_0$, p–i and i–n junctions are still biased in the forward direction[2]) and the excess carrier concentration exceeds the thermally equilibrium value. The stage I ends at the instant $t = t_0$ when the concentrations of electrons and holes at the space-charge boundaries lower to their equi-

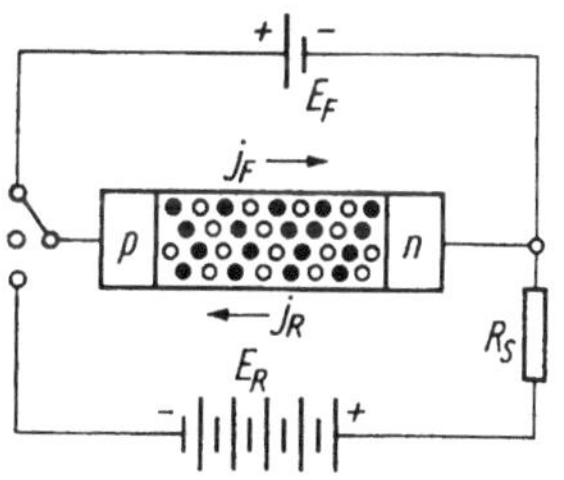

Fig. 2. Switching of a p–i–n diode from forward to reverse direction showing the process of excess carrier extraction out of the base by the reverse current

[1]) It does not follow from this that the large series resistance causes the fact that $j_R \gg j_F$. This resistance is intended to keep (together with the EMF source) the reverse current j_R constant during the period of carrier extraction. "To avoid mathematical complications", it is said in [5], "we shall concern here the limiting case of $j_R = \text{const} \gg j_F$ which is realized in principle with infinitely large EMF and infinitely large series resistance, although the results obtained are also true for $j_R \approx j_F$." It is also stated there that "just as the reverse-biased collector p–n junction of a transistor allows very high currents to be injected by the emitter until the external circuit hinders them, the blocking condition of a reverse-biased rectifier itself does not hinder the high current."

[2]) The fact that during the period $0 < t < t_0$ the junctions are forward-biased although a reverse current flows through them is related to the fact the concentrations of carriers at the boundaries of the p- and n-regions are high as compared to the their thermal-equilibrium values. This follows directly from equations (7) given below; according to them at $0 < t < t_0$ $p(0, t) > n_i$ and $p(w, t) > n_i$ if V_{p-i}, $V_{i-n} > 0$, i.e., the junctions are forward-biased.

librium values and the biases at the junctions become equal to zero. Stage II is featured by the expansion of space-charge layers due to further extraction of majority carriers through the contacts when the junctions begin to become reverse biased. In stage III, the recombination of non-extracted carriers occurs and, as a result, carrier concentrations reach their steady-state values.

Let us consider only the first two stages since only a small part of the reverse voltage relaxes at stage III. Consequently, almost all reverse resistance recovers during the first and second stages. Using the equations for electron and hole currents and assuming that recombination in the i-region is monomolecular and that the condition of quasi-neutrality is valid at stage I (i.e., electric field divergence is small compared to the density of excess-carrier space charge), we may write the diffusion approximation of the continuity equation [5] as

$$L^2 \frac{\partial^2 p}{\partial x^2} = p + \tau \frac{\partial p}{\partial t}, \tag{1}$$

where $p(x, t) \approx n(x, t)$ are the concentrations of holes and electrons, $L^2 = D\tau \equiv \equiv 2b/(b+1) D_p \tau$, τ is the lifetime, D_p the hole diffusion factor, b the ratio of electron and hole mobilities.

As initial condition for equation (1), we use the stationary solution which is of the following form for any relationship between recombination rates in the base and the heavily doped regions [6]:

$$p(x, 0) \equiv p_1(x, 0) + p_2(x, 0) =$$
$$= \frac{p_0(0) + p_0(w)}{2} \frac{\cosh [x/L - w/(2L)]}{\cosh w/(2L)} - \frac{p_0(0) - p_0(w)}{2} \frac{\sinh [x/L - w/(2L)]}{\sinh w/(2L)}. \tag{2}$$

Here, $p_0(0)$ and $p_0(w)$ are the steady-state concentrations of excess carriers in the base at the boundaries of the p- and n-regions with forward current which generally may be found from the following equations [2]:

$$\frac{j_F}{b+1} - \frac{q D p_0(w)}{L} \coth \frac{w}{L} + \frac{q D p_0(0)}{L} \operatorname{cosech} \frac{w}{L} = \frac{q \vartheta_p^* p_0^2(w)}{n_i}, \tag{3}$$

$$\frac{b j_F}{b+1} + \frac{q D p_0(w)}{L} \operatorname{cosech} \frac{w}{L} - \frac{q D p_0(w)}{L} \operatorname{cotanh} \frac{w}{L} = \frac{q \vartheta_n^* p_0^2(0)}{n_i}. \tag{4}$$

The boundary conditions of the problem are the equations for the electron-leakage current through the p–i junction and the hole-leakage current through the i–n junction [2, 6]:

$$j_n(0, t) = \frac{b}{b+1} j(t) + q D \left(\frac{\partial p}{\partial x}\right)_{x=0} = q \vartheta_n^* \left[p(0, t) \exp \frac{q V_{p\text{-}i}(t)}{kT} - n_i\right], \tag{5}$$

$$j_p(w, t) = \frac{1}{b+1} j(t) - q D \left(\frac{\partial p}{\partial x}\right)_{x=w} = q \vartheta_p^* \left[p(w, t) \exp \frac{q V_{i\text{-}n}(t)}{kT} - n_i\right], \tag{6}$$

the boundary concentrations of minority carriers satisfying the Boltzmann relations

$$p(0, t) = n_i \exp \frac{q V_{p\text{-}i}(t)}{kT}, \qquad p(w, t) = n_i \exp \frac{q V_{i\text{-}n}(t)}{kT}. \tag{7}$$

Taking into account equation (7), we may write the boundary conditions (5) and (6) in the following form:

$$\left(\frac{\partial p}{\partial x}\right)_{x=0} - \frac{\vartheta_n^*}{D} \frac{p^2(0, t)}{n_i} = -\frac{j(t)}{2\, q\, D_p}, \tag{8}$$

$$\left(\frac{\partial p}{\partial x}\right)_{x=w} + \frac{\vartheta_p^*}{D} \frac{p^2(w, t)}{n_i} = \frac{j(t)}{2\, b\, q\, D_p}. \tag{9}$$

The relationships (8) and (9) differ from the corresponding expressions in [5] by the presence of the second terms on their left-hand sides. These terms increase with injection level proportional to the squares of boundary concentrations and they become most important in the expressions (8) and (9) with sufficiently high forward currents.

In general it follows from (8) and (9) that electron and hole diffusion and drift currents at the boundaries of the i-layer give unequal contributions to the total current. They turn out to be equal only at a relatively low injection level corresponding to the Hall current range $(j_F \sim \exp[q\,V/(2\,kT)])$.

Since the leakage-dependent terms in (8) and (9) are of field nature [6], the current diffusion component is not essential in the region of very high forward currents ($j_F \sim V^2$) and the total current is practically determined by its drift component. For this reason, parallel with the change in signs of concentration gradients at the i-region boundaries [5], the boundary concentrations must change abruptly with abrupt (at the instant $t = 0$) reversal of the current from forward to reverse direction. Because of Boltzmann's relation (7) the voltage drops across the junctions must also change abruptly. The higher the forward current density before the current reversal and hence the higher the excess carrier leakage rate through the contacts was, the larger are the abrupt changes. It is natural that the above effect is negligible for the current range $j_F \sim \sim \exp[q\,V/(2\,kT)]$ because charge carriers in this region are not extracted through junctions under the action of the field component of the current (in this case the electric field near the junctions is of diffusion origin and results from the Dember effect).

However, it is quite noticeable for those current ranges where the non-ideality of one or both injecting junctions is essential. From the mathematical point of view, voltage jumps across the junctions result from the fact that the right-hand sides of expressions (8) and (9) become positive and negative, respectively, after abrupt reversal of the current from j_F to $-j_R$. In order to obtain left-hand sides of the same sign, it is necessary to change not only the signs and values of concentration gradients but also the values of boundary concentrations. In the current ranges featured by an appreciable deviation of the junction injection coefficients from unity, the terms containing the square of boundary concentration are much larger than their gradients, and the higher the injection level is, the greater is the difference. The signs of both sides of expressions (8) and (9), therefore, are the same only if the values $p(0, t)$ and $p(w, t)$ decrease after current reversal and the gradients increase. The latter fact means that one may neglect the leakage-dependent terms in (8) and (9) at reverse currents, although this is impossible at very high forward currents. This indicates that the following relationships [5] are valid at any values of j_F and j_R:

$$\left(\frac{\partial p}{\partial x}\right)_{x=0} = \frac{j_R}{2\,q\,D_p}, \qquad \left(\frac{\partial p}{\partial x}\right)_{x=w} = -\frac{j_R}{2\,b\,q\,D_p}. \tag{10}$$

It should be noted that under the conditions considered, the electric field can change abruptly the carrier concentration immediately after current reversal only within narrow boundary layers of the i-region whose thicknesses are of the order of magnitude of the Debye length $L_D = \sqrt{(\varepsilon\, kT)/(4\,\pi\, q^2\, p_B)}$. At high injection level ($p_B \gg n_i$) it is $L_D \ll w$ and therefore the transient stage I may be considered to proceed under the conditions of quasi-neutrality of practically the whole base as in the current range $j_F \sim \exp\,[q\,V/(2\,kT)]$ [5].

3. Calculation of Non-Stationary Carrier Distribution at the Transient Stage I

Since the boundary conditions (10) are non-uniform, we shall try to find the solution of equation (1) in the form

$$p(x,t) = p(x,0) + \alpha_1\,[p_1(x,t) - p_1(x,0)] + \alpha_2\,[p_2(x,t) - p_2(x,0)]\,, \tag{11}$$

where the values $p(x,0)$, $p_1(x,0)$, and $p_2(x,0)$ are determined by equation (2) and the arbitrary constants α_1 and α_2 have to be chosen such that

$$\left.\frac{\partial p_{1,2}(x,t)}{\partial x}\right|_{x=0} = \left.\frac{\partial p_{1,2}(x,t)}{\partial x}\right|_{x=w} = 0\,. \tag{12}$$

According to (2) and (10) to (12) we have for α_1 and α_2:

$$\left.\begin{aligned} \alpha_1 &= 1 + \frac{(b+1)\,j_R\,L\coth w/(2\,L)}{2\,b\,q\,D_p\,[p_0(0) + p_0(w)]}\,, \\ \alpha_2 &= 1 + \frac{(b-1)\,j_R\,L\tanh w/(2\,L)}{2\,b\,q\,D_p\,[p_0(0) - p_0(w)]}\,. \end{aligned}\right\} \tag{13}$$

In particular, for the Hall current range where according to (3) and (4)

$$p_0(0) = p_0(w)\,\frac{b\cosh w/L + 1}{\cosh w/L + b} = \frac{j_F\,L}{2\,b\,q\,D_p}\,\frac{b\cosh w/L + 1}{\sinh w/L} \tag{14}$$

it follows from (13) that

$$\alpha_1 = \alpha_2 = \alpha = 1 + \frac{j_R}{j_F} \tag{15}$$

and subsequently we obtain

$$p_{\text{Hall}}(x,t) = p(x,0) + \alpha\,[p'(x,t) - p(x,0)] \tag{16}$$

instead of (11).

It is evident that the function $p'(x,t)$ like $p(x,t)$ satisfies equation (1) and the initial condition (2).

With $L^2\, \mathrm{d}^2 p_{1,2}(x, 0)/\mathrm{d}x^2 = p_{1,2}(x, 0)$, $\alpha_1 \neq 0$, $\alpha_2 \neq 0$, and also $p_1(x, t)$ and $p_2(x, t)$ being independent of each other and assuming that the reverse current does not change with time in stage I ($j_\mathrm{R} = \mathrm{const}$), we find that the functions $p_1(x, t)$ and $p_2(x, t)$ satisfy equation (1), the initial condition (2), and the uniform boundary conditions (12). Let us try to determine the distribution $p_{1,2}(x, t)$ from equation (1) using Fourier's method for the separation of variables [8, 9]. Presenting the function to be determined in the form of a product of two functions, one of which depends only on x and the other only on t, we find for particular solutions of $p_{1,2}(x, t)$:

$$p_{1\,\mathrm{p}}(x, t) = a \cos\left[\frac{\lambda_1}{L}\left(x - \frac{w}{2}\right)\right] \exp\left[-(1 + \lambda_1^2)\frac{t}{\tau}\right], \tag{17}$$

$$p_{2\,\mathrm{p}}(x, t) = \bar{b} \cos\left[\frac{\lambda_2}{L}\left(x - \frac{w}{2}\right)\right] \exp\left[-(1 + \lambda_2^2)\frac{t}{\tau}\right], \tag{18}$$

while according to (12)

$$\sin\frac{\lambda_1 w}{2L} = 0; \qquad \lambda_1 = 2\nu\pi\frac{L}{w}; \qquad \nu = 0, 1, 2, \ldots, \infty\,, \tag{19}$$

$$\cos\frac{\lambda_2 w}{2L} = 0; \qquad \lambda_2 = (2\nu + 1)\pi\frac{L}{w}; \qquad \nu = 0, 1, 2, \ldots, \infty\,. \tag{20}$$

We shall construct the general solution of equation (1) on the basis of the particular solutions (17) and (18), because according to (10) and (20) there is an infinite number of such solutions satisfying the boundary conditions (12). We find this solution using the series expansion

$$p_1(x, t) = \frac{a_0}{2}\,\mathrm{e}^{-t/\tau} + \sum_{\nu=1}^{\infty} a_\nu \cos\left[\frac{2\pi\nu}{w}\left(x - \frac{w}{2}\right)\right] \exp\left\{-\left[1 + \left(\frac{2\pi\nu L}{w}\right)^2\right]\frac{t}{\tau}\right\}, \tag{21}$$

$$p_2(x, t) = \sum_{\nu=0}^{\infty} \bar{b}_\nu \sin\left[\frac{(2\nu + 1)\pi}{w}\left(x - \frac{w}{2}\right)\right] \exp\left\{-\left[1 + \left(\frac{(2\nu + 1)\pi L}{w}\right)^2\right]\frac{t}{\tau}\right\}. \tag{22}$$

It is seen from (21) and (22) that the constants a_ν and $\bar{b}_\nu$ are the factors in the Fourier series for $p_1(x, 0)$ and $p_2(x, 0)$, i.e., according to (2)

$$a_\nu = \frac{2L}{w}\,[p_0(0) + p_0(w)] \tanh\frac{w}{2L}\,\frac{(-1)^\nu}{1 + \left(\dfrac{2\nu\pi L}{w}\right)^2}, \tag{23}$$

$$\bar{b}_\nu = -\frac{2L}{w}\,[p_0(0) - p_0(w)] \coth\frac{w}{2L}\,\frac{(-1)^\nu}{1 + \left[\dfrac{(2\nu + 1)\pi L}{w}\right]^2}. \tag{24}$$

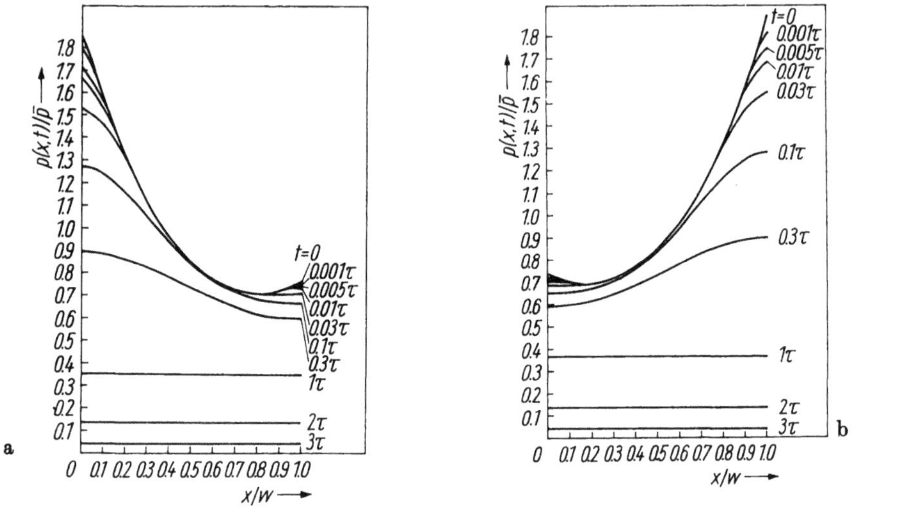

Fig. 3. Time dependence of the distribution $p(x, t)$ for a Si p–i–n diode at $j_R = 0$ for a) $\vartheta_n^* < \vartheta_p^*$ and b) $\vartheta_n^* > \vartheta_p^*$

Using (11), (2), (13), and (21) to (24) we obtain the expression for the excess carrier distribution in the base:

$$p(x, t) = \left(\bar{p} + \frac{j_R \tau}{q w}\right) \left\{ e^{-t/\tau} + 2 \sum_{\nu=1}^{\infty} \frac{(-1)^\nu \cos\left[\frac{2 \nu \pi}{w}\left(x - \frac{w}{2}\right)\right]}{1 + \left(\frac{2 \nu \pi L}{w}\right)^2} \times \right.$$

$$\left. \times \exp\left\{-\left[1 + \left(\frac{2 \nu \pi L}{w}\right)^2\right] \frac{t}{\tau}\right\}\right\} - 2 \left[\frac{p_0(0) - p_0(w)}{w/L} \coth \frac{w}{2 L} + B \frac{j_R \tau}{q w}\right] \times$$

$$\times \sum_{\nu=0}^{\infty} \frac{(-1)^\nu \sin\left[\frac{(2 \nu + 1) \pi}{w}\left(x - \frac{w}{2}\right)\right]}{1 + \left[(2 \nu + 1) \frac{\pi L}{w}\right]^2} \exp\left\{-\left\{1 + \left[(2 \nu + 1) \frac{\pi L}{w}\right]^2\right\} \frac{t}{\tau}\right\} -$$

$$- \frac{j_R \tau}{2 q L} \left[\frac{\cosh\left(\frac{x}{L} - \frac{w}{2 L}\right)}{\sinh \frac{w}{2 L}} - B \frac{\sinh\left(\frac{x}{L} - \frac{w}{2 L}\right)}{\cosh \frac{w}{2 L}}\right], \tag{25}$$

where $B = (b - 1)/(b + 1)$ and $\bar{p}$ is the average concentration of charge carriers in the base which equals, according to (2),

$$\bar{p} = \frac{1}{w} \int_0^w p(x, 0)\, dx = \frac{p_0(0) + p_0(w)}{w/L} \tanh \frac{w}{2 L}. \tag{26}$$

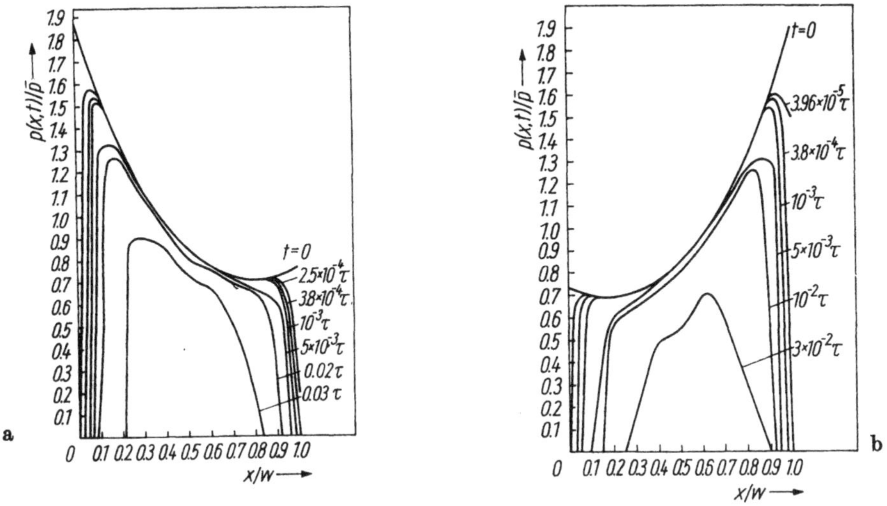

Fig. 4. Theoretical curves for charge carrier distribution in the base (formula (25)) in stage I of extraction under the action of a high reverse current $j_R = 10\, j_F = 5 \times 10^2$ A/cm² for a) $\vartheta_n^* < \vartheta_p^*$ and b) $\vartheta_n^* > \vartheta_p^*$

An essential feature of this distributions $p(x, t)$ consists in its universality, i.e., it is valid for any relationship between the carrier recombination rates in the base and the heavily doped regions; therefore, the distribution (25) is more general than that obtained in [5]. Since in the stage I of the switching process, with carriers being extracted by the reverse current, it is possible that $j_R \gg j_F$, the reverse-current value with essential leakage of carriers through one or both contacts can be much larger than in the Hall case. As a result the period of the initial development of the reverse voltage at very high forward currents is much shorter than for the current range $j_F \sim \exp\,[q\, V/(2\, kT)]$ [5].

Mathematically this means that the distribution $p(x, t)$ varies with forward current not only because of changes in j_R, $\bar{p}$, $p_0(0)$, and $p_0(w)$ and the effect of j_F on them, but also because of the fact that more and more expansion terms must be used in the series (25).

A plot of (25) calculated for a Si diode at $j_R = 0$ and $w/L = 2$ is shown in Fig. 3 and 4. The distribution curves shown in Fig. 3 correspond to $j_R = 0$, $j_F = 50$ A/cm², and a lifetime $\tau = 11$ μs at $\vartheta_n^* = 5 \times 10^{-3}$ cm/s, $\vartheta_p^* = 10^{-2}$ cm/s (Fig. 3a) and $\vartheta_n^* = 0.2$ cm/s, $\vartheta_p^* = 10^{-2}$ cm/s (Fig. 3b).

The curves shown in Fig. 4a and b were calculated for the same values of parameters as the curves shown in Fig. 3a and b, respectively, but for the reverse current $j_R = 10\, j_F = 5 \times 10^2$ A/cm². Since at $j_R = 0$ the extraction of excess carriers occurs during a relatively long time ($t \approx \tau$), the distribution $p(x, t)$ was calculated with a small number of terms in (25) (1 to 10 terms were used depending on the value of t/τ). In the case of high reverse currents, however, we had to use up to 20 to 30 terms in (25).

The comparison of results given in Fig. 3 a with the respective data given in [5] shows that for the parameter values chosen they do not differ much from each other both quantitatively and qualitatively. The data presented in Fig. 3 b differ from the results given in [5] mainly by the fact that the maximum value of $p(x, t)$ corresponds not to $x = 0$ but to $x = w$, and that the minimum of the distribution $p(x, t)$ is located near the point $x = 0$ and not at $x = w$.

In the case of $j_R \neq 0$ (Fig. 4 a and b) the current range $j_F \sim V^2$ is featured not only by the above-mentioned shift of maximum and minimum values of $p(x, t)$ but also by a stage I much shorter than for the current range $j_F \sim \sim \exp [q\, V/(2\, kT)]$. Indeed, as shown in Fig. 4 a and b, the times for the initial decrease of the excess concentration at the left (t_{01}) and right (t_{02}) boundaries of the i-layer are, respectively, $t_{01} = 2.5 \times 10^{-4}\, \tau$, $t_{02} = 3.8 \times 10^{-4}\, \tau$ and $t_{01} = 4 \times 10^{-5}\, \tau$, $t_{02} = 3.8 \times 10^{-4}\, \tau$, i.e., $t_{02}/t_{01} = 1.52$ for Fig. 4 a, whereas $t_{02}/t_{01} = 9.6$ for Fig. 4 b. As it follows from the data of [5], the above values for the current range $j_F \sim \exp [q\, V/(2\, kT)]$ are $t_{01} = 10^{-2}\, \tau$ and $t_{02} = 3 \times 10^{-2}\, \tau$, and therefore they differ from those for the range of very high forward currents by almost two orders of magnitude.

References

[1] A. YU. LEIDERMAN, Thesis for a condidate's degree, Physico-Technical Inst., Acad. Sci. UzbekSSR, Tashkent 1965.

[2] YA. P. KOTOV, Thesis for a candidate's degree, Physico-Technical Inst., Acad. Sci. UzbekSSR, Tashkent 1968.

[3] YU. K. BARSUKOV, Soviet Phys. — Solid State **1**, 1659 (1959).

[4] Z. S. GRIBNIKOV and YU. A. TKHORIK, Ukr. fiz. Zh. **9**, 637 (1964); **9**, 864 (1964); **9**, 972 (1964); **9**, 1201 (1964); Symp. "Fizika p–n perekhodov", Izd. "Zinatne", Riga 1966 (p. 59).

[5] A. HOFFMAN and E. SPENKE, Solid State Electronics **8**, 693 (1965).
K. SCHUSTER and E. SPENKE, ibid. **8**, 881 (1965).
H. BENDA, A. HOFFMAN, and E. SPENKE, ibid. **8**, 887 (1965).
H. BENDA and E. SPENKE, Proc. IEEE **55**, 1331 (1967).

[6] A. YU. LEIDERMAN and P. M. KARAGEORGII-ALKALAEV, Radiotekhnika i Elektronika **9**, 1869 (1964); **10**, 720 (1965).

[7] W. SHOCKLEY, Bell Syst. tech. J. **28**, 435 (1949).

[8] A. N. TIKHONOV and A. A. SAMARSKII, Uravneniya matematicheskoi fiziki, GITTL, Moskva 1953.

[9] KH. S. KOSHLYAKOV, E. B. GLINER, and M. M. SMIRNOV, Osnovnye differentsialnye uravneniya matematicheskoi fiziki, GIFML, Moskva 1962.

(Received October 12, 1972)

See Short Note by H.-J. SCHNABEL and W. STORCH

Fig. 2. Small angle bevel through a selective diffused slice of GaAs

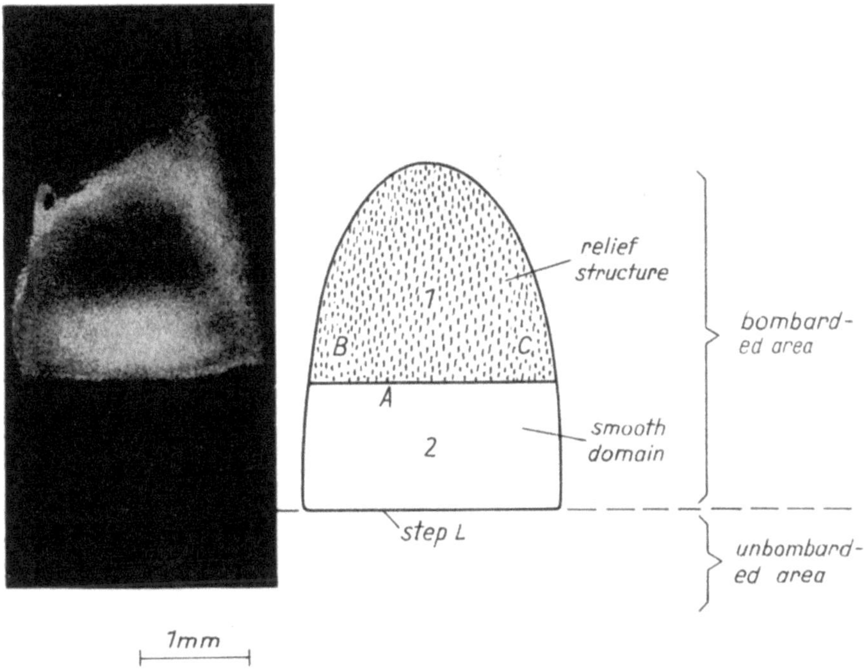

a b

Fig. 2. a) Optical micrograph of the sputtered Si surface, b) drawing of the domains observed in Fig. 2a

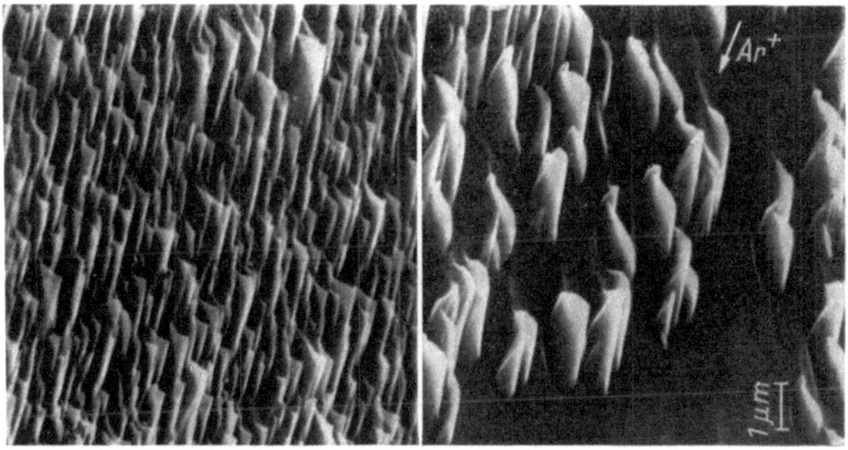

a b

Fig. 3. Relief structure on Si(111) after bombardment with 10 keV Ar^+ ions in the marked direction. The macroscopic surface is viewed at 45° in the scanning electron microscope (tilt axis parallel to the μm scale). a) Domain 1, b) border A between the domains 1 and 2 (see Fig. 2b)

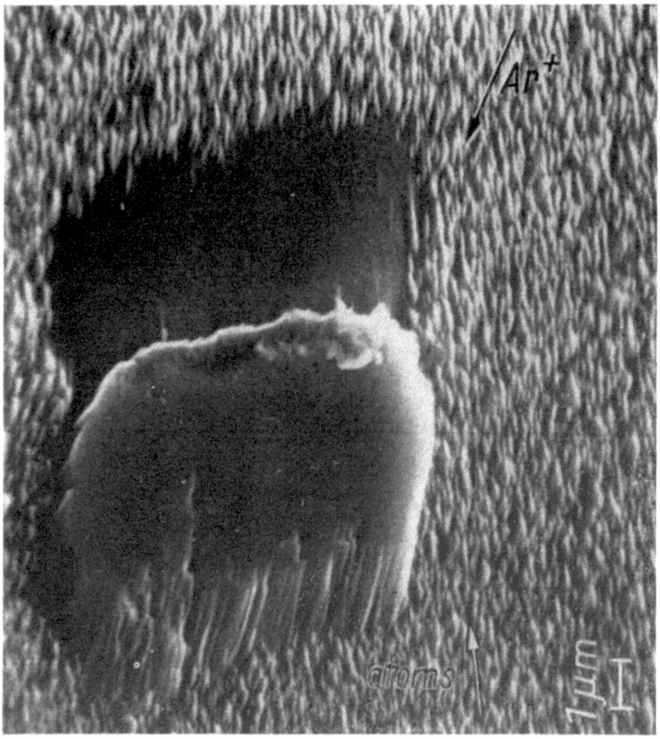

Fig. 4 b. Scanning electron micrograph of the relief structure around a dust particle

a 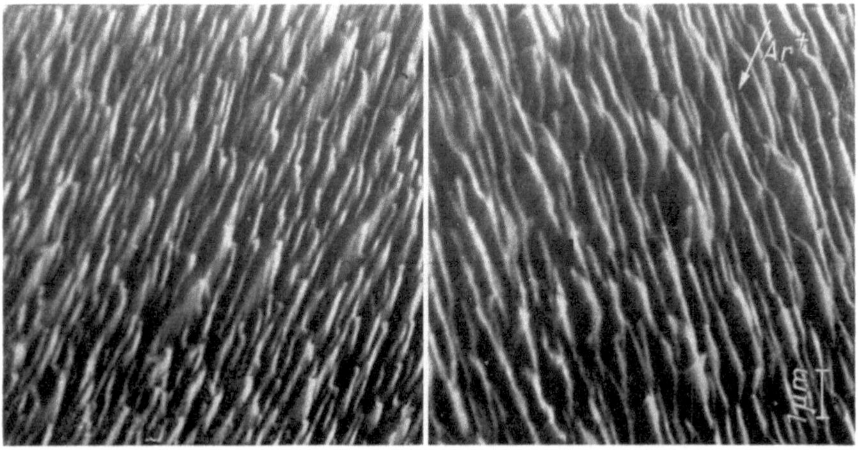 b

Fig. 5. Orientation of the relief structure at different directions of the impinging metal atoms. a) Relief structure at B (see Fig. 2 b); the metal atoms strike the Si surface from the right, b) relief structure at C; the metal atoms strike the surface from the left

See Short Note by R. Ogden

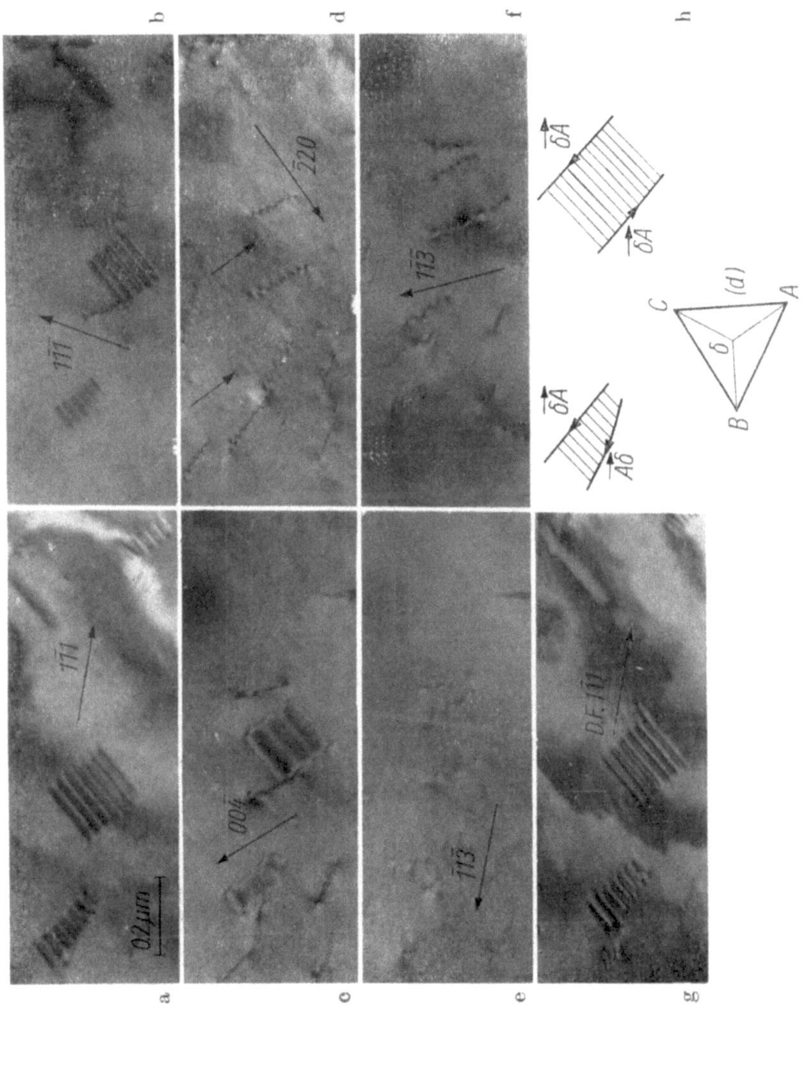

Fig. 1. Diffraction contrast analysis in 110 zone

See Short Note by R. Ogden

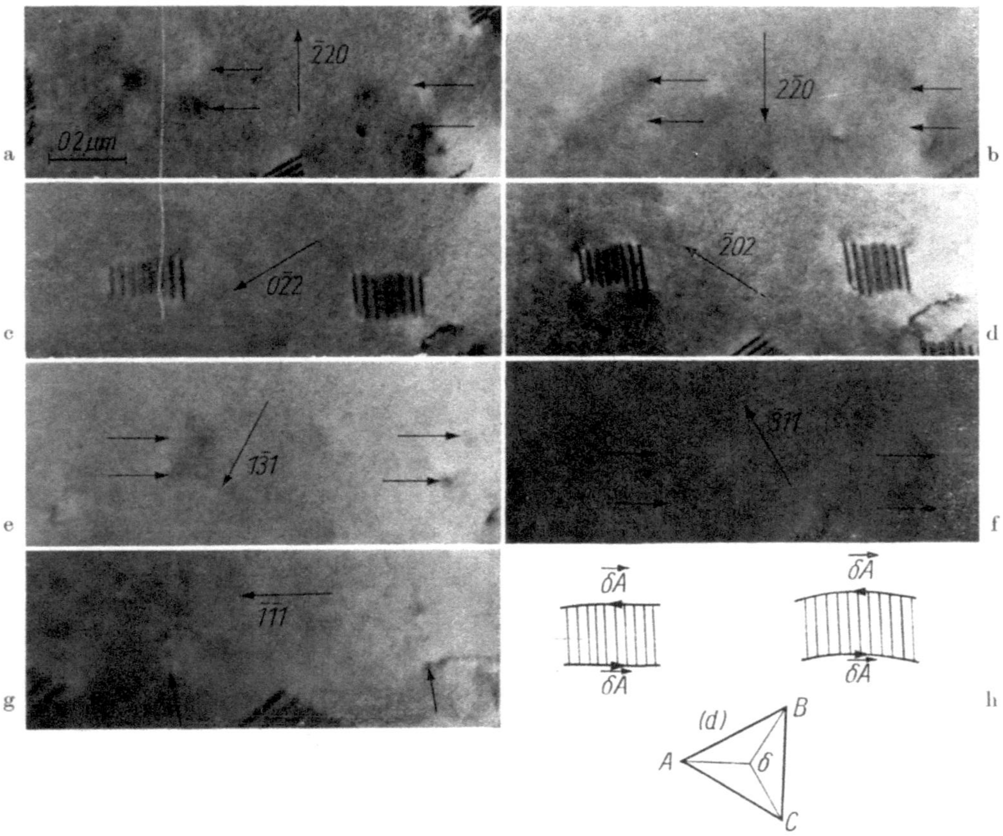

Fig. 2. Diffraction contrast analysis in 111 zone. h) $\vec{CA} = 1/2\,[011]$, $\vec{BA} = 1/2\,[101]$, $\vec{BC} = 1/2\,[1\bar{1}0]$. $\vec{\delta A} = 1/6\,[112]$, $\vec{\delta B} = 1/6\,[\bar{2}1\bar{1}]$, $\vec{\delta C} = 1/6\,[1\bar{2}\bar{1}]$; viewed from outside tetrahedron

Short Notes

phys. stat. sol. (a) 14, K95 (1972)
Subject classification: 9 and 14.3.3; 22.2.1

VEB Halbleiterwerk Frankfurt/Oder, Hauptbereich Forschung, Stahnsdorf

A Peculiar Method for Selective Diffusion of Zn into GaAs

By

H.-J. SCHNABEL and W. STORCH

p-n junctions were produced in Sn doped GaAs substrates of (100) orientation with carrier concentrations ranging from $5x10^{17}$ to $5x10^{18}$ cm^{-3} by the convenient diffusion technique of doped pyrolytic silicon dioxide. Zinc was the doping element (1 to 5). The SiO_2 layers were prepared by oxidation of monosilane with oxygen at 450 ^oC.

It was established that GaAs substrates with SiO_2 layers of ≈ 3000 Å thickness can be annealed under normal pressure in H_2 or N_2 for several hours at 850 ^oC without changing crystal perfection, resistivity, mobility, or carrier concentration. This made diffusion possible from doped SiO_2 layers into GaAs in an open system without influencing substrate properties by heating.

ZnO doped SiO_2 films were deposited by simultaneous reaction of monosilane and zincdiethyl-bis-tetrahydrofurane[1)] with oxygen at 450 ^oC. Zinc concentrations varied in the range of 1 to 25 $\mu g/cm^2$ in layers of 1000 Å thickness. Heat treatment in H_2 at 850 ^oC for 20 min produced junction depths of ≈ 10 μm whereas they were always less than 1 μm in N_2 under the same conditions.

The mechanism of diffusion was assumed to proceed via reduced ZnO. Pure ZnO layers were instable in N_2 at temperatures above 650 ^oC. A reduction of ZnO in doped layers of SiO_2 in H_2 at 650 ^oC was already achieved so that after pre-treating in this way junction depths of ≈ 10 μm were produced in N_2 at 850 ^oC. These results offer the possibility of determining different surface concentrations by achieving ZnO doped SiO_2 layers of constant composition by heat treatment in

---

1) We are very grateful to Prof. Dr. K.-H. Thiele, TH Merseburg, for recommending and preparing.

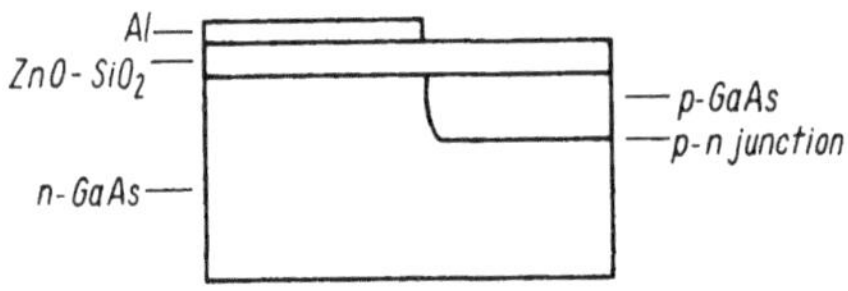

Fig. 1. Model for selective diffusion of Zn into GaAs by use of Al protected doped SiO_2 layers

H_2 at 650 °C when a subsequent diffusion will be carried out in N_2. Aluminium layers vapour-deposited on ZnO doped SiO_2 were qualified for preventing the attack of H_2 at 650 °C. The Al deposited parts of ZnO doped SiO_2 layers were not a source for Zn diffusion. Penetration of Zn into GaAs and the formation of a p-n junction could not be identified (Fig. 1 and 2). Consequently it is possible to produce selective diffused GaAs structures by using photo-lithographically etched aluminium as masks on doped SiO_2 for thermal treatment in H_2 and N_2. Layers of pure SiO_2 on ZnO doped silicon dioxide were unqualified for protecting against H_2 at 650 °C.

References

(1) W. VON MUENCH, Z. angew. Phys. 18, 129 (1964).

(2) W. VON MUENCH, Solid State Electronics 9, 619 (1966).

(3) W. VON MUENCH, Solid State Electronics 9, 939 (1966).

(4) W. VON MUENCH, Symp. on GaAs, 1968, Paper 13 (p. 77).

(5) W. KERN and J.P. WHITE, RCA Rev. 31, 715 (1970).

(Received September 25, 1972)

phys. stat. sol. (a) 14, K97 (1972)

Subject classification: 1.6 and 11; 22.1.2

Sektion Physik der Technischen Universität Dresden

Development of Relief Structures on Si Surfaces by Ion Bombardment

By

J. PUNZEL and W. HAUFFE

Introduction The development of the relief structure on Si single crystals caused by He^+ ion bombardment in an emission electron microscope has been described by one of the authors (1). In the paper quoted a characteristic rough relief structure was shown after bombardment at room temperature with ions produced in a canal ray discharge tube. This relief structure has been found also after bombardment with Ar^+ ions under these conditions. Recent investigations carried out on Si(111)-surfaces bombarded in a special ion beam apparatus with a separated beam of 10 keV Ar^+ ions (no beam of a canal ray discharge tube as in (1)) from different directions of incidence do not exhibit the characteristic structure mentioned above. No relief formation could be observed in an electron microscope (2).

One possible explanation of this discrepancy might be that in (1) foreign atoms have produced this structure. This would be similar to the observation of Wehner (3), who found the development of cones on sputtered copper surfaces induced by simultaneously impinging molybdenum atoms. Therefore we have studied the influence of foreign atoms if Si is sputtered by ions at room temperature.

Experiments We modified Wehner's experiment according to Fig. 1. A separated parallel beam of Ar^+ ions strikes the Si(111)-surface under an angle of incidence of 63° to the normal. Part of the ion beam strikes a metal cylinder C mounted near the silicon surface. Thus Ar^+ ions as well as metal atoms sputtered from the cylinder strike the silicon surface. The density of sputtered metal atoms varies in the bombarded range on the silicon surface from zero on the line L to a maximum.

Further experimental conditions were: total pressure 8×10^{-5} Torr, ion beam density 3 $\mu A/mm^2$, bombarding time 60 min. The cylinder material was stainless steel, copper, molybdenum, aluminium,and in one experiment even silicon.

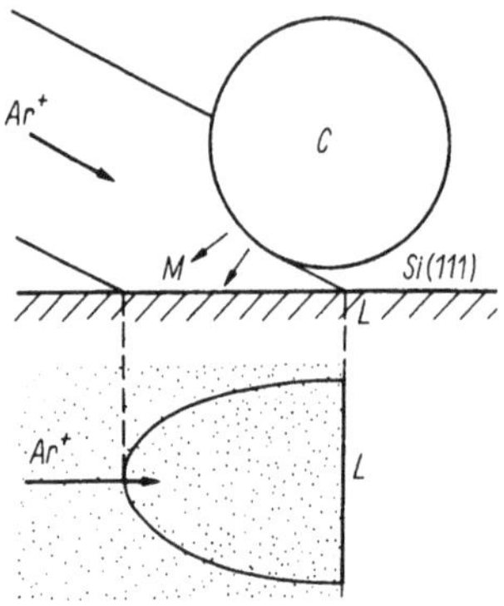

Fig. 1. Sputtering of a Si(111)-surface under the influence of foreign atoms. C metal cylinder, L border of the bombarded area, M atoms sputtered from the cylinder

Experimental results and discussion Fig. 2a shows an optical micrograph of the bombarded silicon surface (cylinder material in this case is stainless steel). In the bombarded are two characteristic domains can be seen. Domain 1 (see Fig. 2b) shows a strong relief formation, domain 2 is sputtered without any structure development. As an example Fig. 3a shows the microrelief of the domain 1 displayed in the scanning electron microscope (JEOL). No cones were found there. Only on the border between 1 and 2 relief structures with single cones could be observed in some cases (Fig. 3b).

The formation of the relief structure is connected with the striking of atoms from the cylinder. This can be seen clearly at dust particles, which produce a shadow for the ions at one direction as well as for the metal atoms at an other direction (see Fig. 4a). In this case the relief formation can be seen in Fig. 4b. The area F which could not be struck by metal atoms has been sputtered without any relief structure. The absence of the structure in the domain 2 of Fig. 2b up to the sharp border A suggests that the density of foreign atoms strongly influences the structure formation (Ar^+ ions reflected at the cylinder can have also an influence).

A further new observation concerns the dependence of the shape of the relief structure on the direction of the impinging metal atoms. In Fig. 5a and 5b there can be seen the structures at B and C (see Fig. 2b) at which the metal atoms strike the silicon surface preferentially from the right at B and from the left at C.

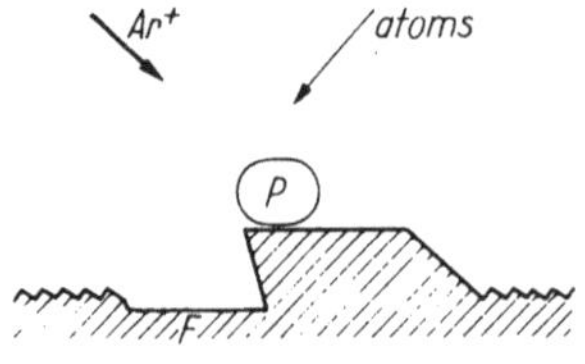

Fig. 4. a) Relief structure at a dust particle P on the initial surface (schematically) F area screened for metal atoms

With a cylinder of copper and molybdenum we observed a result qualitatively equal to that for steel as described above. With a cylinder of aluminium and silicon no relief formation was found under these conditions as in the experiment without any cylinder.

We wish to thank Prof. A. Recknagel for valuable help in this work.

References

(1) J. PUNZEL, phys. stat. sol. 24, K1 (1967).

(2) W. HAUFFE, unpublished.

(3) G.K. WEHNER, J. appl. Phys. 42, 1145 (1971).

(Received July 4, 1972)

phys. stat. sol. (a) 14, K101 (1972)

Subject classification: 10.2; 1.5; 22.1.2

Allen Clark Research Centre, The Plessey Company Limited, Caswell Towcester Northants

Shockley Loops in Epitaxial Silicon

By

R. OGDEN

Many studies have been made of stacking fault structures in epitaxial silicon (1 to 5). These structures have almost invariably been nucleated at the substrate: :layer interface and form closed configurations such as tetrahedra composed of stacking faults intersecting at stair-rod dislocations. Ribbon stacking faults similar to those shown here in Fig. 1 and 2 have been reported occasionally (3, 4) but have not previously been analysed fully. For the present diffraction contrast analysis in the electron microscope, specimens were prepared from $\{111\}$ epitaxial silicon layers grown at low temperatures ($< 900\,^{o}C$) from silane deliberately contaminated with very low levels (≈ 100 ppm) of ethylene. A high-tilt cartridge was used in the microscope to enable micrographs to be taken using reflections in both the 110 and 111 zones. Care was taken to select faults for analysis which were not overlapping and all micrographs were taken close to the 2 beam s = 0 position. Indexing has been chosen so that the faults to be analysed lie on $(11\bar{1})$ planes in both series shown in Fig. 1 and 2 which were taken from different specimens. Closely following an earlier analysis (6), the ribbon defects were confirmed as stacking faults by observation of fringe contrast in Fig. 1a to f. Examination of outer fringe contrast in dark field (7), Fig. 1g shows that both stacking faults are extrinsic. Similar micrographs have shown that the two stacking faults in Fig. 2 are also extrinsic. The possible Burgers vectors (6) of the bounding partials are $\pm 1/3\,[11\bar{1}]$, $\pm 1/6\,[\bar{1}\bar{1}\bar{2}]$, $\pm 1/6\,[\bar{1}21]$, or $\pm 1/6\,[2\bar{1}1]$. The partials show strong contrast in Fig. 1e, f but no contrast in Fig. 1a, b. Reference to Table 1 eliminates the Shockley partials $\pm 1/6\,[\bar{1}21]$ and $\pm 1/6\,[2\bar{1}1]$ since each of these partials should be visible for one set of $\{111\}$ and $\{311\}$ reflections were $\vec{g}\cdot\vec{b} = \pm 1$ or $\pm 2/3$ and invisible for the other set where $\vec{g}\cdot\vec{b} = 0$ or $\pm 1/3$.

The $1/3\,[11\bar{1}]$ Frank partial is eliminated using the $\bar{2}20$ reflection since very little contrast is observed at the partials. The black contrast patches in Fig. 1d

Table 1

$\vec{g}\cdot\vec{b}$ values of possible partials bounding stacking faults on $(11\bar{1})$ planes

$\vec{g}$ \ $\vec{b}$	$\pm 1/3\,[11\bar{1}]$	$\pm 1/6\,[\bar{1}\bar{1}\bar{2}]$	$\pm 1/6\,[\bar{1}21]$	$\pm 1/6\,[\bar{2}11]$
$1\bar{1}1$	±1/3	±1/3	±1/3	±2/3
$1\bar{1}\bar{1}$	±1/3	±1/3	±2/3	±1/3
$00\bar{4}$	±4/3	±4/3	±2/3	±2/3
$\bar{1}1\bar{3}$	±1	±1	0	±1
$1\bar{1}\bar{3}$	±1	±1	±1	0
$\bar{2}20$	0	0	±1	±1
$0\bar{2}2$	±4/3	±1/3	±1/3	±2/3
$\bar{2}02$	±4/3	±1/3	±2/3	±1/3
$1\bar{3}1$	±1	0	±1	±1
$\bar{3}11$	±1	0	±1	±1
$\bar{1}\bar{1}1$	±1	0	0	0

and the black and white contrast patches in Fig. 2a, b (shown arrowed) are not characteristic of the "m" term contrast which the Frank partial would display. Although $\vec{g}\cdot\vec{b}$ and m are both zero for the remaining $\pm 1/6\,[\bar{1}\bar{1}\bar{2}]$ Shockley partial, the patchy contrast may be expected since Head (8) has shown that inclined dislocations in β-brass can produce such contrast under these conditions.

The series of micrographs Fig. 2c to g confirms the $\pm 1/6\,[\bar{1}\bar{1}\bar{2}]$ Shockley partial since in no case is dislocation contrast observed at the faults although faint lines, possible due to decoration, are visible along part of the bounding partials in Fig. 2e. Table 1 shows that the Frank partial should be visible in all these micrographs and the previously eliminated Shockley partials should also be visible for most of them. Clareborough (9) has shown that partials may be invisible for $\vec{g}\cdot\vec{b} = \pm 2/3$ under certain conditions but Table 1 shows that there are sufficient cases where $\vec{g}\cdot\vec{b}$ is zero or an integer to establish the Burgers vectors of partials without using this value.

Fig. 1h and 2h represent the stacking fault ribbons showing the Burgers vectors of the partials using the Thompson notation (10, 11). Plane (d) is the $(11\bar{1})$ fault plane and the stacking faults can be regarded as loops $\overrightarrow{\delta A}$ which are truncated by the foil surfaces. The majority of ribbon faults found were similar to those shown in Fig. 1 and 2 although some intrinsic loops occurred. In both cases the partials were predominantly in screw orientation with Burgers vectors of magnitude $A\beta$, $A\gamma$, and $A\delta$ lying on inclined planes b, c, and d respectively. However occasional examples of extended dislocations such as $\overrightarrow{B\delta} + \overrightarrow{\delta A}$ were found. Full results and a proposed mechanism for the nucleation and growth of the ribbon faults based on the incorporation of carbon atoms on specific lattice sites are to be published separately.

The author is indebted to Dr. W.J. Tunstall and Mr. R.W. Bickness for useful discussion, to the Plessey Company for permission to publish this note, and to the Ministry of Defence (Naval Department) who supported this work.

References

(1) J.M. CHARIG, B.A. JOYCE, D.J. STIRLAND, and R.W. BICKNELL, Phil. Mag. 7, 1847 (1962).

(2) G.R. BOOKER and R. STICKLER, J. appl. Phys. 33, 3281 (1962).

(3) R.H. FINCH, H.J. QUEISSER, G. THOMAS, and J. WASHBURN, J. appl. Phys. 34, 406 (1963).

(4) G.H. SCHWUTTKE and V. SILS, J. appl. Phys. 34, 3127 (1963).

(5) G.R. BOOKER and B.A. UNVALA, Phil. Mag. 11, 11 (1965).

(6) G.R. BOOKER and W.J. TUNSTALL, Phil. Mag. 13, 71 (1966).

(7) R. GEVERS, A. ART, and S. AMELINCKX, phys. stat. sol. 3, 1563 (1963).

(8) A.K. HEAD, Austral. J. Phys. 20, 557 (1967).

(9) L.M. CLAREBOROUGH, Austral. J. Phys. 24, 79 (1971).

(10) N. THOMPSON, Proc. Phys. Soc. B66, 481 (1953).

(11) J.P. HIRTH and J. LÖTHE, Theory of Dislocations, Ser. Mat. Sci. & Engng., McGraw-Hill Publ. Co., 1968.

(Received September 15, 1972)

phys. stat. sol. (a) 14, K105 (1972)

Subject classification: 14.4; 18.3; 22.8.2

Research Centre of Physics, Iassy

Some Observations on the Dielectric Behaviour of $Ca_xNi_{1-x}Fe_2O_4$ Compounds

By

A. VASILIU, E. CUCIUREANU, N. REZLESCU, and E. LUCA

Some new results obtained on the dielectric behaviour of ferrites in the $Ca_xNi_{1-x}Fe_2O_4$ system are given in this note.

Measurements were made on the same polycrystalline samples used in papers (1) and (2), where some magnetic and electrical properties of the Ni-Ca ferrites were studied. The ferrite samples were sintered at different temperatures according to the concentration of Ca ions.

Interesting results were obtained concerning the dependence of the dielectric parameters on temperature and concentration of Ca ions.

ε and tgδ were measured at a frequency of 1 kHz by means of a capacity bridge described in (3).

Fig. 1 shows the variation of the dielectric constant ε and dielectric loss factor tgδ as a function of the concentration of Ca ions for samples with $0 \leqq x \leqq 0.7$, sintered at 1175 °C for 16 h and cooled in two ways; slowly, in the furnace, with approximately 80 deg/h, and quickly, with approximately 500 deg/min, by quenching in water. For the slowly cooled samples, it may be observed that the dielectric constant increases with the concentration of Ca ions at the beginning and has a maximum for x = 0.2; then it decreases to a value approached by the Ni ferrite. In the case of the quenched samples, the shape of the $\varepsilon(x)$ curve is similar, but the maximum appears at about x = 0.3 and is much flatter.

The dielectric loss factor tgδ has a minimum in the region where ε is maximum, for both slowly and quickly cooled samples.

Fig. 2 shows the dependence of the dielectric constant ε on the concentration of Ca ions for smaller concentrations (x < 0.1). In this case, the samples were sintered at 1200 °C (curves 1) and 1250 °C (curves 2); the curves a correspond to slowly cooled samples and curves b to quickly cooled ones. It may be observed that for small concentrations of Ca ions, ε generally increases with increasing x;

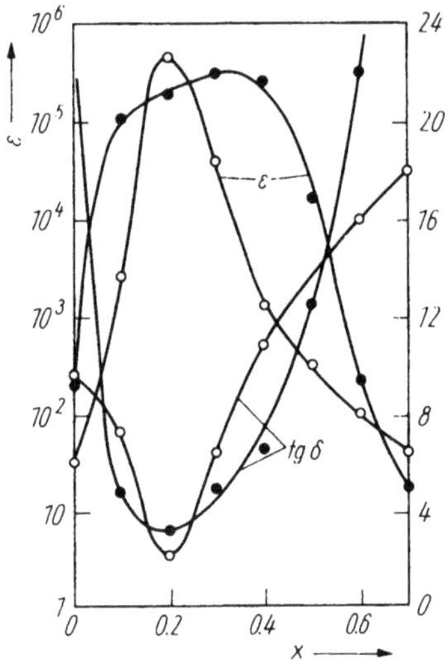

Fig. 1

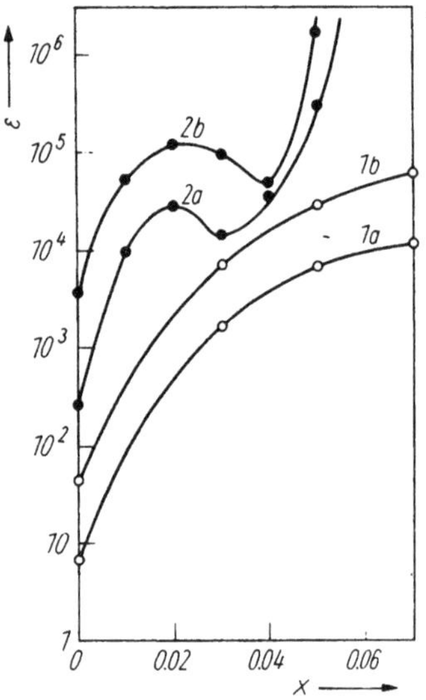

Fig. 2

Fig. 1. ε and tg δ as a function of Ca concentration for the $Ca_xNi_{1-x}Fe_2O_4$ ferrites sintered at 1175 °C:
o o o slowly cooled samples, • • • quickly cooled samples

Fig. 2. Dielectric constant ε as a function of Ca concentration for $Ca_xNi_{1-x}Fe_2O_4$ ferrites:
o o o samples sintered at 1200 °C, • • • samples sintered at 1250 °C

for samples sintered at 1250 °C a minimum occurs in the range $0.03 \leqq x \leqq 0.04$, but this minimum has a value higher than that for the Ni ferrite. Very small concentrations of Ca ions in the Ni ferrite give rise to a remarkable increase of the dielectric cónstant.

A significant fact is that the dielectric constant ε vs. Ca ion concentration curves (Fig. 1 and 2) are the inverse images of those of the resistivity dependence on Ca concentration, given by us in (2). This is a confirmation of the correlation between permittivity ε and resistivity ϱ (4).

The temperature dependence of the dielectric parameters was also studied in the temperature range from -170 to +150 °C. In Fig. 3 and 4 the results of the measurements are given. Fig. 3 shows the variation of ε and tg δ as a function of temperature for samples with different Ca concentrations at the frequency of

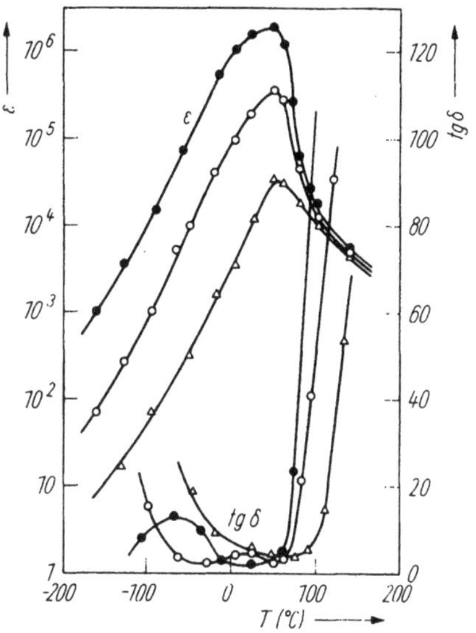

Fig. 3

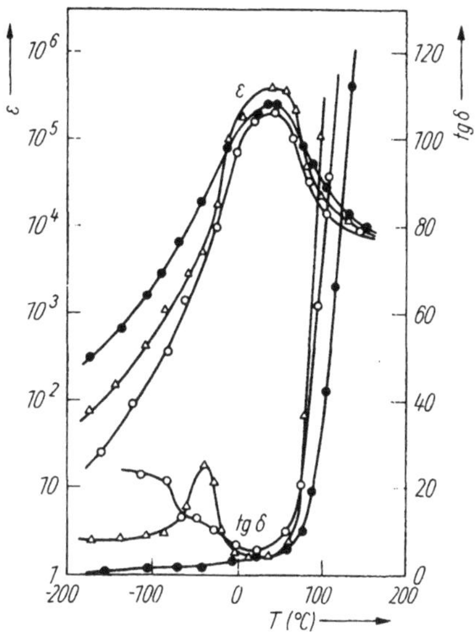

Fig. 4

Fig. 3. ε and tg δ as a function of temperature for samples sintered at 1175 °C: ● ● ● $Ca_{0.2}Ni_{0.8}Fe_2O_4$ ferrite slowly cooled, ○ ○ ○ $Ca_{0.2}Ni_{0.8}Fe_2O_4$ ferrite quickly cooled, △ △ △ $Ca_{0.3}Ni_{0.7}Fe_2O_4$ ferrite quickly cooled

Fig. 4. ε and tg δ as a function of temperature at different frequencies for $Ca_{0.1}Ni_{0.9}Fe_2O_4$ ferrite sintered at 1175 °C and quickly cooled: ● ● ● 0.1 kHz, ○ ○ ○ 1 kHz, △ △ △ 10 kHz

kHz. In Fig. 4 are plotted three curves obtained at various frequencies for $Ca_{0.1}Ni_{0.9}Fe_2O_4$ ferrite. It was found that the dielectric constant varies with temperature in a different manner in comparison with other spinel ferrites. ε increases with temperature for the majority of ferrites (5), but for these ferrites it increases only as far as +50 °C where it has a maximum, and then it decreases. These results are similar to those obtained by (6) for other ferrites. The position of the maximum is independent of the composition (Fig. 3) and the frequency (Fig. 4).

The losses in these ferrites are small in the vicinity of the maximum of ε and increase very strongly above +50 °C.

References

(1) A. VASILIU, GH. MAXIM, M.L. CRAUS, and E. LUCA, phys. stat. sol. (a) 13, 371 (1972).

(2) A. VASILIU, N. REZLESCU, and E. LUCA, Rev. Roum. Phys. 17 (1972).

(3) V. PETRESCU, E. CUCIUREANU, and N. REZLESCU, Rev. Roum. Phys. 15, 731 (1970).

(4) L.I. RABKIN and Z.I. NOVIKOVA, Ferrites, Minsk 1960 (p. 146).

(5) V.A. IOFFE, G.I. KHVOSTENKO, and Z.N. ZONN, Soviet Phys. - J. tech. Phys. 27, 1985 (1957).

(6) F. HABEREY and H.P.J. WIJN, phys. stat. sol. 26, 231 (1968).

(Received August 3, 1972)

phys. stat. sol. (a) 14, K109 (1972)

Subject classification: 18.3; 22.8.2

Ingenieurhochschule Köthen

Induced Magnetic Anisotropy in Iron-Deficient Manganese Ferrites at Low Temperatures

By

J. BIEHOUNEK

In manganese-ferrites $Mn_xFe_{2-x}O_4$ there exists a relaxation phenomenon (effect C) between 90 ^oK and room temperature for $x > 0.6$. With increasing x, its intensity increases. Moreover, it shifts to higher temperatures. Measurements on single crystals showed a predominant contribution of the F-term to the induced magnetic anisotropy. The results obtained in (1 to 3) agree in essential points. A uniform opinion on the explanation of the relaxation mechanism is not yet obtained. It is considered to explain effect C by electron hopping between bivalent ions on B-sites with participation of the Jahn-Teller effect. Závěta et al. (1) developed a model for the effect C that ascribes an essential role to Fe^{2+}. Mizushima (2) supposes on the contrary an electron exchange between bivalent ions of manganese. With a new method measurements of the instable induced anisotropy in magnetically saturated iron-deficient manganese ferrites were carried out. The method is based on the frequency dependence of the transversal susceptibility with the appearance of an after-effect. The circle-cylindrical sample ($d \approx 10$ mm, $h \approx 3$ mm) rotates in a coil whose magnetic field is directed transversally to the direct field of an electromagnet. The coil determines the frequency of the oscillator. If the sample rotates with a varying number of rotations per second then the change of susceptibility causes a change $\Delta\omega$ of the oscillator frequency. Without consideration of the time-constant distribution follows

$$\Delta\omega = A \frac{K_u}{H_e J_s} \left(\frac{1}{1 + (2\Omega\tau)^2} - \frac{1}{1 + (2\Omega_o\tau)^2} \right),$$

where A apparatus constant, Ω angular velocity of the sample, Ω_o smallest angular velocity, H_e magnetic field strength, J_s spontaneous magnetization, τ time constant of diffusion, K_u constant of the induced anisotropy.

Advantages of the method are the large variation range of the number of rotations per second (from 0.4 to 100 Hz) and that the measurements practically cannot be influenced by other anisotropies.

$Mn_{1.5-x}Me_xFe_{1.5}O_4$ ferrites (x = 0, 0.2, 0.4, Me: Ni or Cr) were examined. The Ni-containing ferrites are of special interest because Ni^{2+} replaces Fe^{2+}. Ni^{2+} and the trivalent portion of Mn are on B-sites (4). Fe^{3+} and Mn^{2+} may occupy the A-sites (5). The cation distribution follows from measurements of the specific magnetization and from data about the cation vacancies.

Table 1

sample	x	
1	0	$(Mn^{2+}_{0.62}Fe^{3+}_{0.38})\left[Mn^{3+}_{0.89}Fe^{2+}_{0.35}Fe^{3+}_{0.76}\right]O_{4+0.02}$
2.1	0.2	$(Mn^{2+}_{0.74}Fe^{3+}_{0.26})\left[Ni^{2+}_{0.2}Mn^{3+}_{0.56}Fe^{2+}_{0.02}Fe^{3+}_{1.22}\right]O_{4+0.018}$
2.2	0.4	$(Mn^{2+}_{0.54}Fe^{3+}_{0.46})\left[Ni^{2+}_{0.4}Mn^{3+}_{0.56}Fe^{2+}_{0.06}Fe^{3+}_{0.98}\right]O_{4+0.015}$

Some open questions (for instance Fe^{3+} may be replaced by Mn^{2+} on B-sites) do not affect the fact that the substitution of Mn by Ni strongly influences the Fe^{2+} concentration.

The measurements show a completely different influence of Ni and Cr on the effect C. The Ni substitution shifts the effect C towards lower temperatures with increasing x (Fig. 1). At the same time its intensity increases in contradiction to the results of Gerber et al. (3). This behaviour is explained by the fact that K_u strongly increases with decreasing temperature. The Cr substitution makes the effect C completely disappear already at x = 0.2. Cr replaces Mn^{3+}. Its importance

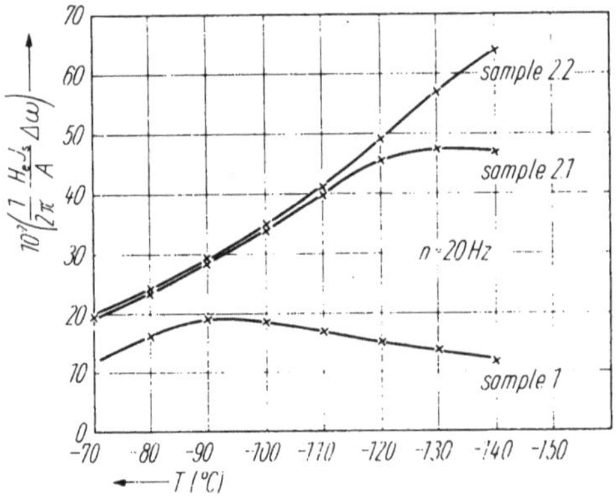

Fig. 1. Dependence of frequency change of the oscillator (with the change of the number of rotations per second from 0.4 to 20 Hz) on the temperature for some sample composition (see Table 1) of the $Mn_{1.5-x}Ni_xFe_{1.5}O_4$ ferrites

for the origin of the effect C becomes clear. Simultaneously it may be concluded that Fe^{2+} is without any influence.

Acknowledgement

The author is very much obliged to Dr. Dietzmann of the Karl-Marx-University of Leipzig for his kind advice and helpful discussions.

References

(1) K. ZÁVĚTA, Z. ZOUNOVÁ, and E.J. TRINKLER, Czech. J. Phys. B18, 1314 (1968).

(2) M. MIZUSHIMA, J. Phys. Soc. Japan 24, 825 (1968).

(3) R. GERBER, L. MICHALOWSKY, K. MOTZKE, and E. STEINBEISS, phys. stat. sol. 16, 795 (1966).

(4) A. MILLER, J. appl. Phys. Suppl. 30 (1959).

(5) M. KÖNIG and C. CHOL, J. appl. Cryst. 1, 124 (1968).

(Received September 6, 1972)

phys. stat. sol. (a) 14, K113 (1972)
Subject classification: 14.3.4 and 16; 22.4.1

Faculty of Physics, University of Sofia

Effect of the Electric Field on the Optical Quenching in CdS Single Crystals

By
K. GERMANOVA, K. MARINOVA, and M. MLADENOVA

The steep decrease of the free electron density in photoconducting CdS with ield is due to transfer of holes from slow to fast recombination centres (1). The ptical and temperature quenching of the photoconductivity is connected with the ame centres. Consequently, an effect of a high electric field on the latter phenomena s to be expected (1, 2). The object of this paper is to discuss field quenching as nechanism of infrared quenching decrease with increasing electric field in CdS ingle crystals.

Photosensitive single crystals of CdS grown by resublimation from the gaseous hase were investigated. The optical quenching spectra were measured at 293 $^{\circ}$K. he samples were illuminated with intrinsic light ($\hbar\omega$ = 2.46 eV). Equal energies f quenching light were used at each wavelength (820 to 1800 nm).

At sufficiently high field strengths $E \geqq E_{cr}$ ($E_{cr} \approx$ 3 kV/cm), the field distribution ithin all samples investigated became inhomogeneous: stationary or travelling high- ield domains were formed, indicating the occurrence of a negative differential con- uctivity (NDC).

The experiments were performed under the following conditions:

The intensity of exciting light was chosen within that region of the photoconductivity- ight intensity characteristic in which the sample was sensitized (at low fields); ie intensity of quenching light was so chosen that the quenching was far from aturation.

) The electric field strength was lower than the critical one ($E < E_{cr}$). In this ase the sample was electrically homogeneous, the homogeneity being verified both y electrooptical and probe methods.

The percent quenching Q was determined by the ratio

$$Q = \frac{\Delta j(E)}{j_o(E)} = \frac{\Delta n(E)}{n(E)} , \qquad (1)$$

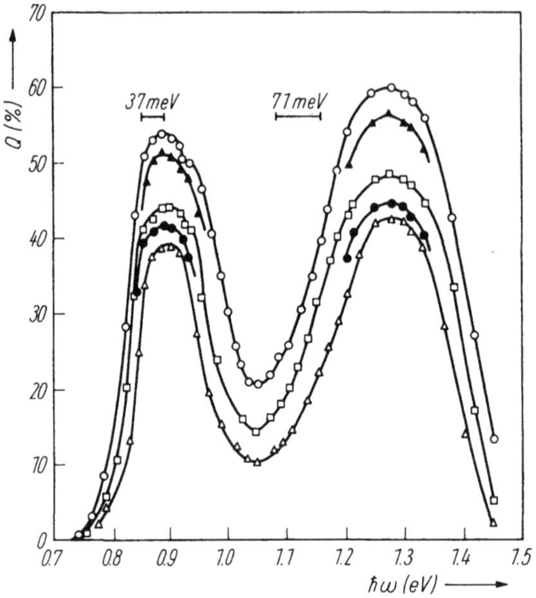

Fig. 1. Spectral distribution of the optical quenching at various strengths of the average field in the sample. ○ E = 3 kV/cm, ▲ E = 3.5 kV/cm, □ E = 5 kV/cm, ● E = 7 kV/cm, △ E = 10 kV/cm

where $j_o(E)$ and n(E) are the bias current and the bias free electron density resulting from the illumination of the sample within the intrinsic absorption range at a given field strength; $\Delta j(E)$ and $\Delta n(E)$ are the corresponding changes produced by additional illumination with quenching light. The spectral distributions of Q at various strengths of the average field in the sample (10^3 V/cm $\leqq E \leqq 10^4$ V/cm) is presented in Fig. 1. In Fig. 2, curve a is shown a typical current-voltage characteristic; in Fig. 2, curve b - the dependence Q = Q(E) in accordance with Fig. 1. It can be seen that: 1) The percent quenching drops monotonically at $E \geqq 3$ kV/cm. 2) The field strength at which the drop of Q occurs is lower than the one necessary for establishing of NDC.

The experimental results can be explained on the basis of the field-quenching model (1), giving the most probable mechanism for the occurrence of an NDC in CdS crystals. In steady state, the free electron density is determined by the equation (reference (1))

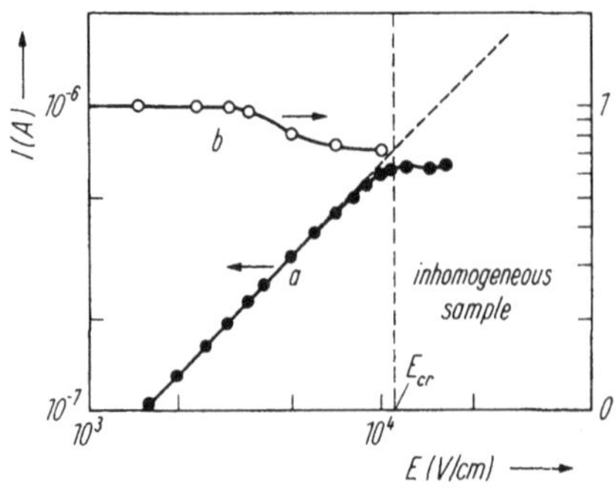

Fig. 2.(a) A typical current-voltage characteristic;
(b) field dependence of the ratio Q(E)/Q(0), where Q(0) is the percent quenching at low field

$$n = \frac{g}{\gamma_1(N_1 - n_1) + \gamma_2(N_2 - n_2)} , \qquad (2)$$

where g is the optical excitation rate per unit volume, γ_i (i = 1, 2) are the electron capture coefficients of the fast and the slow centres respectively; N_i is the density of centres i, while n_i is the density of electrons at centres i. According to reference (1) γ_i are independent of the field. Consequently, the decrease of n with E will be possible only when holes are transferred from slow to fast centres, so that

$$\gamma_1(N_1 - n_1) \gtreqless \gamma_2(N_2 - n_2) . \qquad (3)$$

The decrease of $(N_2 - n_2)$ with the field leads to a drop in the number of holes, which can be transferred into the valence band by additional illumination with quenching light. These considerations provide a qualitative explanation of the observed drop with field of the optical quenching.

From equation (2) one obtains for quenching

$$Q = \frac{\Delta n}{n} = \frac{n}{g}\left[\gamma_1 \Delta(N_1 - n_1) - \gamma_2 \Delta(N_2 - n_2)\right], \qquad (4)$$

where $\Delta n = n - n_q$; $\Delta(N_1 - n_1) = (N_1 - n_1)_q - (N_1 - n_1)$; $\Delta(N_2 - n_2) = (N_2 - n_2) - (N_2 - n_2)_q$. The subscript q denotes the steady state value of the corresponding quantity in the presence of a quenching light. The following simplifying assumption can be made provided the optical and field quenching are not too large:

$$\Delta(N_1 - n_1) \approx \Delta(N_2 - n_2) = f_q S_o (N_2 - n_2). \qquad (5)$$

Here $f_q S_o (N_2 - n_2)$ is the steady state density of holes, excited optically from sensitizing centres, S_o is the optical cross section for quenching photon absorption, f_q is the number of incident photons per unit area.

From (4) and (5) one gets

$$Q(E) = c\, R_2(E) , \qquad (6)$$

where $R_2 = \gamma_2 n(N_2 - n_2)$ is the field dependent rate of electron recombination at the sensitizing centres, c is a field independent quantity. According to the model assumed, R_2 decreases when the field is increased. The critical field at which the decrease of R_2 starts is lower than the one at which NDC is established; this is to be expected inasmuch as the decrease of n requires that condition (3) is also satisfied. Having in mind the character of the field dependence of R_2 (1), it is clear that equation (6) fairly explains the basic experimental results.

References

(1) G.A. DUSSEL and K.W. BÖER, phys. stat. sol. 39, 391 (1970).

(2) L.A. KLIMKA and K.D. GLINCHUK, Fiz. Tekh. Poluprov. 4, 587 (1970).

(Received October 6, 1972)

phys. stat. sol. (a) 14, K117 (1972)

Subject classification: 20.1 and 20.3; 22.4.3

Institute of Physics, Polish Academy of Sciences, Warsaw

Absorption and Photoluminescence of Single Crystals of ZnTe Doped with Cu

By

E.A. GIARO

The influence of copper impurities on the fundamental absorption edge of ZnTe has been investigated by Kurik and Savitskii (1). They have found that crystals doped to an acceptor concentration of 3.52×10^{19} cm^{-3} with resistivity 10^4 to 10^3 Ωcm do not change their exciton spectrum and the absorption edge in comparison with pure ones. The measurements have been carried out at temperatures of 300, 77, and 20.4 K.

In this note the first results of our absorption and photoluminescence measurements on ZnTe:Cu single crystals are presented. Sharp absorption lines we found near the absorption edge at liquid helium temperature. Some of these lines correspond to emission lines in the photoluminescence spectrum. The influence of Cu impurities on the absorption spectrum can be already observed at liquid nitrogen temperature in the form of two steps superposed on the edge.

Single crystals of cubic ZnTe were obtained from the vapour phase in a horizontal furnace heated to 1050 $^{\circ}$C. All samples were mechanically polished and etched in 1% volume solution of bromine in methanol.

The results of the measurements performed on "pure" crystals are in agreement with previous data (1, 2). The absorption edge has an exponential shape

$$\alpha = \alpha_0 \exp\left[\frac{b}{kT}(E - E_0)\right].$$

The temperature shift of the absorption edge measured at $\alpha = 200$ cm^{-1} is linear within the temperature range 80 to 370 K with $\frac{dE}{dT} = -6.5\times10^{-4}$ eV/deg. The concentration of holes in these crystals amounts to 5×10^{15} cm^{-3} at room temperature.

The Cu dopant was incorporated into ZnTe in two ways: a) by addition of a small amount of Cu to the polycrystalline starting material, b) by diffusion at elevated temperature (about 800 $^{\circ}$C). The results of the measurements were identical in both cases. The doped crystals were polished and etched like the pure ones.

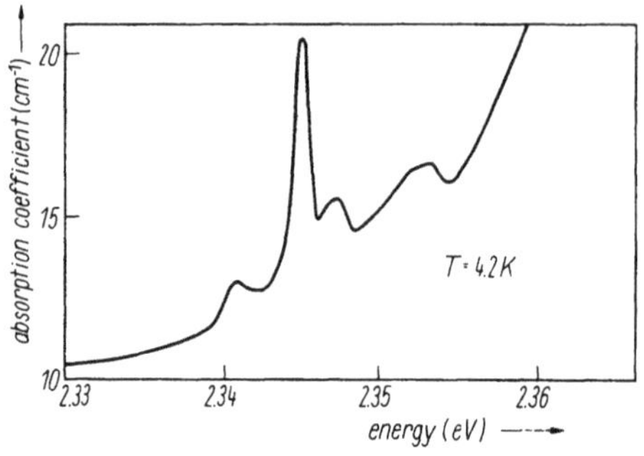

Fig. 1

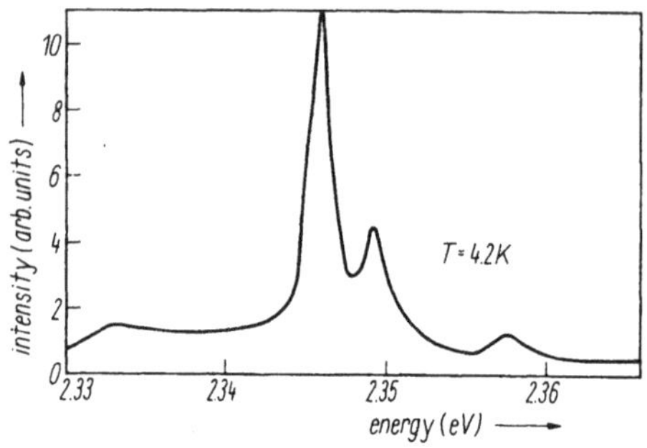

Fig. 2

Fig. 1. Absorption coefficient plotted against photon energy

Fig. 2. Intensity of luminescence vs. photon energy

The absorption spectrum of Cu doped ZnTe near the absorption edge is shown in Fig. 1. Six narrow absorption lines were found. The absorption maxima occur at energies equal to 2.3534, 2.3476, 2.3451, 2.3420, 2.3286, and 2.2750 eV (the last two ones are not shown in the figure). The absorption coefficient α changes in this region from 10 to 20 cm^{-1} and depends on the sample.

The luminescence spectrum was also measured at liquid helium temperature in the same region in which the absorption lines were found. An argon ion laser was used as source of excitation. The results of this measurement are shown in Fig. 2. One can see that the absorption line at 2.3451 eV coincides with the appropriate emission line.

These absorption and emission lines are interpreted as due to zero-phonon transitions in the complex formed by the neutral copper acceptor and the exciton bound to it. The energy of the free exciton in ZnTe was found to be 2.381 eV (3). The binding energy of the exciton in the copper-acceptor complex will therefore be 0.0359 eV and its ratio to the known ionization energy of the Cu acceptor 0.41 eV (4) will be 0.088. This value is close to 0.1 and so is in good agreement with the empirical rule demonstrated by Halsted and Aven (5).

The author wishes to thank Prof. W. Wardzyński for helpful discussions and critical comments and Miss I. Turksa for her photoluminescence measurements.

References

(1) M.V. KURIK and A.V. SAVITSKII, Optika i Spektroskopiya 20, 297 (1966).

(2) M. AVEN and J.S. PRENER (Eds.), Physics and Chemistry of II-VI Compounds, North-Holland Publ. Co., Amsterdam 1967 (chapter 7.6.3)

(3) A.C. ATEN, C.Z. VAN DOORN, and A.T. VINK, Proc. Internat. Conf. Phys. Semicond., Exeter 1962 (p. 696).

(4) R.J. CHERRY and J.W. ALLEN, Proc. Internat. Conf. Phys. Semicond., Exeter 1962 (p. 384).

(5) R.E. HALSTED and A. AVEN, Phys. Rev. Letters 14, 64 (1965).

(Received November 1, 1972)

phys. stat. sol. (a) 14, K121 (1972)
Subject classification: 16; 22.4.1

Research Institute for Technical Physics, Hungarian Academy of Sciences, Budapest (a), and United Incandescent Lamp and Electrical Co. Ltd., Tungsram, Budapest (b)

Photoresponse Measurements on ZnS Polytypes

By

M. SOMOGYI (a), M. FARKAS-JAHNKE (a), and G. SZÉKÁCS (b)

It is well known that ZnS crystals of mixed structure show an anomalous photovoltaic effect (APE), i.e. the wavelength dependence of the photovoltage in the vicinity of the absorption edge varies from sample to sample, and often a change of sign of the photovoltage can be observed (1, 2).

The crystals used in the measurements were grown from the vapour phase in the presence of HCl by a modified Frerichs method. The structure of 0.2 to 0.4 mm wide regions and also a few regions of mixed structure were selected. Indium amalgam contacts were painted on the surface of the crystals along the boundaries of the selected regions (2, 3). The wavelength dependence of the short circuit current between the electrodes was measured in the 315 to 430 nm region.

The photoresponse-wavelength curves measured at atmospheric conditions showed a great variety. On many crystals high currents were observed without any illumination; and also other memory effects could be registered. Generally in the curves the following three effects add up: (1) the high currents of the unilluminated samples by which the curve can be displaced along the ordinata considerably, (2) high responses in the low wavelength region from 345 to 430 nm due mainly to surface effects, (3) the responses characteristic of structural properties in the vicinity of the absorption edge. As only the latter part is the subject of the present investigation the first two effects had to be reduced. This could be achieved by storing the sample in dark and a vacuum of 10^{-5} Torr for 3 to 6 d prior to measurement.

The curves obtained in this way could be divided into three groups according to their shape as follows:

a) Curves showing a simple edge with a small maximum: on crystals of pure 3C, 6H, 4H, and a long periodic polytype structure. (Here the expression "edge" refers to the shape of the curve and is not identical with the optical absorption

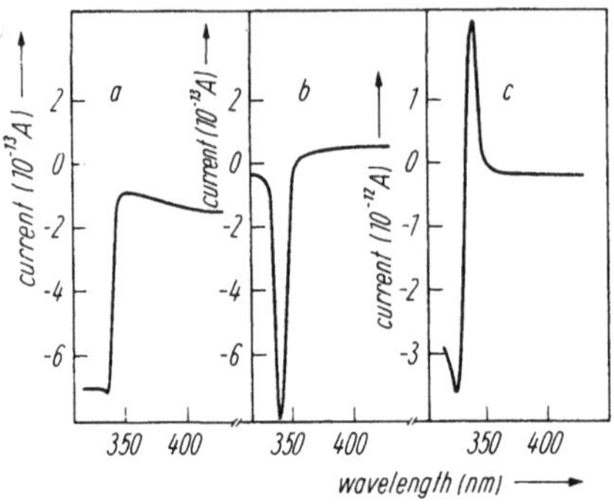

Fig. 1. Photoresponse curves measured at 10^{-5} Torr on three crystals having different structures.
a) Pure polytype structure of long periodicity mainly cubic,
b) 4H structure mixed with a few regions of 3C,
c) 2H parent lattice with a great number of stacking faults and a few cubic region

edge.)

b) Curves having one sharp peak: on crystals with 4H structure mixed with 3C.

c) Curves having an edge and a sharp peak with sign reversal between them: on crystals with 2H structure containing stacking faults or 2H mixed with 3C.

An example for all three types of curves is shown in Fig. 1. According to the convention given in the literature (2, 3) the first extremum regarded from the high energy side is to be taken negative. The wavelength values of the extrema of the photoresponse curves together with the data of optical absorption found in the literature are presented in Table 1.

A simple edge appears not only on the curves measured on crystals of pure cubic, but of pure polytype structures as well. One of these polytype structures had a very long, 150 layer periodicity and a low hexagonality value. The photoresponse edges measured on crystals of pure cubic and polytype structures do not coincide with the optical absorption edges reported previously (4, 5) but are displaced about 5 nm towards shorter wavelengths. The sharp peak appearing on the curves belonging to the classes b and c is situated always between 338 to 340 nm for both groups; this agrees exactly with the data of other authors (2, 6) who measured the wavelength dependence of the photoresponse under similar conditions. It is interesting that the position of this peak coincides with the wavelength given for the optical absorption edge of the pure cubic lattice. According to X-ray data in all hexagonal crystals stacking faults of various amounts were present, so the properties of the pure 2H structure could not be derived from the measured data. All curves of our faculted 2H samples showed the peak mentioned above at about 338 to 340 nm and an edge of opposite polarity at 326 nm. As the magnitude of the

Table 1

structure	wavelength data of the photoresponse extrema edge (nm)	peak (nm)	optical absorption edge (nm)
3C	336	-	339 (4) 341 (5)
long periodic "cubic" polytypes	335 335	- -	
6H	332.5	-	‖ 337.5 ⊥ 338.8 (5)
4H	330.5	-	‖ 336.1 ⊥ 337.8 (5)
4H + 3C	-	340	
2H + stacking faults	326	338	
2H + 3C + + stacking faults	326	339	
2H			‖ 331.8 ⊥ 328 (4) ‖ 332.5 ⊥ 328 (5)

340 nm peak seemed to show a correlation with the amount of structural disorder we concluded that the appearence of the 340 nm peak is always connected with the presence of stacking faults in the 2H lattice, whereas the 326 nm edge is the consequence of band-to-band transitions in the pure 2H regions but should not be identified with the onset of it, i.e. with the optical absorption edge.

As an explanation of the APE in ZnS the existence of built-in fields was supposed by several authors (3, 7). On the other hand, photoresponse curves with similar sharp peaks and sign reversal were measured on semiconductor heterojunctions (8) and were explained by transitions taking place on the opposing sides of the barriers.

It seems very reasonable to conclude that these built-in fields are the consequence of barriers due to the stacking faults present in the parent lattice.

In the light of the new facts the rough correlation found earlier (3) between the APE and the hexagonality of the crystal has the following meaning: it is the parent lattice which has to have a high hexagonality value. The most pronounced APE was found in 2H structures containing faulted regions of low hexagonality.

References

(1) W.J. MERZ, Helv. Phys. Acta 31, 625 (1958).

(2) A. LEMPICKI, Phys. Rev. 113, 1204 (1959).

(3) G. SHACHAR and Y. BRADA, J. appl. Phys. 41, 3127 (1970).

(4) M. CARDONA and G. HARBEKE, Phys. Rev. 137, 1467 (1965).

(5) O. BRAFMAN and I.T. STEINBERGER, Phys. Rev. 143, 501 (1966).

(6) T. OGAWA, Y. FUKUSE, and Y. KAWAI, Japan. J. appl. Phys. 4, 948 (1965).

(7) G.F. NEUMARK, Phys. Rev. 125, 38 (1962).

(8) C. VAN OPDORP and J. VRAKKING, Solid State Electronics 10, 955 (1967).

(Received November 2, 1972)

phys. stat. sol. (a) 14, K125 (1972)
Subject classification: 12.2; 22.6; 22.9

Université des Sciences et Techniques du Languedoc, Montpellier

Mesure de variations de constantes élastiques par diffusion Brillouin stimulée

Par
M. BOISSIER, J. PELOUS et C. VEDEL

Une méthode de mesure des faibles variations des constantes élastiques dans les milieux transparents par hétérodynage optique en diffusion Brillouin stimulée a été mise au point et décrite dans une publication antérieure (1).

Cette technique a déjà appliquée à la détermination des modifications d'une constante élastique en fonction de la pression pour le polyméthyl-métacrylate (P.M.M.A.) (2). Les fréquences des battements de la lumière diffusée par un cristal témoin avec celle diffusée par le cristal soumis à une contrainte extérieure conduisent aux variations de vitesse élastique en différenciant la relation de Brillouin

$$\frac{\Delta\nu}{\nu} = 2\,n\,\frac{V}{c}$$

($\Delta\nu$ est la fréquence Brillouin, ν la fréquence de la radiation excitatrice, n l'indice du milieu, V la vitesse de l'onde sonore, c la vitesse de la lumière). On obtient ensuite les variations des constantes élastiques par la relation

$$\varrho V^2 = C$$

(ϱ masse volumique, C constante élastique).

Il faut tenir compte des variations de n et de ϱ sous l'influence des contraintes extérieures (pression ou température).

Nous présentons ici les résultats obtenus pour des effets de pression sur le quartz et de température sur le P.M.M.A.

Dans le cas du quartz (X axe binaire, Z axe ternaire) on applique une pression uniaxiale selon 0Y; quand on observe le faisceau rétrodiffusé selon 0X, la polarisation du laser incident étant selon 0Z, on obtient

$$\left(\frac{\partial C_{11}^{D}}{\partial P}\right)_T = 4,8 \pm 1,2 \quad ,$$

si l'observation se fait suivant 0Z, on obtient

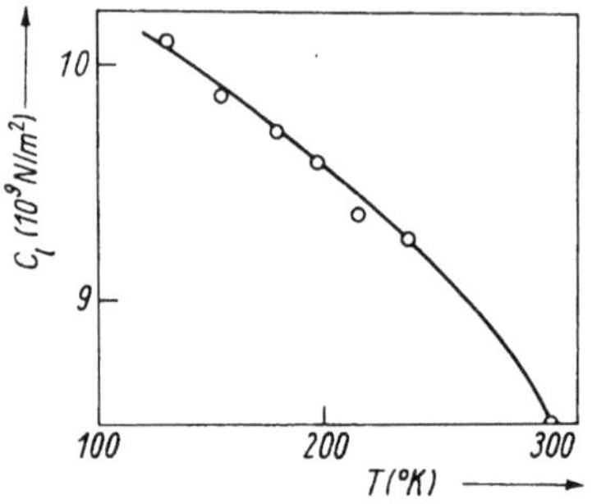

Fig. 1. Variation de la constante élastique C_l, relative aux ondes longitudinales, en fonction de la température T, pour le P.M.M.A.

$$\left(\frac{\partial C_{44}^{E}}{\partial P}\right)_T = 1,9 \pm 0,7 \ .$$

A cause de la piézoélectricité du quartz, il faut distinguer (voir par exemple (3)) les constantes élastiques à induction électrique ou à champ électrique constants, ce que nous avons indiqué respectivement par les indices D et E; l'indice T marque qu'il s'agit de coefficient de pression isotherme - la température étant environ 300 $^{\circ}$K; enfin, le domaine de pression utilisé s'étend de la pression atmosphérique normale à $P = 5\text{x}10^8$ N/m^2.

Pour le P.M.M.A. les variations de la constante élastique principale C_l (liée aux ondes de compression) en fonction de la température dans l'intervalle 130 à 300 $^{\circ}$K sont représentées sur la Fig. 1.

On en déduit un coefficient moyen de variation de la constante élastique en fonction de la température

$$\left(\frac{\partial C_l}{\partial T}\right)_P = -9,6\text{x}10^6 \ \text{Nm}^{-2}\text{deg}^{-1}$$

pour l'intervalle de température considéré; l'indice P indique que la pression est constante (sensiblement nulle).

Ces résultats étant obtenus en rétrodiffusion Brillouin stimulée présentent l'intérêt de mettre en jeu des fréquences sonores (de l'ordre de 25 GHz pour le quartz et 12 GHz pour le P.M.M.A.) plus élevées que celles intervenant dans les méthodes ultrasonores (4) par exemple).

Bibliographie

(1) J. PELOUS, C. VEDEL, M. BOISSIER et L. CECCHI, Phys. Letters (Netherlands) 38A, 73 (1972).

(2) C. VEDEL, M. BOISSIER, J. PELOUS et L. CECCHI, C.R. Acad. Sci. (France) B274, 998 (1972).

(3) R. VACHER, Thèse, Montpellier 1972.

(4) H.J. MCSKIMIN, P. ANDREATCH et R.N. THURSTON, J. appl. Phys. 36, 1624 (1965).

(Received October 18, 1972)

phys. stat. sol. (a) 14, K129 (1972)

Subject classification: 17; 13.4; 22.9

Institute of Organic and Physical Chemistry, Wrocław Technical University

Effect of Impurities and Defects on the Photoemission from Molecular Solids

By

J. SWORAKOWSKI[1)]

Introduction In a previous paper (1) (hereafter referred to as I) a model was put forward explaining the shape and characteristic features of photoelectron emission curves in molecular solids. It has been assumed that there may occur changes in energies of the valence levels of molecules connected with changes of intermolecular distances due to lattice vibrations whose vibration periods are greater than the ionization time. Local changes in the polarization energy may, however, also be due to structural defects occurring in crystals (cf. (2, 3)). Moreover, impurity molecules having lower ionization energies, present in the crystal lattice, may cause changes in shapes of photoemission curves. The main aim of the present note is to discuss the influence of the above mentioned factors on the photoemission from molecular solids.

In order to take into consideration the effect of local levels occurring in crystals, it is necessary to introduce a modified normalized density-of-states function (f(E)) which is proportional to a probability of generation of an electron having an energy E. The function used in this paper can be written in the form

$$f(E) = n(\Delta E) g(E_t) . \tag{1}$$

Here $n(\Delta E)$ is a function depending on a fluctuation of the energies of the valence levels (ΔE) and $g(E_t)$ describes energetic distributions of defect levels in solids.[2)] Similarly as in I, the function $n(\Delta E)$ is assumed to be given by the equation

$$n(\Delta E) = \frac{1}{\sigma\sqrt{2\pi}} \exp\left(-\frac{\Delta E^2}{2\sigma^2}\right), \tag{2}$$

where σ is a parameter being of the order of a few kT.

---

1) Present address: Edward Davies Chemical Laboratories, University College of Wales, Aberystwyth, Cards. U.K.

2) Both E and E_t (E_d, E_l, E_u) are measured with respect to the energy of valence levels of molecules in an ideal rigid crystal (hereafter referred to as E_v).

As shown in I, changes of the energy of valence levels in the near-surface region (corresponding to the "band bending" if the band model can be applied) should cause only energy shifts of photoemission curves without changing other characteristic features (e.g. shapes of the curves, slopes of the straight line approximations etc.). Therefore it is possible to neglect the above-mentioned effect in the subsequent discussion, all other assumptions made in I remaining valid in this note. Under those conditions the equation describing the photoelectron emission curve will take the form[3)]

$$Y = A \int_{E_l}^{E_v} \int_{\Delta E_l}^{\Delta E_u} n(\Delta E) g(E_t) T(E) d\Delta E dE_t \ . \qquad (3)$$

Here A is a constant factor including solutions of integrals over linear and angular coordinates and being assumed independent of the energy of the incident light (cf. I), and T(E) is the surface reflection factor which is assumed after (4) to be of the form

$$T(E) = \begin{cases} 0 & \text{for} \quad E_k \leqq 0 \ , \\ E_k^{1/m} & \text{for} \quad E_k > 0 \ , \end{cases} \qquad (4)$$

E_k denoting the kinetic energy of an electron just outside the surface of a solid, the latter parameter can be derived from the energy balance:

$$E_k = h\nu - I_c + E_t + \Delta E \ . \qquad (5)$$

Results of numerical computations a) monoenergetic local levels Let us consider local monoenergetic levels occurring in the crystal, their energies as measured with respect to valence levels in an ideal crystal amounting to E_d, and their density being H. In this case the function $g(E_t)$ is given by the equation

$$g(E_t) = \delta(E_g, 0) + \frac{H}{N_o} \delta(E_t, E_d), \qquad (6a)$$

where N_o is the total density of states and δ denotes the Kronecker delta-function.

---

3) For the sake of simplicity it will be assumed that such parameters as the electron (exciton) mean free path etc. have the same values both for host and for impurity molecules in the solid.

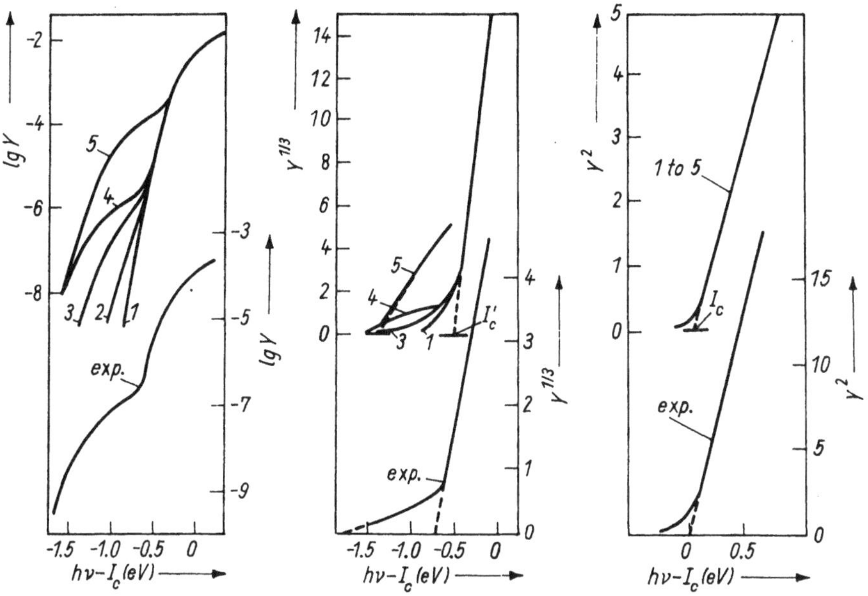

Fig. 1. Photoemission curves in the presence of discrete local levels; (1) a crystal without local states, (2) $E_d = 0.6$ eV, (3) and (5) $E_d = 0.9$ eV, (4) $E_d = 1.2$ eV (curves 1 to 4 were calculated for $H/N_o = 10^{-4}$, curve 5 for $H/N_o = 10^{-2}$), exp.: the experimental curve for the autocomplex AK - 4 (after reference (5) - right-hand scales)

The photoemission curves calculated for several different depths and densities of impuritiy molecules have been shown in Fig. 1. For comparison, an experimental curve obtained for the autocomplex AK - 4 by Alexandrov et al. (5) has been shown. It reveals all characteristic features of the calculated curves. From the plot one may determine the ionization energy of the pure AK - 4 complex to be 6.26 eV. The value of δ, as estimated from the difference of threshold energies in cubic-root and square plots $((I'_c - I_c)/3)$ amounts to 0.23 eV. The low-energy tail in the curve may be attributed to local levels due to undefined impurities, their ionization energy being close to 5.2 eV and their mole fraction amounting to 10^{-4} to 10^{-3}. Similar characteristics have been observed for tetracene-doped anthracene (6), anthracene and tetracene samples containing surface impurities (4, 7), phthalocyanines (8, 9), as well as for several other substances.

b) Continuous distributions of local levels Let us suppose that the energetic distribution of local states within an energy range (E_l, E_v) can be approximated either by an exponential or Gaussian function. The function $g(E_t)$ in these cases can be written as

$$g(E_t) = \delta(E_t, 0) + [1 - \delta(E_t, 0)] \frac{H}{N_o} \alpha(E_t) , \tag{6b}$$

where $\alpha(E_t)$ is equal to either $\varepsilon^{-1} \exp(-|E_t|/\varepsilon)$ or $(\varepsilon\sqrt{2\pi})^{-1} \exp(-E_t^2/2\varepsilon^2)$, ε being an energetic parameter of distribution. Numerical computations have been performed for several different values of H/N_o, ε, and E_l, and the photoemission curves obtained for the exponential distribution of local levels have been presented in Fig. 2. Essentially the same shapes of curves have also been obtained for the Gaussian distribution function. It becomes evident that, contrary to the monoenergetic levels, local levels distributed within an energy region weakly influence the shape of photoemission curves, resulting only in broadening of low-energy tails and leading to higher values of δ. This may be the reason for discrepancies between the values estimated from X-ray data (cf. (3)) and those derived by the method given in I. One may expect that there should be a correlation between the number of structural defects in a crystal and the value of δ, the greatest δ being observed in amorphous films.

Acknowledgements

The author thanks Prof. K. Pigon for helpful discussions and the critical reading of the manuscript. This work was partly supported by the Institute of Physics of the Polish Academy of Sciences.

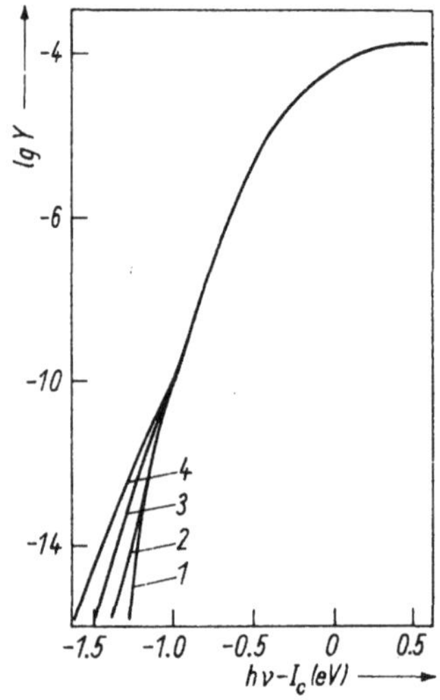

Fig. 2. Photoemission curves in the presence of exponentially distributed local levels; (1) a crystal without local states, (2) and (4) $E_l = 0.5$ eV, (3) $E_l = 0.8$ eV. ε amounts to 4 kT for (2) and (3), and to 8 kT for (4); $H/N_o = 10^{-4}$

References

(1) J. SWORAKOWSKI, phys. stat. sol. (a) 13, 381 (1972).

(2) J. SWORAKOWSKI, Mol. Crystals and Liq. Crystals 11, 1 (1970); in the press.

(3) E.A. SILINSH, phys. stat. sol. (a) 3, 817 (1970); private communication.

(4) M. KOCHI, Y. HARADA, T. HIROOKA, and H. INOKUCHI, Bull. Chem. Soc. Japan 43, 2690 (1970).

(5) S.B. ALEXANDROV et al., Izv. Akad. Nauk Latvian SSR, Ser. fiz. tekh. No. 4, 64 (1970).

(6) M. BATLEY and L.E. LYONS, Mol. Crystals 3, 357 (1968).

(7) S.B. ALEXANDROV, A.I. BELKIND, V.V. ALEXANDROV, V.V. GREKHOV, and YA.D. NEISHLOS, in: Semiconductors and Their Applications to Electrotechnics, Vol. 5, Zinatne, Riga 1971.

(8) V.A. BENDERSKII, A.I. BELKIND, M.I. FEDOROV, and S.B. ALEXANDROV, Fiz. tverd. Tela 14, 790 (1972).

(9) R.J. KOMP and T.J. FITZSIMMONS, Photochem. Photobiol. 8, 419 (1968).

(Received November 16, 1972)

phys. stat. sol. (a) 14, K135 (1972)

Subject classification: 13.3 and 14.3.1; 22.1.3

Institute of Physics, Department of Experimental Physics,
University M. Curie-Skłodowska, Lublin

Measurements of the Work Function and the Fermi Level in Thin Tellurium Films

By

M. JAŁOCHOWSKI, P. MIKOŁAJCZAK, and M. SUBOTOWICZ

The purpose of the present note is to get some information on the energy level structure in thin Te films with decreasing thickness from measurements of the work function by use of the CPD method and other physical characteristics of these films. There are no theoretical papers concerning the energy band structure of tellurium thin films.

The thermoelectric work function φ_{Te} in tellurium bulk crystalline samples is φ^{t}_{Te} = 4.73 eV (1, 2). The photoelectric work function is φ^{f}_{Te} = (4.65±0.05)eV (3).

Experimental Tellurium thin films of thickness from 100 to 1200 Å were evaporated on glass substrates with condensation rate 80 to 120 Å/min. During the condensation the temperature of the glass substrate was held at 300 ^{o}C, the pressure in the evaporation chamber was less than 5x10^{-6} Torr. The chemical purity of the obtained tellurium was 99.8%. After double destillation in vacuum the purity of tellurium was 99.999%. The dimensions of the investigated samples were 10x x20 mm^{2}. The resistivity of the obtained films was in the broad range from 5000 Ω to 1 MΩ. The thickness of most thin films was determined with a precision of up to 10 to 15%. According to (4) the tellurium layers evaporated comparatively quickly on oblique substrates are well crystallized.

CPD was measured by the Kelvin method (5, 6). One plate of the capacitor was the measured sample, the second one - a vibrating reference tungsten plate. Its dimensions were 4x4 mm^{2}; it was placed at a distance of about 0.2 mm from the investigated sample. The amplitude of the vibrating electrode was some hundredths of a millimeter. The rotating table with the investigated sample, made of thick copper plate and rod to enable cooling up to liquid nitrogen temperature, was placed into the vacuum chamber. The electronic circuit enabling the measurement of small contact voltages allowed the steady recording of the measured changes. This appa-

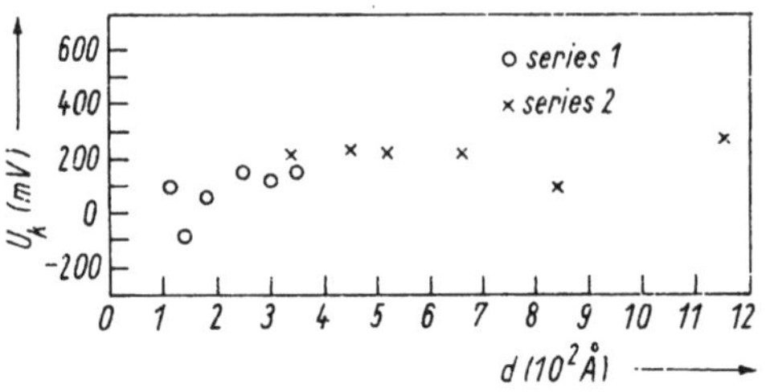

Fig. 1. Dependence of the CPD (U_k) on thickness (d) of the Te films

ratus could respond to changes of CPD higher than ±3 to 4 mV. We have found that the CPD in different points on the surface of the sample differed by no more than ±8 mV. To avoid the presence of water vapour or other undesirable gases, the measurements were performed in the presence of flowing Ar gas at a pressure of 10^{-1} Torr.

Results of measurements and discussion The results of CPD measurements are given in Fig. 1. They prove the shift of the Fermi level in thin Te films with decreasing thickness. Thick layers (above 500 Å) indicate a CPD equal to about 220 mV. The thermoelectric work function φ_W^t of polycrystalline tungsten is $\varphi_W^t = 4.52$ eV (2, 7). In (7) the thermoelectric φ_W^t and photoelectric φ_W^f work functions were determined also for polycrystalline samples.

From the given data and our measurements of the CPD we can determine the tellurium work function, φ_{Te}^t. In tellurium films of thickness 1000 Å $\varphi_{Te}^t = (4.76 \pm 0.01)$eV, and for films of thickness 200 Å $\varphi_{Te}^t = (4.62 \pm 0.01)$eV. Our results do not differ significantly from those of (1) as well as (3) for tellurium bulk samples.

To find a proper interpretation of our results, we have performed measurements of the Hall coefficient.

The Hall coefficient raised from $(5.5 \pm 0.8)cm^3/C$ for the films of thickness 410 Å to $(9.2 \pm 1.1)cm^3/C$ for the film of thickness 2100 Å. According to (8) changes of the energetic position of the acceptor levels with decreasing thickness of the Te films are responsible for the exponential dependence of the Hall carrier mobility on 1/T.

The simultaneous measurements of the work function, that changes from 4.76 eV (for 1000 Å films) to 4.62 eV (for 200 Å films) correspond to the results of (8). We can say even more: the Fermi level is shifted with the change of the acceptor levels in thin Te layers, if the change of the acceptor level position (as suggested in (8)) is responsible for the shift of the Fermi level.

Another possible explanation of the observed shift of the Fermi level in thin Te films with layer thickness is the increased splitting of the valence band into two subbands.

The decreasing value of the Hall coefficient with decreasing thickness of the Te films means an increase of the carrier concentration. The conductivity becomes more intrinsic with decreasing thickness of the tellurium films or the density of levels (probably of the acceptors) increased significantly, raising the Fermi level.

The full explanation of the above presented hypothesis of the Fermi level shift with decreasing film thickness requires further experimental and theoretical investigation.

References

(1) B.M. KAREV, Kontaktnaya raznost potentsiala, Moskva 1955.

(2) R.G. WILSON, J. appl. Phys. 37, 3170 (1966).

(3) L.D. LAUDE, B. FITTON, and M. ANDEREGG, Phys. Rev. Letters 26, 637 (1971).

(4) I. AGARBICEANU, P. CRISTEA, I. TEODORESCU, and C. GHEORGHIȚĂ-OANCEA, phys. stat. sol. (a) 1, K131 (1970).

(5) BAXTER, J. appl. Phys. 42, 2682 (1971).

(6) I.A. KARPOVITS and A.N. KALINIM, Fiz. tverd. Tela 12, 1490 (1970).

(7) S.Y. LUKIANOV, Fotoelementy, Izd. Akad. Nauk SSSR, Moskva 1948.

(8) C. ALBERS, phys. stat. sol. (a) 1, K105 (1970).
C. ALBERS and B. HEINZE, phys. stat. sol. (a) 6, K109 (1971).
C. ALBERS and R. LINK, phys. stat. sol. (a) 12, 89 (1972).

(Received November 15, 1972)

phys. stat. sol. (a) 14, K139 (1972)

Subject classification: 20.2; 22.2.1

Institute of Radiophysics and Electronics,
University of College of Science and Technology, Calcutta

An Alternative Form of the Condition of Population Inversion in Semiconductors at Relatively Higher Temperatures

By

A.N. CHAKRAVARTI and D.P. PARUI

It is well known that in semiconductor lasers or in bulk-excited specimens of semiconductors, the carrier degeneracy in the active region increases with increasing excitation. It is, therefore, felt that certain important features of degenerate semiconductors may be studied from the emission properties of semiconductor lasers or of strongly excited bulk specimens of semiconductors. In this communication, an alternative form of the condition of gain or population inversion in semiconductors at relatively higher temperatures is derived. This condition is then used, together with the fact that the energy of peak spontaneous emission from bulk specimens of semiconductors under strong excitation by an electron beam at relatively higher temperatures increases with increasing impurity concentration, to study qualitatively the doping dependence of the diffusivity-mobility ratio in degenerate semiconductors at such temperatures.

In semiconductor lasers or in bulk specimens of semiconductors under strong excitation, the carriers within a band form a degenerate distribution and create a condition of thermal equilibrium with respect to each other and hence, according to the principle of detailed balance, we can express (reference (1)) the diffusivity-mobility ratio of the electrons as

$$\frac{D_n}{\mu_n} = \frac{1}{e} n_o \frac{d\zeta'_n}{dn_o} , \tag{1}$$

where e is the electron change, n_o the electron concentration, μ_n the mobility, D_n the diffusion coefficient, and ζ'_n the quasi-Fermi energy. In degenerate semiconductors as used in lasers or in bulk specimens under strong excitation the tail states are mostly unoccupied at relatively higher temperatures and, therefore, we can write (reference (2))

$$n_o = N_c F_{1/2}(\eta) , \tag{2}$$

where $N_c = 2(2\pi m_e kT/h^2)^{3/2}$, m_e being the effective electron mass, $\eta\,(=(\zeta'_n - E_c)/kT)$ is the normalized difference between the quasi-Fermi energy and the lowest energy in the conduction band and F is the symbol for Fermi-Dirac integrals. Thus, using equations (1) and (2), we get

$$\frac{D_n}{\mu_n} = \frac{kT}{e}\,\frac{F_{1/2}(\eta)}{F_{-1/2}(\eta)} . \tag{3}$$

For strong degeneracy, we can also write (reference (3))

$$\begin{aligned} n_o &= N_c\,F_{1/2}(\eta) \\ &\approx N_c\,\frac{4\,\eta^{3/2}}{3\,\pi^{1/2}}\,, \qquad \eta > 0 . \end{aligned} \tag{4}$$

Hence, we get from equations (3) and (4)

$$\frac{D_n}{\mu_n} \approx \frac{2}{3}\,\frac{kT}{e}\,\eta \approx \frac{2}{3}\,\frac{1}{e}\,\Delta F_n . \tag{5}$$

ΔF_n being the electron quasi-Fermi energy measured from the edge of the conduction band. Similarly, for holes, we can write

$$\frac{D_p}{\mu_p} \approx \frac{2}{3}\,\frac{1}{e}\,\Delta F_p . \tag{6}$$

Moreover, we note that the condition of gain or population inversion in semiconductors is given (reference (4)) by

$$\zeta'_n - \zeta'_p > h\nu_p . \tag{7}$$

$h\nu_p$ being the energy of the emitted photon. This condition can only be satisfied under conditions of degeneracy. Therefore, combining equations (5) to (7), we can express the condition of gain or population inversion in semiconductors at relatively higher temperatures in an alternative form as

$$\frac{3e}{2}\left(\frac{D_n}{\mu_n} + \frac{D_p}{\mu_p}\right) > h\nu_p - E_g\,, \tag{8}$$

where E_g is the band gap.

It is apparent then that certain features of degenerate semiconductors may be studied with the help of condition (8). For example, since the energy of peak spon-

Fig. 1. Doping dependence of the energy of peak spontaneous emission from n-type GaAs excited by a 40 kV electron beam at 300 ^{o}K as experimentally observed elsewhere (5)

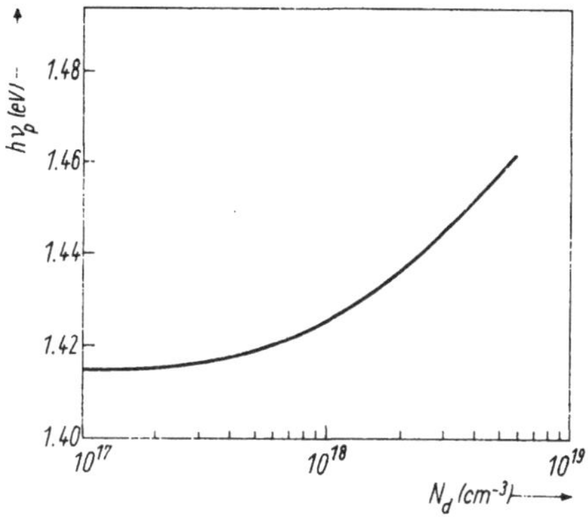

taneous emission increases with increasing impurity concentration as has been experimentally observed (5) in electron-beam-excited degenerate specimens of n-type GaAs at 300 ^{o}K (Fig. 1), the diffusivity-mobility ratio in degenerate semiconductors at relatively higher temperatures should increase according to condition (8) with increasing impurity concentration. At relatively higher temperatures, the impurity states are much less occupied and, therefore, the band gap may be assumed to remain more or less independent of doping. Thus the nature of the doping dependence of the diffusivity-mobility ratio at relatively higher temperatures would be somewhat similar to that of the energy of peak spontaneous emission as shown in Fig. 1. This type of doping dependence of the ratio has in fact been observed elsewhere (6) in Ge and Si at 300 ^{o}K.

It may be added further from equation (5) that the doping dependence of the diffusivity-mobility ratio should also closely follow that of the linewidth ΔE of spontaneous emission preceding laser action since $\Delta E \approx \Delta F_n + kT$ in the absence of filling of the tail states at relatively higher temperatures. In fact the nature of the doping dependence of the diffusivity-mobility ratio in degenerate semiconductors as reported by Lindholm and Ayers (6) is very similar to that of ΔE observed (5) in electron-beam-excited n-type GaAs at 300 ^{o}K. Thus the ratio should also increase with increasing excitation in semiconductor lasers or in bulk-excited semiconductors in more or less the same manner as with increasing doping since the carrier degeneracy increases with increasing excitation. It may be pointed out in this connection that the present work is based on the fact that the Einstein relation for the diffusivity-mobility ratio is not valid for degenerate semiconductors as established by Lindholm and Ayers (6).

The authors are indebted to Prof. J.N. Bhar for his keen interest in the work.

References

(1) E. SPENKE, Electronic Semiconductors, McGraw-Hill. Publ. Co., New York 1958 (p. 293 and 384).

(2) J.S. BLAKEMORE, Semiconductor Statistics, Pergamon Press, London 1962 (p. 79).

(3) J.S. BLAKEMORE, ibid. (p. 91).

(4) M.G.A. BERNARD and G. DURAFFOURG, phys. stat. sol. 1, 699 (1961).

(5) H.C. CASEY, Jr., and R.H. KAISER, J. Electrochem. Soc. 114, 149 (1967).

(6) F.A. LINDHOLM and R.W. AYERS, Proc. IEEE 56, 371 (1968).

(Received October 11, 1972)

phys. stat. sol. (a) 14, K143 (1972)
Subject classification: 10.1; 21.6; 21.7

Department of Mechanical Engineering, University of Salford

The Strength of the Zener Relaxation in Silver Solid Solutions

By
B. MILLS

In a recent note Robinson (1) assumes that the relaxation strength (Δ_m) of alloys of the silver-indium solid solution is controlled by a thermally activated process so that

$$\Delta_m = \frac{A}{T} \exp(-Q_f/kT) .$$

On this assumption Robinson (1) attempted to explain the composition dependence of Δ_m for a range of silver-indium alloys reported recently by Mills (2). Robinson (1) makes no attempt to present a physical model on which his assumptions are based; the purpose of this note is to give reasons why his analysis does not adequately describe the results of Mills (2). Further it will be shown, using existing theories of the Zener relaxation and considering additional features of the temperature dependence of the Zener relaxation, that a thermally activated process cannot control the composition dependence of the relaxation strength. The experimental results of the composition dependence of Δ_m for alloys of silver-indium are shown in (2) together with the result of Williams and Turner (3). It is important to note that the curve passes through the origin which means that Δ_m approaches zero as the solute concentration (c) approaches zero. This observation is characteristic of all reported studies of the composition dependence of Δ_m for the Zener relaxation, e.g. see the work of Berry and Nowick (4) on dilute aluminium-copper alloys. This point appears to have been missed in Robinson's (1) analysis where a finite Δ_m exists at zero solute concentration. Further Robinson (1) suggests that it would be worthwhile testing his analysis by measuring Δ_m at $c = 0$, but this is impossible, and further the magnitude of the Zener relaxation in the silver-indium system is so low that measurements were not possible in alloys containing less than approximately 3 at% indium.

In terms of the re-orientation of adjacent pairs of solute atoms Zener (5) obtained the following expression for the relaxation strength (Δ_m):

$$\Delta_m = \frac{\gamma M_u \lambda^2 n}{kT} = \frac{\gamma z N_v M_u \lambda^2 c^2}{2\,kT}\,, \tag{1}$$

where γ is a numerical constant depending on the orientation and lattice type, n is the number of pairs/volume, N_v is the number of atoms per unit volume of crystal, c is the atom fraction of solute, z is the coordination number, k is the Boltzmann constant, T is the absolute temperature,

$$\lambda = \left(\frac{\partial \varepsilon}{\partial n_p}\right)_{\sigma T}\,,$$

where ε is the appropriate strain, n_p is the excess of solute pairs in the preferred direction when the distribution is random. The relaxation strength based on the LeClaire and Lomer (6) theory of directional short range order is given by

$$\Delta_m = \frac{\alpha N_v a^2 f(p,c)\,\eta^2 c^2 (1-c)^2}{M_u kT}\,, \tag{2}$$

where α is a dimensionless constant, a is the interatomic spacing, f(p,c) is a function of the short range order parameter and the composition c, $\eta = dw/dA$, where $w = w_{AA} + w_{BB} - 2w_{AB}$ which is the difference between like and unlike bond energies.

For very dilute alloys the c^2 dependence of Δ_m (equation (1)) has been confirmed by Berry and Nowick (4). In more concentrated solid solutions the situation appears to be more complicated. An analysis of the detailed experimental study of the Zener relaxation in silver-zinc alloys by Li and Nowick (7) is given below. They show that for three alloys containing 24.2, 30.2, and 33.6 at% zinc a Curie-Weiss temperature dependence of Δ_m is followed, given by

$$\Delta_m = \frac{T_o}{T - T_c}\,,$$

where T_o is a constant, T_c is a critical temperature at which ordering would occur spontaneously if the necessary atomic movements were possible. Further, de Morton and Leak (8) have also shown that $\Delta_m \sim 1/(T - T_c)$ for a number of copper-gold alloys. Consequently Robinson's (1) assumption that $\Delta_m \sim \exp(-Q_f/kT)$ is in conflict with the established Curie-Weiss type of behaviour exhibited by the Zener relaxation Using the data of Li and Nowick (7) for the silver-zinc alloy system and plotting the

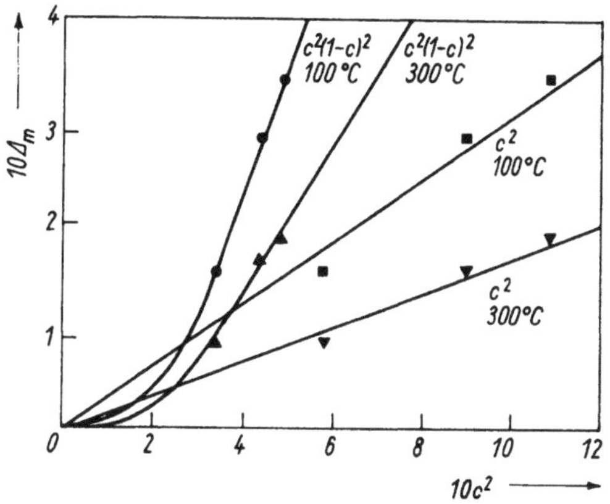

Fig. 1. Relaxation strength Δ_m as a function of c^2 and $c^2(1 - c)^2$ for silver-zinc alloys at 100 and 300 °C (7)

composition dependence of Δ_m using both c^2 and $c^2(1 - c)^2$ for the temperatures 100 and 300 °C the results shown in Fig. 1 are obtained. From Fig. 1 it can be seen that the experimental results of Li and Nowick (7) are in excellent agreement with a c^2 dependence of Δ_m since the straight lines pass through the origin. However, whilst the fit of the experimental points of the $c^2(1 - c)^2$ dependence (equation (2)) is excellent at high concentrations, at low solute concentrations the $c^2(1 - c)^2$ dependence is not obeyed. This behaviour for silver-zinc alloys is very similar to that reported by Mills (2) for silver-indium alloys where the three experimental points for 15.6, 18.2, and 18.95 at% indium alloys lie on a straight line. At lower solute concentration a deviation from a $c^2(1 - c)^2$ relation is observed for silver-indium alloys in agreement with the measurements on the silver-zinc system (7).

In conclusion the theory of LeClaire and Lomer (6) based on directional short range order is in good agreement with the experimental results for concentrated alloys of both silver-zinc and silver-indium. At low solute concentrations the experimental results and the LeClaire-Lomer theory are not in agreement. No measurements were made for the silver-zinc system below 24.2 at% zinc, however for the 3.65, 7.5, and 15.6 at% indium alloys better agreement with a c^2 rather than a $c^2(1 - c)^2$ dependence is found which further supports the Zener (5) theory at low concentrations. The work of Berry and Nowick (4) on dilute copper aluminium alloys has shown excellent agreement with a c^2 dependence of Δ_m. It would appear somewhat fortuitous that the c^2 dependence of Δ_m is also obeyed by concentrated alloys of the silver-zinc system since the concept of isolated pairs of atoms is no longer valid.

Presently a combination of the Zener (5) theory for dilute alloys and the LeClaire-Lomer (6) theory for concentrated alloys is reasonably satisfactory in explaining the composition dependence of Δ_m for the Zener relaxation in a wide range of alloy systems.

References

(1) W.H. ROBINSON, phys. stat. sol. (a) 11, K59 (1972).

(2) B. MILLS, phys. stat. sol. (a) 6, 55 (1971).

(3) G.P. WILLIAMS, Jr., and T.J. TURNER, phys. stat. sol. 26, 645 (1968).

(4) B.S. BERRY and A.S. NOWICK, in: Physical Acoustics, Vol. IIIA, Ed. W.P. MASON, Academic Press, New York 1966.

(5) C. ZENER, Elasticity and Anelasticity of Metals, Univ. Chicago Press, Chicago 1948.

(6) A.D. LECLAIRE and W.M. LOMER, Acta metall. 2, 731 (1954).

(7) C.Y. LI and A.S. NOWICK, Acta metall. 9, 49 (1961).

(8) M.E. DE MORTON and G.M. LEAK, Met. Sci. J. 1, 182 (1967).

(Received October 23, 1972)

phys. stat. sol. (a) 14, K147 (1972)

Subject classification: 19; 22.8

Department of Physics, Monash University, Clayton, Victoria

Paramagnetic Resonance of Fe^{3+} in Adamite

By

D.R. HUTTON, J. LE MARSHALL, and G.J. TROUP

Introduction In a previous paper (1) the analysis of the paramagnetic resonance spectra of Cu^{2+} in adamite $(ZnOH)(ZnAsO_4)$ showed that these ions substitute in both the high and low symmetry zinc sites of this mineral. Some crystals from the Durango (Mexico) sample show further very weak EPR lines and the analysis presented here concludes that these are due to Fe^{3+} ions also substituting into zinc sites.

Paramagnetic resonance results Detailed EPR spectra were recorded at X-band (≈ 9.2 GHz) and Q-band (≈ 35 GHz) at temperatures of 290, 77 , and 4.2 K. Because of the low intensity of the spectra, angular variation was analysed from Hall probe controlled plots, with proton NMR giving final calibration and measurement on magnetic and crystal axes.

Three sets of Fe^{3+} sites were identified and Fig. 1 shows the directions of the magnetic axes with respect to the crystal axes. All spectra could be fitted to the spin Hamiltonian

$$\mathcal{H} = \beta \vec{H} \cdot g \cdot \vec{S} + B_2^0 O_2^0 + B_2^2 O_2^2$$

with g-value $g = 2.004 \pm 0.001$, and crystal field parameters shown in Table 1. Site III accuracy was limited to that obtainable by a perturbation calculation based on the one line observed in the y_3-direction.

Discussion The site I spectra have been associated with occupation of the almost octahedral site by Fe^{3+}. Although the

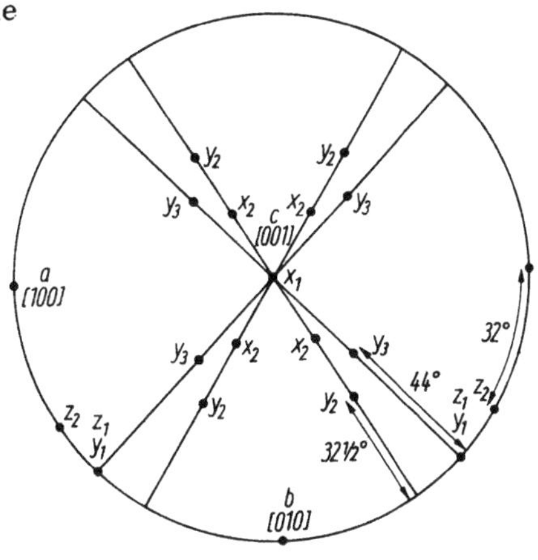

Fig. 1. The stereogram relating Fe^{3+} magnetic axes and crystallographic axes in adamite

Table 1

Crystal field parameters for Fe^{3+} in adamite

site set	no. of sites	B_2^2/B_2^o	B_2^o (GHz)
I	2	0.087 ± 0.003	5.03 ± 0.01
II	4	0.486 ± 0.003	7.92 ± 0.01
III	4	0.34 ± 0.02	4.1 ± 0.6

Cu^{2+} spectra from the same site (1) showed axial symmetry, the non zero B_2^2/B_2^o ratio for the Fe^{3+} spectra was expected since Fe^{3+} is more sensitive to local distortions caused by the nearby protons or charge compensators. Fe^{3+} substituted into the similar site of isostructural andalusite (2, 3), had a low B_2^2/B_2^o ratio but very different B_2^o and B_2^2 values. Although deviations for ostensibly similar sites can be satisfactorily explained for near cubic cases by the work of Sharma et al. (4), no theoretical analysis has yet proved appropriate for low symmetry sites.

From the results of (1) and (3), the site II and site III spectra are expected to be due to substitution into the 5 coordinated site with different charge compensation. The unambiguous assignment of substitutional position and perhaps the charge compensation mechanism might be achieved by detailed analysis of samples richer in iron to determine the pseudo-symmetry axes with the method of Michoulier and Gaite (5).

References

(1) T.J. SEED, J. LE MARSHALL, D.R. HUTTON, and G.J. TROUP, phys. stat. sol. (a) 10, 255 (1972).

(2) V.M. VINOKUROV, M.M. ZAPIROV, YU.E. POLSKII, V.G. STEPANOV, G.K. CHIRKIN, and L.YA. SHEKUN, Soviet Phys. - J. exper. theor. Phys. 12, 1078 (1961).

(3) F. HOLUJ, J.R. THYER, and N.E. HEDGECOCK, Canad. J. Phys. 44, 509 (1966).

(4) R.R. SHARMA, T.P. DAS, and R. ORBACH, Phys. Rev. 171, 378 (1968).

(5) J. MICHOULIER and J.M. GAITE, J. chem. Phys. 56, 5205 (1972).

(Received October 27, 1972)

phys. stat. sol. (a) 14, K149 (1972)

Subject classification: 17; 22.8.1

Institut für Strahlenschutz, GSF, Neuherberg

Energy Distribution of Thermally Stimulated Exo-Electrons from $BaSO_4$

By

D.D. PETERSON

During the investigation of the operation of an energy analyser to be used in exo-electron studies, $BaSO_4$ was chosen as sample because it is known to be a strong exo-electron emitter and also, the energy distribution was previously measured by Kriegseis and Scharmann (1). However the results to be reported in this note show that the maximum exo-electron energy from $BaSO_4$ is at most 1 eV, and not 7.5 eV as previously reported.

The experimental apparatus is shown schematically in Fig. 1. Grids G_1 and G_2 are hemispherically shaped, have diameters of 4 and 5 cm, respectively, and are positioned concentrically with respect to each other and to the sample. The distance from the planar grid G_3 to grid G_2, and to the first dynode of the detector is 5 and 8 mm, respectively.

The sample is made as in (1) by suspending $BaSO_4$ powder in methyl alcohol and placing the mixture onto a thin copper plate. After the alcohol evaporates, the powder adheres to the plate which can then be mounted onto the sample platform. Sample irradiations are done in air with an X-ray machine and an Am^{241} alpha particle source. Following the irradiation, the sample is transfered to the measuring chamber and evacuated. Measurements are taken at a pressure of $4x10^{-6}$ Torr, with a background signal of 0.1 counts/s.

The energy analyser system measures either the retarding potential curve or, using the ac modulation technique, e.g. Eden (2), the derivative of the retarding potential curve. A non-linear heating rate is used to extend the duration of the signal at the peak of the glow curve (i.e. 165 $^{\circ}C$) which allows the energy measurement to be carried out under constant signal conditions. During the derivative curve measurement, in addition to the dc voltage on grid G_2, a 3300 Hz, 400 mA peak-to-peak sine wave is also put on G_2. The shape of the retarding grid and the position of the sample ensure the maximum energy resolution. For a spherical grid retarding

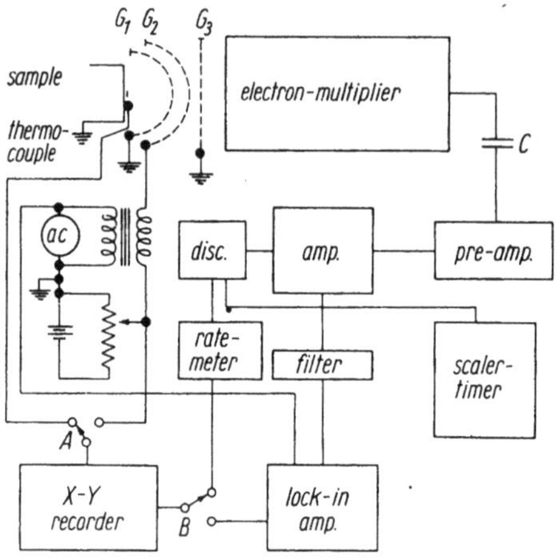

Fig. 1

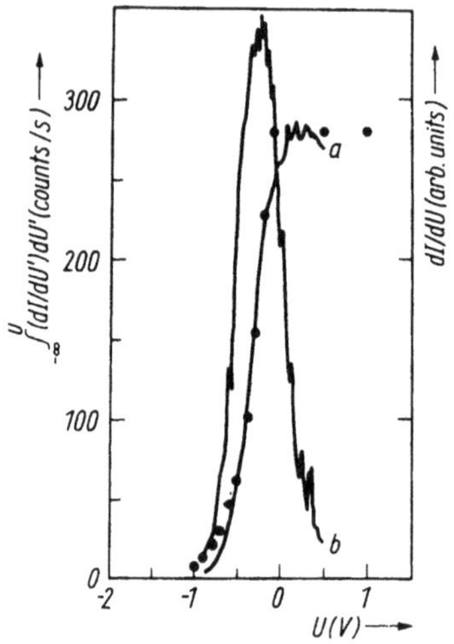

Fig. 2

Fig. 1. Schematic of the energy analyser and the associated electronics

Fig. 2. Measured retarding potential curve (a) and derivative curve (b) of thermally stimulated exo-electrons from irradiated $BaSO_4$ as a function of the voltage U on retarding grid G_2. The filled circles show the theoretical fit to curve a

analyser, Lukirsky (3) calculated the maximum energy error of $\Delta E = (a/b)^2 E_o$, where a and b are the radii of the sample and retarding grid, respectively, and E_o is the energy of the electrons that enter the retarding region. The radius of the sample is 1.5 mm which gives a $\Delta E/E_o < 1\%$. The effective retarding potential of the analyser can be discussed by the theory of electron tubes, e.g. Schintlmeister (4), and is given by $U_{eff} = U_g + DU_a$, where U_g and U_a are the voltages on grids G_2 and the first dynode of the detector, respectively, and D is the reciprocal of the amplification factor. The factor D (i.e. Durchgriff in German) is the product of the individual "Durchgriff" factors for grids G_2 and G_3, and from Kamke (5) (see Fig. 16) $D < 0.1\%$. Thus $U_{eff} = U_g$. As a test, the analyser was calibrated with an electron gun which showed that the retarding is given by the voltage on G_2.

The results of the energy measurements are shown in Fig. 2. Curve a shows the retarding potential curve (i.e. the integral energy spectrum) as a function of the voltage on grid G_2. The maximum exo-electron energy is clearly at most 1 eV.

Curve b shows the measured derivative curve (i.e. the differential energy spectrum) and is characterized by a maximum near 0.3 eV and a halfwidth of 0.5 eV. These results are the same for alpha and X-ray irradiations in air.

The rather slow approach of curve a to saturation has been suggested by Kortov and Myasnikov (6) to be due to localized surface regions with high electric field strength that suppress exo-electron emission from centres with low work function. Neglecting this region, curve a can be fitted reasonably well with the following expression used by Gaprindoshvily et al. (7):

$$I(U) = I_o \exp\left[(U - U_c)/U_m\right],$$

where I_o is the saturation current, U_c is the contact potential between sample and grid G_2, and U_m is the mean exo-electron energy. With $U_c = +0.15$ V, and $U_m = 0.25$ V, I(U) is given by the filled circles in Fig. 2. These maximum and mean energy values of 1 and 0.25 eV, respectively, compared with the previous values of 7.5 and 1.1 eV (1), are easier to explain in terms of thermal stimulation in this temperature region.

References

(1) W. KRIEGSEIS and A. SCHARMANN, phys. stat. sol. 33, K41 (1969).

(2) R.C. EDEN, Rev. sci. Instrum. 41, 252 (1970).

(3) P. LUKIRSKY, Z. Phys. 22, 351 (1924).

(4) J. SCHINTLMEISTER, Die Elektronenröhre als physikalisches Meßgerät, Springer-Verlag, 1943 (p. 9).

(5) D. KAMKE, Hdb. der Physik, Vol. XXXIII.

(6) V.S. KORTOV and I.E. MYASNIKOV, phys. stat. sol. (a) 9, K119 (1972).

(7) A.Y. GAPRINDOSHVILY, V.S. KORTOV, H. KÄAMBRE, R.I. MINTS, and I.E. MYASNIKOV, phys. stat. sol. (a) 4, K231 (1971).

(Received November 6, 1972)

phys. stat. sol. (a) 14, K153 (1972)

Subject classification: 18.3; 22.8.2

Research Centre of Physics, Iassy

The Influence of Oxidation Degree on the Redistribution Rate of Cations in $Mg_{0.9}Fe_{2.1}O_4$ Ferrite

By

E. LUCA, GH. MAXIM, and M.L. CRAUS

In the spinel ferrites, the Mg ions show a preference for octahedral (B) sites, but because the energy difference between normal and inverse spinel is not more than 0.1 eV (1), at high temperature they are distributed statistically. By quenching, one can freeze the high temperature cation distribution. This distribution is not at equilibrium if the temperature is lower than the quenching temperature. Our aim is to investigate the process of the isothermal redistribution of the Mg ions.

The investigations were done on polycrystalline samples of chemical formula $Mg_{0.9}Fe_{2.1}O_4$ obtained by standard ceramic technology. The samples were sintered at 1300 ^{o}C in air for four hours and afterwards quenched in water from 1300, 1200, and 1100 ^{o}C.

The chemical analysis by the method (2) shows that the Fe^{2+} content decreases if the quenching temperature decreases, while the oxidation degree and the concentration of the cation vacancies increase.

To describe the cation distribution we used the saturation magnetization (σ_s) which depends linearly on the Mg ions on the A sites (3). The magnetization measurements were performed at H = 5 kOe and room temperature on disc-shaped samples (d = 14 mm, h = 2 to 3 mm) with a vibrating sample magnetometer. In Fig. 1 is plotted the decrease of σ_s against the time of the treatment at 400 ^{o}C. The curve shape shows that the redistribution process cannot be described by a pure decay curve, but by means of a sum of decay curves.

By means of the curves from Fig. 1, the average time constants of the redistribution process were determined. The results are compiled in Table 1.

One can see an increase of the time constants with the Fe^{2+} concentration, e.g. at a decrease of the concentration of cation vacancies.

Table 1

treatment atmosphere	quenching temperature (^{o}C)	Fe^{2+} content (%)	σ_s after quenching ($G\ cm^3/g$)	average time constant (s)	
				cation redistribution	relaxation of the anisotropy
air	1300	4.8	48.6	9.5×10^3	150
air	1200	4.4	48.2	5.5×10^3	70
air	1100	3.7	45.5	1.9×10^3	40
CO_2	1000	6.4	55.2		

To put in evidence this dependence, in Fig. 1 the $\sigma_s(t)$ curve was plotted for a sample reduced in CO_2 at 1000 ^{o}C and quenched in water. The cation content of this sample is very small. The redistribution process of this cations is very slow in this case.

The sample annealed in CO_2 and quenched in water has a saturation magnetization higher than those treated in air at the same temperature and afterwards quenched in water. That means that there is some influence of the oxidation degree on the

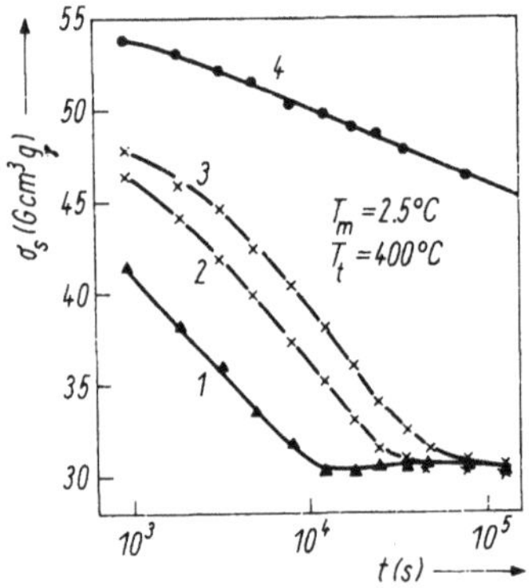

Fig. 1

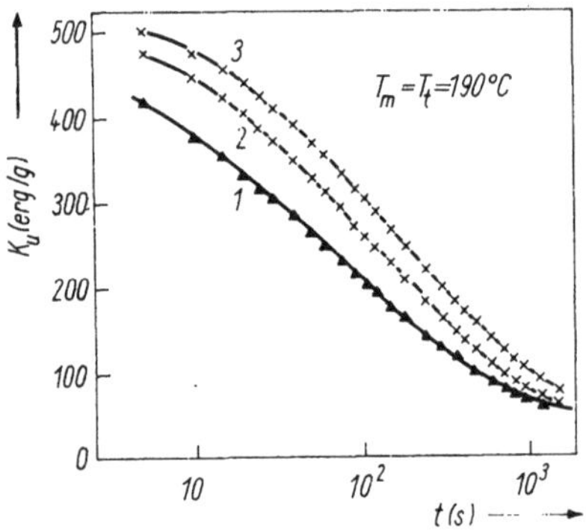

Fig. 2

Fig. 1. The change of the saturation magnetization at 25 ^{o}C with the time of treatment at 400 ^{o}C. (1, 2, 3) specimens annealed in air and quenched in water from 1300, 1200, and 1100 ^{o}C, respectively; (4) specimen annealed in CO_2 at 1000 ^{o}C and quenched in water

Fig. 2. The curves of the isothermal relaxation of the induced anisotropy at 190 ^{o}C for samples annealed in air at 1300, 1200, and 1100 ^{o}C, respectively, curves 1, 2, 3

equilibrium distribution of the cations for strongly reduced samples.

In Fig. 2 are plotted the isothermal relaxation curves of the induced anisotropy (K_u) at 190 ^oC, measured at H = 5 kOe. The average time constant of the process of the directional ordering varies with the Fe^{2+} concentration in the same way as the cation redistribution.

The dependence of the time constant of the directional ordering on the oxidation degree was shown by other authors (4, 5). It led to the hypothesis that the directional ordering is due to cation migration on vacancies.

Our results give the possibility to suppose that the redistribution process of the cations proceeds via vacancies. The Mg ions can jump from the A sites on the B vacancies from the neighbourhood, by leaving a free A site. The A vacancies can be occupied by the Fe ions from the neighbouring B sites obtaining a change of the cation distribution. We suppose that the cation vacancies are distributed not only on the B sites, but on both the octahedral and tetrahedral sites.

References

(1) D.J. EPSTEIN and B. FRANCKIEWICZ, J. appl. Phys. 29, 375 (1958).

(2) T. MERCERON, Ann. Phys. (France) 10, 121 (1965).

(3) L. NEÉL, Ann. Phys. (France) 3, 137 (1948).

(4) K. MOTZKE, phys. stat. sol. 2, K307 (1962).

(5) A. BRAGINSKI, phys. stat. sol. 11, 603 (1965).

(Received October 31, 1972)

phys. stat. sol. (a) 14, K157 (1972)

Subject classification: 18.2; 22

Institute for Low Temperature and Structure Research, Polish Academy of Sciences, Wrocław (a), International Laboratory for High Magnetic Fields and Low Temperatures, Wrocław (b), and University of Pittsburgh, Pennsylvania (c)

Magnetic Properties of Some Cubic Uranium Compounds

By

W. SUSKI (a), T. MYDLARZ (b), and V.U.S. RAO (c)

UTe, U_3Te_4, and U_3Se_4 are the least investigated group of the cubic compounds of uranium. U_3Te_4 and U_3Se_4 have been reported to have a b.c.c. Th_3P_4 structure (1, 2). UTe has an NaCl-type structure (2). Magnetic measurements on UTe and U_3Te_4 performed in the temperature range above 77 K have shown them to be ferromagnetic below 104 and 125 K (3, 4) respectively with a maximum in the temperature dependence of the magnetization (4). Wedgwood and Kuznietz (5) have concluded on the basis of neutron diffraction experiments UTe to have a simple ferromagnetic structure with a magnetic moment 2.2 BM.

U_3Se_4 has been found to be a simple ferromagnet below about 160 K (6) but this conclusion is not confirmed in our measurements. Now we present the results of magnetic measurements in the temperature range from 4.2 up to several degrees above the temperature of the magnetic transition and in magnetic fields up to 80 kOe.

The samples of U_3Te_4 and U_3Se_4 were prepared from elements as described previously (7). UTe was obtained by the volatile metal process according to (8). The lattice constants are given below:

UTe a = 6.146 Å, U_3Te_4 a = 9.412 Å, U_3Se_4 a = 8.760 Å.

These values are close to those reported elsewhere (1, 9, 10). Magnetic measurements were carried out on powder samples using the equipment and technique now standard in our laboratories (11, 12). The Curie points were determined according to the Belov-Goryaga method.

Both tellurides exhibit ferromagnetic properties, at least in stronger magnetic fields. As it follows from Fig. 1, uranium monotelluride shows considerable remanence and seems to be almost saturated in a magnetic field of 80 kOe. However, the magnetic moment calculated on the basis of the highest measured value of the magnetization is equal to 1.21 BM per uranium atom against 2.2 BM obtained in

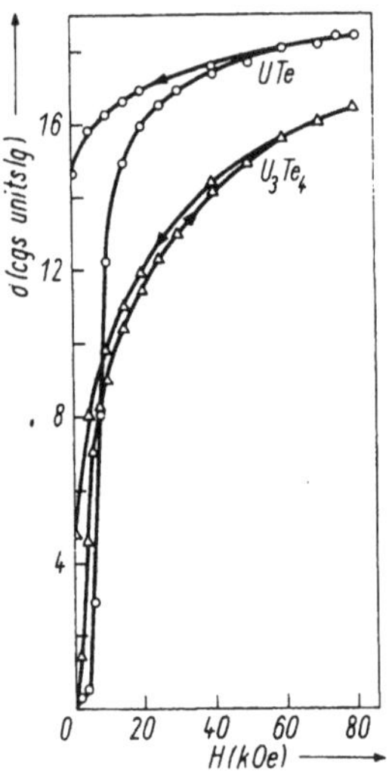

Fig. 1

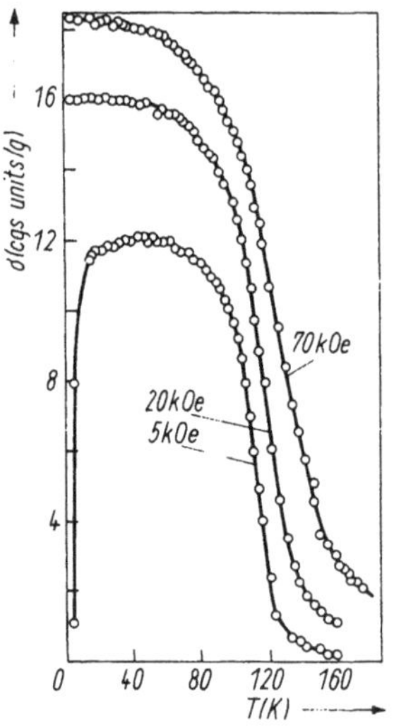

Fig. 2

Fig. 1. The magnetization of UTe and U_3Te_4 as a function ot magnetic field at 4.2 K

Fig. 2. The magnetization of UTe as a function of temperature

neutron diffraction experiments (5). This difference is probably caused by the high anisotropy of UTe confirmed by the orthorhombic distortion below the Curie point (13), as mentioned above. Also, the temperature dependence of the magnetization (see Fig. 2) seems to prove the existence of a marked anisotropy being however lower than, for example, in the case of USe (14). The Curie point is equal to 102 K and is in fair agreement with the previously published data (3, 4). The ferromagnetic properties of uranium monochalcogenides can be explained in terms of RKKY interactions (15), and the lowest Curie point of UTe in the series of monochalcogenides is probably due to the longest U-U distance in the monotelluride. Contrary to UTe, U_3Te_4 does not attain magnetic saturation in the investigated region (Fig. 3), but the magnetic moment calculated from the highest measured value of the magnetization is very close to that for UTe and equal to 1.20 BM per uranium atom. The temperature dependence of the magnetization is different than that for UTe, showing a maximum at 10 K insensitive to the strength of the magnetic field up to 25 kOe. Above this field strength the maximum disappears. The Curie point of U_3Te_4 ((38 ∓ 2)K) is lower than reported previously in (4) (105 K). The differences in both

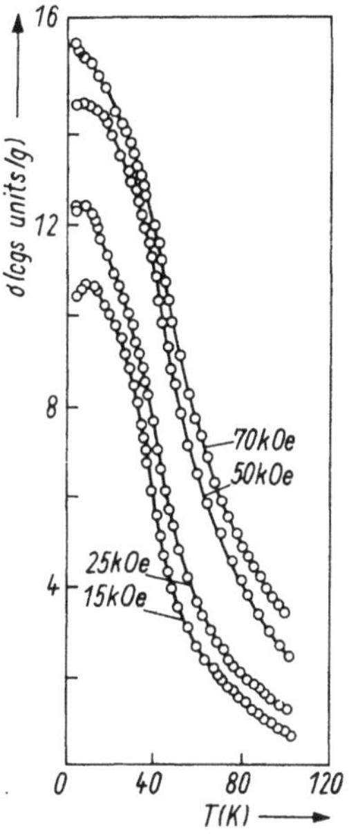

Fig. 3

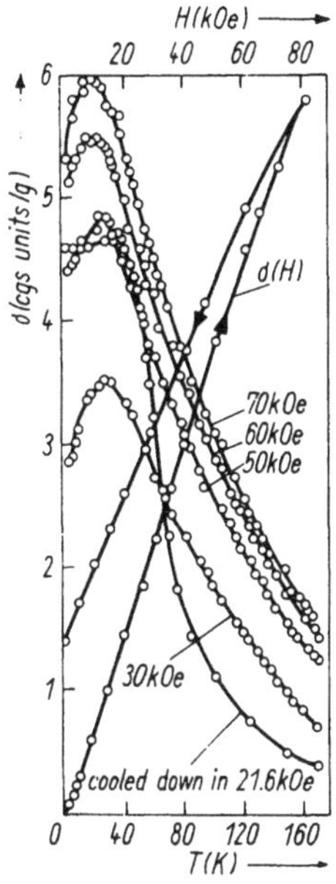

Fig. 4

Fig. 3. The magnetization of U_3Te_4 as a function of temperature

Fig. 4. The magnetization of U_3Se_4 as a function of magnetic field at 4.2 K and of temperature

the Curie point values and the character of the temperature dependence of the magnetization are not yet clear.

An additional indication that the Curie point of U_3Te_4 is lower than 105 K is the Weiss constant equal to +40 K (7).

The dependence of the magnetization of U_3Se_4 on the magnetic field presented in Fig. 4 is almost linear but with considerable remanence (20%).

The temperature dependence of the magnetization of U_3Se_4 (see Fig. 4) exhibits a maximum at about 35 K insensitive to the strength of the magnetic field in the range 0.4 to 30 kOe. As the magnetic field increases above 30 kOe the maximum shifts toward lower temperatures, but only slightly. When the sample was cooled down in a magnetic field of 21.6 kOe and then the measurements performed in the same field, the maximum almost completely disappeared. The neutron diffraction experiments done at 78 K have shown no magnetic order at this temperature (16). The results presented above are different from those reported by Chechernikov et al. (4, 6) for U_3Te_4 and U_3Se_4. These discrepancies are probably due to different purity of samples, the more so as no phase analysis of samples was given in (4, 6).

References

(1) P. KHODODAD, C.R. Acad. Sci. (France) 250, 3998 (1960).

(2) R. FERRO, Z. anorg. allg. Chem. 6, 275 (1954).

(3) W. TRZEBIATOWSKI and A. SEPICHOWSKA, Bull. Acad. Polon. Sci., Ser. Sci. Chim. 8, 457 (1960).

(4) V.I. CHECHERNIKOV, A.V. PECHENNIKOV, M.E. BARYKIN, V.K. SLOVYANSKICH, E.I. YAREMBASH, and G.V. ELLERT, Zh. eksper. teor. Fiz. 52 854 (1967).

(5) F.A. WEDGWOOD and M. KUZNIETZ, Conference Diggest No. 3, Rare Earths and Actinides, Durham 1971, Institute of Physics, London (p. 168).

(6) V.I. CHECHERNIKOV, A.V. PECHENNIKOV, E.I. YAREMBASH, L.F. MARTYNOVA, and V.K. SLOVYANSKICH, Zh. eksper. teor. Fiz. 53, 498 (1967).

(7) R. TROĆ, J. MULAK, and W. SUSKI, phys. stat. sol. (b) 43, 147 (1971).

(8) G.H.B. LOVELL, D.R. PERELS, and E.J. BRITZ, J. nuclear Mater. 39, 303 (1971).

(9) O.L. KRUGER and J.B. MOSER, J. Phys. Chem. Solids 28, 2321 (1967).

(10) W. TRZEBIATOWSKI, Internat. Conf. Magnetismus, Dresden 1966, VEB Deutscher Verlag für Grundstoffindustrie, Leipzig 1967 (p. 88).

(11) E. TROJNAR, C. BAZAN, and J. NIEMIEC, Bull. Acad. Polon. Sci., Ser. Sci. Chim. 13, 481 (1965).

(12) R.A. BUTERA, R.S. CRAIG, and L.V. CHERRY, Rev. sci. Instrum. 32, 708 (1961).

(13) J.A.C. MARPLES, private communication quoted after (5).

(14) W. SUSKI, V.U.S. RAO, R. TROĆ, and W.E. WALLACE, J. Solid State Chem. 4, 223 (1972).

(15) J. GRUNZWEIG-GENOSSAR, M. KUZNIETZ, and F. FRIEDMAN, Phys. Rev. 173, 562 (1968).

(16) A. SZYTUŁA, private communication.

(Received November 7, 1972)

Erratum

To the Short Note by J.D. STEPHENSON, N.G. ALEXANDROPOULOS, and H. WAGENFELD

A Note Concerning the Temperature Dependence of the Linear Absorption Coefficient of X-Rays in a Perfect Single Crystal of Silicon

phys. stat. sol. (a) 13, K71 (1972)

the authors request the following rectifications:

On page K72 in the equation read ln instead of log.

On page K73 in column 4 of the table the figures must read 2.6, 11.1, and 5.3% instead of 1.5, 2.4, and 3.9%, respectively.

Pre-printed Titles

Papers to be published

in "physica status solidi (a) - applied research"

Vol. 15, No. 1

Review Article

D.N. NASLEDOV and V.YA. SHEVCHENKO, Leningrad

Semiconducting $A^{II}B^{V}$ Compounds

Original Papers

D.A. ARONOV and R. MAMATKULOV, Tashkent

On the Theory of Transient Process after Reversal of a p-i-n Diode Current from Forward to Reverse Direction (II)

J. BENOIT et P. BENALLOUL, Paris

Observation des transitions ($^5D_3 \rightarrow {}^7F_J$) dans la cathodoluminescence de chouches minces de TbF_3, ZnS/TbF_3 et ZnS/Tb^{3+}

F.M. VOROBKALO, K.D. GLINCHUK, A.V. PROKHOROVICH, and G. JOHN, Kiev

Effect of Heat Treatment on the 0.93, 1.0, and 1.28 eV Luminescence Bands in n-GaAs

E. MONTRIMAS and B. PETRĖTIS, Vilnius

The Structure of Vacuum-Evaporated Selenium Layers

B. SEEBER and J. NICKL, München

Progress in Single Crystal Growth of V_3Si

K. SOMOGYI, Budapest

A Method for the Determination of the Band Gap from Hall Data

B. REPPICH and G. STREB, Erlangen

Steady State Deformation and Dislocation Structure of Pure and Mg-Doped LiF Single Crystals (I)

L.N. ALEKSANDROV, V.M. ZALETIN, E.A. KRIVOROTOV, and YU.G. SIDOROV, Novosibirsk

Investigation of the Growth Surface of GaAs Epitaxial Films by Chemical Decoration and Small-Angle Shadowing Technique

H. HILDEBRAND, Freiberg/Sa.

The Effect of Low Nickel Contents on the Work-Hardening Characteristics of Copper Single Crystals

Y. OHMURA, Y. ZOHTA, and M. KANAZAWA, Kawasaki, Tokyo

Shallow Donor Formation in Si Produced by Proton Bombardment

G. SAID and P.A. LEE, Brighton

Electrical Conduction Mechanisms in Tin Disulphide

M.A. NICOLET and J. GOLDER, Pasadena, California

Thermal Noise Calculation of Single-Carrier Space-Charge-Limited Current in a Non-Insulating Solid

A. ADDUCI, M. FERRARA, P. TANTALO, A. CINGOLANI, and A. MINAFRA, Bari

Photoelectric Properties of GaSe

G.A. ANDREEV and B.I. SMIRNOV, Leningrad

Precipitation in Strontium-Doped Sodium Chloride Crystals

P. THERNQUIST and J. WANG, Gainesville, Florida

Thermotransport of Tracer Silver in Solid Indium

A. FONTELL, E. ARMINEN, and M. TURUNEN, Helsinki and Otaniemi

Application of the Backscattering Method for the Measurement of Diffusion of Zinc in Aluminium

V.A. SABLIKOV, Moscow

Ambipolar Mobility and Injection Currents in Semiconductors with Deep Traps

B. PŘENOSIL, M. KOUTNÍK, and J. HOLUB, Prague

Dislocation Substructure of a Fatigued Low-Carbon Steel Carbonitrided at 620 °C

A.I. LANDAU, Kharkov

Kinetics of the Dislocation Motion in a Crystal Containing a Spectrum of Local Obstacles (Stoppers)

G. WELSCH, T.E. MITCHELL, and R. GIBALA, Cleveland, Ohio

Dislocation Relaxation in Germanium Single Crystals

S.I. RADAUTSAN, N.N. SYRBU, V.E. TEZLEVAN, K.F. SHERBAN, and E.E. STRUMBAN, Kishinev

Optical and Photoelectrical Properties and the Band Structure of Single Crystals of the Solid Solutions of the System $(CdS)_{3x}-(In_2S_3)_{1-x}$

S.S. LI and F.A. LINDHOLM, Gainesville, Florida

Quantum Yield of p-i-n Photodiodes

H. KUZMANY and W. LIEDERER, Wien

Propagation of Electron Domains in Tellurium

Y. MONOFORT, A. MAISSEU, G. ALLAIS, A. DESCHANVRES, and P. DELAVIGNETTE, Caën

Structure and Morphology of Niobium Suboxides

G.P. SRIVASTAVA, P.C. MATHUR, K.N. TRIPATHI, S.K. LOMASH, A.K. DHALL, Delhi

Magnetic Field Influence on the Gunn Effect Threshold

R. LAIHO, M. NATARAJAN, and M. KAIRA, Turku

Some Electrical and Optical Studies in $CsCuCl_3$ Crystals

G. WALDECKER, H. MEINHOLD, and U. BIRKHOLZ, Karlsruhe

Thermal Conductivity of Semiconducting and Metallic $FeSi_2$

D.I. JONES and M.R. HOPKINS, Dundee and Swansea

The Lorenz Number for Iron and Nickel at High Temperatures

L. FRANK, Brno

True Diffusion Current of Substrate Ions as the Rate-Controlling Factor of Surface Film Growth

Papers to be published

in "physica status solidi (b) - basic research"

Vol. 55, No. 1

Review Article

J.E. FISCHER and D.E. ASPNES, Philadelphia, Pennsylvania, and Murray Hill, New Jersey

Electroreflectance: A Status Report

Original Papers

R. BRUNO, Hamilton, Ontario

The Effective Lattice Theory

S.L. TRUBATCH, Long Beach, California

Self-Consistent Normal Mode Analysis of the One-Mode Polaron

W. BERKHAHN, H.G. KAHLE, L. KLEIN, and H.C. SCHOPPER, Karlsruhe

The Specific Heat of Terbium Arsenate

E. HEINER and J. SCHNEIDER, Dresden

Renormalization of the Bandwidth by Interatomic Interaction

J.M. MACFARLANE, I.S. MCLINTOCK, and J.C. ORR, Dundee

Antiferromagnetism in Amorphous Carbon Films

S. SUGA, T. KODA, and T. MITANI, Tokyo

Interband and Exciton Effects on Faraday Rotation near the Absorption Edge of CuCl

C. NEUMANN and W. HIRSCHWALD, Berlin-Dahlem

Impurity Diffusion in F.C.C. Metals

F. PLACIDO and K. HISANO, Edinburgh

Anharmonic Effects on Surface Mode Absorption in NaCl and KBr Slabs

S. TRIMPER, Leipzig

Perturbation Theory for a Spin System

A.P. GUIMARÃES and A.A. GOMES, Rio de Janeiro

Hyperfine Fields in Ferromagnetic Rare-Earth Alloys

I.V. SVECHKAREV and V.B. PLUZHNIKOV, Kharkov

Pressure Effect on the Fermi Surface of Molybdenum, Tungsten, and Rhenium

T. MATSUKAWA, K. MURATA, and R. SHIMIZU, Suita, Osaka

Investigation of Electron Penetration and X-Ray Production in Solid Targets

E. YANG, Omaha, Nebraska

Electric Field Effect on Two-Photon Transitions near Thresholds in Solids (I)

N.E. CHRISTENSEN, Lyngby

The Band Structure of Rhodium and Its Relation to Photoemission Experiments

A.B. WOLBARST, Hanover, New Hampshire

On the Effects of Applied Electrostatic Fields and Hydrostatic Pressures upon F-Centre ENDOR Spectra

M. KURZYŃSKI, Poznań

Phenomenological Description of Aggregates of Exchange-Bonded Paramagnetic Ions in Crystalline Environment (I)

Y. TAHARA and T. SUGENO, Tokyo

A Theory of Magnetic After-Effect for the Irreversible Magnetization Process in Rayleigh Region

T. PLEFKA, Darmstadt

Theory of Resolved and Unresolved Crystal Field Structure of ESR Spectra in Dilute Magnetic Alloys

G. GERVAIS, B. PIRIOU, and F. CABANNES, Orléans, La Source

Anharmonicity of Infrared Vibration Modes in the Nesosilicate Be_2SiO_4

V.M. BABICH, P.I. BARANSKII, S.L. KOROLYUK, A.G. SAMOILOVICH, and T.N. CHELOVA, Kiev and Chernovtsy

Transverse Magnetoresistance of n-Type Ge in Quantizing Magnetic Fields

H. KISHISHITA, Osaka

Optical Properties of Cu-Compley Centres in KCl Single Crystals

S. RADHAKRISHNA and K.P. PANDE, Madras
Optical Properties of CrO_4^{2-} Ions in Orthorhombic Sulphates

J. HEBER, W. PLATZ, and S. LEUTLOFF, Darmstadt
Higher Energy Levels of Cr^{3+} Ion Pairs in Ruby and Lanthanum Aluminate

S.K. BANSAL and V.P. DUGGAL, Delhi
Temperature Dependence of Magnetoconductivity Tensor Coefficients in Pure Antimony Single Crystals at Low Temperatures

W. HILLMER, Karlsruhe
Investigation of Pair Interactions of Er^{3+} Ions in YPO_4 and $YAsO_4$ by Paramagnetic Resonance

E.J. SIEGEL, East Lansing, Michigan
Tenfold d-Electron Degeneracy in the Hubbard Model and Transition-Metal Magnetism (II)

G. MATSUMARA and G.C. KUCZYNSKI, Notre Dame, Indiana
New Color Centers in KCl Crystals

D. FRÖHLICH, CH. UIHLEIN, and E. MOHLER, Dortmund and Frankfurt/Main
Polariton Fusion in CuCl

E. HESS, I. TOPOL, K.-R. SCHULZE, H. NEUMANN, and K. UNGER, Leipzig
Band Structure and Pseudopotential Form Factors for AlAs

J.L. REGOLINI, S. LEWONCZUK, J. RINGEISSEN, S. NIKITINE, and C. SCHWAB, Strasbourg
Optical Properties of $CuGaS_2$ near the Fundamental Absorption Edge

R. SERNEELS and R. GEVERS, Antwerp and Mol
A Discussion and Generalization of Bethe's Second-Order Approximation

Å. DANEMAR, D.O. WELCH, and B.S.H. ROYCE, Princeton, New Jersey
A Born Model Calculation of the Energies of Schottky Pair Formation and Cation Migration in Potassium Azide

B.V. THOSAR, R.G. LAGU, V.G. KULKARNI, and GIRISH CHANDRA, Bombay
Dependence of Positron Lifetimes on the Temperature and Phase of Some Organic Media

G. BABONAS, A. RĖZA, and A. ŠILEIKA, Vilnius
Birefringence of CdS under Uniaxial Stress

K. LEIBLER, W. GIRIAT, K. CHĘCIŃSKI, Z. WILAMOWSKI, and R. IWANOWSKI, Warsaw
EPR of Mn^{2+} Ions in Mixed Crystals $Zn_wHg_{1-w}Se$ and $Zn_mHg_{1-m}Te$

B. KRISHNAMURTHY and K.P. SINHA, Bangalore
Mössbauer Lineshape in the Presence of Radio-Frequency Fields

Y. RUEDIN, P.-A. SCHNEGG, C. JACCARD, and M.A. AEGERTER, Neuchâtel
EPR Optical Detection of F-Centre Pairs in Alkali Halides (II)

V.S. BOIKO, Kharkov
Dislocation Description of Twin Dynamic Behaviour

H.-W. STREITWOLF and J. MERTSCHING, Berlin
Single-Site Approximation in Alloys with Short-Range Order

H. FREESE and U. KREY, Hamburg
Spin Wave Spectrum and Magnetization of Magnetic Thin Films with a Rough Surface

H. IBACH and W. MÖNCH, Aachen
Surface Phonons in Spectral Oscillations in the Photoconductivity on Silicon (111) Surfaces

F.K. NAEHRING, Berlin
Fluorine Superhyperfine Structure in the EPR of $CdF_2:Gd^{3+}$

M.Z. ZGIERSKI, Cracow
Vibronic Coupling and Problem of Bulk Photoconductivity in Molecular Crystals

O. GOEDE and F. FINK, Berlin
Piezoemission Study of CdS:Te

19. Magnetic Resonance
20. Optical Properties
    - 20.1 Spectra. Optical Constants (X-Ray Spectra See 20)
    - 20.2 Lasers
    - 20.3 Luminescence (X-Ray Spectra See 20)
21. Metals. Alloys
    - 21.1 Transition Metals (Group Sc to Zn) and Their Alloys
        - 21.1.1 Fe and Fe Alloys
    - 21.2 Alkali Metals
    - 21.3 Alkali-Earth Metals
    - 21.4 Rare-Earth Metals
    - 21.5 Actinides
    - 21.6 Noble Metals
    - 21.7 Semi-Metals (Semiconducting Alloys See 22.7)
22. Semiconductors and Ionic Crystals
    - 22.1 Elements
        - 22.1.1 Germanium
        - 22.1.2 Silicon
        - 22.1.3 Group VI Elements
    - 22.2 III–V Compounds
        - 22.2.1 Arsenides
        - 22.2.2 Phosphides
        - 22.2.3 Antimonides
        - 22.2.4 Mixed Crystals
    - 22.3 IV–IV, II–V, and III–VI Compounds
        - 22.3.1 Mixed Crystals
    - 22.4 II–VI Compounds
        - 22.4.1 Sulphides
        - 22.4.2 Selenides
        - 22.4.3 Tellurides
        - 22.4.4 Mixed Crystals
    - 22.5 Halides
        - 22.5.1 Silver Halides
        - 22.5.2 Alkali Halides
        - 22.5.3 Alkali-Earth Halides
        - 22.5.4 Mixed Crystals
    - 22.6 Simple Oxides (CdO, ZnO See 22.4)
        - 22.6.1 Mixed Crystals
    - 22.7 Semiconducting Intermetallic Compounds
    - 22.8 Tertiary and Higher Compounds
        - 22.8.1 Salts of Oxy-Acids (Carbonates, Phosphates, Silicates Including Mica etc.)
        - 22.8.2 Spinels, Garnets, Ferrites
    - 22.9 Organic Semiconductors
23. Solidified Gases

Attention

The preceding classification scheme is closely akin to the scheme used by Physics Abstracts to classify the various fields of solid state physics. A paper will either be classified under one or more of the subsections or, if none of these are relevant, it will just appear under the general section heading (i. e. sections and subsections are equivalent in this sense).

The *subject classification* given at the head of each paper begins with the main subject matter of the paper, and this is followed, if necessary, by figures referring to subsidiary subject matter and the type of solids investigated.

The papers published in this issue are listed in the order of their subject classification at the end of the table of contents.

Bei Fragen zur Produktsicherheit wenden Sie sich bitte an:
If you have any questions regarding product safety,
please contact:

Walter de Gruyter GmbH
Genthiner Straße 13
10785 Berlin
productsafety@degruyterbrill.com